INTRODUCTION TO
PROBABILITY AND
STATISTICS
FOR SCIENTISTS AND ENGINEERS

McGraw-Hill Series in Probability and Statistics

Brillinger: *Time Series: Data Analysis and Theory*
De Groot: *Optimal Statistical Decisions*
Dixon and Massey: *Introduction to Statistical Analysis*
Milton and Arnold: *Introduction to Probability and Statistics: Principles and Applications for Engineering and the Computing Sciences*
Milton: *Statistical Methods in the Biological and Health Sciences*
Mood, Graybill, and Boes: *Introduction to the Theory of Statistics*
Morrison: *Multivariate Statistical Methods*
Rosenkrantz: *Introduction to Probability and Statistics for Scientists and Engineers*
Steel, Torrie, and Dickey: *Principles and Procedures of Statistics: A Biometrical Approach*

McGraw-Hill Series in Industrial Engineering and Management Science

CONSULTING EDITORS

Kenneth E. Case, *Department of Industrial Engineering and Management, Oklahoma State University*
Philip M. Wolfe, *Department of Industrial and Management Systems Engineering, Arizona State University*

Barnes: *Statistical Analysis for Engineers and Scientists: A Computer-Based Approach*
Bedworth, Henderson, and Wolfe: *Computer-Integrated Design and Manufacturing*
Black: *The Design of the Factory with a Future*
Blank: *Engineering Economy*
Blank: *Statistical Procedures for Engineering, Management, and Science*
Bridger: *Introduction to Ergonomics*
Denton: *Safety Management: Improving Performance*
Grant and Leavenworth: *Statistical Quality Control*
Hicks: *Industrial Engineering and Management: A New Perspective*
Hillier and Lieberman: *Introduction to Mathematical Programming*
Hillier and Lieberman: *Introduction to Operations Research*
Huchingson: *New Horizons for Human Factors in Design*
Juran and Gryna: *Quality Planning and Analysis: from Product Development through Use*
Khoshnevis: *Discrete Systems Simulation*
Kolarik: *Creating Quality: Concepts, Systems, Strategies, and Tools*
Law and Kelton: *Simulation Modeling and Analysis*
Marshall and Oliver: *Decision Making and Forecasting*
Moen, Nolan, and Provost: *Improving Quality through Planned Experimentation*
Nash and Sofer: *Linear and Nonlinear Programming*
Nelson: *Stochastic Modeling: Analysis and Simulation*
Niebel, Draper, and Wysk: *Modern Manufacturing Process Engineering*
Pegden: *Introduction to Simulation Using SIMAN*
Riggs, Bedworth, and Randhawa: *Engineering Economics*
Rosenkrantz: *Introduction to Probability and Statistics for Scientists and Engineers*
Taguchi, Elsayed, and Hsiang: *Quality Engineering in Production Systems*
Wu and Coppins: *Linear Programming and Extensions*

INTRODUCTION TO
PROBABILITY AND
STATISTICS
FOR SCIENTISTS AND ENGINEERS

Walter A. Rosenkrantz

Department of Mathematics and Statistics

University of Massachusetts at Amherst

The McGraw-Hill Companies, Inc.
New York   St. Louis   San Francisco   Auckland   Bogotá   Caracas
Lisbon   London   Madrid   Mexico City   Milan   Montreal   New Delhi
San Juan   Singapore   Sydney   Tokyo   Toronto

McGraw-Hill

*A Division of The **McGraw·Hill** Companies*

INTRODUCTION TO PROBABILITY AND STATISTICS FOR SCIENTISTS AND ENGINEERS

This book is printed on acid-free paper.

domestic 1 2 3 4 5 6 7 8 9 DOC DOC 9 0 0 9 8 7

international 1 2 3 4 5 6 7 8 9 DOC DOC 9 0 0 9 8 7

ISBN 0-07-053988-X

This book was set in Palatino by Publication Services.
The editors were Eric Munson and Maggie Rogers; the interior designer was Elizabeth Williamson; the production supervisor was Michelle Lyon; the cover designer was Nadja Lazansky.

R. R. Donnelley & Sons, Crawfordsville, was the printer and binder.

Library of Congress Cataloging-in-Publication Data
Rosenkrantz, Walter A.
    Introduction to probability and statistics for scientists and
engineers / Walter A. Rosenkrantz.
        p.   cm.
    Includes index.
    ISBN 0-07-053988-X
    1. Engineering–Statistical methods.   2. Probabilities.
I. Title.
TA340.R6   1997
519.5′024′5–dc21   96-50182
                                                            CIP

INTERNATIONAL EDITION

When ordering this title, use ISBN 0-07-114666-0

http://www.mhcollege.com

CONTENTS

PREFACE

OBJECTIVES

My primary goal in writing *Introduction to Probability and Statistics for Scientists and Engineers* was to present the basic ideas of probability theory and statistics within the context of interesting applications in the engineering, physical, biological, and computer sciences. For many students, this will be their only statistics and probability course before entry into the competitive workplace of engineers and scientists. This text has been carefully designed so that upon successful completion of a course based on this book, students will be well equipped for the task of modeling, displaying, interpreting, and collecting data for a variety of scientific and engineering applications. In addition, this text gives students a solid foundation for more advanced studies in probability and statistics. It is intended for a two-semester, calculus-based course for undergraduate students majoring in engineering, computer science, or mathematics.

ORGANIZATION OF THE BOOK

The book consists of 14 chapters. Chapter 1 is an introduction to exploratory data analysis, a powerful set of graphical methods for organizing and visualizing the distribution of data. Throughout the text, I emphasize that a major role of data analysis is to help the experimenter visualize the distribution of the data. Consequently, beginning in Chapter 1, the text highlights the fundamental role of the distribution function (frequency, empirical, sampling, and theoretical) in statistics and probability. This early coverage helps explain how the empirical distribution function plays a vital role in many areas of statistics, including normal probability plots and the bootstrap methodology. In addition, it is a concrete example of the more abstract notion of a distribution function of a random variable, discussed in the chapters on probability. Chapter 2 successfully links probability theory and statistical inference by carefully constructing a mathematical model of random sampling. Chapters 3 through 6 cover the basic concepts of probability theory, including the most important distributions: the hypergeometric, binomial, Poisson, normal, gamma, chi-square, and others. Also included are the central limit theorem and the law of large numbers, including an application to the fundamental theorem of mathematical statistics. Chapters 7 through 14 take students through the traditional topics of a first course in statistics, including estimation, hypothesis testing, regression, analysis of variance, and a brief introduction to statistical quality control.

ORGANIZATION OF EACH CHAPTER

Each chapter begins with an orientation section that is a brief overview and outline of the chapter. An abundance of applied examples are worked out in every chapter to give students relevant applications to a variety of science and engineering disciplines. I have included real data sets from referenced sources, in addition to simulated data sets, where appropriate. There are a large number of problems for students to work after each section, and answers to odd-numbered problems are included at the back of the textbook. The problems, which range from the routine to the challenging, help students master the basic concepts and give them a glimpse of the vast range of applications to a variety of disciplines. In chapters of a more computational nature, I have included a section titled "Mathematical Details and Derivations" at the end for additional reference. For students interested in learning more about the topics covered in the text, suggested readings are listed in the last section of each chapter under the heading "To Probe Further."

PEDAGOGICAL FEATURES OF THE BOOK

Data Sets.   The data sets themselves have been carefully selected to illustrate current trends in statistical methods. Instead of using a large number of data sets, this book analyzes a limited number of data sets using a variety of statistical methods. This approach enables students to learn that careful data analysis means applying a broad spectrum of techniques to a single set of data.

Examples.   The main focus of this text is to introduce students to the most important concepts in statistics and probability through worked examples that relate to students' chosen fields of study in science, engineering, or computer science. Each chapter contains a variety of applied examples, both real and simulated, that are intended to get students motivated and thinking about abstract concepts in a concrete setting. In most cases, applied examples are presented directly after the mathematical concepts have been developed. Some of the useful applications of statistics and probability that are stressed include applications to queuing theory, reliability theory, acceptance sampling, and computer performance analysis. These applications are representative of some of the simpler modeling problems that an engineer is likely to encounter in industry and finance. Many of the examples included throughout the text use both statistical software packages and scientific calculators and include fully worked solutions to help reinforce students' mastery of the basic concepts.

Probability and Statistical Inference.   Because probability theory is often a difficult topic for students to understand, I introduce it as a tool for constructing a mathematical model of a random sample drawn from a population. This distinct focus enables me to explain and clarify the link between probability theory and statistical inference. This clarification paves the way for the introduction of important concepts of sampling from a distribution, randomization, and probability modeling.

Hypothesis Testing.   Besides classical hypothesis testing, equal coverage is given to the confidence interval approach to hypothesis testing. This approach gives students a useful tool for their chosen disciplines because the confidence interval approach is most commonly used by working engineers.

Reliability Theory. Data sets and concepts from reliability theory, acceptance sampling, and queuing theory are introduced early in the text, beginning in Chapter 1, with additional material in Chapters 3 and 5.

Statistical Software. I strongly recommend the use of a statistical software package (although it is not required) with this textbook because it lets the instructor and students spend time focusing on the interpretation of the output, not just working on the tedious computations involved. In addition, many of the problems and examples in the book can be solved numerically using a statistical software package or scientific calculator.

To help familiarize students with modern methods for graphically displaying data, I included output from several standard statistical software packages, including MINITAB and SAS. The output is annotated to help students learn how to effectively read and interpret the types of graphical displays they can expect to encounter as working scientists and engineers.

SUPPLEMENTS

Instructor's Solutions Manual. A manual containing worked solutions to all the problems in the textbook is available to adopters of the textbook.

Data Sets. Selected data sets are available electronically to adopters of the textbook.

ACKNOWLEDGMENTS

The publication of a new book is not possible without the cooperation of a large number of highly talented individuals, so it is a great pleasure for me to have this opportunity of thanking them. First, I want to thank my editors at McGraw-Hill: Eric Munson for initiating this project and Maggie Rogers for gently guiding and prodding me to its completion. While writing this book I benefited greatly from the criticisms and the many helpful suggestions from my colleagues, including Professors John Buonacorsi, Joe Horowitz, H. K. Hsieh, Ramesh Korwar, A. Rudvalis, Morris Skibinsky, Sundar Subramanian, and Mike Sutherland. I owe a special thanks in particular to Professor Craig Zirbel of Bowling Green State University, who read several chapters with great care and made many useful suggestions for improving them. I am particularly indebted to Eva Goldwater and Trina Hosmer for valuable assistance with SAS and MINITAB. I am also grateful to Mark and Elizabeth Rosenkrantz for correcting several mistakes and for providing a student's perspective on an earlier version of this book.

The task of preparing the original version using the LaTeX document preparation system was made considerably easier thanks to the support and assistance of the Research Computing Facility (Scott Conti, Director) and his excellent technical staff, Volker Ecke, Gordon Kieffer, Mark Stowell, and Rick Tuthill. Thanks also go to Karen Hawk and her staff at Publication Services for their assistance in producing the final textbook.

It is also a pleasure to thank the numerous reviewers, including Saul Blumenthal, Ohio State University; Kenneth E. Case, Oklahoma State University; Jeffery K. Cochran, Arizona State University; Peyton Cook, University of Tulsa; Laura Raiman DuPont; Muhammad El-Taha, University of Southern Maine; John R. English, University of Arkansas; Kwang-Jae Kim, Pennsylvania State University; Mansooreh Mollaghasemi, University of Central Florida; Emily S. Murphree, Miami University; Roger Nelsen, Lewis and Clark College; William I. Notz, Ohio State University; Dennis M. O'Brien, University of Wisconsin–LaCrosse; Eric V. Slud, University of Maryland; Alice E. Smith, University of Pittsburgh; James R. Smith, Tennessee Technological University; and Gary S. Wasserman, Wayne State University. Their criticisms were always valuable, even if they were not always accepted.

I am also grateful to the American Association for the Advancement of Science, the American Statistical Association, the Biometrika Trustees, Elsevier Science, Iowa State University Press, Richard D. Irwin, McGraw-Hill, Prentice Hall, Routledge, Chapman & Hall, the Royal Society of Chemistry, and John Wiley & Sons for permission to use copyrighted material. *A Handbook of Small Data Sets*, edited by D. J. Hand, F. Daly, A. D. Lunn, K. J. McConway, and E. Ostrowski (Chapman & Hall, 1994), was a great help to me in finding interesting data sets for inclusion in this text. I have made every effort to secure permission from the original copyright holders for each data set, and I would be grateful to my readers for calling my attention to any omissions so they can be corrected by the publisher.

Finally, I dedicate this book to my wife, Linda, and children, Mark and Liz, for their patience and support while I was sometimes busy, frequently busy, and always busy working on this book.

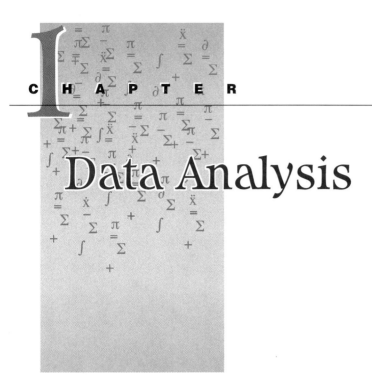

CHAPTER 1

Data Analysis

Information, that is imperfectly acquired, is generally as imperfectly retained; and a man who has carefully investigated a printed table, finds, when done, that he has only a very faint and partial idea of what he has read; and that like a figure imprinted on sand, is soon totally erased and defaced.

William Playfair (1786), *Commercial and Political Atlas*

1.1 ORIENTATION

Statistics is the science and art of collecting, displaying, and interpreting data in order to test theories and make inferences concerning all kinds of phenomena. Scientists and engineers use statistics to summarize and interpret data so that they can make inferences. Statistical software packages such as MINITAB[1] and SAS[2] produce graphical displays that are very useful for visualizing and interpreting the data. It is therefore of no small importance that we be able to understand and interpret the current methods of graphically displaying statistical data. However, it is worth keeping in mind that the primary purpose of a statistical analysis is insight and understanding—not mindless, formal calculations using statistical software packages.

In this chapter we develop the concepts and techniques of data analysis in the context of examples taken from science and engineering.

---

[1]MINITAB is a registered trademark of Minitab, Inc.

[2]SAS is a trademark of the SAS Institute, Inc., Cary, North Carolina.

Organization of Chapter

1.2 THE ROLE AND SCOPE OF STATISTICS IN SCIENCE AND ENGINEERING

A scientific theory, according to the philosopher of science Karl Popper, is characterized by "its falsifiability, or refutability, or testability" (*Conjectures and Refutations: The Growth of Scientific Knowledge*, New York, Harper Torchbooks, 1965, p. 37). In practical terms, this means that a scientist tests a theory by performing an experiment and observing and recording the values of one or more variables of interest. The variable of interest for a chemist might be the atomic weight of an element; for an engineer, it might be the lifetime of a TV monitor; for an epidemiologist, it might be the prevalence of a disease in a given population; for the Bureau of the Census, it may be the population of each state, which is used to reapportion political power in accordance with the U.S. Constitution. The next three examples illustrate how we use statistical models to validate scientific theories, interpret data, and grapple with the variability inherent in all experiments.

---

example 1.1 Einstein's special relativity theory, published in 1905, asserts that the speed of light is a constant value, independent of position, inertial frame of reference, direction, or time. The physicists A. A. Michelson and E. W. Morley (1881 and 1887) and D. Miller (1924) performed a series of experiments to determine the velocity of light. They obtained contradictory results. Michelson and Morley could not detect significant differences in the velocity of light, in support of relativity theory, but Miller's experiments led him to the opposite conclusion. The most plausible explanation for the different results is that the experimental apparatus had to detect differences in the velocity of light as small as one part in 100,000,000, and, unfortunately, temperature changes in the laboratory as small as 1/100 of a degree could produce an effect three times as large as the one Miller was trying to detect. In addition, Miller's interferometer was so sensitive that "a tiny movement of the hand, or a slight cough, made the interference fringes so unstable that no readings were possible" (R. Clark, *Einstein: The Life and Times*, New York, World Publishing, 1971, p. 329). In spite of their contradictory results, all of the scientists were rewarded for their work. Michelson won the Nobel Prize for Physics in 1907, and Miller received the American Association for the Advancement of Science Award in 1925. Reviewing this curious episode in the history of science, H. Collins and T. Pinch wrote, "Thus, although the famous Michelson-Morley experiment of 1887 is regularly taken as the first, if inadvertent, proof of relativity, in 1925, a more refined and complete version of the experiment was widely hailed as, effectively, disproving relativity" (*The Golem: What Everyone Should Know about Science*, New York, Cambridge University Press, 1993, p. 40).

Although the debate between Michelson and Miller continued—they confronted each other at a scientific meeting in 1928 and agreed to differ—the physics community now accepts special relativity theory as correct. This demonstrates that *the validity of a scientific experiment strongly depends upon the theoretical framework within which the data are collected, analyzed, and interpreted.* The British astronomer Sir Arthur Eddington (1882–1944) perhaps put it best when he said that "no experiment should be believed until confirmed by theory." From the statistician's perspective, this shows how experimental error can invalidate experimental results obtained by highly skilled scientists using the best available state-of-the-art equipment; *if the design of the experiment is faulty, then no reliable conclusions can be drawn from the data.*   ■

---

example **1.2**   In his famous breeding experiment the geneticist Mendel classified peas according to their shape [round (r) or wrinkled (w)] and color [yellow (y) or green (g)]. Each seed was classified into one of four categories: ry = (round, yellow), rg = (round, green), wy = (wrinkled, yellow), and wg = (wrinkled, green). According to Mendelian genetics, the frequency counts of seeds of each type produced in this experiment occur in the following ratio:

Mendel's predicted ratio: $ry : rg : wy : wg = 9 : 3 : 3 : 1$

For example, the ratio of the number of ry peas to the number of rg peas should be 9 : 3. An unusual feature of Mendel's model should be noted: It predicts a set of frequency counts (called a *frequency distribution*) instead of a single number. The actual and predicted counts obtained by Mendel for $n = 556$ peas appear in Table 1.1. The noninteger values appearing in the third column result from dividing the 556 peas into four categories in the ratio 9 : 3 : 3 : 1. This means that the first category has $556 \times (9/16) = 312.75$ peas, the next two categories each have $556 \times (3/16) = 104.25$ peas, and the last category has $556 \times (1/16) = 34.75$ peas.

TABLE 1.1 MENDEL'S DATA

Seed type	Observed frequency	Predicted frequency
ry	315	312.75
wy	101	104.25
rg	108	104.25
wg	32	34.75

From the results in Table 1.1 we see that the observed counts are close but not exactly equal to those predicted by Mendel. This leads to one of the most fundamental scientific questions: How much agreement is there between the collected data and the scientific model that explains it? Are the discrepancies between the predicted counts and the observed counts small enough to be attributable to chance, or are they so large as to cast doubts on the theory itself? One problem with Mendel's data, first pointed out by Sir R. A. Fisher (*Annals of Science*, 1936, pp. 115–137), is that every one of his data sets fits his predictions extremely well—too well, in fact. Statisticians measure how well the experimental data fit a theoretical model by using the χ^2 (*chi-square*) statistic (which we study in Chap. 9). A large χ^2 value provides strong evidence against the model. Fisher noted that the χ^2 values for Mendel's numerous experiments were smaller than would be expected from random sampling; that is, *Mendel's data were too good to be true.*[3] Fisher's method of analyzing Mendel's data is a good example of *inferential statistics*, which is the science of making inferences from the data based upon the theory of probability. This theory is presented in Chaps. 2 to 6. ∎

The scope of modern statistics extends far beyond the validation of scientific theories, as important as this task might be. Statistics is also used to search for mathematical models that explain the given data and that are useful for predicting new data. In brief, it is a highly useful tool for inductive learning from experimental data. The next example illustrates how misinterpreting statistical data can seriously harm a corporation, its reputation, and its employees.

example **1.3**   The lackluster performance of the American economy in the last quarter of the twentieth century has profoundly affected corporations, workers, and families. Massive layoffs, the household remedy of America's managers for falling sales and mounting losses, do not appear to have aided the basic mission of the modern corporation, which is to produce high-quality products at competitive prices. Even the ultimate symbol of success, the CEO of a major American corporation, was no longer welcome at many college commencements, being "too unpopular among students who cannot find entry level jobs" (*New York Times*, 29 May 1995). W. E. Deming (1900–1993), originally trained as a physicist but employed as a statistician, became a highly respected, if not always welcome,

---

[3]In spite of the fudged data, Mendelian genetics has survived and is today regarded as one of the outstanding scientific discoveries of all time.

consultant to America's largest corporations. "The basic cause of sickness in American industry and resulting unemployment," he argued, "is failure of top management to manage" (*Out of the Crisis*, Cambridge, Mass., MIT Press, 1982, p. ix). This is a sharp departure from the prevalent custom of blaming workers when poor product quality leads to a decline in sales, profits, and dividends. To illustrate this point he devised a simple experiment that is really a parable, a simple story with a simple moral lesson.

Deming's Parable of the Red Bead Experiment

(This example is adapted from W. Edwards Deming, *Out of the Crisis*, Cambridge, Mass., MIT Press, 1982, pp. 109–112.) Consider a bowl filled with 800 red beads and 3200 white beads. The beads in the bowl represent a shipment of parts from a supplier; the red beads represent the defective parts. Each worker stirs the beads and then, while blindfolded, inserts a special tool with which he draws out exactly 50 beads. The aim of this process is to produce white beads; red beads, as previously noted, are defective and will not be accepted by the customers. Table 1.2 shows the results for six workers.

TABLE 1.2 DATA FROM DEMING'S RED BEAD EXPERIMENT

Name	Number of red beads produced
Mike	9
Peter	5
Terry	15
Jack	4
Louise	10
Gary	8
Total	51

It is clear that the workers' skills in stirring the beads are irrelevant to the final results; the observed variation between them is due solely to chance. As Deming writes,

> It would be difficult to construct physical circumstances so nearly equal for six people, yet to the eye, the people vary greatly in performance.

For example, from the data in Table 1.2 we are tempted to conclude that Jack, who produced only 4 red beads, is an outstanding employee and that Terry, who produced nearly four times as many red beads as Jack, is incompetent and should be immediately dismissed. To explain the data, Deming goes on to calculate what he calls *the limits of variation attributable to the system*. Omitting the technical details—we will explain his mathematical model, methods, and results in Chapter 4 (Example 4.16)—Deming concludes that 99 percent of the observed variation in the workers' performance is due to chance. He continues:

The six employees obviously all fall within the calculated limits of variation that could arise from the system that they work in. There is no evidence in these data that Jack will in the future be a better performer than Terry. Everyone should accordingly receive the same raise. . . . It would obviously be a waste of time to try and find out why Terry made 15 red beads, or why Jack made only 4.

In other words, the quality of the final product will not be improved by mass firings of poorly performing workers. The solution is to improve the system. Management should remove the workers' blindfolds or switch to a supplier that will ship fewer red beads. ∎

This example illustrates one of the most important tasks of the engineer, which is *to identify, control, and reduce the sources of variation* in the manufacture of a commercial product.

1.3 DATA DEFINED ON A POPULATION

The key to data analysis is to construct a mathematical model so that information can be efficiently organized and interpreted. Formally, we begin constructing the mathematical model by stating the following definition.

▬ DEFINITION 1.1

1. A *population* is a set, denoted S, of well-defined distinct objects, the elements of which are denoted by s.
2. A *sample* is a subset of S, denoted by $A \subset S$. We represent the sample by listing its elements as a sequence $A = \{s_1, \ldots, s_n\}$. ∎

To illustrate these concepts, we now consider an example of a sample drawn from a population.

example 1.4 The federal government requires manufacturers to monitor the amount of radiation emitted through the closed door of a microwave oven. One manufacturer measured the radiation emitted by 42 microwave ovens; the recorded values are shown in Table 1.3 and constitute a *data set*. Statisticians call these *raw data*—a list of measurements whose values have not been manipulated in any way. The set of microwave ovens produced by this manufacturer is an example of a population; the subset of 42 ovens whose emissions were measured is a sample. We denote its elements by $\{s_1, \ldots, s_{42}\}$.

Defining a Variable on a Population. We denote the radiation emitted by the microwave oven s by $X(s)$, where X is a function with domain S. It is worth noting that the observed variation in the emitted radiation values comes from the sampling procedure, and the observed value of the radiation is a function of the particular oven tested. From this point of view, Table 1.3 lists the values of the function X evaluated on the sample $\{s_1, \ldots, s_{42}\}$. In particular, $X(s_1) = 0.15$, $X(s_2) = 0.09, \ldots, X(s_{42}) = 0.05$. Less formally, X can be described as a *variable defined on a population*. Engineers are usually interested in several variables defined on the same population. For example, the manufacturer also measured the amount of radiation emitted through the open door of each of the 42 microwave ovens. (This data set is presented later, in Prob. 1.1.) In this text we will denote the ith item in the data set by the symbol x_i, where $X(s_i) = x_i$, and the data set of which it is a member by $X = \{x_1, \ldots, x_n\}$. Note that we use the same symbol X to denote both the data set and the function $X(s)$ that produces it. The sequential listing corresponds to reading from left to right across the successive rows of Table 1.3. In detail,

$$x_1 = 0.15, \ x_2 = 0.09, \ \ldots, \ x_8 = 0.05, \ \ldots, \ x_{42} = 0.05$$

We summarize these remarks in the following formal definition.

■ **DEFINITION 1.2**

1. The elements of a *data set* are the values of a real-valued function $X(s_i) = x_i$ defined on a sample $A = \{s_1, \ldots, s_n\}$.
2. The number of elements in the sample is called the *sample size*. ■

TABLE 1.3 RAW DATA FOR THE RADIATION EMITTED BY 42 MICROWAVE OVENS

0.15	0.09	0.18	0.10	0.05	0.12	0.08
0.05	0.08	0.10	0.07	0.02	0.01	0.10
0.10	0.10	0.02	0.10	0.01	0.40	0.10
0.05	0.03	0.05	0.15	0.10	0.15	0.09
0.08	0.18	0.10	0.20	0.11	0.30	0.02
0.20	0.20	0.30	0.30	0.40	0.30	0.05

Source: R. A. Johnson and D. W. Wichern, *Applied Multivariate Statistical Analysis,* 3d ed., Englewood Cliffs, N.J., 1992, Prentice Hall, p. 156. Used with permission. ■

1.3.1 Other Types of Data

In many situations of practical interest the objective is to compare two populations with respect to some numerical characteristic. For example, we might be interested in determining which of two gasoline blends yields more miles per gallon (mpg). Problems of this sort lead to data sets with a more complex structure than those considered previously. We now look at an example where the data have been obtained from two populations.

example 1.5   A light bulb manufacturer tested 10 light bulbs containing filaments of type A and 10 light bulbs with filaments of type B. The values for the observed lifetimes of the 20 light bulbs are recorded in Table 1.4. The purpose of the experiment was to determine which filament had longer lifetimes.

TABLE 1.4 LIFETIMES (IN HOURS) OF LIGHT BULBS WITH TYPES A AND B FILAMENTS

A	B	A	B
1293	1061	1643	1138
1380	1065	1466	1143
1614	1092	1627	1094
1497	1017	1383	1270
1340	1021	1711	1028

In this example we have two samples taken from two populations, each consisting of $n = 10$ light bulbs. The lifetimes of light bulbs containing type A and type B filaments are denoted by the variables X and Y, where X = lifetime of type A light bulb and Y = lifetime of type B light bulb. ∎

Multivariate Data. In the previous section we defined the concept of a single variable X defined on a population S. Multivariate data sets arise in the study of values of two or more variables defined on the same population. Data collected in public health studies, as discussed in the next example, are frequently of this sort.

example 1.6 The original StatLab population consisted of 1296 member families of the Kaiser Foundation Health Plan living in the San Francisco Bay area during the years 1961–1972. These families participated in a Child Health and Development Studies project conducted under the supervision of the School of Public Health, University of California, Berkeley. The purpose of the study was "to investigate how certain biological and socioeconomic attributes of parents affect the development of their offspring." The data for the 36 families listed in Table 1.5 come from a much larger data set of 648 families to which a baby girl had been born between 1 April 1961 and 15 April 1963. Ten years later, between 5 April 1971 and 4 April 1972, measurements were taken on 32 variables for each family, only 10 of which are listed here.

Key to the Variables. From left to right in the headings of Table 1.5, the 10 variables are girl's height (*height*), girl's weight (*weight*), girl's score on the Peabody Picture Vocabulary Test (*PEA*), girl's score on the Raven progressive matrices test (*RAV*), mother's height (*mheight*), mother's weight (*mweight*), mother's education level (*meduc*), father's height (*fheight*), father's weight (*fweight*), and father's education level (*feduc*). The level of education was coded according to the following scheme:

Code	Education of parent
0	Less than 8th grade
1	8th to 12th grade
2	High school graduate
3	Some college
4	College graduate

TABLE 1.5 STATLAB DATA SET CONSISTING OF 36 OBSERVATIONS
ON 10 VARIABLES

height	weight	PEA	RAV	mheight	mweight	meduc	fheight	fweight	feduc
55.7	85	85	34	66.0	130	1	70.1	171	3
48.9	59	74	34	62.8	159	3	65.0	130	1
54.9	70	64	25	66.1	138	2	70.0	175	2
53.6	88	87	43	61.8	123	2	71.8	196	3
53.4	68	87	40	62.8	146	2	68.0	163	2
59.9	93	83	37	63.4	116	2	74.0	180	3
53.1	72	81	33	65.4	220	2	68.1	173	2
52.2	84	74	37	62.3	120	2	72.0	150	3
56.8	68	72	21	65.4	141	3	71.0	150	2
53.8	76	64	31	66.4	184	2	75.0	235	3
49.0	51	72	29	64.8	107	1	70.0	145	3
51.9	59	87	38	66.9	114	2	72.8	197	3
55.7	78	72	19	65.6	151	1	74.0	204	2
53.4	73	78	27	63.6	123	2	71.0	220	3
50.3	60	77	35	65.4	143	2	64.0	150	2
55.8	70	75	29	63.3	150	2	69.0	180	3
55.2	78	71	25	63.7	149	2	69.0	170	3
58.8	106	60	20	64.3	205	2	72.0	235	2
53.6	76	79	36	58.3	121	3	75.0	190	3
53.3	64	63	32	63.3	147	0	70.0	190	2
49.9	50	80	42	65.3	134	2	68.5	160	3
53.8	56	66	28	66.4	117	2	65.5	130	1
53.9	68	65	38	65.6	151	3	69.0	160	3
53.1	68	69	15	63.8	150	3	73.5	154	2
56.1	64	63	24	62.0	110	2	70.8	183	3
51.7	56	70	14	59.6	138	2	64.9	169	1
48.0	64	51	40	65.1	128	2	67.0	146	4
51.3	58	80	45	62.0	132	2	71.0	165	2
50.5	60	75	29	65.4	144	2	71.0	190	3
57.3	79	96	38	68.1	143	3	73.5	185	3
52.8	64	94	25	66.3	147	2	71.0	180	3
51.3	56	82	34	61.0	113	2	71.0	149	2
50.1	56	82	39	64.6	116	2	66.0	145	3
48.7	50	74	35	62.8	132	2	67.8	166	3
57.7	89	90	35	66.8	134	3	70.5	200	2
54.8	65	64	32	67.0	131	2	73.0	175	2

Source: J. L. Hodges, Jr., D. Krech, R. S. Crutchfield, *StatLab,* New York, McGraw-Hill, 1975. Used with permission.

The education level of the parent is a *categorical variable;* that is, the value of the variable is nonnumerical. Other examples of categorical variables are responses to a public opinion poll (*support, oppose, undecided*), quality control of a product (*defective, nondefective*), and Mendel's genetics data.

In this and subsequent chapters we examine several widely used graphical and analytical methods for studying multivariate data.   ∎

1.4   THE FREQUENCY DISTRIBUTION OF A VARIABLE DEFINED ON A POPULATION

In this section we first examine the basic concept of the *frequency distribution of a variable defined on a population*. We then introduce a variety of tabular, graphical, and analytic methods that statisticians use to answer the question of how the data are distributed. Among the most widely used graphical displays are *bar charts, dot plots, stem-and-leaf displays, histograms*, and *box plots*. These methods are elementary in the sense that no knowledge of calculus is required. Because our primary focus is on understanding and interpreting these graphs, we deemphasize the tedious details of their construction and produce these displays using statistical software packages.

1.4.1   Organizing the Data

We organize and display data in order to extract information from them. Because the purpose of collecting the data in Table 1.3 was to determine the proportion of microwave ovens that emit radiation below an acceptable level, the easiest way to obtain this information is to first sort the data in *increasing* order of magnitude, as in Table 1.6.[4] The technique of *sorting* or *ranking* the observations of a data set in increasing order of magnitude is widely used in statistics, and the following terminology is standard.

TABLE 1.6   ORDER STATISTICS OF THE RADIATION DATA OF TABLE 1.3

0.01	0.01	0.02	0.02	0.02	0.03	0.05
0.05	0.05	0.05	0.05	0.07	0.08	0.08
0.08	0.09	0.09	0.10	0.10	0.10	0.10
0.10	0.10	0.10	0.10	0.10	0.11	0.12
0.15	0.15	0.15	0.18	0.18	0.20	0.20
0.20	0.30	0.30	0.30	0.30	0.40	0.40

■ **DEFINITION 1.3**

When the observations $\{x_1, x_2, \ldots, x_n\}$ are sorted in increasing order,

$$\{x_{(1)} \leq x_{(2)} \leq \cdots \leq x_{(n)}\}$$

the elements of this set are called the *order statistics* and $x_{(i)}$ is called the *ith order statistic*.   ■

Looking at Table 1.6, we see that it is easier to extract useful information from sorted data. For example, the smallest observed value is 0.01, which appears as

---

[4]Both MINITAB and SAS have "commands" or "procedures" that will sort the data in increasing order of magnitude.

the thirteenth entry in the original data set (Table 1.3); thus $x_{(1)} = 0.01 = x_{13}$. The largest observed value is 0.40, which appears as the twentieth entry, so $x_{(42)} = 0.40 = x_{20}$. This shows that, in general, $x_{(i)} \neq x_i$.

The difference between the smallest observed value and the largest observed value, which equals $0.40 - 0.01 = 0.39$, is called the *sample range*. It is a crude measure of the amount of variability in the data. We can express the smallest and largest values in terms of the order statistics as follows:

$$x_{(1)} = \min\{x_1, \ldots, x_n\}$$
$$x_{(n)} = \max\{x_1, \ldots, x_n\}$$

The primary focus of a statistical study is the frequency distribution of the data set. Next we explain how to use order statistics to compute it.

Frequency Distribution of a Variable. We compute the frequency distribution of a variable by counting the number of times that the value x appears; this number is denoted $f(x)$. The smallest observed value is 0.01 and occurs twice, so $f(0.01) = 2$; similarly, $f(0.02) = 3$, $f(0.03) = 1$, etc. The value 0.10 has the highest observed frequency, $f(0.10) = 9$; it is called the *mode*. In general, the mode is the value that occurs most often. A data set can have two or more modes.

Cumulative Frequency Distribution. The cumulative frequency distribution is a count of the number of observed values *less than or equal to x*. It is denoted $F(x)$. Looking at Table 1.6, we see that there are six values less than or equal to 0.03, so $F(0.03) = 6$. The cumulative frequency $F(x)$ is the sum of the frequencies of the values less than or equal to x. For example,

$$F(0.03) = f(0.01) + f(0.02) + f(0.03) = 2 + 3 + 1 = 6$$

All major statistical software packages have programs that not only compute these distributions but display the results in attractive visual formats. We study several of these formats in the next section.

1.4.2 Graphical Displays

The art of displaying data in an informative and visually attractive format is called *exploratory data analysis,* which is the name statisticians give to a collection of methods for constructing graphical displays of data. One very informative display is the *horizontal bar chart*. It is a compact graphical display of the order statistics, frequency distribution, and cumulative frequency distribution. Figure 1.1 is the horizontal bar chart for the radiation data in Table 1.3.

The information in the horizontal bar chart is laid out in columns as follows:

- **Column 1.** The entries in the leftmost column, where R = emitted radiation, are the distinct observed values sorted in increasing order.
- **The bar chart.** Between the first and second columns is a visual display of the frequency of each observation, where the width of the bar to the right of the observed value x is proportional to the frequency $f(x)$.

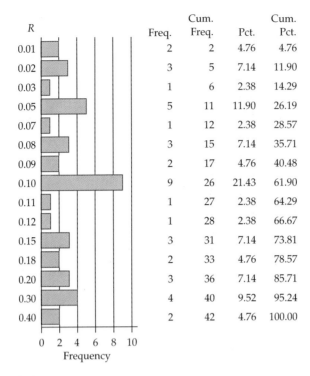

R	Freq.	Cum. Freq.	Pct.	Cum. Pct.
0.01	2	2	4.76	4.76
0.02	3	5	7.14	11.90
0.03	1	6	2.38	14.29
0.05	5	11	11.90	26.19
0.07	1	12	2.38	28.57
0.08	3	15	7.14	35.71
0.09	2	17	4.76	40.48
0.10	9	26	21.43	61.90
0.11	1	27	2.38	64.29
0.12	1	28	2.38	66.67
0.15	3	31	7.14	73.81
0.18	2	33	4.76	78.57
0.20	3	36	7.14	85.71
0.30	4	40	9.52	95.24
0.40	2	42	4.76	100.00

0   2   4   6   8   10
Frequency

Figure 1.1 Frequency distribution of radiation data (Table 1.3)

- **Column 2 (Freq.).** This is the frequency function $f(x)$. For example, to compute $f(0.09)$, scan down the first column until you locate 0.09 and then read across to the second column to find a value of 2; that is, $f(0.09) = 2$.
- **Column 3 (Cum. Freq.).** This is the cumulative frequency distribution $F(x)$. To determine, for instance, how many microwave ovens emitted radiation in amounts less than or equal to 0.20, scan down the first column until you reach 0.20 and then read across to the entry 36 in column 3.
- **Columns 4 (Pct.) and 5 (Cum. Pct.).** Columns 4 and 5 contain the values in columns 2 and 3 restated in terms of percentages. For example, to determine the percentage of microwave ovens whose emitted radiation level equals 0.10, scan down the first column until you locate 0.10 and then read across to 21.43 in column 4; that is, 21.43 percent of the ovens emitted radiation in the amount of 0.10. To determine the percentage of microwave ovens whose emitted radiation level is less than or equal to 0.10, read across to column 5 to find 61.90; that is, 61.90 percent of the ovens emitted radiation at a level less than or equal to 0.10.

Notation. We use the symbol $\#(A)$ to denote the number of elements in the set A.

example 1.7

Refer to the StatLab data set (Table 1.5). Let X denote the Peabody test scores in the third column and let $A = \{x_i : x_i\} = 64$. Then, using the $\#(A)$ symbol, we have

$$\#\{x_i : x_i = 64\} = 3$$
$$\#\{x_i : x_i \leq 64\} = 7$$
$$\#\{x_i : x_i > 64\} = 29$$ ∎

The Empirical Distribution Function. The *empirical distribution function* $\hat{F}_n(x)$ is the proportion of the observed values less than or equal to x, so it equals $F(x)/n$, where $F(x)$ is the cumulative frequency distribution. The formal definition follows.

■ **DEFINITION 1.4**

The *empirical distribution function* $\hat{F}_n$ of the data set $\{x_1, x_2, \ldots, x_n\}$ is the function defined by

$$\hat{F}_n(x) = \frac{\#\{x_i : x_i \leq x\}}{n}$$

That is, the empirical distribution function $\hat{F}_n(x)$ is the proportion of values that are less than or equal to x. ∎

The function $100 \times \hat{F}_n(x)$ denotes the percentage of values that are less than or equal to x; this is the function displayed in the column labeled Cum. Pct. in Fig. 1.1.

The empirical distribution and the frequency distribution functions are related to one another by the equation

$$\hat{F}_n(x) = \frac{1}{n}\sum_{y \leq x} f(y) \tag{1.1}$$

The empirical frequency distribution, denoted $\hat{f}(x)$, is the frequency distribution divided by the total number of observations; that is,

$$\hat{f}(x) = \frac{f(x)}{n}$$

Note: Some authors call $\hat{F}_n(x)$ the *sample distribution function.*

example 1.8

Suppose the goal of the manufacturer in Example 1.4 is to produce microwave ovens whose level of radiation emissions does not exceed 0.25. What proportion of the ovens tested meet this standard?

Solution. Looking at the order statistics in the horizontal bar chart (Fig. 1.1), we see that 36 microwave ovens emitted radiation less than or equal to 0.25; consequently, the proportion of ovens meeting the standard is $\hat{F}_n(0.25) = 36/42 = 0.857$. Notice that

$$\hat{F}_n(0.25) = \hat{f}(0.01) + \hat{f}(0.02) + \cdots + \hat{f}(0.18) + \hat{f}(0.20)$$
$$= 0.0476 + 0.0714 + \cdots + 0.0476 + 0.0714 = 0.857$$ ∎

example 1.9   Plot the empirical distribution function of the microwave oven radiation data
(Table 1.3).

Solution. The graph of $\hat{F}_n(x)$ is shown in Fig. 1.2. Notice that this graph resembles
a staircase with steps located at the order statistics. Looking at this staircase, we
see that the height of each step at x—that is, the size of the jump j—equals $f(x)/n$.
In particular, the graph of the empirical distribution function of the microwave
radiation data starts with the first order statistic $x_{(1)} = 0.01$, whose frequency
is $f(0.01) = 2$ (see Table 1.6) and plots the point $(0.01, 2/42) = (0.01, 0.0476)$.
The next distinct order statistic is 0.02, and its frequency is $f(0.02) = 3$; so the
point $\{0.02, [f(0.01) + f(0.02)]/42\} = (0.02, 5/42) = (0.02, 0.119)$ is plotted. This
procedure continues until the last two points, $(0.30, 40/42) = (0.30, 0.9524)$ and
$(0.40, 42/42) = (0.40, 1)$, are plotted. Note that $f(0.10) = 9$, so the jump at 0.10
equals $9/42 = 0.214$, and that $f(0.25) = 0$, so the jump at 0.25 is zero; that is, $\hat{F}_n$
is continuous at $x = 0.25$.   ∎

The empirical distribution function is an example of a *piecewise-defined
function*; that is, the formula which produces the output $\hat{F}_n(x)$ depends on the
interval where the input x is located.

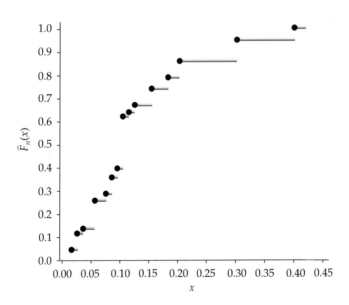

Figure 1.2 Empirical distribution function of radiation data

Dot Plots. Another useful way to display the radiation data in Table 1.3 is the
dot plot, in which each observation x_i is displayed as a "dot" above the point x_i
on the x axis, as shown in Fig. 1.3. The frequency of x_i is the number of dots
above x_i.

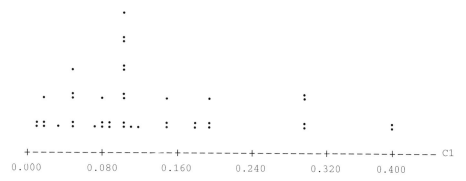

Figure 1.3 Dot plot (produced by MINITAB) of emitted microwave radiation data of Table 1.3

The dot plot is more than an alternative visual display to the horizontal bar chart; it is also a useful first glance at the *shape* and *variability* of the distribution. Looking at Fig. 1.3, we see, for example, that the values are unevenly distributed along the x axis; there is a single *peak* at $x = 0.10$, around which most of the data seem to be concentrated, and a *tail* extending rightward out to the value $x = 0.40$. A *skewed* distribution is one that has a single peak and a long tail in one direction. A distribution is *skewed to the right*, or *positively skewed*, when the tail goes to the right, as is the case here. This skewed distribution suggests that a few of the microwave ovens are emitting excessive amounts of radiation.

example **1.10** The father's height (*fheight*) in the StatLab data set (Table 1.5) is an example of a variable with a *symmetric distribution*. Looking at the dot plot shown in Fig. 1.4, we see that the data seem to have a well-defined single peak at approximately 71 inches, about which they appear to be symmetrically distributed. ∎

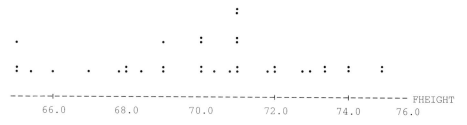

Figure 1.4 Dot plot of the fathers' heights from Table 1.5

Stem-and-Leaf Plots. One of the quickest ways of visualizing the shape of a distribution with a minimum of computational effort is the stem-and-leaf plot.

Assume that each observation x_i consists of at least two digits, as in the radiation data (Table 1.3). Then break each observation into a *stem* consisting of the leading digit(s) and a *leaf* consisting of the remaining digits; decimal points

Stem	Leaf	Frequency
0	95858721215359825	17
1	5802000000505801	16
2	000	3
3	000	4
4	00	2

Figure 1.5 Stem-and-leaf plot of the radiation data in Table 1.3

are ignored in a stem-and-leaf display. For the data in Table 1.3 look only at the digits after the decimal point—we ignore not only the decimal point but the leading digit 0 as well. The observation 0.15 has a stem of 1 and a leaf of 5, 0.09 has stem 0 and leaf 9, and 0.10 has stem 1 and leaf 0. In detail:

Data value	Split	Stem	Leaf
0.15 $\rightarrow$	1 | 5	1	5
0.09 $\rightarrow$	0 | 9	0	9
0.10 $\rightarrow$	1 | 0	1	0

We then form two columns: column 1 lists the stems, and column 2 contains the leaves corresponding to the given stem. The result is displayed in Fig. 1.5.

Sorting the leaves in increasing order for each stem yields an *ordered* stem-and-leaf plot. An ordered stem-and-leaf plot produced by MINITAB is displayed in Fig. 1.6.

The stem-and-leaf plot displays in condensed form all the information contained in the horizontal bar chart in Fig. 1.1, including a rough look at the variability of the data.

```
0    11222355555788899
1    0000000001255588
2    000
3    0000
4    00
```

Figure 1.6 Ordered stem-and-leaf plot of radiation data in Table 1.3

Scatter Plots. Horizontal bar charts, dot plots, and stem-and-leaf plots help us to understand the distribution of a single variable. *Scatter plots* are useful for studying the relationship between two variables X and Y. In the following example we explore the possible existence of a link between girls' test scores on the Raven test (Y) and on the Peabody test (X) (Table 1.5).

example **1.11**   For this example, refer to the StatLab data set in Table 1.5. We are particularly interested in exploring the relationship between the girls' scores on the Raven test (Y) and the corresponding scores on the Peabody test (X). The Peabody test measures verbal facility; the Raven test measures a different sort of mental ability, related to geometrical intuition and spatial organization. The Raven test uses no words, only geometrical figures. It is an interesting problem in psychology to determine whether there is any relationship between the verbal facility measured by the Peabody test and the geometrical-spatial abilities measured by the Raven test. The scatter plot is a useful first step for exploring this question. We construct the scatter plot by successively plotting the points (x_i, y_i) $(i = 1, \ldots, 36)$, where x_i and y_i denote the girls' scores on the Peabody and Raven tests, respectively. The first girl scored 85 on the Peabody test and 34 on the Raven test, so we plot the point $(85, 34)$; the second girl scored 74 on the Peabody test and 34 on the Raven test, so we plot the point $(74, 34)$; and so on until all 36 points are plotted as shown in Fig. 1.7.

The scatter plot suggests that there is a weak *positive association* between the scores on the two tests; that is, low Peabody scores appear to be associated with low Raven scores, and high Peabody scores appear to be associated with high Raven scores. There are, however, a few data points that do not follow this pattern; for instance, girl number 27, who scored the lowest on the Peabody test (51) achieved a 40 on the Raven test, which was the fourth highest! Girl number

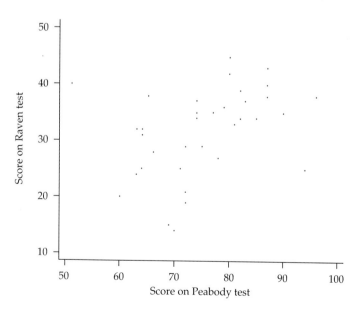

Figure 1.7 Scatter plot of Raven text scores (Y) versus Peabody verbal test scores (X)

31, on the other hand, scored a 94 on the Peabody test (the maximum score was 96) and a relatively low 25 on the Raven test. Consequently, at this stage the precise relationship between these two variables is not clear. We cannot say more at this time because we do not have at our disposal the tools statisticians use to study the association between two variables. We will return to this topic in Chapter 10, where we examine the association between two variables using the tools of regression and correlation analysis.   ■

1.4.3   Histograms

In this section we demonstrate another way to visualize the distribution of a data set, called the *histogram;* it is, in essence, a vertical bar chart constructed in a special way.

Although it is possible to draw a histogram by hand, it is easier, and more reliable, to construct it via a suitable software package. Figure 1.8 is a computer-generated histogram for the radiation data in Table 1.3.

Looking at this histogram, we see that the x axis represents the values of the emitted radiation and the y axis represents the frequency values. The data have been grouped into seven nonoverlapping intervals, called *class intervals.* Their midpoints, called the *class marks,* are

$$0.03, \; 0.09, \; 0.15, \; 0.21, \; 0.27, \; 0.33, \; 0.39$$

and are clearly marked on the x axis.

To describe the class intervals in more detail we use the symbol $[a, b)$ to denote the set of points on the line that are less than b and greater than or equal

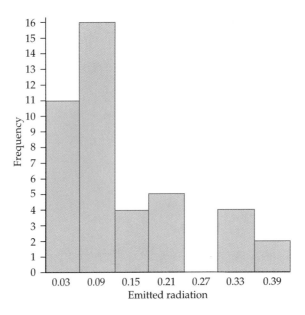

Figure 1.8 Histogram of radiation data (Table 1.3)

to a; that is,

$$[a, b) = \{x : a \le x < b\}$$

The seven class intervals in Fig. 1.8 are

$$[0.0, 0.06), [0.06, 0.12), \ldots, [0.36, 0.42]$$

The lengths of the class intervals are called the *class widths,* which in this case equal 0.06 since the class intervals are of equal length.

Above each class interval is a rectangle whose base has length equal to the class width and whose height equals the class frequency; consequently, the areas of the rectangles are proportional to the class frequencies when the class widths are equal. Looking at the order statistics displayed in Table 1.6, we see that the frequency of the class interval $[0.0, 0.06)$ equals 11. The *relative (class) frequency* is the proportion of the values that are in the class interval, so the relative frequency of the class interval $[0.0, 0.06)$ equals $11/42 = 0.26$.

In the general case we construct a histogram by first dividing the sample range into a small number of nonoverlapping intervals, usually at least 5 but no more than 20, and then counting the number of values in each interval. In detail, we proceed as follows:

1. Choose k nonoverlapping intervals that cover the sample range; they are denoted

$$[u_0, u_1), [u_1, u_2), \ldots, [u_{k-1}, u_k]$$

   The numbers u_i are called the *class boundaries.* The midpoints of the class intervals are denoted by m_j; that is,

$$m_j = (u_j + u_{j-1})/2$$

2. Count the number of observations in the jth class. This number is called the jth class frequency and is denoted by f_j, so that

$$f_j = \#(x_i : u_{j-1} \le x_i < u_j) \quad \text{and} \quad f_n = \#(x_i : u_{n-1} \le x_i \le u_n)$$

   The relative frequency of the jth class is the proportion of observations in that class; it is denoted $\hat{f}_j = f_j/n$.

3. The computer draws the frequency histogram by constructing above each class interval a rectangle with height proportional to the class frequency f_j.

4. The *relative frequency histogram* corresponds to constructing a rectangle whose area equals the relative frequency of the jth class.

Drawing a Histogram by Hand

Although we usually use computers to construct histograms, we can also construct them by hand, an exercise that can give some insight into their computation. There are no hard and fast rules for this kind of construction, but there are some steps and rules of thumb we can follow. We now give a method for determining the number of classes, the class marks, and the class frequencies. We then apply this method to show how to construct the histogram shown in Fig. 1.8.

1. **Determining the number of classes.** The first problem in constructing a histogram is to determine the number of classes. The number of classes should be at least 5 but no more than 20; the exact number to use depends on the sample size n and computational convenience. One rule[5] that has been suggested is this: Choose the number of classes k so that

$$2^{k-1} < n \le 2^k \tag{1.2}$$

Thus, for the radiation data (Table 1.3) we would choose $k = 6$ since $n = 42$ and

$$2^5 = 32 < 42 \le 64 = 2^6$$

2. **Choosing the class marks.** The next rule is that the class intervals should be of equal length, although this is not always possible, and that the class marks should be easily computable. We now describe how these rules were applied to produce the frequency distribution in Table 1.7.

   Looking at the order statistics in Fig. 1.1, we see that the sample range equals $x_{(42)} - x_{(1)} = 0.40 - 0.01 = 0.39$. We divide the sample range into equal subintervals of a length that is easily computable. According to the rule stated in Eq. (1.2), we should divide the data into six classes. This suggests that we divide the sample range into six equal parts, with the length of each interval being equal to $0.39/6 = 0.065$. Since this is not a particularly convenient number to work with, we round it down to the more convenient number 0.06; this leads to the choice of seven class intervals of total length 0.42, which covers the sample range. This leads to the class intervals

$$[0.0, 0.06), [0.06, 0.12), \ldots, [0.36, 0.42]$$

produced by the computer. To avoid any ambiguity in the classification procedure, an observation, such as 0.12, that falls on a class boundary is assigned to the interval on the right; thus 0.12 is placed in the interval $[0.12, 0.18)$.

TABLE 1.7 FREQUENCY TABLE OF RADIATION DATA

Class boundaries	Tally	Frequency	Cumulative frequency
$0 \le x < 0.06$	ℍℍℍℍ /	11	11
$0.06 \le x < 0.12$	ℍℍ ℍℍ ℍℍ /	16	27
$0.12 \le x < 0.18$	////	4	31
$0.18 \le x < 0.24$	ℍℍ	5	36
$0.24 \le x < 0.30$		0	36
$0.30 \le x < 0.36$	////	4	40
$0.36 \le x \le 0.42$	//	2	42

---

[5]This is called *Sturges' rule*; see H. A. Sturges, "The Choice of Class Interval," *Journal of the American Statistical Association*, vol. 21, March 1926, pp. 65–66.

3. **Computing the class frequencies.** The *frequency table* shown in Table 1.7 is formed by counting the number of values that fall into each class interval and recording the counts in the *tally* column. The frequency f_i of the ith class is recorded in the next column; the last column indicates the cumulative frequency.

example **1.12**   The histogram of the fathers' heights recorded in Table 1.5 is shown in Fig. 1.9. It is an example of a distribution that is nearly symmetric about a single well-defined peak.   ■

example **1.13**   A histogram that displays two or more clearly defined peaks suggests that the population from which the data were drawn is *heterogeneous*; that is, the population consists of two or more distinct subpopulations. Figure 1.10 displays the histogram obtained by combining the heights of the girls and of their fathers into one data set consisting of 72 observations.

  Looking at Fig. 1.10, we clearly see two peaks, one centered at 54 inches and the other at 70 inches. Of course, this is not surprising because we knew beforehand that the population consisted of two dissimilar subpopulations. In a typical engineering study this will not be known. To illustrate this point, let us pretend that Fig. 1.10 is the histogram of the lengths of a metal rod (measured in inches) shipped from some manufacturer. To an experienced statistician this histogram suggests that the rods came from two different plants or were produced by two different machines, for example. In other words, there is a serious variation in the quality of the output that the engineers must reduce, if not eliminate.   ■

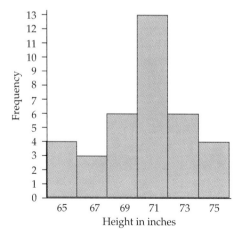

Figure 1.9 Histogram of heights of fathers (Table 1.5)

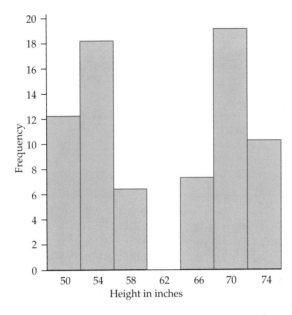

Figure 1.10 Histogram of combined heights of fathers and daughters (Table 1.5)

PROBLEMS

1.1 The manufacturer of microwave ovens (see Example 1.4) also measured the radiation emitted through the open doors of the 42 ovens and recorded the values in the following table.

0.30	0.09	0.30	0.10	0.10	0.12	0.09
0.10	0.09	0.10	0.07	0.05	0.01	0.45
0.12	0.20	0.04	0.10	0.01	0.60	0.12
0.10	0.05	0.05	0.15	0.30	0.15	0.09
0.09	0.28	0.10	0.10	0.10	0.30	0.12
0.25	0.20	0.40	0.33	0.32	0.12	0.12

Source: R. A. Johnson and D. W. Wichern, *Applied Multivariate Statistical Analysis,* 3d ed., Englewood Cliffs, N.J., Prentice Hall, p. 169. Used with permission.

(a) Construct an ordered stem-and-leaf plot of this data set.

(b) Draw the histogram of this data set using six intervals with class marks at $0.05, 0.15, 0.25, \ldots, 0.55$ and with class width 0.1. Describe the shape of the distribution.

1.2 Statisticians prefer to work with distributions that are symmetrically distributed about a single well-defined peak. When the data are skewed, as is the case for the radiation data (see Fig. 1.8), the statisticians Box and

Cox showed that transforming the data by means of the function $y = g(x)$, where

$$g(x) = \begin{cases} \dfrac{x^\lambda - 1}{\lambda} & \lambda \neq 0 \\ \ln y & \lambda = 0 \end{cases}$$

can produce a distribution that is more nearly symmetric for well-chosen λ. With respect to the radiation data in Table 1.3, Johnson and Wichern showed that $\lambda = 0.25$ is a good choice. The transformed data are shown in the following table.

−1.51	−1.81	−1.39	−1.75	−2.11	−1.65	−1.87
−2.11	−1.87	−1.75	−1.94	−2.50	−2.74	−1.75
−1.75	−1.75	−2.50	−1.75	−2.74	−0.82	−1.75
−2.11	−2.34	−2.11	−1.51	−1.75	−1.51	−1.81
−1.87	−1.39	−1.75	−1.33	−1.70	−1.04	−2.50
−1.33	−1.33	−1.04	−1.04	−0.82	−1.04	−2.11

(a) Make an ordered stem-and-leaf plot of the data.
(b) Draw the histogram of the data. Use the class marks

$$-2.7, \ -2.4, \ -2.1, \ -1.8, \ -1.5, \ -1.2, \ -0.9$$

and class width 0.30. Compare the histogram for the transformed data with the histogram for the original data set (Fig. 1.8). Is the distribution of the transformed data symmetric or nearly so?

1.3 Refer to the open-door radiation data in Prob. 1.1. To obtain a data set with a more symmetric distribution, Johnson and Wichern used the same function as in Prob. 1.2; that is, the transformed data are given by

$$y_i = \frac{x_i^{0.25} - 1}{0.25}$$

The transformed data set is shown in the following table.

−1.04	−1.81	−1.04	−1.75	−1.75	−1.65	−1.81
−1.75	−1.81	−1.75	−1.94	−2.11	−2.74	−0.72
−1.65	−1.33	−2.21	−1.75	−2.74	−0.48	−1.65
−1.75	−2.11	−2.11	−1.51	−1.04	−1.51	−1.81
−1.81	−1.09	−1.75	−1.75	−1.75	−1.04	−1.65
−1.17	−1.33	−0.82	−0.97	−0.99	−1.65	−1.65

Draw the histogram. Is its distribution more symmetric than that for the original data in Prob. 1.1?

1.4 Refer to the lifetimes of type B filaments in Table 1.4.
(a) Compute the order statistics.
(b) Compute the cumulative frequency distribution $F(x)$ for $x = 1065, 1100, 1143$.
(c) Graph the empirical distribution function $\hat{F}_n(x)$.

1.5 The following data are measurements of the capacity (in ampere-hours) of 10 batteries.

$$157, 152, 146, 153, 151, 154, 149, 148, 153, 147$$

(a) Compute the order statistics.

(b) Compute the empirical distribution function $\hat{F}_n(x)$ for

$$x = 145, 150, 150.5, 152.9, 153.1, 158$$

(c) Graph the empirical distribution function $\hat{F}_n(x)$.

1.6 The following table lists 17 measurements of the burst strength of a rocket motor chamber.

15.30	17.10	16.30	16.05	16.75	16.60
17.10	17.50	16.10	16.10	16.00	16.75
17.50	16.50	16.40	16.00	16.20	

Source: S. Weerahandi and R. A. Johnson, "Testing Reliability in a Stress–Strength Model When x and y Are Normally Distributed," *Technometrics,* vol. 34, no. 1, 1992, pp. 83–91. Used with permission.

(a) Make a horizontal bar chart.

(b) Make the dot plot and describe the shape of the distribution as belonging to one of the following types: symmetric, skewed left, or skewed right.

(c) Compute the order statistics.

(d) What percentage of the values are less than or equal to 16.10? Greater than or equal to 16.10?

(e) What percentage of the values are less than or equal to 16.40? Greater than or equal to 16.40?

(f) What percentage of the values are less than or equal to 16.75? Greater than or equal to 16.75?

(g) Draw the graph of the empirical distribution function.

1.7 The following table lists 24 measurements (psi) of the operating pressure of a rocket motor.

7.7401	7.7749	7.7227	7.77925	7.96195	7.4472
8.0707	7.89525	8.0736	7.4965	7.5719	7.7981
7.8764	8.1925	8.01705	7.9431	7.71835	7.87785
7.2904	7.7575	7.3196	7.6357	8.06055	7.9112

Source: I. Guttman, R. Johnson, G. Bhattacharyya, and B. Reiser, "Confidence Limits for Stress–Strength Models with Explanatory Variables," *Technometrics,* vol. 30, no. 2, 1988, pp. 161–168. Used with permission.

(a) Compute the order statistics.

(b) What percentage of the values are less than or equal to 7.78? Greater than or equal to 7.78?

(c) What percentage of the values are less than or equal to 7.65? Greater than or equal to 7.65?

(d) What percentage of the values are less than or equal to 7.95? Greater than or equal to 7.95?

(e) Make the frequency histogram; use Sturges' rule, Eq. (1.2), to guide you in choosing the number of classes.

(f) Comment on the shape of the distribution. Is it symmetric, skewed left, or skewed right?

1.8 Carbon dioxide (CO_2) emissions are believed to be associated with the phenomenon of global warming. The following data come from a sample consisting of the 20 countries with the highest CO_2 emissions in 1989.

CO_2 EMISSIONS DATA FOR THE 20 COUNTRIES WITH THE HIGHEST EMISSIONS IN 1989

Country	1989 emis- sions, X_1	1989 emis- sions per capita, X_2	1987 GNP per capita, X_3	1987 emis- sions per GNP, X_4	Country	1989 emis- sions, X_1	1989 emis- sions per capita, X_2	1987 GNP per capita, X_3	1987 emis- sions per GNP, X_4
United States	1,328.3	5.37	18,529	2.8	GDR	88.1	5.40	11,300	4.9
Soviet Union	1,038.2	3.62	8,375	4.3	Mexico	87.3	1.01	1,825	5.4
China	651.9	0.59	294	19.0	South Africa	76.0	2.20	1,870	12.5
Japan	284.0	2.31	15,764	1.3	Australia	70.3	4.22	11,103	3.6
India	177.9	0.21	311	6.1	Czecho- slovakia	61.8	3.95	9,280	4.5
FRG	175.1	2.86	14,399	2.1	ROK	60.3	1.42	2,689	4.4
UK	155.1	2.70	10419	2.6	Rumania	57.9	2.50	6,030	4.2
Canada	124.3	4.73	15,160	2.9	Brazil	56.5	0.38	2,021	1.9
Poland	120.3	3.15	1,926	17.6	Spain	55.5	1.42	5,972	2.1
Italy	106.4	1.86	10,355	1.7					
France	97.5	1.74	12,789	1.4					

Source: UN Conference on Trade and Development UNCTAD/RDP/DFP/1, New York, United Nations, 1992.
Note: The data for USSR, FRG, GDR, and Czechoslovakia were obtained prior to the breakup of the Soviet Union and Czechoslovakia and prior to the reunification of Germany.

In this data set the countries are ranked (in decreasing order) according to their total 1989 CO_2 emissions data; this is simply a display of the order statistics corresponding to the variable X_1.

(a) Looking at the data, we see that the United States leads the world in total CO_2 emissions. Now rank the countries according to the variable X_2 (1987 emissions per capita). What is the rank of the United States? Of China?

(b) Rank the countries according to the variable X_4 (1987 emissions per GNP). What is the rank of the United States? Of China?

(c) How would you explain the differences in the rankings of the various countries according to the variables X_2 and X_4, respectively?

1.9 Refer to the StatLab data set in Table 1.5.

(a) Make an ordered stem-and-leaf plot corresponding to the variable *height*.

(b) Draw the frequency histogram for the variable *height*. Use a class width equal to 2.0 in. and the following class marks (midpoints):

$$49, \ 51, \ 53, \ 55, \ 57, \ 59$$

(c) Make an ordered stem-and-leaf plot corresponding to the variable *mheight*.

(d) Draw the frequency histogram for the variable *mheight*; use a class width equal to 1.5 in. and the following class marks (midpoints):

$$58.5, \ 60.0, \ 61.5, \ 63.0, \ 64.5, \ 66.0, \ 67.5$$

(e) Compare the shapes of the histograms obtained in parts (*b*) and (*d*); comment on their similarities and differences.

1.10 Sir Francis Galton (1822–1911), author of *Hereditary Genius* (1869) and a cousin of Charles Darwin (1809–1882), was interested in finding scientific laws governing the inheritance of physical traits. He was particularly interested in how the heights of parents are passed on to their children. Let us study this question using the StatLab data set of Table 1.5.

(a) Draw the scatter plot of the daughter's height (Y) against the mother's height (X).

(b) Galton also considered the "midparent" height, obtained by taking the average of the heights of the mother and father. Define the new variable

$$\text{Midparent height} = \frac{mheight + fheight}{2}$$

and draw the scatter plot of the daughter's height (Y) against the midparent height (X).

(c) Comparing these scatter plots, can you determine which variable— mother's height or midparent height—is most positively associated with the daughter's height?

1.11 Refer again to the StatLab data set of Table 1.5.

(a) Make an ordered stem-and-leaf plot for the variable *fweight*.

(b) Draw the histogram for the variable *fweight*.

(c) We expect that taller men tend to weigh more than shorter men. To check the accuracy of this inference, draw the scatter plot of *fweight* (Y) against *fheight* (X). What can you conclude from the scatter plot?

1.12 The following table records the results of a study of lead absorption in children of employees who worked in a factory where lead is used to make batteries. In this study the authors matched 33 such children (exposed) from different families to 33 children (control) of the same age and from the same neighborhood whose parents were employed in industries not using lead. The purpose of the study was to determine if children in the exposed group were at risk from lead inadvertently brought home by their parents. The column titled "Difference" records the differences in lead levels between a child in the exposed group and one in the control group.

Pair	Exposed, X	Control, Y	Difference, D = X − Y	Pair	Exposed, X	Control, Y	Difference, D = X − Y
1	38	16	22	18	10	13	−3
2	23	18	5	19	45	9	36
3	41	18	23	20	39	14	25
4	18	24	−6	21	22	21	1
5	37	19	18	22	35	19	16
6	36	11	25	23	49	7	42
7	23	10	13	24	48	18	30
8	62	15	47	25	44	19	25
9	31	16	15	26	35	12	23
10	34	18	16	27	43	11	32
11	24	18	6	28	39	22	17
12	14	13	1	29	34	25	9
13	21	19	2	30	13	16	−3
14	17	10	7	31	73	13	60
15	16	16	0	32	25	11	14
16	20	16	4	33	27	13	14
17	15	24	−9				

Source: D. Morton et al., "Lead Absorption in Children of Employees in a Lead-Related Industry," *American Journal of Epidemiology,* vol. 115, 1982, pp. 549–555. Used with permission.

For each of the three variables ("exposed," "control," and "difference"), draw the frequency histogram and describe the shape of the distribution.

1.13 Cadmium (Cd) is a bluish-white metal used as a protective coating for iron, steel, and copper. Overexposure to cadmium dust can damage the lungs, kidney, and liver. The federal standard for cadmium dust in the workplace is 200 $\mu g/m^3$. The following (simulated) data were obtained by measuring the levels of cadmium dust at 5-min intervals over a 4-h period.

181	190	192	197	201	204
199	189	200	198	192	189
192	204	199	196	195	199
198	202	193	194	197	189
197	195	209	201	195	202
190	190	193	196	200	196
206	192	197	198	195	198
200	193	195	193	198	195

(a) Make an ordered stem-and-leaf plot of the data using an appropriate stem and leaf.

(b) Draw the frequency histogram. Do the data appear to be symmetrically distributed?

(c) Comparing these data with the federal standard, would you work in this factory? Justify your answer.

1.14 The following table records the measured capacitances of 40 capacitors rated at 0.5 microfarads (μF). The data are simulated.

0.5121	0.5059	0.4838	0.4981
0.4851	0.4964	0.4955	0.4990
0.5092	0.5075	0.4905	0.5026
0.5010	0.4863	0.5100	0.5042
0.4944	0.5084	0.5031	0.5072
0.5170	0.4957	0.5056	0.5038
0.5066	0.4935	0.5113	0.5186
0.4986	0.5040	0.4967	0.5001
0.4920	0.5001	0.4972	0.5061
0.5080	0.4842	0.4878	0.4948

(a) Make an ordered stem-and-leaf plot using the first two nonzero digits as the stem values (ignore the decimal point). Compute the sample range.

(b) Make a dot plot and describe the shape of the frequency distribution.

(c) Make the frequency histogram; use Sturges' rule [Eq. (1.2)] to guide you in choosing the number of classes. Comment on the shape of the distribution.

1.5   QUANTILES OF A DISTRIBUTION

In the previous sections we introduced a variety of graphical displays for studying the distribution of the data. In this section and the following one we examine several *numerical* measures that also describe important features of the data.

For example, for some engineers it may be enough to know that the median lifetime (defined in Sec. 1.5.1) of type A light bulbs (see Table 1.4) is 1481.5 h with a minimum lifetime of 1293 h and a maximum lifetime of 1711 h. We can summarize much of the information contained in a data set by computing suitable *percentiles* of the empirical distribution function. When we say, for instance, that a student's class rank is in the 90th percentile, we mean that 90 percent of her classmates rank below her and 10 percent rank above. Similarly, in reliability engineering the 10th percentile of a data set indicates the length of time for which approximately 90 percent of the systems will operate without failure. This is sometimes called the *safe life*.

For many purposes, including mathematical convenience, it is easier to work with the p *quantile* of a distribution, where $0 < p < 1$; it corresponds to the $(100p)$th percentile of the distribution. Quantiles are a useful tool for comparing two distributions via *quantile-quantile plots* or Q-Q plots, which some authors call *probability plots*. We examine this important data analysis tool in Sec. 8.4. We begin our discussion by considering the important special case of the 50th percentile, which is also called the *median*.

1.5.1 The Median

A median of the data set $\{x_1, \ldots, x_n\}$ is a number denoted by $\tilde{x}$ that divides the data set in half, so that at least half the data values are greater than or equal to $\tilde{x}$ and at least half the data values are less than or equal to $\tilde{x}$.

We now describe a simple procedure for computing a median. In Example 1.14 we consider two cases: (1) n even and (2) n odd.

example 1.14   Given (1) the lifetimes of type A filaments in Table 1.4 and (2) the failure times of Kevlar/epoxy pressure vessels in Table 1.8, find the median lifetime of the type A filaments and the median failure time of the pressure vessels.

TABLE 1.8 ORDERED LIST OF FAILURE TIMES OF KEVLAR/EPOXY PRESSURE VESSELS AT 86 PERCENT STRESS LEVEL (4300 PSI)

2.2	4.0	4.0	4.6	6.1
6.7	7.9	8.3	8.5	9.1
10.2	12.5	13.3	14.0	14.6
15.0	18.7	22.1	45.9	**55.4**
61.2	87.2	98.2	101.0	111.4
144.0	158.7	243.9	254.1	444.4
590.4	638.2	755.2	952.2	1108.2
1148.5	1569.3	1750.6	1802.1	

Source: R. E. Barlow, R. H. Toland, and T. Freeman, "A Bayesian Analysis of the Stress-Rupture Life of Kevlar/Epoxy Spherical Pressure Vessels," in *Accelerated Life Testing and Experts' Opinions in Reliability,* C. Clarotti and D. Lindley, eds., Amsterdam, North Holland, 1988, pp. 203–236. Used with permission.

1. *Solution.* Looking at the lifetimes of type A filaments in Table 1.4, we see that $n = 10$, so the sample size is even. The order statistics for the light bulbs with type A filaments have two values (shown here in boldface) occupying the middle two positions:

   $$1293 < 1340 < 1380 < 1383 < \mathbf{1466} < \mathbf{1497} < 1614 < 1627 < 1643 < 1711$$

   We define the median to be the average of the two middle terms, so that

   $$\tilde{x} = \frac{1466 + 1497}{2} = 1481.5$$

2. *Solution.* Looking at the failure times of the pressure vessels in Table 1.8, we see that $n = 39$, so the sample size is odd. The data in this table are already arranged in increasing order; they are, in fact, the order statistics of the sample. The median is the twentieth term in the ordered array since there are 19 values less than $x_{(20)}$ and 19 values greater than $x_{(20)}$. Consequently, $\tilde{x} = x_{(20)} = 55.4$. ∎

Equations (1.3) and (1.4) are formulas for the median in terms of the order statistics.

$$\tilde{x} = \tfrac{1}{2}(x_{(n/2)} + x_{(n/2+1)}) \qquad \text{if } n \text{ is even} \tag{1.3}$$

$$\tilde{x} = x_{((n+1)/2)} \qquad \text{if } n \text{ is odd} \tag{1.4}$$

---

example 1.15   1. Compute the median lifetime of type A filaments in Table 1.4 via Eq. (1.3).

Solution. Here $n = 10$ is even, so Eq. (1.3) yields

$$\tilde{x} = \tfrac{1}{2}(x_{(5)} + x_{(6)}) = \tfrac{1}{2}(1466 + 1497) = 1481.5$$

which is in agreement with the result obtained earlier.

2. Compute the median of the data in Table 1.8 via Eq. (1.4).

Solution. Here $n = 39$ is odd, so Eq. (1.4) yields

$$\frac{n+1}{2} = \frac{40}{2} = 20, \qquad \text{so}$$

$$\tilde{x} = x_{((n+1)/2)} = x_{(40/2)} = x_{(20)} = 55.4 \qquad \blacksquare$$

The median is an example of a *measure of location* because it locates the *center* of a distribution. For this reason it is also called a measure of *central value* (see Sec. 1.6).

1.5.2 Quantiles of the Empirical Distribution Function

The inherent variability in a typical data set cannot be reduced to a single number such as the median. This is why economists use *quintiles* (the 20th, 40th, 60th, and 80th percentiles) to more accurately describe the income distribution of a country, for example. Similarly, many colleges and universities now report the SAT scores of students accepted for admission as a *range* of values instead of reporting, as they used to, the average SAT score. This minor shift in policy was deemed sufficiently newsworthy to be published in the *New York Times* (5 Aug 1990, p. 4A).

> Syracuse and Colgate Universities have joined 50 other colleges and universities in pledging to release the SAT scores as a *range* rather than a single average score. Such scores are published by college guides and often requested by families of prospective students.
> In the standardized testing agreement the schools pledge not to provide an average test score. Instead, the pledge suggests presenting the range of scores of the middle 50% of students accepted.

For example, the University of Florida reported in 1995 that the SAT scores of its entering freshman class had a middle 50 percent range of [1080, 1250]. This means that 50 percent of those accepted had SAT scores in the interval [1080, 1250]; in addition, 25 percent of those accepted had SAT scores below

1080 and 25 percent had SAT scores above 1250. The numbers 1080 and 1250 are examples of what are called the *lower* and *upper quartiles* of a distribution. As this example makes clear, the quantiles of a distribution provide additional information concerning the distribution of the data.

A p quantile $(0 < p < 1)$ of the empirical distribution function is a number $Q(p)$ with the property that at least $100p$ percent of the data values are less than or equal to $Q(p)$ and at least $100(1 - p)$ percent of the data values are greater than or equal to $Q(p)$. We first give a geometrical method for computing a p quantile. Then we give an analytic formula for computing it.

---

example **1.16**   Find the median of the radiation data (Table 1.3) from the empirical distribution function of the data.

Solution. Figure 1.11 is the graph of the empirical distribution function of the radiation data (Table 1.3), modified by connecting the jumps by vertical line segments. We obtain the median, which equals $Q(0.5)$, by drawing a horizontal line starting at the point $(0, 0.5)$ on the y axis and going to the right (or left) until it intersects the graph of the empirical distribution function. The x coordinate of this point is the median. Looking at Fig. 1.11, we see that the point of intersection has coordinates $(Q(0.5), 0.5) = (0.10, 0.5)$, so $\tilde{x} = 0.10$. We obtain any quantile $Q(p)$ from the graph of the empirical distribution function in this way; we draw a horizontal line starting at the point $(0, p)$ on the y axis and going to the right (or left) until it intersects the graph of the empirical distribution function. The x coordinate of this point is the p quantile $Q(p)$; the point of intersection has

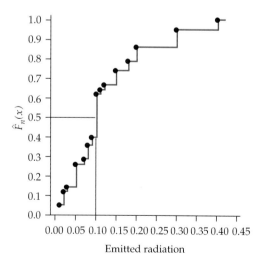

Figure 1.11 Graphical determination of the median

coordinates $(Q(p), p)$. When the empirical distribution function has a flat interval at the level p, we define the p quantile to be the midpoint of the flat interval. Since the height of the empirical distribution function at a point x gives the proportion of the data less than or equal to x, defining the p quantile $Q(p)$ in this way means that at least $100p$ percent of the data will be less than or equal to $Q(p)$ and at least $100(1 - p)$ percent of the data will be greater than or equal to $Q(p)$.   ∎

The Upper and Lower Quartiles. The 25th and 75th percentiles are called the *lower* and *upper quartiles* and are denoted by $Q_1 = Q(0.25)$ and $Q_3 = Q(0.75)$, respectively.

Because drawing the graph of the empirical distribution function can be quite tedious, we now give a formula for $Q(p)$ using the order statistics.

A Formula for Computing $Q(p)$**.** To calculate the quantile $Q(p)$, we proceed in a manner analogous to that used to calculate the median [Eqs. (1.3) and (1.4)]. We begin by computing np. There are two cases to consider, according to whether or not np is an integer.

1. np **is an integer.** In this case compute the npth and $(np + 1)$th order statistics $x_{(np)}$ and $x_{(np+1)}$; $Q(p)$ is defined to be their average:

$$Q(p) = \tfrac{1}{2}(x_{(np)} + x_{(np+1)}) \tag{1.5}$$

2. np **is not an integer.** In this case there exist two consecutive integers r and $r + 1$ such that np lies between them; that is, $r < np < r + 1$. We define $Q(p)$ as

$$Q(p) = x_{(r+1)} \tag{1.6}$$

Note: Equation (1.5) corresponds to computing the median when the sample size is even, and Eq. (1.6) corresponds to computing the median when the sample size n is odd.

---

example 1.17

1. Compute $Q(0.10)$, or the safe life, for the lifetimes of the type A filaments (Table 1.4).

   Solution. Here $n = 10$ and $p = 0.1$, so $np = 1$, an integer. Consequently,

   $$Q(0.10) = \tfrac{1}{2}(x_{(1)} + x_{(2)}) = 1316.5$$

2. Compute the lower and upper quartiles of the failure times of the pressure vessels in Table 1.8.

   Solution. Here $n = 39$ and $p = 0.25$; therefore, $9 < np = 9.75 < 10$. Thus,

   $$Q_1 = Q(0.25) = x_{(10)} = 9.1$$

   Similarly, for $n = 39$ and $p = 0.75$, we have $29 < np = 29.25 < 30$. Thus,

   $$Q_3 = Q(0.75) = x_{(30)} = 444.4$$

Order Statistics and Quantiles. It is worth pointing out that the ith order statistic $x_{(i)}$ is the $(i - 0.5)/n$ quantile; that is,

$$x_{(i)} = Q\left(\frac{i - 0.5}{n}\right) \qquad (1.7)$$

We leave the derivation of Eq. (1.7) as Prob. 1.23.

The Sample Range and Interquartile Range. Order statistics and percentiles are often used to measure the variability of a distribution. The *sample range* is the length of the smallest interval that contains all of the observed values; the *interquartile range* (IQR) is the length of the interval that contains the middle half of the data.

The sample range is defined as

$$\text{Sample range} = x_{(n)} - x_{(1)} \qquad (1.8)$$

and the interquartile range is defined as

$$\text{Interquartile range (IQR)} = Q_3 - Q_1 \qquad (1.9)$$

The interval $[Q_1, Q_3]$ is called *the middle 50 percent range.*

An *outlier* is any data value that lies outside the interval

$$(Q_1 - 1.5 \times \text{IQR}, Q_3 + 1.5 \times \text{IQR}) \qquad \blacksquare$$

---

example **1.18**   Compute (1) the interquartile range and (2) the outliers for the variable *weight* in the StatLab data set (Table 1.5).

Solution

1. *IQR:* The lower quartile is $Q_1 = 59$ and the upper quartile is $Q_3 = 77$. Therefore, the interquartile range is IQR $= 77 - 59 = 18$.
2. *Outliers:* We note that $1.5 \times$ IQR $= 27$; therefore,

$$Q_3 + 1.5 \times \text{IQR} = 77 + 27 = 104$$
$$Q_1 - 1.5 \times \text{IQR} = 59 - 27 = 32$$

Looking at Table 1.5, we see that there is one outlier, identified as girl 18, with weight 106 lb. $\qquad \blacksquare$

Box Plots. The *box plot* is a graphic display of the median, quartiles, and interquartile range of a data set.

---

example **1.19**   Figure 1.12 is the computer-generated box plot for the radiation data (Table 1.3). The plus sign (+) locates the median, and the left and right edges of the box, printed as "I", locate the lower and upper quartiles. In this case $Q_1 = 0.05$, $Q_3 = 0.18$, and IQR $= 0.13$. The box represents the middle 50 percent range of the data. The *whiskers* are the lines that extend outward from the ends of the box to a distance of at most 1.5 units of IQR. More precisely, the whisker on the

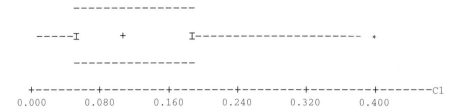

Figure 1.12 Box plot of the radiation data

right starts at Q_3 and terminates at the number $\min(x_{(n)}, Q_3 + 1.5 \times IQR)$. In this case $x_{(42)} = 0.40$ and $Q_3 + 1.5 \times IQR = 0.18 + 0.195 = 0.375$, so

$$\min(x_{(42)}, Q_3 + 1.5 \times IQR) = \min(0.40, 0.375) = 0.375$$

Similarly, the whisker on the left starts at Q_1 and terminates at the number $\max(x_{(1)}, Q_1 - 1.5 \times IQR)$. In this case $x_{(1)} = 0.01$ and $Q_1 - 1.5 \times IQR = -0.145$, so

$$\max(x_{(1)}, Q_1 - 1.5 \times IQR) = \max(0.01, -0.145) = 0.01$$

Any value beyond these limits is marked with an asterisk (*). In Fig. 1.12 there is an asterisk at 0.40. Referring back to the original data set, we see that there are two outliers, both equal to 0.40. ∎

The following example demonstrates how to use the box plot to compare two populations.

example 1.20 Figure 1.13 is a side-by-side box plot of the lifetimes of types A and B filaments (Table 1.4). This display shows two things: (1) Type A filaments have significantly longer lifetimes because the endpoint of the left whisker for the type A filaments

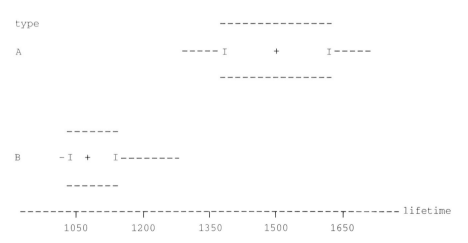

Figure 1.13 Side-by-side box plot of the lifetimes (in hours) of type A and type B filaments

(1293) is greater than the endpoint of the right whisker for the type B filaments (1270), and (2) type A filaments have a longer interquartile range. This reveals an excessive variability in the lifetimes of type A filaments that the quality control engineers cannot ignore. ■

PROBLEMS

1.15 The 1989 total charges (in dollars) for 33 female patients, ages 30–49, admitted for circulatory disorders are recorded in the following table. The data were collected by an insurance company interested in the distribution of claims.

2337	2179	2348	4765	2088	2872	1924	2294
2182	2138	1765	2467	3609	2141	1850	3191
3020	2473	1898	7787	6169	1802	2011	2270
3425	3558	2315	1642	5878	2101	2242	5746
3041							

Source: E. W. Frees, "Estimating Densities of Functions of Observations," *Journal of the American Statistical Association,* vol. 89, no. 426, 1994, pp. 517–525. Used with permission.

(a) Draw the histogram and describe the shape of the distribution.
(b) Compute the median and the upper and lower quartiles, and determine the outliers.
(c) Draw the box plot.

1.16 Refer to the lifetimes of type B filaments in Table 1.4.
(a) Compute the sample range and median of the lifetimes.
(b) Compute the lower and upper quartiles and the IQR. Are there any outliers?

1.17 The annual incomes (in thousands of dollars) of 30 families in a metropolitan region are listed in the following table. The data were collected to determine the profitability of locating a shopping mall in the region.

19	51	43	43	51	23
23	23	31	43	67	31
23	99	67	51	91	55
31	27	35	31	35	35
39	27	47	71	111	63

(a) Compute the quintiles of the income distribution; that is, compute the 20th, 40th, 60th, and 80th percentiles.
(b) Draw the box plot.

1.18 The lifetimes (in hours) of 20 bearings are recorded in the following table.

6,278	3,113	9,350	5,236	11,584
12,628	7,725	8,604	14,266	6,215
3,212	9,003	3,523	12,888	9,460
13,431	17,809	2,812	11,825	2,398

Source: R. E. Schafer and J. E. Angus, "Estimation of Weibull Quantiles with Minimum Error in the Distribution Function," *Technometrics,* vol. 21, no. 3, 1979, pp. 367–370. Used with permission.

The 10th percentile, $Q(0.10)$, of the empirical distribution function is called, as noted earlier, the safe life.
(a) Compute $Q(0.10)$ and the median, $\tilde{x} = Q(0.50)$.
(b) Compute the sample range, the upper and lower quartiles, and the IQR.
(c) Identify any outliers.
(d) Draw the box plot.

1.19 Refer to the StatLab data set (Table 1.5).
(a) Compute the sample range, median, Q_1, Q_3, and IQR for the variable *fweight.*
(b) Identify all outliers.

1.20 Refer to the StatLab data set (Table 1.5).
(a) Compute the sample range, median, Q_1, Q_3, and IQR of the girls' scores on the Peabody vocabulary test.
(b) Draw the box plot.

1.21 Refer to the StatLab data set (Table 1.5).
(a) Compute the sample range, median, Q_1, Q_3, and IQR of the girls' scores on the Raven progressive matrices test.
(b) Draw the box plot.

1.22 In Example 1.16 we used the graph of the empirical distribution function (Fig. 1.11) to compute the median of the radiation data (Table 1.3).
(a) Use the same graphical method to compute the lower and upper quartiles.
(b) Compute the lower and upper quartiles using Eq. (1.6). Your answers to parts (*a*) and (*b*) should be the same.

1.23 Show that the *i*th order statistic $x_{(i)}$ is the $(i - 0.5)/n$ quantile by deriving Eq. (1.7).

1.24 Refer to the battery data of Prob. 1.5.
(a) Compute the sample range, median, Q_1, Q_3, and IQR.
(b) Draw the box plot.

1.25 Refer to the failure times data of Table 1.8.
(a) Compute the safe life, $Q(0.10)$.
(b) Compute the sample range and the IQR.
(c) Identify all outliers.
(d) Draw the box plot.

1.26 Refer to the StatLab data set (Table 1.5).

(a) Compute the sample range, the median, the upper and lower quartiles, and the IQR for the variable *mweight*. Are there any outliers?

(b) Draw the box plot.

1.27 The times between successive failures of the air-conditioning systems of Boeing 720 jet airplanes are shown in the following table. It contains 97 values taken from 4 airplanes. This is an edited version of the original data set, which contained the maintenance data on a fleet of 13 airplanes and 212 values. The purpose of the data collection was to obtain information on the distribution of failure times and to use this information for predicting reliability, scheduling maintenance, and providing spare parts. Notice that the number of failures is not the same for each airplane; periods (.) represent missing values. The first column (OBS) lists the order in which the data were recorded.

OBS	B7912	B7913	B7914	B8045	OBS	B7912	B7913	B7914	B8045
1	23	97	50	102	16	12	54	36	34
2	261	51	44	209	17	120	31	22	.
3	87	11	102	14	18	11	216	139	.
4	7	4	72	57	19	3	46	210	.
5	120	141	22	54	20	14	111	97	.
6	14	18	39	32	21	71	39	30	.
7	62	142	3	67	22	11	63	23	.
8	47	68	15	59	23	14	18	13	.
9	225	77	197	134	24	11	191	14	.
10	71	80	188	152	25	16	18	.	.
11	246	1	79	27	26	90	163	.	.
12	21	16	88	14	27	1	24	.	.
13	42	106	46	230	28	16	.	.	.
14	20	206	5	66	29	52	.	.	.
15	5	82	5	61	30	95	.	.	.

Source: F. Proschan, "Theoretical Explanation of Observed Decreasing Failure Rate," *Technometrics*, vol. 5, no. 3, 1963, pp. 375–383. Used with permission.

Compute the sample range, median, lower and upper quartiles, IQR, and all outliers for:

(a) Airplane B7912; then draw the box plot.

(b) Airplane B7913; then draw the box plot.

(c) Airplane B7914; then draw the box plot.

(d) Airplane B8045; then draw the box plot.

1.28 Refer to the data on lead absorption in children of employees who worked in a factory where lead is used to make batteries (see Prob. 1.12). Compare the exposed group with the control group by drawing side-by-side box plots. What inferences can you make from these plots concerning the effect on children of the parents' exposure to lead in their workplace?

1.29 Refer to the data on lead absorption in children of employees who worked in a factory where lead is used to make batteries (see Prob. 1.12). Compare the exposed group with the control group by drawing the box plot for the difference variable, $d_i = x_i - y_i$ ($i = 1, \ldots, 33$). What inferences can you make concerning the effect on children of the parents' exposure to lead in their workplace?

1.30 Refer to the data in Prob. 1.14.
(a) Compute the median, the lower and upper quartiles, and the IQR. Are there any outliers?
(b) Draw the box plot.

1.31 The manufacturer's cost and the selling price (in dollars) of light bulbs are denoted by C_1 and C_2, respectively. The bulb is guaranteed for H hours; if the bulb fails before then, the manufacturer guarantees a total refund. Let R denote the manufacturer's net revenue. Using the data for type A filaments in Table 1.4, compute:
(a) R for $C_1 = 1$, $C_2 = 2$, and $H = 1400$.
(b) R for $C_1 = 1$, $C_2 = 2$, and $H = 1500$.

1.6 MEASURES OF LOCATION (CENTRAL VALUE) AND VARIABILITY

In addition to the numerical measures such as the sample range, median, and the lower and upper quartiles already discussed, we now introduce the *(arithmetic) mean, variance,* and *standard deviation.* The mean defines what physicists call the *center of mass* of the distribution. The standard deviation, like the interquartile range, is a measure of the variability of the distribution.

1.6.1 The Sample Mean

We use the term *average* in everyday speech to denote the arithmetic mean of a set of numbers. In statistics we call this the *sample mean,* and it is, except for the median, the most widely used numerical measure of the central value of a distribution.

DEFINITION 1.5

The *sample mean* $\bar{x}$ of the set of numerical data $\{x_1, \ldots, x_n\}$ is defined by

$$\bar{x} = \frac{1}{n} \sum_{1 \le i \le n} x_i \tag{1.10}$$

∎

example 1.21   1. The mean time before failure (MTBF) of a device, such as a light bulb, plays an important role in reliability engineering. Compute the MTBF of light bulbs with type A filaments (Table 1.4).

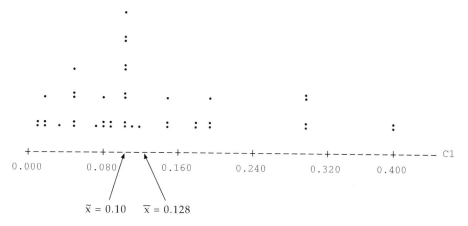

Figure 1.14 Dot plot of the emitted microwave radiation data of Table 1.3 displaying the sample mean $\bar{x} = 0.128$ and the median $\tilde{x} = 0.10$

Solution. The sum of the 10 lifetimes of the type A filaments equals 14,954.00. Consequently, the MTBF is given by

$$\bar{x} = \frac{1293 + 1380 + \cdots + 1711}{10} = \frac{14,954.00}{10} = 1495.4$$

2. Compute the mean amount of radiation emitted by the microwave ovens (Table 1.3).

Solution. When the sample size is large, as it is in this example, we use a statistical software package because it is too tedious to compute the mean by hand. We obtain the value $\bar{x} = 0.128$.

Looking at the dot plot of the microwave radiation data in Fig. 1.14, we notice that the mean is greater than the median. This occurs because the distribution is skewed to the right. In particular, the two outliers (observations $x_{20} = x_{40} = 0.40$) increase the value of the mean but do not have any effect on the median. For instance, if the largest observed value were increased to 100, the median would remain unchanged but the mean would be sharply increased.

It is helpful to think of the values displayed in Fig. 1.14 as consisting of $n = 42$ point masses placed along a rod, with each mass weighing $1/n = 1/42$ grams, say. Thus, we place 9/42 grams at $x = 0.10$ since $\hat{f}(0.10) = 9/42$. The sample mean $\bar{x}$ can now be interpreted as the *center of mass* of this distribution. ∎

Mean, Median, or Mode?

We noted in Example 1.21 that the sample mean is larger than the sample median; this is fairly typical when the data, as is the case here, are skewed to the right. This raises the question: Which measure of central value is appropriate? The answer depends on the shape of the distribution and the variable type, that is, whether or not the variable is numerical or categorical. For example, it makes no sense to talk about the average birth month, but it is reasonable to identify the mode, which is the month in which the most births occur. In other words,

1. The mode is a reasonable measure of central value for a categorical variable.
2. The median is an appropriate measure of central value when there are outliers skewing the distribution either to the right or to the left. For instance, income data are skewed to the right, and this explains why economists prefer to use the median income instead of the average income.
3. The sample mean is the most appropriate choice for the central value when the distribution appears to be symmetric about its mean.

1.6.2   Sample Variance and Sample Standard Deviation

The *sample variance* is used to measure the spread of a distribution about its sample mean. In particular, a small variance implies that a high proportion of the observations are clustered about the sample mean.

■ DEFINITION 1.6

The *sample variance* s^2 and its square root, the *sample standard deviation s*, are defined by

$$s^2 = \frac{1}{n-1} \sum_{1 \le i \le n} (x_i - \bar{x})^2 \tag{1.11}$$

$$s = \sqrt{\frac{1}{n-1} \sum_{1 \le i \le n} (x_i - \bar{x})^2} \tag{1.12}$$

■

To compute s^2, one proceeds as follows: For each data value x_i compute $(x_i - \bar{x})^2$, then calculate the *sum of squares* $\sum_{1 \le i \le n}(x_i - \bar{x})^2$, and, finally, divide this sum by $n - 1$.

---

example 1.22 Compute the sample variance for the weights of the first seven girls in the StatLab data set (Table 1.5).

Solution. The sample mean equals 76.43. The remaining steps in the calculation are displayed in Table 1.9.

TABLE 1.9

$x_i - \bar{x}$		$(x_i - \bar{x})^2$
$85 - 76.43 =$	8.57	73.44
$59 - 76.43 =$	-17.43	303.80
$70 - 76.43 =$	-6.43	41.34
$88 - 76.43 =$	11.57	133.86
$68 - 76.43 =$	-8.43	71.06
$93 - 76.43 =$	16.57	274.56
$72 - 76.43 =$	-4.43	19.62

$$s^2 = 917.71/6 = 152.95$$
$$s = 12.37$$

For tedious computations such as these we should use a calculator. Indeed, most calculators, even the least expensive ones, have a program that computes the sample mean and standard deviation. If the computations must be done by hand, the drudgery can be eased by using Eq. (1.13), the *shortcut formula*. It can be shown (we omit the details) that Eq. (1.11) for the variance is algebraically equivalent to Eq. (1.13).

$$s^2 = \frac{1}{n-1}\left(\sum_{1 \le i \le n} x_i^2 - \left(\sum_{1 \le i \le n} x_i\right)^2 / n\right) \qquad (1.13)$$

∎

example **1.23**   Compute, using the shortcut formula (1.13), the sample variance for the weights of the first seven girls in the StatLab data set (Table 1.5).

Solution. We first calculate the sum of the weights and the sum of the squares of the weights of the first seven girls. We obtain the following results:

$$\sum_{1 \le i \le 7} x_i = 535 \qquad \text{and} \qquad \sum_{1 \le i \le 7} x_i^2 = 41{,}807$$

Therefore,

$$s^2 = \frac{1}{6}\left(41{,}807 - \frac{(535)^2}{7}\right) = 152.95 \qquad \blacksquare$$

Although Eqs. (1.11) and (1.13) are algebraically equivalent, Eq. (1.13) is computationally more efficient; it can be shown that Eq. (1.13) requires approximately $2n$ arithmetic operations, whereas Eq. (1.11) requires approximately $3n$ operations.

Computing the Sample Mean and Sample Variance from the Frequency Distribution. The frequency distribution contains all the information in the sample, and we can compute all numerical measures of location and variability from the frequency distribution itself. Equations (1.14) and (1.15), for example, give formulas for the sample mean and sample variance directly in terms of the frequency distribution.

$$\bar{x} = \frac{1}{n}\sum_x xf(x) \qquad (1.14)$$

$$s^2 = \frac{1}{n-1}\sum_x (x - \bar{x})^2 f(x)$$

$$= \frac{1}{n-1}\left(\sum_x x^2 f(x) - n\bar{x}^2\right) \qquad (1.15)$$

We derive Eqs. (1.14) and (1.15) by noting that

$$\sum_{1 \le i \le n} x_i = \sum_x xf(x)$$

and

$$\sum_{1 \le i \le n} (x_i - \bar{x})^2 = \sum_{x} (x - \bar{x})^2 f(x)$$

The right-hand sides of the preceding equations are obtained by first grouping identical summands, x; multiplying them by their frequency, $f(x)$; and then performing the summation. We give the details of these computations in the following example.

---

example **1.24**   Compute (1) the sample mean and (2) the sample variance for the radiation data (Table 1.3) using Eqs. (1.14) and (1.15).

1. Solution. To compute the sample mean, we apply Eq. (1.14) to the frequency distribution shown in the Freq. column of Fig. 1.1. The computations yield

$$\bar{x} = \frac{1}{42}(0.01 \times 2 + 0.02 \times 3 + \cdots + 0.30 \times 4 + 0.40 \times 2)$$

$$= \frac{5.39}{42} = 0.128$$

which agrees with the value obtained earlier (see Example 1.21).

2. Solution. To compute the sample variance, we apply Eq. (1.15) to the frequency distribution shown in the Freq. column of Fig. 1.1. The computations yield

$$s^2 = \frac{1}{41}[(0.01)^2 \times 2 + (0.02)^2 \times 3 + \cdots + (0.40)^2 \times 2 - 42 \times (0.128)^2]$$

$$= \frac{1}{41}(1.1016 - 0.6881) = \frac{0.4135}{41} = 0.01$$ ∎

Population Mean and Variance. Sometimes a sample is not enough. The Constitution of the United States, for example, requires that every 10 years a census be taken of the entire population. The basic purpose of the census is to reapportion the membership of the House of Representatives among the states. For us a *census* is a sample that consists of the entire population; that is, we measure the variable for *every* element of the population. The numerical measures obtained from a census, such as the median, mean, variance, and standard deviation, are called *population parameters*. The *population mean* and *population variance* are particularly important population parameters.

▮ DEFINITION 1.7

Let $\{x_1, \ldots, x_N\}$ denote the data obtained from a census of a population consisting of N elements. We denote the population mean and population variance by the Greek letters μ (mu) and σ^2 (sigma squared). They are defined by

$$\text{Population mean:} \quad \mu = \frac{1}{N} \sum_{1 \le i \le N} x_i$$

$$\text{Population variance:} \quad \sigma^2 = \frac{1}{N} \sum_{1 \le i \le N} (x_i - \mu)^2$$

$$\text{Population standard deviation:} \quad \sigma = \sqrt{\frac{1}{N} \sum_{1 \le i \le N} (x_i - \mu)^2} \quad \blacksquare$$

▨ REMARK 1.1

Note that the divisor in the definition of the population variance is the population size N and *not* $N - 1$. $\blacksquare$

When the size N of the population is large, however, the task of computing μ may be next to impossible. We therefore use the sample mean $\bar{x}$ and sample variance s^2, based on a sample size $n < N$, to estimate μ and σ^2. Example 1.25 helps us to understand the difference between a sample and a census.

---

example **1.25**   To simplify the discussion we consider the 36 girls and their weights recorded in Table 1.5 as our population, so $N = 36$. We take the first 7 girls listed in the table as our sample, so $n = 7$. Compute (1) the population mean, population variance, and population standard deviation; and (2) the sample mean, sample variance, and sample standard deviation (using the first 7 girls as the sample).

1. Solution. The values for the population mean, population variance, and population standard deviation are given by

$$\frac{\sum_{1 \le i \le 36} x_i}{36} = 68.92 = \mu$$

$$\frac{\sum_{1 \le i \le 36} (x_i - 68.92)^2}{36} = 163.86 = \sigma^2$$

$$\sigma = \sqrt{163.86} = 12.80$$

2. Solution. We first list the weights (in pounds) of the sample consisting of the first seven girls of the StatLab data set:

$$85, 59, 70, 88, 68, 93, 72$$

The mean, variance, and standard deviation for this data set were computed earlier as

$$\bar{x} = 76.43, \qquad s^2 = 152.95, \qquad s = 12.37 \qquad \blacksquare$$

1.6.3   Linear Transformations of Data

It is sometimes necessary to rewrite observations in a different system of units, for instance, transforming pounds to kilograms, inches to centimeters, and degrees Fahrenheit to degrees Celsius. We use the subscript notation s_x and s_y to distinguish between the sample standard deviations for the data sets X and Y. The transformed data set, denoted $Y = \{y_1, \ldots, y_n\}$, is obtained from the original one by computing $y_i = g(x_i)(i = 1, \ldots, n)$, where $g(x)$ is a function. The most frequently used functions are the *linear functions* defined by

$$y = g(x) = ax + b$$

We now give some examples.

1. To transform x, in pounds, to y, in kilograms, we set $a = 1/2.2$ and $b = 0$.
2. To transform x, in inches, to y, in centimeters, we set $a = 2.54$ and $b = 0$.
3. To transform x, in degrees Celsius, to y, in degrees Fahrenheit, we set $a = 9/5$ and $b = 32$.

We compute the mean, variance, and standard deviation for data transformed by the linear function $g(x) = ax + b$ by applying the following proposition. The proof is given in Sec. 1.7.

▦ PROPOSITION 1.1

The sample mean, sample variance, and sample standard deviation of the transformed data $y_i = ax_i + b(i = 1, 2, \ldots, n)$ are given by

$$\bar{y} = a\bar{x} + b$$
$$s_y^2 = a^2 s_x^2$$
$$s_y = |a|s_x$$

■

---

example 1.26   Compute the sample mean, sample variance, and sample standard deviation for the girls' weights (in kilograms) in the StatLab data set (Table 1.5).

Solution. The weight y in kilograms is given by $y = x/2.2$, so $a = 1/2.2$ and $b = 0$. The sample mean, sample variance, and sample standard deviation of the original data are given by (see Example 1.25)

$$\bar{x} = 68.92, \qquad s_x^2 = 168.54, \qquad s_x = 12.92$$

Therefore, the sample mean, sample variance, and sample standard deviation of the transformed data (in kilograms) are given by

$$\bar{y} = \frac{68.92}{2.2} = 31.33, \qquad s_y^2 = \frac{168.54}{(2.2)^2} = 34.82, \qquad s_y = \frac{12.92}{2.2} = 5.90 \quad ■$$

PROBLEMS

1.32 The Analytical Methods Committee of the Royal Society of Chemistry reported the following results on the determination of tin in foodstuffs. The samples were boiled with hydrochloric acid under reflux for different times.

Refluxing time (min)	Tin found (mg/kg)					
30	55	57	59	56	56	59
75	57	55	58	59	59	59

Source: "The Determination of Tin in Organic Matter by Atomic Absorption Spectrometry," *The Analyst*, vol. 108, 1983, pp. 109–115. Used with permission.

Compute the sample means, sample variances, and sample standard deviations for the 30- and 75-min refluxing times.

1.33 Eleven samples of the outflow from a sewage treatment plant were divided in two parts, with one half going to a private commercial laboratory and the other half sent to a state laboratory. Each laboratory measured the biochemical oxygen demand (BOD) and the suspended solids (SS). The data appear in the following table.

Commercial lab		State lab		Commercial lab		State lab	
BOD	SS	BOD	SS	BOD	SS	BOD	SS
6	27	25	15	28	26	42	30
6	23	28	13	71	124	54	64
18	64	36	22	43	54	34	56
8	44	35	29	33	30	29	20
11	30	15	31	20	14	39	21
34	75	44	64				

Source: R. A. Johnson and D. W. Wichern, *Applied Multivariate Statistical Analysis,* 3d ed., Englewood Cliffs, N.J., Prentice Hall, 1992, p. 223. Used with permission.

(a) Compute the sample means, sample variances, and sample standard deviations for the BOD data obtained by each laboratory. Compare the precision of their results. Are the mean BOD levels determined by the two laboratories in agreement with one another? Use the appropriate data analysis tools to guide your thinking.

(b) Compute the sample means, sample variances, and sample standard deviations for the SS data for each laboratory. Compare the precision of their results. Are the mean SS levels determined by the two laboratories in agreement with one another?

1.34 Find the sample means, variances, and standard deviations of the lifetimes of the type B filaments in Table 1.4.

1.35 Find the sample means, variances, and standard deviations of the burst strengths of the rocket motor from Prob. 1.6.

1.36 One measure of the effect of sewage effluent on a lake is the concentration of nitrates in the water. Assume that the following data, which are simulated, were obtained by taking 12 samples of water, dividing each sample into two equal parts, and measuring the nitrate concentration for each part by

Method 1, x_i	Method 2, y_i	Difference, $d_i = x_i - y_i$	Method 1, x_i	Method 2, y_i	Difference, $d_i = x_i - y_i$
119	95	24	61	124	-63
207	159	48	376	360	16
213	174	39	314	350	-36
232	344	-112	55	34	21
208	256	-48	215	235	-20
193	175	18	101	198	-97

two different methods. We are interested in determining whether the two methods yield similar results.

(a) Compute the sample means and sample standard deviations for each variable (x, y, d) in the data set.

(b) Compute the box plots for each variable (x, y, d) in the data set.

(c) Using the results from parts (a) and (b), answer, as best you can, the following question: Do the data indicate a difference between the two methods of measuring the nitrate concentration? Justify your answer.

1.37 The frequency distribution for the fathers' heights listed in Table 1.5 is given in the following table, where height $= x$ and freq. $= f(x)$.

Height	64.9	65	65.5	66	67	67.8	68	68.1	68.5	69	70
Freq.	1	2	1	1	1	1	1	1	1	3	3
Height	70.1	70.5	70.8	71	71.8	72	72.8	73	73.5	74	75
Freq.	1	1	1	6	1	2	1	1	2	2	2

Compute the sample mean and sample standard deviation of the heights of the fathers. [*Hint:* Use Eqs. (1.14) and (1.15).]

1.38 The frequency distribution for the girls' scores on the Raven progressive matrices test (Table 1.5) is given in the following table, where score $= x$ and freq. $= f(x)$.

Score	14	15	19	20	21	24	25	27	28	29	31	32
Freq.	1	1	1	1	1	1	3	1	1	3	1	2
Score	33	34	35	36	37	38	39	40	42	43	45	
Freq.	1	3	3	1	2	3	1	2	1	1	1	

Compute the sample mean and sample standard deviation of the girls' scores. [*Hint:* Use Eqs. (1.14) and (1.15).]

1.39 Find the sample mean, variance, and standard deviation of the bearing lifetimes from Prob. 1.18.

1.40 With reference to the failure times of the air-conditioning systems from Prob. 1.27, find the sample mean, variance, and standard deviation for:

(a) Airplane B7912

(b) Airplane B7913

(c) Airplane B7914

(d) Airplane B8045

(e) How do you explain the fact that in each case the sample mean is greater than the sample median?

1.41 Refer to the battery data of Prob. 1.5. Compute the sample mean and sample standard deviation.

1.42 Compute the sample mean, sample variance, and sample standard deviation for the capacitances listed in Prob. 1.14.

1.43 Refer to the data in Prob. 1.12.

    (a) Compare the lead levels in the blood of the exposed and control groups of children by computing the sample mean, sample variance, and sample standard deviation of the amount of lead in the blood of each group.

    (b) What inferences can you draw from this comparison?

1.44 The data in Prob. 1.12 are from an experiment where the goal is to *control* all factors except the one being tested. The factors of age and neighborhood were controlled by pairing children of the same age and neighborhood; they differed only with respect to their parents' exposure to lead. This is an example of *paired data.* One way of comparing the blood lead levels of the two populations is to compute the sample mean and sample standard deviation of D, the difference between the exposed and control groups of children. Calculate the sample mean and sample standard deviation for the variable D and interpret your results.

1.45 Compute the sample mean, sample variance, and sample standard deviation for the cadmium levels displayed in Prob. 1.13.

1.46 Suppose each observation x_i is bounded above and below by M and m, respectively; that is, assume that $m \le x_i \le M, i = 1, \ldots, n$.

    (a) Show that $m \le \bar{x} \le M$.

    (b) The writer Garrison Keillor has described the mythical town of Lake Wobegon as a place "where all the women are strong, the men are good-looking, and the children above average." Let x_i denote the grade point average (GPA) of the ith student and $\bar{x}$ denote the average GPA for all the students in Lake Wobegon High School. Explain why it is impossible for every student to have a GPA that is above the average. In other words, show that it is mathematically impossible to have $x_i > \bar{x}$ for every x_i. [*Hint:* Use the result from part (*a*).]

1.47 The following data come from an experiment that measured the weight loss for a group of 30 men who were randomly assigned to three different weight loss programs labeled A, B, and C. The data are simulated.

A	B	C	A	B	C
15.3	3.4	11.9	10	5.2	8.8
2.1	10.9	13.1	8.3	2.5	12.5
8.8	2.8	11.6	9.4	10.5	8.6
5.1	7.8	6.8	12.5	7.1	17.5
8.3	0.9	6.8	11.1	7.5	10.3

    (a) As a first step toward comparing the three diets, compute the mean and standard deviation of the weight losses for each of the three diets.

    (b) Construct side-by-side box plots of the weight losses for each of the three diets. Are there any outliers?

(c) Using the results of parts (*a*) and (*b*), comment on the similarities and differences in the weight loss data among the three diets. Which diet is most effective? Least effective?

1.48 Thirty-five items were inspected and the number of defects per item recorded:

2	3	0	1	0	0	0
0	0	2	3	3	1	3
3	1	0	0	2	0	0
2	1	2	0	1	0	3
1	4	0	0	1	0	4

(a) Compute the mean and median number of defects per item.
(b) Compare the mean and median and explain why they differ.

1.49 Suppose the data set $Y = \{y_1, \ldots, y_n\}$ is obtained from $X = \{x_1, \ldots, x_n\}$ by means of the linear transformation $y_i = ax_i + b$, where a and b are constants and $a > 0$.

(a) Show that the order statistics for the data set Y are given by

$$y_{(i)} = ax_{(i)} + b$$

(b) Let $Q_x(p)$ and $Q_y(p)$ denote the quantiles of order p for the X and Y data sets, respectively. Show that

$$Q_y(p) = aQ_x(p) + b$$

(c) Let $\hat{F}_n$ and $\hat{G}_n$ be the empirical distribution functions of the data sets $\{x_1, \ldots, x_n\}$ and $\{y_1, \ldots, y_n\}$, respectively. Show that

$$\hat{G}_n(y) = \hat{F}_n\left(\frac{y - b}{a}\right)$$

1.7   MATHEMATICAL DETAILS AND DERIVATIONS

We prove Proposition 1.1 using the standard rules for manipulating sums:

$$\sum_{1 \le i \le n} y_i = \sum_{1 \le i \le n} (ax_i + b)$$

$$= \sum_{1 \le i \le n} ax_i + \sum_{1 \le i \le n} b$$

Therefore,

$$\sum_{1 \le i \le n} y_i = a\left(\sum_{1 \le i \le n} x_i\right) + nb$$

We complete the proof by dividing both sides by n. This yields

$$\bar{y} = a\bar{x} + b$$

To prove $s_y^2 = a^2 s_x^2$, we proceed as follows:

$$s_y^2 = \sum_{1 \le i \le n} (y_i - \overline{y})^2$$

$$= \sum_{1 \le i \le n} [(ax_i + b) - (a\overline{x} + b)]^2$$

$$= \sum_{1 \le i \le n} a^2 (x_i - \overline{x})^2$$

$$= a^2 \sum_{1 \le i \le n} (x_i - \overline{x})^2 = a^2 s_x^2$$

1.8 CHAPTER SUMMARY

Scientists and engineers analyze their data by studying the frequency distribution of a variable defined on a population. They model data by postulating a theoretical distribution; they test their model by comparing the observed frequency distribution with the theoretical one. Consequently, the two most important concepts in this chapter are the (1) frequency distribution and (2) empirical distribution function of a variable defined on a population. Graphical methods, such as horizontal bar charts, dot plots, stem-and-leaf plots, histograms, and box plots are among the most effective methods for visualizing the shape of the distribution. Also useful are numerical measures such as quantiles, the sample range, the mean, the variance, and the standard deviation.

To Probe Further. The role of William Playfair (1759–1826) in the development of graphical methods for visualizing complex data is recounted in P. Costigan-Eaves and M. Macdonald-Ross, "William Playfair and Graphics," *Statistical Science*, vol. 5, no. 3, 1990, pp. 318–326.

The importance of identifying the sources of variation in the quality of a product or service is emphasized repeatedly in W. E. Deming, *Out of the Crisis*, Cambridge, Mass., MIT Press, 1982.

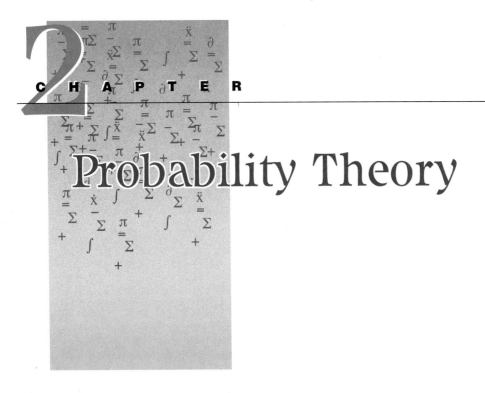

CHAPTER 2

Probability Theory

Millionaires should always gamble, poor men never.

John Maynard Keynes (1883–1946), British economist

2.1 ORIENTATION

After the distribution of the data has been analyzed via the methods of exploratory data analysis, the next step is *statistical inference,* which is a scientific method for making inferences about a population based on the data collected from a sample. The reliability of these inferences—and this is a point whose importance cannot be exaggerated—depends crucially on how one chooses the sample. The theoretical ideal, which is often quite difficult to implement, is to select the sample via *random sampling,* that is, *every population subset of the same size has an equal chance (probability) of being selected.* If the sample is selected based on a method that deviates significantly from random sampling, then the inferences will tend to be *biased* because, in some sense, the sample gives a distorted picture of the population. In their book *Statistical Methods* (7th ed., Ames, Iowa State University Press, 1980, p. 6) G. W. Snedecor and W. G. Cochran describe an interesting classroom experiment that illustrates how biased sampling can occur.

In a class exercise, a population of rocks of varying sizes was spread out on a large table so that all the rocks could be seen. After inspecting the rocks and lifting any that they wished, students were asked to select samples of five rocks such that their average weight would estimate as accurately as possible the average weight of the whole population of rocks. The average weights of the students' samples *consistently overestimated* the population mean.

The students' consistent overestimation of the population mean is an example of *biased sampling.* Snedecor and Cochran offer the following reasonable explanation for the observed bias: "A larger and heavier rock is more easily noticed than a smaller one."

Since different students will, in general, select different rocks for their samples, it follows that the value of the average weight of the rocks in the sample will also vary. This is an example of a *sampling experiment*; it is also an example of an experiment whose outcomes are subject to chance. In this chapter our primary goal is to construct a mathematical model of random sampling based on the theory of probability.

The basic concepts of probability theory originated in the analysis of games of chance by such renowned mathematicians as Blaise Pascal (1623–1662) and Pierre de Fermat (1601–1665). Indeed, one of the first books published on probability theory was entitled *The Doctrine of Chances: or, A Method of Calculating the Probability of Events in Play* (1718) by the mathematician Abraham de Moivre (1667–1754).[1] In the dedication of his book de Moivre, anticipating the charge "that the Doctrine of Chances has a tendency to promote Play," refuted it by arguing "that this Doctrine is so far from encouraging Play, that it is rather a guard against it, by setting in a clear light, the advantages and disadvantages of those games wherein chance is concerned."

It is an interesting fact, and one that we take advantage of, that the concepts and techniques used by de Moivre and his contemporaries to analyze games of chance are precisely the tools we need to construct a mathematical model of random sampling. For these reasons we begin our introduction to probability theory with the analysis of some simple games of chance.

Organization of Chapter

---

[1]The word *play* for de Moivre is synonymous with a game of chance.

2.2   SAMPLE SPACE, EVENTS, AND AXIOMS OF PROBABILITY THEORY

Before shipping a mass-produced product to its customers, a manufacturer typically inspects a small proportion of the output for defects. The number of defective items will vary from sample to sample. In Deming's red bead experiment (see Example 1.3) a worker selects 50 beads from a bowl filled with 800 red beads and 3200 white beads. The red beads represent the defective items. This is an important and well-known example of a sampling experiment. The number of red beads produced by six workers were 9, 5, 15, 4, 10, and 8 (see Table 1.2). The workers' performance, as previously noted, appears to vary greatly, with one worker producing 4 red beads and another producing 15, nearly four times as many! The problem for the quality control engineer is to account for this observed variation in workers' performance: Is it to be found in the raw materials used, or is it due to differences among the workers themselves? We will see later, as a result of the theory presented here and in the following chapters, that most of the observed variation in workers' performance is due to random sampling and not to any intrinsic differences among the workers.

The key to understanding variations in experimental data is to construct a mathematical model of random sampling. *Probability theory* is that branch of mathematics concerned with the construction, analysis, and comparison of probability models of random phenomena. We begin by defining some important terms and basic concepts.

Selecting items for inspection is an example of an *experiment* with unpredictable outcomes. A sample that contains 15 defective items is an example of an *event*. The set of logically possible outcomes of an experiment is called the *sample space* (corresponding to the experiment). The formal definition follows.

Sample Space. The *sample space* Ω (capital "omega") is the set of all logically possible outcomes resulting from some experiment. An element ω (lower-case "omega") of the sample space is called a *sample point* or, more simply, an *outcome*.

These terms are just convenient abstractions of some fairly simple ideas that are nicely illustrated in the intuitive context of gambling games such as craps, roulette, poker, and state lotteries. Indeed, P. S. Laplace (1749–1827), a great mathematician and one of the founders of the theory of probability, wrote, "It is remarkable that a science which commenced with the consideration of games of chance, should be elevated to the rank of the most important subjects of human knowledge." We begin our introduction of probability models with a brief description of simple games of chance since the mathematical techniques used to analyze them are widely used in other scientific contexts and the results are of independent interest.

---

example **2.1**   Describe the sample space corresponding to the experiment of throwing two dice.

Solution. We assume that the two dice can be distinguished by color, with one colored red, say, and the other colored white. We denote the result of throwing the two dice by the ordered pair i, j, so that $1, 3$ means the red die and white die showed a 1 and a 3, respectively. The experiment of throwing two dice has a total of 36 outcomes, listed in Table 2.1.

TABLE 2.1   SAMPLE SPACE FOR THROWING TWO DICE

1, 1	1, 2	1, 3	1, 4	1, 5	1, 6
2, 1	2, 2	2, 3	2, 4	2, 5	2, 6
3, 1	3, 2	3, 3	3, 4	3, 5	3, 6
4, 1	4, 2	4, 3	4, 4	4, 5	4, 6
5, 1	5, 2	5, 3	5, 4	5, 5	5, 6
6, 1	6, 2	6, 3	6, 4	6, 5	6, 6

■

The sample space is said to be *discrete* if it consists of a finite or infinite sequence of sample points.

Notation for Discrete Sample Spaces

$$\Omega = \{\omega_1, \ldots, \omega_N\} \qquad \text{(finite sample space)}$$
$$\Omega = \{\omega_1, \ldots, \omega_n, \ldots\} \qquad \text{(discrete infinite sample space)}$$

We give additional examples of discrete sample spaces later in this chapter (Sec. 2.3) and in the next.

Events. An *event A* is simply a subset of the sample space Ω; in symbols, $A \subset \Omega$. The event that contains no sample points at all is the *impossible event* and is denoted $\varnothing$.

Notation. Unless otherwise stated, Ω denotes a fixed but otherwise unspecified sample space. Subsets of the sample space are denoted by the capital letters A, B, C, etc.

---

example 2.2   The gambler playing casino craps has a choice of outcomes on which to place a bet. Following is a partial list of events on which one may place a bet; the symbol A_i denotes the event that the number i is thrown.

1. $A_{12} = \{(6, 6)\}$
2. $A_{11} = \{(5, 6), (6, 5)\}$
3. $A_7 = \{(3, 4), (4, 3), (2, 5), (5, 2), (1, 6), (6, 1)\}$
4. *Throwing a natural*: The player throws a 7 or an 11; there are eight sample points in a natural:

$$\text{Natural} = \{(3, 4), (4, 3), (2, 5), (5, 2), (1, 6), (6, 1), (5, 6), (6, 5)\}$$

5. *Throwing a point:* The player throws a 4, 5, 6, 8, 9, or 10. This event contains 24 sample points, which are too many to list here; for our purposes the verbal description is satisfactory.

We say that the *event A occurred* if the outcome of the experiment is a sample point $\omega \in A$.

Thus, if $(4, 3)$ is thrown, we say A_7 occurred.   ■

Combinations of Events. Events can be combined in various ways to form more complex events, and, conversely, complex events can be written as combinations of events that are simpler to analyze. An event, as defined earlier, is a subset of the sample space, so it is quite natural to combine events using such basic concepts of set theory as union, complement, and intersection. A familiarity with the basic concepts of set theory is assumed.

The Union of Two Events. The event *A or B or both occurred* is called the *union* of the two events and is denoted by $A \cup B$. It consists of all sample points that are in either A or B or both.

---

example **2.3**

A *natural* occurs, as noted previously, when the dice show a 7 or 11. In the language of probability theory, a natural is the union of the two events A_7 and A_{11} and is written $A_7 \cup A_{11}$.

The operation of union can be extended to more than two events. Consider the event that a shooter throws a natural or a point. This event, denoted A, is a union of eight events and is written

$$A = A_4 \cup A_5 \cup A_6 \cup A_7 \cup A_8 \cup A_9 \cup A_{10} \cup A_{11} \qquad \blacksquare$$

The Complement of an Event. The set of all sample points not in A is also an event; it is called the *complement of A* and is denoted by A'.

---

example **2.4**

Refer to the event A in the preceding example. List the sample points in A'.

Solution. The event A' occurs when the shooter fails to throw a natural or a point; we say that the player has thrown a *craps*. The event A' can be written in the following equivalent form:

$$A' = A_2 \cup A_3 \cup A_{12}$$

The sample points in the set A' are

$$A' = \{(1,1),(1,2),(2,1),(6,6)\} \qquad \blacksquare$$

The Intersection of Two Events. The event *A and B occurred simultaneously*, denoted by $A \cap B$, is called the *intersection* of the two events A and B. It consists of all sample points common to A and B; that is, a sample point $\omega \in A \cap B$ if and only if $\omega \in A$ and $\omega \in B$.

---

example **2.5**

Let $A = \{$an even number is thrown$\}$ and $B = \{$a number greater than 9 is thrown$\}$. Describe the events $A, B,$ and $A \cap B$ in terms of the events $A_2, \ldots, A_{12}$.

Solution. We express each of these events in terms of the A_i as follows:

$$A = A_2 \cup A_4 \cup A_6 \cup A_8 \cup A_{10} \cup A_{12}$$
$$B = A_{10} \cup A_{11} \cup A_{12}$$
$$A \cap B = A_{10} \cup A_{12} \qquad \blacksquare$$

Mutually Disjoint Events. The events A and B are said to be *mutually disjoint* if $A \cap B = \varnothing$. More generally, the events $A_1, A_2, \ldots, A_n$ are mutually disjoint if

$$A_i \cap A_j = \varnothing, \quad i \neq j$$

---

example **2.6**   Here are some examples of disjoint events:

1. Note that one throw of a pair of dice cannot simultaneously yield a 4 and a 9, so $A_4 \cap A_9 = \varnothing$.
2. Every sample point is either in A or in its complement, A', but not in both; thus,

$$\Omega = A \cup A'; \qquad A \cap A' = \varnothing$$

This is the simplest example of a *partition*, which is a decomposition of a sample space into a union of mutually disjoint sets.

Partition of a Sample Space. The events $A_1, A_2, \ldots, A_n$ form a *partition* of the sample space Ω if:

(a) The events are mutually disjoint, and

(b) $\Omega = A_1 \cup A_2 \cup \cdots \cup A_n$

In other words, a partition divides the sample space into mutually disjoint events with the property that every point in the sample space belongs to exactly one of these events.

3. Referring to the sample space corresponding to one throw of a pair of dice (Table 2.1), it is easy to verify that the events $A_2, \ldots, A_{12}$ form a *partition* of the sample space, because they are mutually disjoint, and they are *exhaustive* in the sense that every point in the sample space belongs to one and only one of these events. Using set theory notation, we express this by

$$\Omega = A_2 \cup A_3 \cup \cdots \cup A_{12}$$

Table 2.2 is a visual display of this partition.

TABLE 2.2 A PARTITION OF THE DICE THROWING SAMPLE SPACE DISPLAYED IN THE FORMAT OF A HORIZONTAL BAR CHART

A_i	Sample points in A_i
A_2	$(1,1)$
A_3	$(1,2),(2,1)$
A_4	$(1,3),(2,2),(3,1)$
A_5	$(1,4),(2,3),(3,2),(4,1)$
A_6	$(1,5),(2,4),(3,3),(4,2),(5,1)$
A_7	$(1,6),(2,5),(3,4),(4,3),(5,2),(6,1)$
A_8	$(2,6),(3,5),(4,4),(5,3),(6,2)$
A_9	$(3,6),(4,5),(5,4),(6,3)$
A_{10}	$(4,6),(5,5),(6,4)$
A_{11}	$(5,6),(6,5)$
A_{12}	$(6,6)$

■

A **implies** *B.* We say that the event *A* *implies* *B* when the occurrence of *A* implies the occurrence of *B*. This means that every sample point in *A* is also in *B*; that is, *A* is a *subset* of the event *B* or, in symbols, $A \subset B$.

Venn Diagrams

The set-theoretic operations $A \cup B, A \cap B$, and A' can be represented graphically via *Venn diagrams*. The sample space Ω is represented as a rectangle, and the subsets *A* and *B* are represented as circles contained within the rectangle, as shown in Fig. 2.1.

The Algebra of Sets. We next list, without proof, some facts about the set-theoretic operations of union, intersection, and complementation. We verify them by drawing the Venn diagram representations of the sets that occur on the left- and right-hand sides of the equals sign.

■■■ **PROPOSITION 2.1**

The operations $\cup$ and $\cap$ satisfy the following relations:

$$A \cup B = B \cup A; \qquad A \cap B = B \cap A \qquad \text{(symmetry)} \qquad (2.1)$$

$$A \cap (B \cup C) = (A \cap B) \cup (A \cap C) \qquad \text{(distributive law)} \qquad (2.2)$$

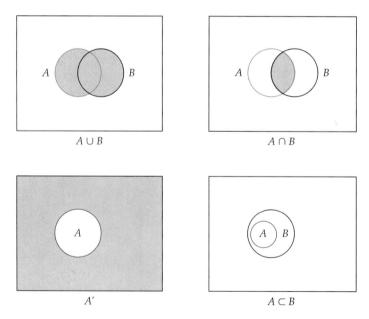

$$A \cup B$$

$$A \cap B$$

$$A'$$

$$A \subset B$$

Figure 2.1 The interior of each rectangle represents the sample space Ω, and the shaded regions are the Venn diagrams for the sets $A \cup B$, $A \cap B$, and A'

$$(A \cup B)' = A' \cap B' \tag{2.3}$$
$$(A \cap B)' = A' \cup B' \tag{2.4}$$

Equations (2.3) and (2.4) are called *De Morgan's laws*.   ∎

---

example **2.7**   Consider the experiment of tossing a coin three times. Let A_i and B_i denote, respectively, the events *exactly i heads are thrown* and *at least i heads are thrown*, for $i = 0, 1, 2, 3$.

1. List the sample points in A_2 and B_2.

Solution. The sample space contains eight points, which are listed in Table 2.3. The symbols H and T stand for heads and tails, respectively.

TABLE 2.3   SAMPLE SPACE FOR COIN-TOSSING EXPERIMENT ($n = 3$)

HHH	HTH	THH	TTH
HHT	HTT	THT	TTT

We now have all the information we need to list the sample points in A_2 and B_2.

$$A_2 = \{(HHT),(HTH),(THH)\}$$
$$B_2 = \{(HHT),(HTH),(THH),(HHH)\}$$

Note that $A_2 \subset B_2$ and, more generally, $A_i \subset B_i$; the latter assertion follows from the fact that if exactly i heads are thrown, then at least i heads are thrown.

2. Express in terms of A_i, B_i the following events:
(a) One tail is thrown.

Solution. If one tail is thrown, then two heads were thrown; consequently,
$$\text{One tail was thrown} = A_2$$

(b) Only tails were thrown.

Solution. This is equivalent to the assertion that no heads were thrown; consequently,
$$\text{Only tails were thrown} = A_0$$

(c) At least two tails were thrown.

Solution. This is equivalent to the statement that at most one head was thrown; thus,
$$\text{At least two tails were thrown} = B_2' = A_0 \cup A_1$$ ∎

---

example **2.8**   Let us now consider these concepts in the context of reliability theory. The *reliability* of a device is defined as the probability that it performs without failure for a specified period of time in a given environment. Consider a system—for example, a chain three links long—consisting of three components, each of which can be in only one of two states: (1) a functioning state and (2) a nonfunctioning or failed state. The state of the system itself is denoted by W (functioning) and W' (nonfunctioning). Similarly, the state of the ith component is denoted by W_i (functioning) and by W_i' (nonfunctioning). Reliability engineers are interested in expressing the functioning state W of the system as a whole in terms of the functioning states of its components W_i ($i = 1, 2, 3$). We will consider two examples: (1) the components are linked in series and (2) the components are linked in parallel.

1. **Components in series**. Linking the components in series means that the system functions if and only if each component is functioning; this can be expressed in set-theoretic language as

$$W = W_1 \cap W_2 \cap W_3$$

2. **Components in parallel**. Linking the components in parallel means that the system functions if at least one of the components is functioning; i.e., the other two components serve as "backups." In this case the functioning state W can be expressed as

$$W = W_1 \cup W_2 \cup W_3$$ ■

2.2.1   Probability Functions

At the beginning of every NFL football game an official flips a coin to decide who will receive the kickoff; this procedure is regarded by everyone as fair because everyone accepts without question that heads (H) and tails (T) are equally likely. The sample space corresponding to this experiment consists of two sample points $\Omega = \{H, T\}$. The assumption that the coin is "fair" means that each of the two sample points is assigned the probability $1/2$. We use the symbol $P(\omega)$ to denote the probability assigned to the sample point ω. With reference to the experiment of flipping a coin once, we write

$$P(H) = \frac{1}{2} \quad \text{and} \quad P(T) = \frac{1}{2}$$

Probability Function Defined on a Discrete Sample Space

■ **DEFINITION 2.1**

A *probability function* is a real-valued function $P(\omega)$ defined on a discrete sample space $\Omega = \{\omega_1, \ldots, \omega_n, \ldots\}$ with the following properties:

$$P(\omega_i) \geq 0 \qquad (2.5)$$

$$\sum_{1 \leq i < \infty} P(\omega_i) = 1 \qquad (2.6)$$

■

The Probability of an Event A

It is natural to define the probability of an event A to be equal to the sum of the probabilities of all the sample points in it; more precisely, the probability of a set A consisting of the distinct sample points $\{s_1, \ldots, s_k, \ldots\}$ is given by

$$P(A) = \sum_{s_i \in A} P(s_i) \tag{2.7}$$

---

example 2.9   *Playing the field* in craps means betting on the event
$$A = \{2, 3, 4, 9, 10, 11, 12\}$$
This is also called a *field bet*. Compute $P(A)$.

Solution. Refer to Table 2.2, which displays the events $A_i = $ (the number i is thrown) and their probabilities in the format of a horizontal bar chart. From this table we see that
$$A = A_2 \cup A_3 \cup A_4 \cup A_9 \cup A_{10} \cup A_{11} \cup A_{12}$$
Since there are 16 sample points in A, each of which is assigned the probability $1/36$, it follows that $P(A) = 16/36$.   ∎

In Example 2.9 we computed the probability of an event A by first counting the number of sample points in A and then dividing it by the total number of sample points in the sample space Ω. This is a special case of the equally likely probability function, which we define next.

Equally Likely Probability Function

▦ **DEFINITION 2.2**

The *equally likely probability function* P defined on a finite sample space $\Omega = \{\omega_1, \omega_2, \ldots, \omega_N\}$ assigns the same probability $P(\omega_i) = 1/N$ to each sample point.   ∎

Computing $P(A)$ **When** P **Is an Equally Likely Probability Function.** When P is the equally likely probability measure on a finite sample space containing N sample points and the set $A = \{s_1, \ldots, s_k\}$ contains k sample points, then $P(A)$ equals k/N. This follows from Eq. (2.7); the details follow.

$$P(A) = P(s_1) + \cdots + P(s_k) = \frac{1}{N} + \cdots + \frac{1}{N} \quad (k \text{ times})$$
$$= \frac{k}{N}$$

Notation. We continue to use the symbol $\#(A)$ to denote the number of sample points (outcomes) in the event A. It is also called the *number of outcomes favorable to A*. Thus,

$$P(A) = \frac{\#(A)}{\#(\Omega)}$$

The Empirical Probability Function

The frequency function $f(x)$ of a data set $X = \{x_1, x_2, \ldots, x_n\}$ generates a probability function $\hat{f}(x)$ defined by the equation

$$\hat{f}(x) = \frac{f(x)}{n} \tag{2.8}$$

We call $\hat{f}(x)$ the *empirical probability function*. The empirical distribution function is obtained from the empirical probability function by noting that

$$\hat{F}_n(x) = \sum_{y \le x} \hat{f}(y)$$

example 2.10   Compute the empirical probability function for the microwave radiation data (Table 1.3).

Solution. To compute the empirical probability function, we use Eq. (2.8) and the second column of the horizontal bar chart (Fig. 1.1), which lists the values of $f(x)$. Table 2.4 displays the values $\hat{f}(x)$.

TABLE 2.4   EMPIRICAL PROBABILITY FUNCTION FOR THE RADIATION DATA SET (TABLE 1.3)

x	0.01	0.02	0.03	0.05	0.07	0.08	0.09	0.10
$\hat{f}(x)$	2/42	3/42	1/42	5/42	1/42	3/42	2/42	9/42

x	0.11	0.12	0.15	0.18	0.20	0.30	0.40
$\hat{f}(x)$	1/42	1/42	3/42	2/42	3/42	4/42	2/42

The empirical probability function plays a fundamental role in the *bootstrap* method, described later in the chapter, which is a useful tool for studying distributions associated with a data set.

The Frequency Interpretation of Probability

If a fair coin is tossed n times, we expect the relative frequency with which a head (H) appears to be nearly equal to its theoretical probability $P(H) = 1/2$; in other words, we expect the following relation to hold:

$$\frac{\text{Number of heads in } n \text{ tosses}}{n} \approx P(H) = \frac{1}{2} \quad \text{for } n \text{ large}$$

example 2.11   R. Wolf (1882) threw a die 20,000 times and recorded the number of times each of the six faces appeared. The results follow.

Face	1	2	3	4	5	6
Frequency	3407	3631	3176	2916	3448	3422

Source: D. J. Hand et al., *Small Data Sets*, London, Chapman & Hall, 1994.

When the die is fair, the expected frequency of occurrence for each face is $20,000 \times (1/6) = 3333.33$. We can now compare the observed frequencies with the expected frequencies to test whether the die is fair. (We omit the details of the analysis since it involves an application of the chi-square test, which will be discussed in Sec. 9.3.) ∎

In general, when an experiment is repeated n times, with n a large number, we expect the relative frequency with which the event A occurs to be approximately equal to its theoretical probability $P(A)$, that is,

$$\lim_{n \to \infty} \frac{\text{number of occurrences of event } A \text{ in } n \text{ repetitions}}{n} = P(A) \qquad (2.9)$$

Equation (2.9) is a mathematical formulation of this result, called the *law of large numbers* (to be given as Theorem 5.2).

Axioms for a Probability Measure

The function $P(A)$ defined by Eq. (2.7) is an example of the more general concept of a *probability measure,* which is a function defined as follows:

■ **DEFINITION 2.3**

Let Ω be a sample space corresponding to some experiment. A *probability measure P* is a function that assigns to each event $A \subset \Omega$ a number $P(A)$ satisfying the following axioms. For every event A

$$P(A) \geq 0 \qquad (2.10)$$
$$P(\Omega) = 1 \qquad (2.11)$$

For any two mutually disjoint events A and B

$$P(A \cup B) = P(A) + P(B) \quad (A \cap B = \varnothing) \qquad (2.12)$$

More generally, for any finite sequence of mutually disjoint events $(A_i \cap A_j = \varnothing, i \neq j)$, the probability of their union equals the sum of their probabilities:

$$P(A_1 \cup A_2 \cup \cdots \cup A_n) = \sum_{1 \leq i \leq n} P(A_i) \qquad (2.13)$$

When the sample space contains an infinite number of outcomes and $A_1, A_2, \ldots, A_n, \ldots$ is an infinite sequence of mutually disjoint events, then we assume that

$$P(A_1 \cup A_2 \cup \cdots \cup A_i \cup \cdots) = \sum_{1 \leq i < \infty} P(A_n) \qquad (2.14)$$
∎

---

example **2.12**   Give an example of an infinite sequence of mutually disjoint events.

Solution. Consider the experiment of tossing a coin until the first head appears. Clearly, this is an experiment that can continue indefinitely. Denote by A_i the event that a head is thrown for the first time at the ith throw; this is a simple example of an infinite sequence of mutually disjoint events. The union $\cup_{1 \leq i < \infty} A_i$ is the event that a head ever appears. ∎

Some Consequences of the Axioms

▨ PROPOSITION 2.2

Probability of the Complement: For every event A,
$$P(A') = 1 - P(A)$$

Proof. It suffices to show that $P(A) + P(A') = 1$. Since $\Omega = A \cup A'$ and $A \cap A' = \varnothing$, it follows from axioms (2.11) and (2.13) (where $n = 2$ and $A_1 = A$, $A_2 = A'$) that

$$1 = P(\Omega) = P(A) + P(A') \tag{2.15}$$

■

In Example 2.13 we show that it is sometimes easier to compute $P(A)$ via the formula $P(A) = 1 - P(A')$ instead of computing it directly.

---

example **2.13**   Consider the experiment of tossing a coin $n = 3$ times (Example 2.7). Compute the probability of getting at least one head.

Solution. Let A denote the event that at least one head is thrown. Its complement A' is the event that no heads are thrown. The event A' contains exactly one sample point, namely, $A' = (TTT)$, so $P(A') = 1/8$. Therefore, by Proposition 2.2,

$$P(\text{at least one head}) = 1 - P(\text{no heads}) = 1 - \frac{1}{8} = \frac{7}{8}$$

Axiom (2.14), called the *axiom of countable additivity*, plays an important role in more advanced treatises on probability theory.   ■

The computation of $P(B)$ is sometimes made easier by first partitioning B into two disjoint events and then using the following proposition, which generalizes Eq. (2.15).

▨ PROPOSITION 2.3

For every pair of events A, B,
$$P(B) = P(B \cap A) + P(B \cap A') \tag{2.16}$$

Proof. Using the partition $\Omega = A \cup A'$ and the distributive law [Eq. (2.2)], we can partition B into two disjoint events as follows:
$$B = B \cap \Omega = B \cap (A \cup A') = (B \cap A) \cup (B \cap A')$$
Thus,
$$P(B) = P((B \cap A) \cup (B \cap A'))$$
$$= P(B \cap A) + P(B \cap A') \quad \text{[by axiom (2.12)]}$$

■

A Formula for $P(A \cup B)$

We next give a formula for computing $P(A \cup B)$, where the events A and B are not necessarily disjoint.

▦ PROPOSITION 2.4

For arbitrary events A and B,

$$P(A \cup B) = P(A) + P(B) - P(A \cap B) \tag{2.17}$$

$$P(A \cup B) \le P(A) + P(B) \tag{2.18}$$

Proof. We begin with the partition $A \cup B = A \cup (B \cap A')$, which can be verified by drawing Venn diagrams for $A \cup B$ and $A \cup (B \cap A')$.

$$A \cup B = A \cup (B \cap A')$$

Therefore,

$$P(A \cup B) = P(A) + P(A' \cap B)$$

Equation (2.16) implies

$$P(A' \cap B) = P(B) - P(A \cap B)$$

Therefore,

$$P(A \cup B) = P(A) + P(B) - P(A \cap B)$$

The right-hand side of Eq. (2.17) is less than the right-hand side of Eq. (2.18) since it is obtained from the latter by subtracting the nonnegative number $P(A \cap B)$. ∎ ■

---

example 2.14   Let A, B denote events for which

$$P(A) = 0.3, \qquad P(B) = 0.8, \qquad P(A \cap B) = 0.2$$

Using only this information and the basic properties of a probability measure, compute $P(A \cup B)$, $P(A' \cap B')$, and $P(A' \cap B)$.

Solutions

$$P(A \cup B) = P(A) + P(B) - P(A \cap B) \quad \text{(Proposition 2.4)}$$
$$= 0.3 + 0.8 - 0.2 = 0.9$$
$$P(A' \cap B') = P([A \cup B]') \quad \text{(De Morgan's laws)}$$
$$= 1 - P(A \cup B) = 1 - 0.9 = 0.1$$
$$P(A' \cap B) = P(B) - P(A \cap B) \quad \text{(Proposition 2.3)}$$
$$= 0.8 - 0.2 = 0.6 \quad ■$$

Proposition 2.4 will now be extended to compute the probability of the union of three events that are not assumed to be mutually disjoint.

▦ PROPOSITION 2.5

$$P(A_1 \cup A_2 \cup A_3) = P(A_1) + P(A_2) + P(A_3) - P(A_1 \cap A_2)$$
$$- P(A_1 \cap A_3) - P(A_2 \cap A_3) + P(A_1 \cap A_2 \cap A_3)$$

We omit the derivation since it is a special case of the inclusion-exclusion relation (see W. Feller, *An Introduction to Probability Theory and Its Applications*, vol. 1, 3d ed., New York, John Wiley & Sons, 1968, p. 99). ■

example **2.15**   Given three events A_1, A_2, A_3 such that
$$P(A_1) = 0.3, \qquad P(A_2) = 0.6, \qquad P(A_3) = 0.4$$
$$P(A_1 \cap A_2) = 0.2, \qquad P(A_1 \cap A_3) = 0.1, \qquad P(A_2 \cap A_3) = 0.3$$
$$P(A_1 \cap A_2 \cap A_3) = 0.05$$
compute the following probabilities:

1. $P(A_1 \cup A_2 \cup A_3)$
2. $P(A_1' \cap A_2' \cap A_3')$

Solutions

1. To compute $P(A_1 \cup A_2 \cup A_3)$ we apply Proposition 2.5:
$$P(A_1 \cup A_2 \cup A_3) = 0.3 + 0.6 + 0.4 - 0.2$$
$$- 0.1 - 0.3 + 0.05 = 0.75$$

2. To compute $P(A_1' \cap A_2' \cap A_3')$ we apply De Morgan's laws or, rather, a straightforward extension of Eq. (2.3) to compute the complement of the union of three events:
$$A_1' \cap A_2' \cap A_3' = (A_1 \cup A_2 \cup A_3)'$$
Consequently,
$$P(A_1' \cap A_2' \cap A_3') = 1 - P(A_1 \cup A_2 \cup A_3) = 1 - 0.75 = 0.25 \qquad \blacksquare$$

PROBLEMS

2.1 The sample space corresponding to the experiment of tossing a coin four times is shown. Assume that all 16 sample points have equal probabilities.

HHHH	HTHH	THHH	TTHH
HHHT	HTHT	THHT	TTHT
HHTH	HTTH	THTH	TTTH
HHTT	HTTT	THTT	TTTT

Let A_i and B_i denote, respectively, the events *exactly i heads are thrown* and *at least i heads are thrown*, for $i = 0, 1, 2, 3, 4$. List the sample points in each of the events and compute their probabilities.

(a) A_0 (b) A_1 (c) B_3
(d) B_4 (e) A_4

2.2 (Continuation of Prob. 2.1) Which of the following statements are true? Which are false?

(a) $A_0' = B_1$
(b) $B_2 = A_2 \cup A_3 \cup A_4$
(c) $B_2 \subset A_2$
(d) $A_2 \subset B_2$

2.3 (Continuation of Prob. 2.1) Express, in terms of the sets A_i and B_i defined earlier, the following events:

(a) No more than two tails are thrown.
(b) Exactly three tails are thrown.
(c) No tails are thrown.
(d) Only tails are thrown.

2.4 A quality control engineer observes that a printed circuit board with i defects ($i = 0, 1, 2, 3, 4$) seems to occur with a frequency inversely proportional to $i + 1$; that is, the sample space corresponding to this experiment consists of the finite set of points $\Omega = \{\omega_0, \omega_1, \omega_2, \omega_3, \omega_4\}$, where ω_i is the event that the circuit board has i defects and

$$P(\omega_i) = c/(i + 1), \quad i = 0, 1, \ldots, 4$$

(a) Compute the constant c. [*Hint:* Use Eq. (2.6) to derive an equation for c.]
(b) What is the probability that a circuit board has no defects? At least one defect?

2.5 In an environmental study of the effect of radon gas on a population, blood samples were drawn from human volunteers, 100 cells were cultured, and the cells that exhibited chromosome aberrations were counted. Denote by ω_i the event that i chromosome aberrations occurred. Based on the observed frequency counts of these events, the microbiologist feels that the following probability model is appropriate:

$$\Omega = \{\omega_0, \omega_1, \omega_2, \omega_3, \omega_4, \omega_5\}$$

$$P(\omega_i) = \frac{c(i + 1)^2}{2^{i+1}}, \quad i = 0, 1, 2, 3, 4, 5$$

(a) Determine the value of the constant c.
(b) What is the probability that a person selected at random from the population has exactly three chromosome aberrations? Three or more?

2.6 Let $W_i, i = 1, 2, 3$, denote the event that component i is working. W_i' denotes the event that the ith component is not working. Express in set-theoretic language the following events: Of the components W_1, W_2, W_3,
(a) Only W_1 is working.   **(b)** All three components are working.
(c) None are working.   **(d)** At least one is working.
(e) At least two are working.   **(f)** Exactly two are working.

2.7 Let Ω denote the sample space arising from the tossing of two dice. Let A_i denote the event *an i is thrown*, where $i = 2, 3, \ldots, 11, 12$. Express each of the following events in terms of the events A_i.
(a) An odd number is thrown.
(b) A number ≤ 6 is thrown.
(c) A number > 6 is thrown.

2.8 Compute the probabilities of each of the events in the previous problem; assume that all sample points have the same probability, $1/36$.

2.9 A and B are events whose probabilities are given by $P(A) = 0.7, P(B) = 0.4, P(A \cap B) = 0.2$. Compute the following probabilities:
(a) $P(A \cup B)$ **(b)** $P(A' \cap B')$
(c) $P(A' \cup B')$ **(d)** $P(A' \cap B)$
(e) $P(A \cap B')$ **(f)** $P(A' \cup B)$
(g) $P(A \cup B')$

2.10 Given: $P(A_1) = 0.5$, $P(A_2) = 0.4$, $P(A_3) = 0.4$, $P(A_1 \cap A_2) = 0.04$, $P(A_1 \cap A_3) = 0.1, P(A_2 \cap A_3) = 0.2, P(A_1 \cap A_2 \cap A_3) = 0.02$. Using this information, Venn diagrams, and Propositions 2.1 and 2.5, compute the following probabilities:

(a) $P(A_1 \cup A_2 \cup A_3)$ (b) $P(A_1' \cap A_2' \cap A_3')$
(c) $P((A_1 \cup A_2) \cap A_3)$ (d) $P((A_1 \cup A_2) \cap A_3')$
(e) $P(A_1' \cap A_2 \cap A_3)$

2.11 Which of the following statements are true? Which are false?

(a) If $A \subset B$, then $P(B') \leq P(A')$.
(b) If $P(A) = P(B)$, then $A \equiv B$.
(c) If $P(A \cap B) = 0$, then either $P(A) = 0$ or $P(B) = 0$.

2.12 Suppose $P(A) = 0.7$ and $P(B) = 0.8$. Show that

$$P(A \cap B) \geq 0.5$$

2.13 Let B denote an event for which $P(B) > 0$. Define a set function P' via the formula

$$P'(A) = \frac{P(A \cap B)}{P(B)}$$

Show that P' satisfies the axioms for a probability function. $P'(A)$ is called "the conditional probability of A given that B has occurred."

2.3 MATHEMATICAL MODELS OF RANDOM SAMPLING

We return now to the main focus of this chapter: to use the theory of probability to construct a mathematical model of the concept of a *random sample* of size k taken from a finite population consisting of n objects. We consider two basic methods for choosing a random sample: *sampling without replacement* and *sampling with replacement*.

- **Sampling without replacement:** A five-card poker hand is an example of a random sample of size 5 taken without replacement from a deck of 52 cards. That is, once a card is drawn, it is removed from the deck.
- **Sampling with replacement:** Spinning a roulette wheel in Monte Carlo is mathematically equivalent to randomly selecting a number, with replacement, from the population $\{0, 1, \ldots, 36\}$; that is, the next spin of the roulette wheel is again a random sample from the same population.

As noted earlier, random sampling means that each subset of a given size has the same probability of being the one that is drawn. This probability equals $1/N$, where N is the number of subsets of size k. Similarly, in order to compute the probability $P(A)$ of an event A, it is necessary to compute $\#(A)/N$; that is, we must compute the number of ways that A occurs and then divide it by the total number of outcomes N. However, when either $\#(A)$ or N, or both, are very large numbers, a complete listing of the sample points in A and in Ω is virtually impossible—and it wouldn't provide much information even if it were possible. The sample space corresponding to the experiment of tossing a coin 10 times, for example, contains $N = 2^{10} = 1024$ sample points, and the event *five heads are thrown* has 252 sample points (these results are particular consequences of more general theorems to be proved shortly). Fortunately, in most cases the computation of N and $\#(A)$ can be simplified by reducing it to an equivalent problem in *combinatorial analysis.* For this reason a brief introduction to this subject is warranted.

The Multiplication Principle

The *multiplication principle* and its variations play a fundamental role throughout this section. Suppose experiment 1 has n_1 outcomes denoted $\{A_1, \ldots, A_{n_1}\}$ and experiment 2 has n_2 outcomes denoted $\{B_1, \ldots, B_{n_2}\}$. Then performing the two experiments in succession has $n_1 \times n_2$ distinct outcomes.

We denote an outcome of this experiment by the ordered pair (A_i, B_j), which indicates that the first experiment produced the outcome A_i and the second experiment produced the outcome B_j. The set of all such outcomes can be displayed as a rectangular array consisting of n_1 rows and n_2 columns, as shown in Fig. 2.2. We see that the array in Fig. 2.2 has n_1 rows and n_2 columns, so it has $n_1 \times n_2$ elements.

It is worth noting that performing the two experiments in succession is itself an experiment called the *product of two experiments;* its sample space is denoted $S_1 \times S_2$, where $S_1 = \{A_1, \ldots, A_{n_1}\}$ and $S_2 = \{B_1, \ldots, B_{n_2}\}$.

Because of its importance in probability and statistics, some simple examples of this computation are worth studying in detail.

---

example **2.16**   Describe the product experiment $S_1 \times S_2$, where:

1. $S_1 = S_2 = \{1, 2, 3, 4, 5, 6\}$. Here $n_1 = n_2 = 6$, so there are $6 \times 6 = 36$ ordered pairs of the form (i, j). The set of ordered pairs is identical to the sample space corresponding to throwing two dice, or one die twice in succession; see Table 2.1.

2. $S_1 = S_2 = \{H, T\}$. Here $n_1 = n_2 = 2$, so there are $2 \times 2 = 4$ ordered pairs of the form $\{(H, H), (H, T), (T, H), (T, T)\}$. This is the sample space corresponding to the experiment of throwing a coin twice.

$$
\begin{array}{cccc}
(A_1, B_1) & (A_1, B_2) & \cdots & (A_1, B_{n2}) \\
(A_2, B_1) & (A_2, B_2) & \cdots & (A_2, B_{n2}) \\
\vdots & (A_i, B_j) & \cdots & \vdots \\
(A_{n1}, B_1) & (A_{n1}, B_2) & \cdots & (A_{n1}, B_{n2})
\end{array}
$$

Figure 2.2 The sample space corresponding to the product of two experiments

3. Consider an experiment in which the yield of a chemical process depends on (a) the operating pressure and (b) the operating temperature. To determine the optimum yield, the experiment is run at $n_1 = 6$ different pressures and $n_2 = 4$ different temperatures. Here

$$S_1 = \{P_1, \ldots, P_6\}$$
$$S_2 = \{T_1, \ldots, T_4\}$$

A *complete trial* is an experiment that is run for all possible combinations, which in this case total $6 \times 4 = 24$. ∎

In the preceding example the last experiment is determined by making two choices out of six possibilities for the operating pressure and four possibilities for the operating temperature. The method of enumeration will now be extended to the case of three or more experiments. The result is intuitively obvious, so we omit the formal proof.

The Generalized Multiplication Principle. Consider a sequence of k experiments where the ith experiment has n_i outcomes (or choices). Then the product experiment, which is the experiment corresponding to performing the k experiments in succession, has $n_1 \times n_2 \times \cdots \times n_k$ outcomes.

---

example 2.17

The following three situations illustrate various applications of the multiplication principle.

1. *Tossing a coin k times.* Each experiment has only two outcomes, H and T; thus $n_i = 2$, $i = 1, 2, \ldots, k$, and therefore the total number of outcomes in the sample space equals $2 \times \cdots \times 2 = 2^k$.

2. *The bootstrap.* A *bootstrap sample*, denoted by $X^* = \{x_1^*, \ldots, x_n^*\}$, is a random sample of size n taken with replacement from the data set $X = \{x_1, x_2, \ldots, x_n\}$. Use of the data generated by repeated resampling from the original data set is the starting point for a computer-intensive method for studying the sampling distribution of certain statistics that are otherwise very difficult to compute. (See B. Efron and R. J. Tibshirani, *An Introduction to the Bootstrap*, New York, Chapman & Hall, 1993.)

3. *License plates.* A license plate consisting of two letters followed by four digits is an ordered 6-tuple of the form (A, B, 1, 2, 3, 4), for example. The two letters can be chosen in any of 26^2 ways, and the four digits can be chosen in any of 10^4 ways. Consequently, the total number of possible license plates is equal to $26^2 \times 10^4 = 6{,}760{,}000$. ∎

Tree Diagrams

When n_1 and n_2 are small, one can visualize the multiplication principle in the format of a *tree diagram*. The tree diagram corresponding to tossing a coin twice is displayed in Fig. 2.3. We construct the tree diagram by listing in a column

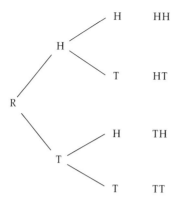

Figure 2.3 Tree diagram for coin-tossing experiment ($k = 2$)

all the elements of S_1 and connecting each of these elements, called *nodes*, by a line segment, called a *branch*, to the point labeled R, called the *root*. Each node in turn serves as the root of another tree consisting of n_2 branches placed to the right of the node. Each branch corresponds to one of the n_2 elements of S_2. Each outcome of the experiment corresponds to a *path* through the tree, which can be identified by listing the nodes through which it passes. Thus, HH denotes the event that two heads were thrown, as represented by the path starting at R and connecting the two nodes labeled H.

example **2.18** An urn contains three objects labeled a, b, c. The experiment consists of successively drawing two objects, without replacement, at random from the urn. Draw the tree diagram and use it to list the elements of the sample space.

Solution. The tree diagram corresponding to this experiment is shown in Fig. 2.4. There are six outcomes in the sample space:

$$S = \{ab, ac, ba, bc, ca, cb\}$$ ∎

Although tree diagrams are a useful visual display of the outcomes of a product of two experiments, they are of little practical value when either n_1 or n_2 is large. In these cases it is much easier to count the number of outcomes in an event, or in the sample space, by reducing it to an equivalent problem in *permutations* and *combinations*.

Permutations

The concept of a *permutation* plays an important role in mathematics (probability theory) and in computer science (theory of algorithms). A *permutation of n objects*

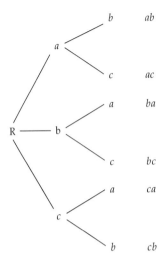

Figure 2.4 Tree diagram for selecting two objects without replacement from the set $\{a, b, c\}$

taken k at a time is an arrangement of k objects taken without replacement from n objects. There are six permutations of the three objects $\{a, b, c\}$ taken two at a time:

$$\{ab, ac, ba, bc, ca, cb\}$$

Our goal here is to use the multiplication principle to count the number of different permutations of n objects taken k at a time.

■ THEOREM 2.1

The number of permutations of k objects taken from n distinct objects is denoted $P_{n,k}$ and is determined from the formula

$$P_{n,k} = n(n-1)\cdots(n-k+1)$$

The special case $k = n$ is particularly important and is denoted $n!$ (pronounced "n factorial"):

$$n! = n(n-1)\cdots 3 \cdot 2 \cdot 1 = P_{n,n}$$

Proof. Denote the n distinct objects by the symbols $O_1, \ldots, O_n$. A permutation is an ordered k-tuple of the form

$$(O_{x_1}, \ldots, O_{x_k})$$

where we have n choices for O_{x_1}, $n-1$ choices for $O_{x_2}, \ldots$, and, finally, $n-k+1$ choices for O_{x_k}; consequently, by the general multiplication principle, the total number of such outcomes equals $n(n-1)\cdots(n-k+1)$. ■

---

example **2.19**   Consider a lottery in which there are three winning tickets, corresponding to the first, second, and third prizes, respectively. Suppose 100 lottery tickets are sold, with the first, second, and third prizes to be determined by a random drawing. Compute the number of different ways of drawing the three winning tickets.

Solution. Each drawing corresponds to a permutation of 3 objects taken from 100 objects. Thus, $k = 3$ and $n = 100$, so the number of such drawings equals

$$P_{100,3} = 100 \times 99 \times 98 = 970{,}200$$

∎

---

example **2.20**   Compute the following quantities: (1) $P_{8,3}$, (2) $P_{9,4}$, (3) 3!, (4) 5!.

Solutions

1. $P_{8,3} = 8 \times 7 \times 6 = 336$
2. $P_{9,4} = 9 \times 8 \times 7 \times 6 = 3024$
3. $3! = 3 \times 2 \times 1 = 6$
4. $5! = 5 \times 4 \times 3 \times 2 \times 1 = 120$ ∎

Notation. It is convenient to define $0! = 1$. With this convention we define $n!$ recursively as

$$n! = n \times (n - 1)!, \quad n = 1, 2, \ldots$$

Thus, $5! = 5 \times 4! = 5 \times 24 = 120$.

---

example **2.21**   The birthday problem, first popularized by W. Feller (*An Introduction to Probability Theory and Its Applications,* vol. 1, 3d ed., New York, John Wiley & Sons, 1968), is an interesting application of the methods of random sampling with results that are at first quite surprising. The problem is this: There are k people in a room; what is the probability that at least two people in the room have the same birthday?

Solution

1. *The sample space.* We begin by recording the birthday of the ith person as an integer $x_i (1 \le x_i \le 365)$; thus, if the ith person was born on January 1, then $x_i = 1$, and so on. Going around the room and recording the k birthdays produces a k-tuple of the form $(x_1, \ldots, x_k)$. The sample space Ω of this experiment is just the set of all such k-tuples, and we assume that each sample has an equal probability of being selected because we assume that each person has an equal chance of being born on any one of the 365 days of the year. This sample space can also be realized by means of the following sampling-with-replacement experiment:

   > Consider an urn containing 365 balls numbered $1, 2, \ldots, 364, 365$. After carefully stirring the balls, pick one, record its value, and then replace it in the urn. Repeat this procedure k times; the sample

space corresponding to this experiment consists of the ordered k-tuples of the form $(x_1, \ldots, x_k)$, where $x_i \in \{1, 2, \ldots, 365\}$.

Applying the generalized multiplication principle, we see that the total number of possible birthdays is equal to $365 \times \cdots \times 365 = 365^k$ (we ignore leap years).

2. We denote by A the event that at least two people in the room have the same birthday. It turns out to be much simpler to first compute the probability of the complement $A' = \{$everyone has a different birthday$\}$ and then compute $P(A)$ via the formula $P(A) = 1 - P(A')$. Since the sample space is equipped with the equally likely probability function, we compute the probability of A' by first counting the number of outcomes favorable to A' and then dividing by the total number of outcomes. The number of ways in which k persons all have different birthdays is equal to the number of k-tuples $(x_1, \ldots, x_k)$ in which all the x_i are different; but this is equivalent to drawing a sample of size k without replacement from $n = 365$ objects. Consequently, the number of sample points in A' equals $P_{365,k}$, and therefore,

$$P(k \text{ persons have different birthdays}) = \frac{P_{365,k}}{365^k}$$

For $k = 23$, a pocket calculator yields the result

$$\frac{P_{365,23}}{365^{23}} = 0.493$$

Thus, if one selects a group of 23 persons at random, the probability that at least 2 of them have a common birthday is equal to $1 - 0.493 = 0.507 > 1/2$, a result that Feller describes as astounding. What do you think?   ∎

Combinations

For many problems in probability and statistics it is important to count the number of distinct subsets of size k taken from a finite population consisting of n distinct objects. We denote this number by $C_{n,k}$, called *the number of combinations of k objects taken from n objects*. Each subset of size k is also called an *unordered sample of size k* taken from the set S. For instance, the subsets of size $k = 2$ taken from the $n = 3$ objects $\{a, b, c\}$ are the three subsets

$$\{a, b\}, \{a, c\}, \{b, c\}$$

On the other hand, the permutations of two objects taken from three objects are the $P_{3,2} = 6$ subsets

$$(a, b), (b, a), (a, c), (c, a), (b, c), (c, a)$$

The difference between a permutation and a combination is that the permutations (a, b) and (b, a) are different, because they are ordered differently. We do not, however, distinguish between the subsets $\{a, b\}$ and $\{b, a\}$ because they contain the same elements.

Another Interpretation of the Binomial Coefficient $C_{n,k}$

Suppose we have n objects $O_1, \ldots, O_n$ and two types of labels: k labels denoted H and $n - k$ labels denoted T. We want to determine how many different ways these objects can be labeled using the label H k times and the label T $n - k$ times. If we think about it for a moment, each such labeling corresponds to first choosing the k objects to be labeled H, and then labeling the remaining objects T. But this labeling process is equivalent to the number of different ways of choosing subsets of size k from n objects; consequently, there are $C_{n,k}$ ways of carrying out this labeling. In particular, there are $C_{n,k}$ ways of getting k heads (H) in n tosses of a coin; each such outcome corresponds to labeling the integers $1, \ldots, n$ with k H's and $n - k$ T's. For instance, the number of ways we can get five heads in 10 tosses of a coin is $C_{10,5}$. Theorem 2.2 gives a formula for computing $C_{n,k}$.

■■■ **THEOREM 2.2**

A population S consisting of n distinct objects contains $C_{n,k}$ subsets of size $k, 0 \leq k \leq n$, where

$$C_{n,k} = \frac{P_{n,k}}{k!} = \frac{n!}{k!(n-k)!}$$

Notation. The following alternative notation is also widely used:

$$C_{n,k} = \binom{n}{k}$$

$C_{n,k}$ is also called a *binomial coefficient*. Before proceeding to the proof, note that

$$\binom{n}{0} = \binom{n}{n} = 1$$

This is a consequence of the convention that $0! = 1$.

Proof. The assertion of the theorem is equivalent to the relation

$$C_{n,k} \times k! = P_{n,k}$$

which is a result of the multiplication principle, as will now be explained. Every permutation can be obtained by first selecting a subset of k objects, which has $C_{n,k}$ outcomes, and then noting that each k-element subset has $k!$ permutations. There are therefore $C_{n,k} \times k!$ ways of performing these two operations in succession. The proof is complete. ■

---

example **2.22**   Compute the following binomial coefficients: (1) $C_{3,2}$, (2) $C_{8,3}$, (3) $C_{10,5}$.

Solution

1. $C_{3,2} = \dfrac{3 \times 2}{2!} = 3$

2. $C_{8,3} = \dfrac{8 \times 7 \times 6}{3!} = 56$

3. $C_{10,5} = \dfrac{P_{10,5}}{5!} = \dfrac{30,240}{120} = 252$ ∎

---

example 2.23   The binomial coefficients are useful for computing the chances of winning in poker or winning a state lottery. Following are some of the possible applications.

1. *Poker hands.* A poker hand consists of 5 cards taken from 52, i.e., a subset of size 5 taken from 52; so $n = 52, k = 5$. The number possible of 5-card poker hands equals

$$C_{52,5} = 2{,}598{,}960$$

Let us compute the probability of getting *four of a kind* in poker, for example, four kings or four aces. There are 13 denominations, and for each occurrence of four aces, say, there are 48 choices for the fifth card. Consequently, there are $13 \times 48 = 624$ ways of getting a hand with four of a kind. The probability of getting a four of a kind is then equal to $624/\binom{52}{5}$.

2. *Bridge hands.* A bridge hand is a subset of size 13 taken from 52 cards; thus, $n = 52, k = 13$. The number of 13-card bridge hands equals

$$C_{52,13} = 635{,}013{,}559{,}600$$

3. *State lotteries.* Consider a lottery in which one has to pick six different digits out of $\{1, 2, \ldots, 40\}$ to win first prize. A winning ticket is a subset of size $k = 6$ taken from $n = 40$ objects. The sample is unordered because the order in which the winning numbers are drawn is not relevant. The probability of winning is $1/\binom{40}{6}$, where $\binom{40}{6} = 3{,}838{,}380$.   ∎

A Recurrence Relation for Binomial Coefficients. A *recurrence relation* is a rule that defines each element of a sequence in terms of the preceding elements. Recurrence relations are particularly useful for computing the elements of the sequence. The binomial coefficients satisfy the following recurrence relation:

$$\binom{n}{k+1} = \frac{n-k}{k+1} \times \binom{n}{k}, \qquad \binom{n}{0} = 1 \qquad (2.19)$$

Proof. The proof is left as an exercise.

---

example 2.24   Use the recurrence relation to compute $\binom{5}{k}$, $k = 1, 2, 3$.

Solution

$$n = 5, k = 1: \qquad \binom{5}{1} = \frac{5}{1} \times \binom{5}{0} = 5$$

$$n = 5, k = 2: \qquad \binom{5}{2} = \frac{5-1}{1+1} \times \binom{5}{1} = 10$$

$$n = 5, k = 3: \qquad \binom{5}{3} = \frac{5-2}{2+1} \times \binom{5}{2} = 10$$ ∎

The Binomial Theorem

In addition to their obvious importance in combinatorial analysis, the binomial coefficients have many interesting properties as well as many applications. They arise frequently, for example, in the analysis of algorithms in computer science (see, for example, Donald E. Knuth, *The Art of Computer Programming: Fundamental Algorithms,* vol. 1, Reading, Mass., Addison-Wesley, 1973). The most useful application for us is the binomial theorem.

■■■ **THEOREM 2.3**

The binomial theorem:

$$(a + b)^n = \binom{n}{0}a^n + \cdots + \binom{n}{k}a^{n-k}b^k + \cdots + \binom{n}{n}b^n \qquad (2.20)$$

Proof. For a detailed proof, see Sec. 2.5. ■

---

example **2.25**   Give the binomial expansion for $(a + b)^n$ for $n = 0, 1, 2, 3, 4$.

Solution

$$n = 0: \quad (a + b)^0 = 1$$
$$n = 1: \quad (a + b)^1 = a + b$$
$$n = 2: \quad (a + b)^2 = a^2 + 2ab + b^2$$
$$n = 3: \quad (a + b)^3 = a^3 + 3a^2b + 3ab^2 + b^3$$
$$n = 4: \quad (a + b)^4 = a^4 + 4a^3b + 6a^2b^2 + 4ab^3 + b^4$$

■

The binomial coefficients displayed in Example 2.25 reveal an interesting pattern that is called the *Pascal triangle* (see Fig. 2.5). The tip of the triangle, which contains the single entry 1, is called the zeroth row; the next row, which contains the two coefficients 1 and 1, is called the first row, and so forth. The number $\binom{n+1}{k}$ in the $(n + 1)$th row is the sum of the two adjacent numbers in the row above; for instance, in the fifth row, $4 = 1 + 3, 6 = 3 + 3$, etc. In the general case we have the following *binomial identity:*

Figure 2.5 The Pascal Triangle

$$\binom{n}{k} + \binom{n}{k-1} = \binom{n+1}{k} \tag{2.21}$$

The Pascal triangle is a visual display of the identity (2.21).

Derivation of the Binomial Identity (2.21). The binomial identity will now be proved using a clever counting argument that avoids tedious algebraic computations. The binomial coefficient appearing on the right-hand side of Eq. (2.21) counts the number of subsets of size k taken from $(n+1)$ objects; the left-hand side also counts the number of k-element subsets, but it does so in a slightly different way. To see this, pick one of the $n+1$ elements and note that a k-element subset either does not contain this element, or it does. There are $\binom{n}{k}$ of the first kind and $\binom{n}{k-1}$ of the second; adding these two terms together yields the identity (2.21).

2.3.1   Multinomial Coefficients

We begin our discussion of multinomial coefficients by considering the following problem: How many distinguishable arrangements of the letters of the word ILLINOIS are there? This is an eight-letter word consisting of three I's, two L's, and one each of the letters N, O, and S. This is a generalization of the labeling problem considered earlier. We have eight labels divided into five subgroups denoted by the letters I, L, N, O, S. The problem is equivalent, then, to counting the number of different ways eight objects can be labeled using three labels of type I, two labels of type L, and one label each of the types N, O, and S. It is a special case of the following general problem:

■ THEOREM 2.4

Suppose we have k_1 labels of type 1, k_2 labels of type 2, ..., and k_m labels of type m, for a grand total of $n = k_1 + k_2 + \cdots + k_m$ labels. Then the total number of distinguishable arrangements is given by

$$\binom{n}{k_1, \ldots, k_m} = \frac{n!}{k_1! \cdots k_m!} \tag{2.22}$$

We give an example before sketching the proof.   ■

---

example **2.26**   Compute the total number of distinguishable arrangements of the letters of the word ILLINOIS.

Solution. ILLINOIS has $n = 8$ letters consisting of $k_1 = 3$ I's, $k_2 = 2$ L's, and one each of the letters N, O, and S, so $k_3 = k_4 = k_5 = 1$. Inserting these values into Eq. (2.22) yields the result

$$\frac{8!}{3! \times 2! \times 1! \times 1! \times 1!} = 3360$$ ■

The term appearing in Eq. (2.22) is called a *multinomial coefficient*. Note that the binomial coefficient is a special case of the multinomial coefficient with $m = 2, k = k_1, k_2 = n - k$.

Proof of Theorem 2.4: It suffices to give the proof for the case $m = 3$ since the proof in the general case is similar. There are $\binom{n}{k_1}$ ways of placing the k_1 labels of type 1, which leaves $n - k_1$ unlabeled positions. There are $\binom{n-k_1}{k_2}$ ways of placing k_2 labels of type 2 over the remaining $n - k_1$ positions. Once these two sets of labels have been placed, the type 3 labels must go into the remaining positions. The total number of ways this can be done then is clearly the product $\binom{n}{k_1} \times \binom{n-k_1}{k_2}$. An easy calculation left to the reader shows that this product equals

$$\frac{n!}{k_1!k_2!k_3!}$$

PROBLEMS

2.14 Compute $P_{n,k}$ and $\binom{n}{k}$ for:
  (a) $n = 10, k = 1, 2, 3$
  (b) $n = 7, k = 4, 5, 6$

2.15 Verify the binomial identity

$$\binom{n}{k} + \binom{n}{k-1} = \binom{n+1}{k}$$

in the following cases:
  (a) $n = 8, k = 3$
  (b) $n = 5, k = 2$

2.16 Using the recurrence formula (2.19), compute

$$\binom{6}{k}, \quad \text{for } k = 1, \dots, 6$$

2.17 The format of a Massachusetts license plate is three digits followed by three letters, as in 736FSC. How many possible such license plates are there?

2.18 To win Megabucks of the Massachusetts state lottery, one must select six different digits out of $\{1, 2, \dots, 36\}$ that match the winning combination. What is the probability of winning?

2.19 How many distinguishable arrangements of the letters of the word *MISSISSIPPI* are there?

2.20 Let S denote the sample space consisting of the 24 distinguishable permutations of the symbols 1, 2, 3, 4 and assign to each permutation the probability $1/24$. Let A_i denote the event that the digit i appears at its natural place, e.g., $\{2134\} \in A_4$. Compute:
  (a) $P(A_1 \cap A_2)$
  (b) $P(A_1)$
  (c) $P(A_2)$

2.21 A group of 10 components contains 3 that are defective.
  (a) What is the probability that a component drawn at random is defective?

(b) Suppose a random sample of size 4 is drawn without replacement. What is the probability that *exactly* 1 of these components is defective?

(c) [Continuation of part (*b*)] What is the probability that *at least* 1 of these components is defective?

2.22 In how many ways can 12 people be divided into three groups of 4 persons (in each group) for an evening of bridge?

2.23 A bus starts with five people and makes 10 stops. Assume that passengers are equally likely to get off at any stop.

(a) Describe the sample space and calculate the number of different outcomes.

(b) Compute the probability that no two passengers get off at the same stop.

2.24 From a batch of 20 radios a sample of size 3 is randomly selected for inspection. If there are 6 defective radios in the batch, what is the probability that the sample:

(a) Contains only defectives?

(b) Contains only nondefectives?

(c) Contains 1 defective and 2 nondefectives?

2.25 Consider the following two games of chance:

(1) Person A throws 6 dice and wins if at least one ace appears.

(2) Person B throws 12 dice and wins if at least two aces appear.

Who has the greater probability of winning?

2.26 Consider a deck of four cards marked 1, 2, 3, 4.

(a) List all the elements in the sample space Ω coming from the experiment *two cards are drawn in succession and without replacement from the deck.*

If Ω is assigned the equally likely probability function, compute the probability that:

(b) The largest number drawn is a 4.

(c) The smallest number drawn is a 2.

(d) If X_i denotes the number drawn on the *i*th draw, compute

$$P(X_1 + X_2 = 5)$$

2.27 An urn contains two nickels and three dimes.

(a) List all elements of the sample space corresponding to the experiment *two coins are selected at random and without replacement.*

(b) Compute the probabilies that the value of the coins selected equals $0.10, $0.15, $0.20.

2.28 A poker hand is a set of 5 cards drawn at random from a deck of 52 cards consisting of four suits (hearts, diamonds, spades, and clubs), where each suit consists of 13 cards with face value (also called denomination) denoted {ace, 2, 3, ..., 10, jack, queen, king}. Find the probability of obtaining each of the following poker hands.

(a) Royal flush (10, jack, queen, king, ace of a single suit).

(b) Full house (one pair and one triple each of the same denomination).

2.29 A lot consists of 15 articles of which 8 are free of defects, 4 have minor defects, and 3 have major defects. Two articles are selected at random without replacement. Find the probability that:

(a) Both have major defects. **(b)** Both are good.

(c) Neither is good. **(d)** Exactly 1 is good.

(e) At least 1 is good. **(f)** At most 1 is good.

(g) Both have minor defects.

2.4 CONDITIONAL PROBABILITY, BAYES' THEOREM, AND INDEPENDENCE

2.4.1   Conditional Probability

In one of the earliest studies (1936) establishing a "link" between smoking and lung cancer, two British physicians reported that of 135 men afflicted with lung cancer, 122, or 90 percent, were heavy smokers. In nontechnical language, the inference is clear: If you have lung cancer, it is much more likely the case than not that you are a heavy smoker. Is the converse true? If you are heavy smoker, does it necessarily follow that your chances of developing lung cancer are much higher than for a nonsmoker? To answer questions like these we need to define the concept of *conditional probability*, which is of fundamental importance in probability theory.

We assume that the data reported by the physicians came from a population Ω of N men, and we are interested in the two subpopulations:

1. $A = \{$men who are heavy smokers$\}$

2. $B = \{$men afflicted with lung cancer$\}$

Denote by $\#(B)$ the number of sample points in the set B. Thus $\#(B) = 135$ and $\#(B \cap A) = 122$. The probability that a person selected at random from the subpopulation B (men afflicted with lung cancer) is also in A (men who are heavy smokers) is clearly

$$\frac{\#(B \cap A)}{\#(B)} = \frac{122}{135} = 0.90 \tag{2.23}$$

In words we say, "The probability of the event A (the man is a heavy smoker) given that the event B (the man has lung cancer) has occurred is equal to 0.9." In symbols this is written

$$P(A \mid B) = \frac{\#(B \cap A)}{\#(B)} = 0.9$$

Now the probability that a person selected at random from the population Ω is both a heavy smoker and afflicted with lung cancer is clearly

$$P(A \cap B) = \#(B \cap A)/N$$

Similarly, the probability that a person selected at random from the population Ω is afflicted with lung cancer is clearly

$$P(B) = \#(B)/N$$

Therefore,

$$P(A\,|\,B) = \frac{\#(B \cap A)}{\#(B)} = \frac{\#(B \cap A)/N}{\#(B)/N} = \frac{P(A \cap B)}{P(B)}$$

The formal definition follows.

■ **DEFINITION 2.4**

Let A and B be two arbitrary events with $P(B) > 0$. The *conditional probability* of the event A given that B has occurred is defined to be

$$P(A\,|\,B) = \frac{P(A \cap B)}{P(B)} \tag{2.24}$$

We frequently use Eq. (2.24) in the equivalent form:

$$P(A \cap B) = P(A\,|\,B) \times P(B) \tag{2.25}$$

Equation (2.25) is called the *product rule*. It is useful because it is often the case that one is given $P(A\,|\,B)$ and $P(B)$ and *not* $P(A \cap B)$, as in Example 2.27, where we analyze the performance of a screening test to detect a disease. ■

---

example 2.27   Consider a population partitioned into two groups distinguished by whether or not individuals are infected with a rare disease. To be specific, denote by D the event that a person selected at random has the disease, and suppose that the *prevalence rate* of this disease is 1 person in 5000; that is, we assume that $P(D) = 0.0002$. We are interested in studying the performance of a screening test for the disease. A screening test is not foolproof; it sometimes yields an incorrect result. These errors are of two types: *false positives* and *false negatives*. A false positive reading simply means that the person tested positive for the disease even though he or she is not infected. Similarly, a false negative reading means that even though the person is infected the test failed to detect it. Let T^+ denote the event that the screening test is positive for a person selected at random. Suppose the manufacturer claims that its screening test is quite reliable in the sense that the rate of false positive readings is $2/100 = 0.02$ and the rate of false negative readings is $1/100 = 0.01$. Compute $P(T^+)$.

Solution. It is interesting to note that the rate of false positive readings and the rate of false negative readings are conditional probabilities, although nowhere in the specifications is the term *conditional probability* actually used. To see this, denote the event that a person taking the test yields a positive (negative) result by T^+ (T^-), and denote by D (D') the event that the person has the disease (does not have the disease). A false positive means that the event T^+ occurred given that D' occurred. We express these specifications as conditional probabilities as follows:

$$P(T^+\,|\,D') = 0.02 \quad \text{and} \quad P(T^-\,|\,D) = 0.01$$

To evaluate the performance of the test, it is necessary to calculate the probability that the test is positive, which is given by

$$P(T^+) = P(T^+ \cap D) + P(T^+ \cap D') \tag{2.26}$$

Proof. Equation (2.26) is a consequence of the fact that a positive response can occur for only one of two reasons: (1) the person being tested actually has the disease, or (2) the person does not (false positive). This yields the partition

$$T^+ = (T^+ \cap D) \cup (T^+ \cap D')$$

Equation (2.26) is a straightforward application of Proposition 2.3 with $T^+ = B$ and $D = A$. Next, using Eq. (2.26) and the product rule (2.25), we obtain the following value for $P(T^+)$:

$$P(T^+) = P(T^+ \cap D) + P(T^+ \cap D') \tag{2.27}$$
$$= P(T^+ \mid D)P(D) + P(T^+ \mid D')P(D') \tag{2.28}$$

Inserting into Eq. (2.28) the values $P(T^+ \mid D') = 0.02$, $P(T^- \mid D) = 0.01$, $P(D) = 0.0002$, $P(D') = 0.9998$ yields the result

$$P(T^+) = 0.99 \times 0.0002 + 0.02 \times 0.9998 = 0.02 \tag{2.29}$$

∎

The reasoning used in this example is a special case of a more general theorem called *the law of total probability.*

▰ **THEOREM 2.5** *The law of total probability.*

Let the events $A_1, A_2, \ldots, A_n$ be a partition of the sample space Ω, and let B denote an arbitrary event. Then

$$P(B) = \sum_{1 \le i \le n} P(B \mid A_i)P(A_i) \qquad ∎$$

Before proving this result, we give a typical application.

---

example **2.28** In a factory that manufactures silicon wafers, machines 1 and 2 produce, respectively, 40 percent and 60 percent of the total output. Suppose that 2 out of every 100 wafers produced by machine 1 and 3 out of every 200 wafers produced by machine 2 are defective. What is the probability that a wafer selected at random from the total output is defective?

Solution. Let

$$D = \{\text{a wafer selected at random is defective}\}$$
$$M_1 = \{\text{a wafer selected at random is produced by machine 1}\}$$
$$M_2 = \{\text{a wafer selected at random is produced by machine 2}\}$$

Figure 2.6 is the tree diagram corresponding to this experiment. The numbers appearing along the branches are the conditional probabilities.

Clearly, the defective wafer is produced by either machine 1 or machine 2; these two possibilities are represented in the tree diagram by the two paths labeled M_1D and M_2D, respectively. This suggests that we compute $P(D)$ by "conditioning" on whether the wafer was produced by machine 1 or machine 2. More precisely, we use Theorem 2.5 to show that

$$P(D) = P(D \mid M_1)P(M_1) + P(D \mid M_2)P(M_2)$$

The probabilities and conditional probabilities are derived from the information contained in the problem description. Specifically, it is easy to verify that

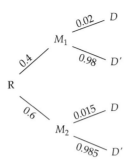

Figure 2.6 Tree Diagram for Example 2.28

$$P(M_1) = 0.4, \qquad P(M_2) = 0.6$$
$$P(D \mid M_1) = 0.02, \qquad P(D \mid M_2) = 0.015$$
$$P(D) = 0.02 \times 0.4 + 0.015 \times 0.6 = 0.017 \qquad ■$$

Proof of Theorem 2.5. The events $A_1, A_2, \ldots, A_n$ form a partition of Ω; consequently,

$$B = (B \cap A_1) \cup (B \cap A_2) \cup \cdots \cup (B \cap A_n)$$

Therefore,

$$P(B) = \sum_{1 \leq i \leq n} P(B \cap A_i) = \sum_{1 \leq i \leq n} P(B \mid A_i) P(A_i)$$

where we have used the product rule (2.25).

■ THEOREM 2.6

The function $P(A \mid B)$, for fixed B, $P(B) > 0$, satisfies axioms (2.10) through (2.14) for a probability function, i.e.,

$$0 \leq P(A \mid B) \leq 1 \tag{2.30}$$
$$P(\Omega \mid B) = 1 \tag{2.31}$$
$$A_1 \cap A_2 = \varnothing \Rightarrow P(A_1 \cup A_2 \mid B) = P(A_1 \mid B) + P(A_2 \mid B) \tag{2.32}$$

If $A_1, A_2, \ldots, A_n, \ldots$ is an infinite sequence of mutually disjoint events, then

$$P(A_1 \cup A_2 \cup \cdots \cup A_n \cup \cdots \mid B) = \sum_{n=1}^{\infty} P(A_n \mid B) \tag{2.33}$$

The proof is left to the reader (see Prob. 2.13). ■

2.4.2 Bayes' Theorem

In Example 2.27 we computed the probability that the screening test for a rare disease is positive for a person selected at random. But this is not what the patient wants to know. What the patient wants to know is, Does the positive test result mean that he or she actually has the disease? Example 2.29 gives a partial and surprising answer.

example **2.29**   This is a continuation of Example 2.27; we use the same notation. Suppose the screening test is positive. Compute the probability that the patient actually has the disease.

Solution. We have to compute the conditional probability that the patient has the disease given that the test is positive; that is, we have to compute the following conditional probability:

$$P(D \mid T^+) = \frac{P(D \cap T^+)}{P(T^+)}$$

The prevalence rate of the disease is 1 in 5000, so $P(D) = 0.0002$. We showed earlier [see Eq. (2.29)] that $P(T^+) = 0.02$. Similarly,

$$P(T^+ \cap D) = P(T^+ \mid D)P(D) = 0.99 \times 0.0002 = 0.000198 \approx 0.0002$$

Therefore,

$$P(D \mid T^+) = \frac{P(D \cap T^+)}{P(T^+)} = \frac{0.0002}{0.02} = 0.01$$

So only 1 percent of those who test positive for the disease actually have the disease! What is the explanation for this result? The answer lies in the fact that the bulk of the positive readings come from the false positives. One should not conclude, however, that the test is without value. Prior to administration of the test, the chances of a person selected at random having the disease were 0.0002; once the test is administered and results in a positive reading, the chances have increased to 0.01. That is, the probability that the person actually has the disease has increased by a factor of 50. Clearly, further testing of the patient is warranted. We may summarize these results in the following way: Before administration of the test, the probability that a person selected at random has the disease is given by the *prior* probability

$$P(D) = 0.0002$$

Similarly, the prior probability that a randomly selected person does not have the disease is

$$P(D') = 0.9998$$

If the test yielded a positive result, then the *posterior* probabilities are

$$P(D \mid T^+) = 0.01, \qquad P(D' \mid T^+) = 0.99 \qquad \blacksquare$$

The reasoning used in Example 2.29 is a special case of a more general result called *Bayes' theorem.*

▧ THEOREM 2.7

Let the events $A_1, A_2, \ldots, A_n$ be a partition of the sample space Ω, and let B denote an arbitrary event satisfying the condition $P(B) > 0$. Then

$$P(A_k \mid B) = \frac{P(B \mid A_k)P(A_k)}{\sum_{1 \leq i \leq n} P(B \mid A_i)P(A_i)}$$

Proof. $P(A_k \mid B) = P(A_k \cap B)/P(B)$ (this is just Definition 2.4). The product rule (2.25) applied to the numerator yields

$$P(A_k \cap B) = P(B \cap A_k) = P(B \mid A_k)P(A_k)$$

Next apply Theorem 2.5 to the term $P(B)$ in the denominator. Putting these results together yields

$$P(A_k \mid B) = \frac{P(A_k \cap B)}{P(B)} = \frac{P(B \mid A_k)P(A_k)}{\sum_{1 \le i \le n} P(B \mid A_i)P(A_i)}$$ ∎

2.4.3   Independence

Consider the following experiment. A card is drawn at random from a deck of 52, and its face value and suit are noted. The event that an ace was drawn is denoted by A, and the event that a club was drawn is denoted by B. There are four aces, so $P(A) = 4/52 = 1/13$, and there are 13 clubs, so $P(B) = 13/52 = 1/4$. $A \cap B$ denotes the event that the ace of clubs was drawn, and since there is only one such card in the deck,

$$P(A \cap B) = 1/52 = (1/13) \times (1/4) = P(A) \times P(B)$$

Thus,

$$P(A \mid B) = \frac{P(A \cap B)}{P(B)} = \frac{1/52}{1/4} = \frac{1}{13} = P(A)$$

In other words, knowing that the card selected was a club did not change the probability that the card selected was an ace. We say that the event A is *independent* of the event B.

■ **DEFINITION 2.5**

The events A and B are said to be *(probabilistically) independent* if

$$P(A \cap B) = P(A) \times P(B) \tag{2.34}$$ ∎

Clearly, if Eq. (2.34) holds, then

$$P(A \mid B) = P(A) \quad \text{and} \quad P(B \mid A) = P(B)$$

Consequently, if A is independent of B, then B is independent of A.

The definition of independence is easily extended to three or more events. To simplify the notation, we consider only the case of three events.

■ **DEFINITION 2.6**

The events A, B, C are probabilistically independent if they are:

1. Pairwise independent, i.e., the events (A, B), (A, C), (B, C) are independent, and
2. The probability of the intersection of the three events equals the product of their probabilities:

$$P(A \cap B \cap C) = P(A) \times P(B) \times P(C)$$ ∎

Independence is often assumed rather than proved, as in the following example.

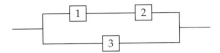

Figure 2.7 System of components for
Example 2.30

---

example 2.30   Consider a system consisting of three components linked as indicated in Fig. 2.7.

Let W_i be the event that component i is working, and denote its probability by $P(W_i) = p$. Let W denote the event that the system is working. We define the reliability of the system to be $P(W)$. Compute $P(W)$.

Solution. It is clear that this system works only if *both* W_1 and W_2 or W_3 are working. Using set theory notation, we can write

$$W = (W_1 \cap W_2) \cup W_3$$

We now assume that the failures of these components are mutually independent. Thus

$$P(W) = P((W_1 \cap W_2) \cup W_3)$$
$$= P(W_1 \cap W_2) + P(W_3) - P(W_1 \cap W_2 \cap W_3) \quad \text{by Proposition 2.4}$$
$$= p^2 + p - p^3 \quad \text{by independence}$$

Thus, if $p = 0.9$ then $P(W) = 0.9^2 + 0.9 - 0.9^3 = 0.981$.   ∎

PROBLEMS

2.30 A and B are events whose probabilities are given by $P(A) = 0.7, P(B) = 0.4$, $P(A \cap B) = 0.2$. Compute the following conditional probabilities.

(a) $P(A \mid B)$ **(b)** $P(A' \mid B')$
(c) $P(B' \mid A')$ **(d)** $P(A' \mid B)$
(e) $P(A \mid B')$ **(f)** $P(B \mid A')$
(g) $P(A \cup B \mid B')$

2.31 The student body of a college is composed of 70 percent men and 30 percent women. It is known that 40 percent of the men and 60 percent of the women are engineering majors. What is the probability that an engineering student selected at random is a man?

2.32 In a bolt factory, machines A, B, and C manufacture, respectively, 20 percent, 30 percent, and 50 percent of the total. Of their output, 3 percent, 2 percent, and 1 percent of the bolts, respectively, are defective.

(a) A bolt is selected at random. Find the probability that it is defective.
(b) A bolt is selected at random and is found to be defective. What is the probability it was produced by machine A?

2.33 An automobile insurance company classifies drivers into three classes: class A (good risks), class B (medium risks), and class C (poor risks).

The percentage of drivers in each class is: class A, 20 percent; class B, 65 percent; and class C, 15 percent. The probabilities that a driver in one of these classes will have an accident within one year are given by 0.01, 0.02, and 0.03, respectively. After purchasing an insurance policy, a driver has an accident within the first year. What is the probability that the driver is a class A risk? A class B risk? A class C risk?

2.34 In Examples 2.27 and 2.29, involving testing for a rare disease, suppose the results of the test were negative; i.e., the event T^- occurred. Compute the posterior probabilities in this case.

2.35 A lot consists of three defective and seven good transistors. Two transistors are selected at random. Given that the sample contains a defective transistor, what is the probability that both are defective?

2.36 (Continuation of previous problem) Consider the following procedure to find and remove all defective transistors: Transistors are randomly selected and tested until all three defective transistors are found. What is the probability that the third defective transistor will be found on:
(a) The third test?
(b) The fourth test?
(c) The tenth test?

2.37 A coin is tossed four times.
(a) What is the probability that the number of heads is greater than or equal to 3 given that one of the tosses is a head?
(b) What is the probability that the number of heads is greater than or equal to 3 given that the first toss is a head?

2.38 A coin is tossed three times.
(a) Let A denote the event that one head appeared in the first two tosses, and let B denote the event that one head occurred in the last two tosses. Are these events independent?
(b) Let A denote the event that one head appeared in the first toss, and let B denote the event that one head occurred in the last two tosses. Are these events independent?

2.39 In the manufacture of a certain article, two types of defects are noted that occur with probabilities 0.02 and 0.05, respectively. If these two defects occur independently, what is the probability that:
(a) An article is free of both kinds of defects?
(b) An article has at least one type of defect?

2.40 Let $W_i, i = 1, 2, 3,$ denote the event that component i is working. W_i' denotes the event that the ith component is not working. Assume that the events W_i are mutually independent and $P(W_i) = 0.9$, $i = 1, 2, 3$. Compute the probability that, of the components W_1, W_2, W_3:
(a) Only W_1 is working.   **(b)** All three components are working.
(c) None are working.   **(d)** At least one is working.
(e) At least two are working.   **(f)** Exactly two are working.

2.41 Show that $P(A \cup B \mid B) = 1$.

2.42 (a) Show that Ω and A are independent for *any A*.
    (b) Show that $\varnothing$ and A are independent for *any A*.

2.43 Show that if A is independent of itself, then $P(A) = 0$ or $P(A) = 1$.

2.44 Show that if $P(A) = 0$ or $P(A) = 1$, then A is independent of every event B.

2.45 Show that if A and B are independent, then so are A and B'.

2.46 Suppose $A, B,$ and C are mutually independent events. Assume that $P(B \cap C) > 0$ (this hypothesis is just to ensure that the following conditional probabilities are defined).
    (a) Show that $P(A \mid B \cap C) = P(A)$.
    (b) Show that $P(A \mid B \cup C) = P(A)$.

2.5   THE BINOMIAL THEOREM (OPTIONAL)

Instead of proving Theorem 2.3 directly, we will prove the following special case first.

▬ THEOREM 2.8

$$(1 + t)^n = \binom{n}{0} + \binom{n}{1}t + \cdots + \binom{n}{k}t^k + \cdots + \binom{n}{n}t^n \tag{2.35}$$

Proof. It is obvious that $(1 + t)^n$ is a polynomial of the nth degree, so it can be written as

$$f(t) = (1 + t)^n = a_0 + a_1 t + a_2 t^2 + \cdots + a_k t^k + \cdots + a_n t^n$$

We will prove the binomial theorem by showing that

$$a_k = \frac{P_{n,k}}{k!} = \binom{n}{k}$$

In the ensuing sequence of calculations $f'(t)$ denotes the derivative of f and $f^{(k)}$ denotes the kth derivative.

$$f'(t) = n(1 + t)^{n-1} = a_1 + 2a_2 t + \cdots + na_n t^{n-1}$$
$$f''(t) = n(n - 1)(1 + t)^{n-2} = 2a_2 + 3 \cdot 2a_3 t + \cdots + n(n - 1)a_n t^{n-2}$$

and so on.

Since these equations are valid for all t, we can set $t = 0$ in each one to conclude

$$f(0) = 1 = a_0$$
$$f'(0) = n = a_1$$
$$f''(0) = n(n - 1) = 2a_2$$

In general,

$$f^{(k)}(0) = n(n - 1) \cdots (n - k + 1) = P_{n,k} = k!a_k$$

Consequently,

$$a_k = \frac{P_{n,k}}{k!}$$

Finally, we show that Theorem 2.8 implies Theorem 2.3. We begin with the algebraic identity

$$(a + b)^n = \left[a\left(1 + \frac{b}{a}\right)\right]^n = a^n\left(1 + \frac{b}{a}\right)^n$$

and then apply Theorem 2.8 to the term $(1 + b/a)^n$ to conclude

$$\left[a\left(1 + \frac{b}{a}\right)\right]^n = a^n\sum_{k=0}^{n}\binom{n}{k}\left(\frac{b}{a}\right)^k$$

Therefore,

$$a^n\left(1 + \frac{b}{a}\right)^n = a^n\sum_{k=0}^{n}\binom{n}{k}a^{-k}b^k$$

$$= \sum_{k=0}^{n}a^n\binom{n}{k}a^{-k}b^k$$

$$= \sum_{k=0}^{n}\binom{n}{k}a^{n-k}b^k$$

The proof of the binomial theorem is complete.   ∎

2.6 CHAPTER SUMMARY

Statistical inference is a scientific method for making inferences about a population based on data collected from a sample. To guard against the scientist's hidden biases, the theoretical ideal is to collect the data via random sampling, which ensures that *every subset of the population of the same size has an equal chance (probability) of being selected.* Probability theory is the mathematical framework within which we analyze the results of random sampling as well as other experiments whose outcomes are due to chance. Although it is not possible to predict in advance the result of one throw of a pair of dice, it is possible to predict the distribution of the outcomes. It is the comparison of the observed frequency distributions with those predicted by the mathematical theory that enables the scientist to make inferences about the population.

To Probe Further. There is still no better introduction to probability theory than W. Feller's *An Introduction to Probability Theory and Its Applications*, vol. 1, 3d ed., New York, John Wiley & Sons, 1968. Permutations, combinations, and recurrence relations also play an important role in computer science, particularly in the analysis of algorithms; see T. H. Cormen, C. E. Leiserson, and R. L. Rivest, *Introduction to Algorithms*, New York, McGraw-Hill, 1990.

CHAPTER 3

Discrete Random Variables and Their Distribution Functions

It is the calculus of probabilities which alone can regulate justly the premiums to be paid for assurances; the reserve fund for the disbursements of pensions, annuities, discounts, etc. It is under its influence that lotteries and other shameful snares cunningly laid for avarice and ignorance have definitely disappeared.

Dominique François Jean Arago (1786–1853), French physicist

3.1 ORIENTATION

In this chapter we study the fundamental concept of the *distribution of a variable X defined on a sample space.* A variable defined on a sample space is called a *random variable.* It is analogous to the concept of a variable defined on a population S studied in Chap. 1. There are other analogies as well. The analogue of the empirical distribution function is called the *distribution function of X,* and the analogues of the sample mean and sample variance are called the *expected value* and *variance of X,* respectively. Random variables are classified as *discrete* or *continuous;* the major difference between these two types is in the level and sophistication of the mathematical tools used to study them. In this chapter we study only discrete random variables; the concept of a continuous random variable, which requires use of the calculus in an essential way, will be treated in Chap. 4.

Organization of Chapter

3.2 DISCRETE RANDOM VARIABLES

The number of heads that appear in n tosses of a coin, the payoff from a bet placed on a throw of two dice, and the number of defective items in a sample of size n taken from a lot of N manufactured items are all examples of functions whose values are determined by chance; such functions are called *random variables.* To illustrate, consider Table 3.1, which lists (1) the eight sample points corresponding to the experiment of tossing a coin three times and (2) the values of the random variable X, which counts the number of heads for each outcome.

TABLE 3.1 THE RANDOM VARIABLE X,
WHICH COUNTS THE NUMBER OF HEADS
IN THREE TOSSES OF A COIN

$X(TTT) = 0$	$X(THT) = 1$	$X(HHT) = 2$	$X(THH) = 2$
$X(TTH) = 1$	$X(HTT) = 1$	$X(HTH) = 2$	$X(HHH) = 3$

■ **DEFINITION 3.1**

A *random variable* X is a real-valued function defined on a sample space Ω. The value of the function at each sample point is denoted by $X(\omega)$. ■

The set of values $\{X(\omega) : \omega \in \Omega\}$ is called the *range* and is denoted R_X. The random variable X of Table 3.1 has range $R_X = \{0, 1, 2, 3\}$. It is an example of a *discrete random variable,* so called because its range is a discrete set. In the general case we say that the random variable X is discrete if its range is the discrete set $R_X = \{x_1, x_2, \ldots, x_n, \ldots\}$. Because it is convenient to do so, we assume that the numbers in the range appear in increasing order; that is, $R_X = \{x_1 < x_2 < \cdots < x_n < \cdots\}$.

The Event $\{X = x\}$

A random variable is often used to describe events. For instance, with respect to the experiment of throwing a coin three times, the expression $X = 2$ is equivalent to saying "two heads are thrown." More generally, $X = x$ means that "x heads are thrown." In detail, we write

$$\{X = 0\} = (TTT)$$
$$\{X = 1\} = (HTT), (THT), (TTH)$$
$$\{X = 2\} = (HHT), (HTH), (THH)$$
$$\{X = 3\} = (HHH)$$

In the general case we use the notation

$$\{X = x\} = \{\omega : X(\omega) = x\} \tag{3.1}$$

The Probability Function of a Discrete Random Variable X

The event $\{X = x\}$ has a probability denoted by $P(X = x)$. If x is not in the range $R_X = \{x_1, x_2, \ldots\}$, then the set $\{X = x\}$ is the empty set $\emptyset$; in this case $P(X = x) = 0$; consequently, it is only necessary to list the values $P(X = x), x \in R_X$. The function

$$f_X(x) = P(X = x), \quad x \in R_X$$
$$f_X(x) = 0, \quad x \text{ not in } R_X \tag{3.2}$$

is called the *probability function* (pf) of the random variable X. Some authors use the term *probability mass function* (pmf).

Distribution Function of a Random Variable X

For computational purposes it is often more convenient to work with the *distribution function* (df), instead of the pf $f_X(x)$.

■ **DEFINITION 3.2**

The distribution function (df) $F_X(x)$ of the random variable X is the real-valued function defined by the equation

$$F_X(x) = P(X \le x) \qquad \blacksquare$$

The distribution function of a discrete random value is given by

$$F_X(x) = \sum_{x_j \le x} f_X(x_j)$$

The distribution function $F_X(x)$ determines the values $f_X(x_j)$ via the equation[1]

$$f_X(x_j) = F_X(x_j) - F_X(x_{j-1}) \tag{3.3}$$

Notation. To simplify the notation we sometimes drop the subscript X and simply write $f(x)$ or $F(x)$.

---

example 3.1

Compute the probability function for the number of heads appearing in three tosses of a fair coin.

Solution. The assumption that the coin is fair means that we assign the equally likely probability measure to the sample space. Since the sample space has eight sample points, each sample point has probability $1/8$.

TABLE 3.2 THE PROBABILITY AND DISTRIBUTION FUNCTIONS FOR THE NUMBER OF HEADS IN THREE TOSSES OF A FAIR COIN, DISPLAYED IN THE FORMAT OF A HORIZONTAL BAR CHART

x	$f_X(x)$	$F_X(x)$	$\{\omega : X(\omega) = x\}$
0	1/8	1/8	TTT
1	3/8	4/8	HTT, THT, TTH
2	3/8	7/8	HHT, HTH, THH
3	1/8	1	HHH

∎

In the preceding example, note that the probability function is nonnegative and sums to 1; that is,

$$f_X(x) \geq 0 \quad \text{and} \quad \sum_{0 \leq x \leq 3} f_X(x) = 1$$

This is just a special case of the following proposition.

■ **PROPOSITION 3.1**

Let X be a discrete random variable with probability function $f_X(x)$. Then

$$f_X(x) \geq 0 \quad \text{and} \quad \sum_x f_X(x) = 1 \tag{3.4}$$

∎

Derivation of Eq. (3.4). This result follows from the fact that the events $A_x = \{X = x\}$ form a partition of the sample space Ω; consequently, the axiom of countable additivity [axiom (2.14)] and the fact that $f_X(x) = P(A_x)$ together imply

---

[1]We assume that $x_j < x_{j+1}$ ($j = 1, \ldots$).

$$1 = P(\Omega) = P(\cup_x A_x)$$

$$= \sum_x P(A_x)$$

$$= \sum_x f_X(x)$$

Any function satisfying the conditions in Eq. (3.4) is also called a probability function.

---

example 3.2   Compute the probability function of the random variable X that records the sum of the faces of two dice.

Solution. The sample space is the set

$$\Omega = \{(i,j) : i = 1,\dots,6; j = 1,\dots,6\}$$

(See Table 2.1.) The random variable X is the function $X(i,j) = i + j$. Thus, $X(1,1) = 2$, $X(1,2) = 3$, ..., $X(6,6) = 12$. The range of X is the set

$$R_X = \{2, 3, \dots, 12\}$$

and its probability function is displayed in Table 3.3.   ■

Probability Histogram

Another way to visualize a pf is by constructing a *probability histogram*, as in Fig. 3.1. Above each integer $x = 2,\dots,12$ we construct a rectangle centered at x and with height proportional to $f_X(x)$.

TABLE 3.3 THE PROBABILITY FUNCTION
FOR THE RANDOM VARIABLE $X(i,j) = i+j$ DISPLAYED IN THE
FORMAT OF A HORIZONTAL BAR CHART

x	$f_X(x)$	$F_X(x)$	$\{\omega : X(\omega) = x\}$
2	1/36	1/36	$(1,1)$
3	2/36	3/36	$(1,2),(2,1)$
4	3/36	6/36	$(1,3),(2,2),(3,1)$
5	4/36	10/36	$(1,4),(2,3),(3,2),(4,1)$
6	5/36	15/36	$(1,5),(2,4),(3,3),(4,2),(5,1)$
7	6/36	21/36	$(1,6),(2,5),(3,4),(4,3),(5,2),(6,1)$
8	5/36	26/36	$(2,6),(3,5),(4,4),(5,3),(6,2)$
9	4/36	30/36	$(3,6),(4,5),(5,4),(6,3)$
10	3/36	33/36	$(4,6),(5,5),(6,4)$
11	2/36	35/36	$(5,6),(6,5)$
12	1/36	1	$(6,6)$

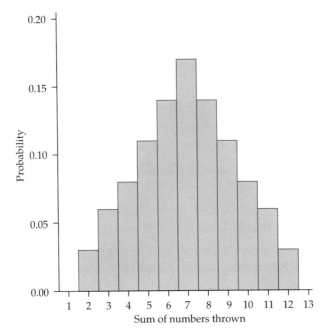

Figure 3.1 Histogram of the probability function of throwing two dice

Interpreting the Probability Histogram as a Relative Frequency Histogram. When we throw a pair of fair dice a large number of times, we expect the relative frequency of the occurrences of the number $x(x = 2, \ldots, 12)$ to be nearly equal to its probability $f_X(x)$; this is the frequency interpretation of probability discussed in the previous chapter. Of course, in any actual experiment of tossing a pair of dice n times, the relative frequency histogram will differ from the theoretical frequency displayed in Fig. 3.1.

example 3.3   Compute and sketch the graph of the distribution function of the random variable X that records the sum of the dots that appear when two dice are thrown.

Solution. The values of the df $F_X(j)$ ($j = 2, \ldots, 12$) are displayed in the third column of Table 3.3. We obtain the values of $F_X(x)$ for all x by noting that

$$F_X(x) = 0, \quad x < 2$$
$$= F_X(j), \quad j \le x < j + 1 \, (j = 2, \ldots, 12)$$
$$= 1, \quad x \ge 12$$

The graph is displayed in Fig. 3.2.   ∎

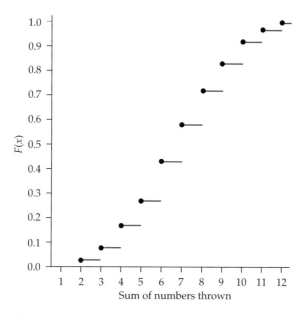

Figure 3.2 Distribution function of throwing two dice

---

example **3.4**

Compute the probability function of the payoff for a one-roll bet on a 12 with a payoff of 30 to 1.

Solution. The probability function of the payoff random variable W, which represents the gambler's *net* winnings, is easily determined. First note that

$$W(6,6) = 30 \quad \text{and} \quad W(i,j) = -1 \quad \text{for all other sample points}$$

The range of W is the set consisting of the two points $\{-1, 30\}$. Its pf is displayed in Table 3.4.

TABLE 3.4   THE PROBABILITY
FUNCTION OF THE PAYOFF FOR A
ONE-ROLL BET ON THE NUMBER 12

x	-1	30
$f_W(x)$	$35/36$	$1/36$

∎

---

example **3.5**

Compute the probability function of the number of hearts in a five-card poker hand.

Solution. The probability that a five-card poker hand contains x hearts is given by

$$f(x) = \frac{\binom{13}{x}\binom{39}{5-x}}{\binom{52}{5}} \qquad x = 0,1,2,3,4,5 \tag{3.5}$$

Derivation of Eq. (3.5). The term in the denominator counts the total number of five-card poker hands, each of which is assigned the same probability. The term in the numerator counts the number of ways of getting x hearts and $5 - x$ nonhearts in a five-card poker hand. Their ratio is the probability of getting x hearts in a five-card poker hand. In detail: The number of ways of getting x hearts out of 13 is $\binom{13}{x}$, and the number of ways of getting $5 - x$ nonhearts from the 39 nonhearts in the deck is $\binom{39}{5-x}$. Consequently, the number of ways of getting x hearts and $5 - x$ nonhearts in the poker hand is their product, $\binom{13}{x}\binom{39}{5-x}$; this is the term in the numerator. There are $\binom{52}{5}$ sample points, and each one is assigned the same probability, $1/\binom{52}{5}$; this is the term in the denominator. ∎

---

example 3.6   For each of the following nonnegative functions $f(x)$, determine the constant c so that it satisfies the condition

$$\sum_x f_X(x) = 1$$

for it to be a probability function.

1. $f(x)$ is given by

$$f(x) = cx, \quad x \in R_X = \{1, 2, 3, 4\}; \qquad f(x) = 0 \quad \text{elsewhere}$$

Solution. The sum

$$\sum_{x \in R_X} cx = \sum_{1 \le x \le 4} cx$$

$$= c \sum_{1 \le x \le 4} x = 10c = 1$$

Consequently, $c = 1/10$.

2. $f(x)$ is given by

$$f(x) = c(5 - x^2), \quad x = -2, -1, 0, 1, 2; \qquad f(x) = 0 \quad \text{elsewhere}$$

Solution. The sum

$$\sum_{-2 \le x \le 2} c(5 - x^2) = c(1 + 4 + 5 + 4 + 1) = 15c = 1$$

Consequently, $c = 1/15$. ∎

Bernoulli Random Variables

A random variable X that assumes only the two values 0 and 1 with $P(X = 0) = 1 - p$ and $P(X = 1) = p$ is called a *Bernoulli random variable*. The probability function for these variables, denoted $f(x; p)$, is given by Table 3.5.

TABLE 3.5 PROBABILITY FUNCTION OF A BERNOULLI RANDOM VARIABLE

x	0	1
$f(x; p)$	$1 - p$	p

The Geometric Distribution

The *geometric distribution* frequently arises in connection with so-called waiting time problems. For example, consider a sequence of coin tosses, with probability p for heads and probability $q = 1 - p$ for tails. Denote the number of trials until the first head is thrown by X. This is also called the *waiting time until the first success*. The probability of the first success at the kth trial equals

$$P(X = k) = q^{k-1}p, \quad k = 1, 2, \ldots$$

since there are $k - 1$ failures before the first success occurs at the kth trial. In detail,

$$
\begin{aligned}
P(X = k) &= P(\mathrm{T}, \ldots, \mathrm{T}, \mathrm{H}) \\
&= P(\mathrm{T}) \cdots P(\mathrm{T})P(\mathrm{H}) \\
&= q \times \cdots \times q \times p = q^{k-1}p, \quad k = 1, 2, \ldots
\end{aligned}
$$

▨ **DEFINITION 3.3**

A random variable X is said to have a *geometric distribution* with parameter p if

$$P(X = k) = pq^{k-1}, \quad k = 1, \ldots \tag{3.6}$$

■

To show that Eq. (3.6) defines a probability function, we must show that it satisfies Eq. (3.4). This is a consequence of the following well-known property of the *geometric series*:

$$\sum_{0 \le k < \infty} r^k = \frac{1}{1 - r}, \quad |r| < 1 \tag{3.7}$$

Thus,

$$
\sum_{1 \le k < \infty} pq^{k-1} = p \sum_{0 \le k < \infty} q^k
$$

$$
= p \times \frac{1}{1 - q} = p \times \frac{1}{p} = 1
$$

Functions of a Random Variable

In physics the kinetic energy of a body of mass m with speed v is defined to be equal to $(1/2)mv^2$. In statistical mechanics the speed of a molecule in a gas is assumed to be given by a random variable X with a df whose precise form does not concern us here. It follows that the kinetic energy of a molecule is given by the random variable $Y = (1/2)mX^2$, where m is the mass of the molecule. We write $Y = \phi(X)$, where $\phi(v) = (1/2)mv^2$. We say that Y is a *function of the random variable X*.

Computing the Probability Function of $\phi(X)$. Since the distribution of the kinetic energy can be related to such measurable quantities as the pressure of the gas, it is useful to have a method for computing the probability function of $\phi(X)$. Rather than giving a general formula, it is more instructive to work out specific examples.

---

example 3.7   The random variable X has a pf given by Table 3.6.

TABLE 3.6

x	-2	-1	0	1	2
$f(x)$	0.1	0.2	0.3	0.3	0.1

1. Compute the pf of $Y = X^2$.

   Solution. Here $\phi(x) = x^2$. Since $R_X = \{-2, -1, 0, 1, 2\}$, it follows that
   $$R_Y = \{-2^2, -1^2, 0^2, 1^2, 2^2\} = \{0, 1, 4\}$$
   Moreover, you can easily verify that
   $$\{X^2 = 0\} = \{X = 0\}$$
   $$\{X^2 = 1\} = \{X = -1\} \cup \{X = 1\}$$
   $$\{X^2 = 4\} = \{X = -2\} \cup \{X = 2\}$$
   Consequently,
   $$P(\{X^2 = 0\}) = P(\{X = 0\}) = 0.3$$
   $$P(\{X^2 = 1\}) = P(\{X = -1\} \cup \{X = 1\}) = 0.5$$
   $$P(\{X^2 = 4\}) = P(\{X = -2\} \cup \{X = 2\}) = 0.2$$
   Therefore, the probability function of $Y = X^2$ is given by Table 3.7.

   TABLE 3.7

y	0	1	4
$f_Y(y)$	0.3	0.5	0.2

2. Compute the pf of $Y = 2^X$.

Solution. Here $\phi(x) = 2^x$, and the range of Y is the set
$$R_Y = \{2^{-2}, 2^{-1}, 2^0, 2^1, 2^2\}$$
$$= \left\{\frac{1}{4}, \frac{1}{2}, 1, 2, 4\right\}$$

Proceeding as in the preceding case, we see that
$$\{2^X = 2^x\} = \{X = x\}$$

Consequently,
$$P(\{2^X = 2^x\}) = P(\{X = x\}) = f(x)$$

Therefore the probability function of $Y = 2^x$ is given by Table 3.8.

TABLE 3.8

y	1/4	1/2	1	2	4
$f_Y(y)$	0.1	0.2	0.3	0.3	0.1

∎

PROBLEMS

3.1 The random variable X has the pf defined by
$$f(x) = \frac{5 - x^2}{15}, \quad x = -2, -1, 0, 1, 2; \qquad f(x) = 0 \quad \text{elsewhere}$$
Compute:
(a) $P(X \le 0)$ (c) $P(X \le 1)$ (e) $P(|X| \le 1)$
(b) $P(X < 0)$ (d) $P(X \le 1.5)$ (f) $P(|X| < 1)$

3.2 With reference to the random variable defined in Prob. 3.1, compute the:
(a) Probability function of X^2
(b) Probability function of 2^X
(c) Probability function of $2X$

3.3 The random variable X has the pf defined by
$$f(x) = c(6 - x), \quad x = -2, -1, 0, 1, 2; \qquad f(x) = 0 \quad \text{elsewhere}$$
Compute:
(a) c
(b) $F(x) = P(X \le x)$ for $x = -2.5, -0.5, 0, 1.5, 1.7, 3$

3.4 The df of a discrete random variable Y is as follows:
$$F(y) = \begin{cases} 0, & y < -1 \\ 0.2, & -1 \le y < 0 \\ 0.5, & 0 \le y < 1 \\ 0.8, & 1 \le y < 3 \\ 1, & y \ge 3 \end{cases}$$

(a) Draw the graph of $F(y)$.
(b) Compute the pf of Y.
(c) Compute $P(0 \leq Y \leq 2)$.

3.5 Consider the experiment of selecting one coin at random from an urn containing four pennies, three nickels, two dimes, and one quarter. Let X denote the monetary value of the coin that is selected. Compute the probability function of X.

3.6 The number of customers who make a reservation for a limousine service to an airport is a random variable X with probability function given by

$$f(x) = \frac{1}{21}(5 - |x - 3|), \quad x = 1, 2, 3, 4, 5, 6$$

The limousine has a capacity of five passengers and the cost is $10 per passenger. Consequently, the revenue R per trip is also a random variable. Compute the probability function of R.

3.7 Consider a deck of five cards marked 1, 2, 3, 4, 5. Two of these cards are picked at random and without replacement; let $W =$ sum of the numbers picked. Compute the pf of W.

3.8 Suppose a lot of 100 items contains 10 defective ones, and a sample of 5 items is selected at random from the lot. Let X denote the number of defective items in the sample; compute $P(X = x)$ for $x = 0, 1, 2, 3, 4, 5$. (*Hint:* This problem is similar to the computation of the probability function of the number of hearts in a five-card poker hand [Eq. (3.5)], with "defective" and "nondefective" substituted for "heart" and "nonheart.")

3.9 Suppose a lot of 50 items contains 4 defective ones, and a sample of 5 items is selected at random from the lot. Let X denote the number of defective items in the sample; compute the probability function of X.

3.10 Let X have the pf given by

$$f(x) = cx \quad \text{for } x = 1, 2, 3, 4, 5, 6; \qquad f(x) = 0 \quad \text{elsewhere}$$

(a) Compute c.
(b) Compute $F(x)$ for $x = 0, 1, 5, 7$.
(c) Compute $P(X \text{ is odd})$.
(d) Compute $P(X \text{ is even})$.
(e) Compute $P(X < 4)$.
(f) Compute $P(2 \leq X \leq 4)$.

3.11 Let X have the pf given by

$$f(x) = cx \quad \text{for } x = 1, 2, \ldots, n; \qquad f(x) = 0 \quad \text{elsewhere}$$

(a) Show that

$$c = \frac{2}{n(n + 1)}$$

(b) Show that

$$F(x) = \frac{x(x + 1)}{n(n + 1)} \quad \text{for } x = 1, 2, \ldots, n$$

Hint:

$$\sum_{1 \le i \le x} i = \frac{x(x+1)}{2}$$

3.12 The random variable X has the pf defined by

$$f(x) = \frac{1}{x(x+1)} \quad \text{for } x = 1, 2, \ldots, n, \ldots$$

(a) Show that

$$\sum_{1 \le x < \infty} \frac{1}{x(x+1)} = 1$$

Hint: Use the algebraic identity

$$\frac{1}{x(x+1)} = \frac{1}{x} - \frac{1}{x+1}$$

(b) Show that

$$F(x) = P(X \le x) = \frac{x}{x+1} \quad \text{for } x = 1, 2, \ldots, n, \ldots$$

3.3 EXPECTED VALUE AND VARIANCE OF A RANDOM VARIABLE

In Chap. 1 we defined the sample mean $\bar{x}$ of a data set $\{x_1, \ldots, x_n\}$ to be the arithmetic mean

$$\bar{x} = \frac{1}{n} \sum_{1 \le i \le n} x_i$$

We then derived an equivalent formula [Eq. (1.14)] for the sample mean, and this formula can be expressed as follows in terms of the empirical probability function [Eq. (2.8)]:

$$\bar{x} = \sum_x x \hat{f}(x)$$

We define the *expected value* of a random variable in exactly the same way, except that we use the probability function $f_X(x)$ in place of the empirical probability function $\hat{f}(x)$. The formal definition follows.

▦ DEFINITION 3.4

The *expected value* of a random variable X with probability function $f_X(x)$ is denoted $E(X)$ and is defined to be

$$E(X) = \sum_x x P(X = x)$$

$$= \sum_x x f_X(x) \tag{3.8}$$

provided that the sum

$$\sum_x |x| f_X(x) < \infty \tag{3.9}$$

converges.   ■

The expected value can be interpreted as the center of mass of a distribution of weights $f(x_i)$ located at the points x_i on the line.

Notation. The expected value of X is also denoted by $\mu = E(X)$; to emphasize its dependence on X, the notation μ_X is also used.

An Interpretation of $E(X)$ in Terms of the Average Win Per Bet. Let X denote the payoff random variable for some gambling game. The gambler wins x_j ($j = 1,\ldots,n$) dollars with probability $P(X = x_j)$. A negative value $x_j < 0$ is interpreted as a loss of x_j dollars. According to the frequency interpretation of probability [i.e., the relative frequency with which the event $\{X = x_j\}$ occurs is approximately equal to its theoretical probability $P(X = x_j)$], we see that after N repetitions of the game the gambler has won x_j dollars $NP(X = x_j)$ times. Consequently, after N plays the total winnings W_N are approximately equal to

$$W_N \approx \sum_{1 \le j \le n} N x_j P(X = x_j)$$

The average win per bet equals W_N/N; consequently,

$$\text{Average win per bet} = \frac{W_N}{N}$$
$$\approx \sum_{1 \le j \le n} x_j P(X = x_j)$$
$$= E(X)$$

---

example 3.8   Compute $E(X)$ for each of the following random variables:

1. A Bernoulli random variable X with $P(X = 1) = p$.

Solution. We show that $E(X) = p$. The computation is simple, but the result is quite useful.

$$E(X) = 0 \times (1 - p) + 1 \times p = p \tag{3.10}$$

2. A geometric random variable X with probability function
$$f(x) = pq^{x-1}, \quad x = 1, 2, \ldots$$

Solution. We show that $E(X) = 1/p$. Recalling the interpretation of X as the waiting time until the first success, we see that this result is intuitively plausible since a small probability of success implies a long waiting time for the first success.

We use the recurrence relation

$$f(x + 1) = qf(x), \quad 1 \le x < \infty$$

Multiplying both sides of the recurrence relation by x and summing over x, we obtain the equation

$$\sum_{1 \le x < \infty} xf(x+1) = \sum_{1 \le x < \infty} qxf(x) = qE(X)$$

On the other hand,

$$\sum_{1 \le x < \infty} xf(x+1) = \sum_{1 \le x < \infty} (x-1)f(x)$$

$$= \sum_{1 \le x} xf(x) - \sum_{1 \le x} f(x)$$

$$= E(X) - 1$$

Therefore,

$$E(X) - 1 = qE(X).$$

Solving for $E(X)$ yields the equation

$$(1 - q)E(X) = 1$$

Thus, $E(X) = 1/(1-q) = 1/p$.

3. Chuck-a-luck is a popular carnival game (so beware!) in which three fair dice are rolled. A bet placed on one of the numbers 1 through 6 has a payoff that depends on the number of times the number appears on the three dice. To fix our ideas, suppose one bets \$1 on the number 4. The payoff random variable X equals $-\$1$ if none of the three dice shows a 4; otherwise $X = \$x$, where x is the number of dice that show the number 4. For instance, the payoff is \$3 if all three dice show the number 4. We now compute $E(X)$.

   Solution. The range $R_X = \{-1, 1, 2, 3\}$. The probability function is given by

$$P(X = -1) = \frac{5^3}{6^3}$$

$$P(X = 1) = \frac{3 \times 5^2}{6^3}$$

$$P(X = 2) = \frac{3 \times 5}{6^3}$$

$$P(X = 3) = \frac{1}{6^3}$$

Consequently, $E(X) = -0.0789$.

   Details of the Computations. The sample space corresponding to throwing three dice has $6^3 = 216$ sample points, all of which have equal probability. The events $X = -1, 1, 2, 3$ contain 5^3, 3×5^2, 3×5, and 1 sample point, respectively (why?).

4. W is the random variable giving the payoff of a one-roll bet on a 12.

   Solution. The payoff of a one-roll bet on a 12 is 30 to 1; its probability function $f(w)$ is given by Table 3.9.

TABLE 3.9

w	-1	30
$f(w)$	$35/36$	$1/36$

Therefore,

$$E(W) = -1 \times \frac{35}{36} + 30 \times \frac{1}{36} = -\frac{5}{36} = -0.1389$$ ∎

▦ DEFINITION 3.5

Let W be the payoff random variable for a gambling game. The game is called:

1. Favorable, if $E(W) > 0$
2. Unfavorable, if $E(W) < 0$
3. Fair, if $E(W) = 0$ ∎

The quantity $100 \times |E(W)|$ is called the *house percentage*.

Comment. A one-roll bet on the number 12 with a 30-to-1 payoff is a sucker bet because the house percentage, which is nearly 14 percent, is so high.

---

example 3.9   Let us now analyze the payoff random variable of a one-roll bet. A one-roll bet pays you $\$b$ if you win; otherwise you lose your stake of $\$a$. If you denote the probabilities of winning and losing by p and $1 - p$, respectively, then the probability function and the expected value of the payoff random variable W of a one-roll bet are given by

$$P(W = -a) = 1 - p \quad \text{and} \quad P(W = b) = p$$

So

$$E(W) = -a(1 - p) + bp = \mu \tag{3.11}$$

To repeat, if the event $B = \{W = b\}$ occurs, you win b dollars; if B does not occur, you lose your stake of a dollars. The ratio b/a ("b to a") is called the *house odds*. A *fair bet* is one for which the offered odds b'/a' are chosen so that $E(W) = 0$; in this case we call the ratio b'/a' the *true odds*.

The condition that $E(W) = -a'(1 - p) + b'p = 0$ implies that the true odds b'/a' must satisfy the condition

$$\textbf{True odds:} \quad \frac{b'}{a'} = \frac{1 - p}{p} \tag{3.12}$$

The true odds for a one-roll bet on a 12 are 35 to 1; the house odds are only 30 to 1. The true odds on a one-roll bet on the number 11 are 34/2, or 17 to 1; the house odds are 15 to 1. It is this difference between the true odds and the house odds that guarantees the fantastic profits of gambling casinos. More generally, if W is the payoff for some gambling game, we call $E(W) = \mu$ the *expected win per bet*. ∎

Doubling the Bet—The Road to Financial Ruin

Over the weekend of 25–26 Feb. 1995, the financial world was startled to learn that Baring Brothers, a highly respected merchant bank that had been in business for 233 years, collapsed into bankruptcy as a result of a series of highly imprudent financial transactions by one of its junior bond traders. For the bank's senior management, the failure, coming as it did just before the annual bonuses were to be distributed, could not have occurred at a more embarrassing time. It was soon revealed that the trader was attempting to recoup his previous losses by committing ever-increasing sums of the bank's capital to high-risk bets on the future value of several indices based on the prices of a complex hodgepodge of stocks and bonds. This practice, to use the currently fashionable terminology, is called *risk management*.[2] The bond trader's method, as we will now see, is a variant of an old and discredited gambling system called *doubling the bet*, which is a gambling strategy to cover mounting losses by placing bigger and bigger bets.

Consider a gambler who has a probability p of winning on any one of a sequence of bets. The probability of winning for the first time at the xth bet is pq^{x-1}. To simplify the calculations, we assume $p = q = 1/2$, so that the probability of winning for the first time at the xth bet is $(1/2)^x$. Suppose the gambler's initial bet is one dollar and that he doubles his bet each time until he wins for the first time. Doubling the bet means that he bets 2^{x-1} dollars on the xth bet, where $x = 1, 2, \ldots$. How large must his initial capital be if he is to sustain this betting system through the xth bet given that he lost his previous $x - 1$ bets? We solve this problem by computing the pf of X, the amount of capital the gambler needs to play this game until he wins for the first time.

To place his first bet he needs $1. If he loses, he then bets $2, so he needs $1 + 2 = 2^2 - 1$ dollars to bet a second time. Repeating this reasoning, we see that the gambler's cumulative losses from the preceding $(x - 1)$ unsuccessful bets followed by a bet of 2^{x-1} on the xth bet is given by the geometric series

$$X = 1 + 2 + \cdots + 2^{x-1} = 2^x - 1$$

If he wins for the first time on the xth bet, his net gain will be $2^x - (2^x - 1) = 1$; that is, when he wins he will be $1 ahead. However, from the preceding remarks it is clear that he will need $2^x - 1$ dollars if he is to play the game through the xth play. Consequently, the random variable X has the probability function given by

$$P(X = 2^x - 1) = \left(\frac{1}{2}\right)^x, \quad x = 1, 2, \ldots$$

---

[2]From a mathematician's perspective, the only difference between gambling and risk management is that the latter activity does not take place in a casino.

The expected value of X is given by

$$E(X) = \sum_{1 \leq x < \infty} (2^x - 1)\left(\frac{1}{2}\right)^x$$

$$= \sum_{1 \leq x < \infty} (1 - 2^{-x})$$

$$= \infty$$

This series diverges to infinity since its xth term, which equals $(1 - 2^{-x})$, converges to 1 and not to zero. In other words, *no finite amount of money is sufficient to sustain this betting system.* In the Baring bank failure, the losses eventually reached a billion dollars. ■

3.3.1 Moments of a Random Variable

The kth moment μ_k, $k = 1, 2, \ldots$, of a random variable X is defined by the equation

$$\mu_k = E(X^k) \quad \text{where } k = 1, 2, \ldots$$

It is clear from the definition that the first moment $\mu_1 = \mu$, the expected value. The moments of a random variable give useful information on the shape and spread of the distribution function of X. They are also used to construct estimators for population parameters via the so-called method of moments, discussed in Chap. 7. For now we focus our attention on how to compute them. Note that $\mu_k = E\phi(X)$, where $\phi(x) = x^k$. The next theorem gives a formula for computing the expected value of a function of a random variable X.

▨ THEOREM 3.1

Let X be a discrete random variable with pf $f_X(x)$ and let $Y = \phi(X)$. Then

$$E(\phi(X)) = \sum_x \phi(x) f_X(x) \tag{3.13}$$

provided that

$$\sum_x |\phi(x)| f_X(x) < \infty \tag{3.14}$$
■

We omit the proof. The significance of the theorem is this: To compute $E(Y)$ it is not necessary to recompute the pf $f_Y(y)$ of $Y = \phi(X)$ and then compute $E(Y)$ using Eq. (3.8). In terms of the probability function, the kth moment is given by

$$\mu_k = \sum_x x^k f(x) \tag{3.15}$$

---

example 3.10   Compute the first and second moments for each of the following random variables.

1. Let X denote the value of a number picked at random from the set of the first N integers $R_X = \{1, 2, \ldots, N\}$. Show that

$$\mu_1 = \frac{N + 1}{2} \quad \text{and} \quad \mu_2 = \frac{(N + 1)(2N + 1)}{6} \tag{3.16}$$

Solution. The probability function is the equally likely measure on the set R_X. To see why, note that since each number is equally likely to be chosen, the probability function is given by

$$f_X(x) = 1/N, \quad x = 1, \ldots, N \qquad f_X(x) = 0 \quad \text{elsewhere} \tag{3.17}$$

The probability function (3.17) is called the *discrete uniform distribution*. The formulas for the moments displayed in Eq. (3.16) are the consequences of the following well-known formulas for the sums of the powers of the first N integers:

$$\sum_{1 \leq x \leq N} x = \frac{N(N + 1)}{2} \tag{3.18}$$

$$\sum_{1 \leq x \leq N} x^2 = \frac{N(N + 1)(2N + 1)}{6} \tag{3.19}$$

Thus,

$$\mu_1 = \sum_{1 \leq x \leq N} x f_X(x)$$

$$= \frac{\sum_{1 \leq x \leq N} x}{N}$$

$$= \frac{N(N + 1)}{2N} = \frac{(N + 1)}{2}$$

Similarly,

$$\mu_2 = \sum_{1 \leq x \leq N} x^2 f_X(x)$$

$$= \frac{\sum_{1 \leq x \leq N} x^2}{N}$$

$$= \frac{N(N + 1)(2N + 1)}{6N}$$

$$= \frac{(N + 1)(2N + 1)}{6}$$

2. Show that the first and second moments of the geometric distribution [see Eq. (3.6)] are given by

$$\mu_1 = \frac{1}{p} \quad \text{and} \quad \mu_2 = \frac{2}{p^2} - \frac{1}{p} \tag{3.20}$$

Solution. Using the recurrence relation

$$f(x + 1) = q f(x) \quad 1 \leq x < \infty$$

we previously derived the result that $\mu_1 = 1/p$. Multiplying both sides of the recurrence relation by x^2 and summing over x, we obtain

$$\sum_{1 \leq x < \infty} x^2 f(x+1) = \sum_{1 \leq x < \infty} qx^2 f(x) = qE(X^2) = q\mu_2$$

On the other hand, the left-hand side of the preceding equation equals

$$\sum_{1 \leq x < \infty} x^2 f(x+1) = \sum_{1 \leq x < \infty} (x-1)^2 f(x)$$

$$= \sum_{1 \leq x < \infty} x^2 f(x) - 2\sum_{1 \leq x < \infty} xf(x) + \sum_{1 \leq x < \infty} f(x)$$

$$= \mu_2 - 2\mu_1 + 1$$

Solving the equation

$$\mu_2 - 2\mu_1 + 1 = q\mu_2$$

for μ_2 and using the result derived earlier that $\mu_1 = 1/p$ yields

$$(1-q)\mu_2 = p\mu_2 = \frac{2}{p} - 1$$

Consequently,

$$\mu_2 = \frac{2}{p^2} - \frac{1}{p}$$

3. The random variable X has the pf given in Table 3.10.

TABLE 3.10

x	-2	-1	0	1	2
$f(x)$	0.1	0.2	0.3	0.3	0.1

Compute (*a*) $E(X^2)$ and (*b*) $E(2^X)$.

Solutions

(a) The random variable $X^2 = \phi(X)$ where $\phi(x) = x^2$. Consequently,

$$E(X^2) = (-2)^2 \times 0.1 + (-1)^2 \times 0.2 + 1^2 \times 0.3 + 2^2 \times 0.1$$
$$= 1.3$$

(b) The random variable $2^X = \phi(X)$ where $\phi(x) = 2^x$. Consequently,

$$E(2^X) = 2^{-2} \times 0.1 + 2^{-1} \times 0.2 + 2^0 \times 0.3 + 2^1 \times 0.3 + 2^2 \times 0.1$$
$$= 1.425$$

∎

The following corollary of Theorem 3.1 is particularly useful.

◼ COROLLARY 3.1

Let X be a random variable with expected value $E(X)$ and let a, b be arbitrary constants. Then

$$E(aX + b) = aE(X) + b \tag{3.21}$$

◼

Before giving the derivation of Eq. (3.21) we note the following special cases:

1. Set $b = 0$ in Eq. (3.21); then

$$E(aX) = aE(X)$$

2. Set $a = 1$ and $b = -\mu_X$ in Eq. (3.21); then

$$E(X - \mu_X) = E(X) - \mu_X = 0$$

Derivation of Eq. (3.21). We choose $\phi(x) = ax + b$ in Eq. (3.13) and obtain

$$
\begin{aligned}
E(aX + b) &= \sum_x (ax + b) f_X(x) \\
&= \sum_x axf_X(x) + \sum_x bf_X(x) \\
&= a \sum_x xf_X(x) + b \sum_x f_X(x) \\
&= aE(X) + b
\end{aligned}
$$

---

example **3.11**   Given that $E(X) = 1.5$, compute the following expected values: (1) $E(2X + 4)$, (2) $E(-3X - 0.5)$, (3) $E(0.1X)$.

Solution

1. $E(2X + 4) = 2 \times 1.5 + 4 = 7$
2. $E(-3X - 0.5) = -3 \times 1.5 - 0.5 = -5$
3. $E(0.1X) = 0.1 \times 1.5 = 0.15$ ∎

3.3.2   Variance of a Random Variable

We defined the population variance σ^2 of a finite population by the equation

$$\sigma^2 = \frac{\sum_{1 \le i \le N}(x_i - \mu)^2}{N}$$

where μ is the population mean (see Definition 1.7). The population variance is a weighted average of the squared deviations of the values x_j from the population mean μ. The weight assigned to the squared deviation $(x_j - \mu)^2$ is $1/N$. We now extend this concept to random variables.

The *variance* of a random variable X is denoted by $V(X)$ and is defined as

$$V(X) = E((X - \mu_X)^2) = \sum_{1 \le j < \infty} (x_j - \mu_X)^2 f_X(x_j) \qquad (3.22)$$

The variance of X is a weighted average of the squared deviations of the values x_j from the expected value μ_X; the weight assigned to $(x_j - \mu_X)^2$ is $f_X(x_j)$. Note also that the variance $V(X) = E(\phi(X))$, where $\phi(x) = (x - \mu_X)^2$.

Notation. The variance is frequently denoted by σ_X^2 or $\sigma^2(X)$; sometimes we drop the subscript X and write σ^2. The *standard deviation* of X is the square root of the variance:

$$\text{Standard deviation } \sigma_X = \sqrt{\sigma_X^2}$$

A Shortcut Formula for $V(X)$

Determining the variance via Eq. (3.22) requires tedious computations. The following formula, which is algebraically equivalent to Eq. (3.22), is called the *shortcut formula* for the variance because it requires fewer arithmetic operations.

$$V(X) = E(X^2) - \mu_X^2$$

$$= \sum_{1 \leq i < \infty} x_i^2 f_X(x_i) - \mu_X^2 \tag{3.23}$$

We can also express $V(X)$ in terms of the first two moments by noting that

$$V(X) = \mu_2 - \mu_1^2$$

Derivation of Eq. (3.23). Noting that

$$(X - \mu_X)^2 = X^2 - 2\mu_X X + \mu_X^2$$

and taking expected values of both sides, we conclude that

$$E((X - \mu_X)^2) = E(X^2) + E(-2\mu_X X) + E(\mu_X^2)$$

Since μ_X is a constant, we have

$$E(-2\mu_X X) = -2\mu_X E(X) = -2\mu_X^2 \quad \text{and} \quad E(\mu_X^2) = \mu_X^2$$

Therefore,

$$E((X - \mu_X)^2) = E(X^2) - 2\mu_X^2 + \mu_X^2 = E(X^2) - \mu_X^2$$

---

example 3.12 **1.** Compute $V(X)$, where X is a Bernoulli random variable.

Solution. Recall that X takes on the two values 0 and 1 with probabilities $P(X = 0) = 1 - p$, $P(X = 1) = p$, respectively. We claim that the variance equals

$$V(X) = p(1 - p) \tag{3.24}$$

To see this, note that $X^2 = X$ (why?); consequently, $E(X^2) = E(X) = p$ and therefore

$$V(X) = E(X^2) - E(X)^2 = p - p^2 = p(1 - p)$$

2. Compute the variance of the random variable with probability function

$$f(x) = \frac{x}{10}, \quad x \in R_X = \{1, 2, 3, 4\}; \qquad f(x) = 0 \quad \text{elsewhere}$$

Solution. Using the shortcut formula, we first compute

$$E(X) = 3 \quad \text{and} \quad E(X^2) = 10$$

Consequently,

$$V(X) = 10 - 3^2 = 1$$

∎

A Formula for $V(aX + b)$

We now derive a formula for computing the variance of the random variable $Y = aX + b$, which is obtained from X by means of the linear transformation $\phi(x) = ax + b$. For instance, if X records the temperature of some object in degrees Celsius, then $Y = (9/5)X + 32$ is its temperature in degrees Fahrenheit. Here $a = 9/5$ and $b = 32$. The following formula for $V(aX + b)$ will prove to be quite useful.

Let X be a random variable with variance $V(X)$, and let a, b denote arbitrary constants; then

$$V(aX + b) = a^2 V(X) \tag{3.25}$$

Before giving the derivation of Eq. (3.25), we give an example.

---

example 3.13 Suppose $Y = (9/5)X + 32$ and $V(X) = 5$. Compute $V(Y)$.

Solution. Apply Eq. (3.25) with $a = 9/5$. Then

$$V(Y) = \left(\frac{9}{5}\right)^2 \times 5 = \frac{81}{5} = 16.2 \qquad \blacksquare$$

Derivation of Eq. (3.25). We begin with the observation that

$$\mu_{aX+b} = E(aX + b) = aE(X) + b = a\mu_X + b$$

and

$$V(aX + b) = E([aX + b - \mu_{aX+b}]^2)$$

Therefore,

$$
\begin{aligned}
V(aX + b) &= E([aX + b - (a\mu_X + b)]^2) \\
&= a^2 E([X - \mu_X]^2) \\
&= a^2 V(X)
\end{aligned}
$$

PROBLEMS

3.13 The number of defects on a printed circuit board is a random variable X with pf given by
$$P(X = i) = c/(i + 1), \quad i = 0, 1, \ldots, 4$$
(a) Compute the constant c.
(b) Compute the mean and variance of X.

3.14 The number of cells (out of 100) that exhibit chromosome aberrations is a random variable X with pf given by
$$P(X = i) = \frac{c(i + 1)^2}{2^{i+1}}, \quad i = 0, 1, 2, 3, 4, 5$$
(a) Determine the value of the constant c.
(b) Compute the mean and variance of X.

3.15 The probability function for the number of heads X in three tosses of a fair coin is given in Table 3.2.
(a) Compute $E(X)$.
(b) Use the shortcut formula to compute $V(X)$.

3.16 Table 3.3 displays the probability function corresponding to the experiment of throwing two dice. Compute $E(X)$ and $V(X)$.

3.17 Compute $E(X)$ and $V(X)$, where X is the number of hearts in a five-card poker hand. [*Hint:* Use Eq. (3.5) to compute $f_X(x)$.]

3.18 The payoff of a one-roll bet on the number 11 is 15 to 1.
(a) Compute the house percentage. (*Note:* The definition of *house percentage* is given in Definition 3.5.)
(b) Compute the true odds. [For the definition of *true odds*, refer to Eq. (3.12).]

3.19 The random variable X has the pf defined by

$$f(x) = \frac{1}{5}, \quad x = 1, 2, 3, 4, 5; \qquad f(x) = 0 \quad \text{elsewhere}$$

Compute:
(a) $E(X)$ (b) $E(X^2)$ (c) $V(X)$
(d) $E(2^X)$ (e) $E(\sqrt{X})$
(f) $E(2X + 5)$ (g) $V(2X + 5)$

3.20 The random variable X has the pf given by the following table:

x	-2	-1	0	1	2
$f(x)$	0.2	0.1	0.1	0.4	0.2

Compute:
(a) $E(X)$ (b) $E(X^2)$ (c) $V(X)$
(d) $E(2^X)$ (e) $E(2^{-X})$ (f) $E(\sin(\pi X))$

3.21 The random variable X has the pf defined by

$$f(x) = \frac{5 - x^2}{15}, \quad x = -2, -1, 0, 1, 2; \qquad f(x) = 0 \quad \text{elsewhere}$$

Compute:
(a) $E(X)$ (b) $E(X^2)$ (c) $V(X)$
(d) $E(2^X)$ (e) $E(2^{-X})$ (f) $E(\cos(\pi X))$

3.22 Let the random variable Y have the distribution function $F(y)$ defined by

$$F(y) = \begin{cases} 0, & y < -1 \\ 0.2, & -1 \le y < 0 \\ 0.5, & 0 \le y < 1 \\ 0.8, & 1 \le y < 3 \\ 1, & y \ge 3 \end{cases}$$

Compute:
(a) $E(Y)$ (b) $E(Y^2)$ (c) $V(Y)$

3.23 The random variable X has the pf defined by

$$f(x) = \frac{6 - x}{30}, \quad x = -2, -1, 0, 1, 2; \quad f(x) = 0 \quad \text{elsewhere}$$

Compute:
(a) $E(X)$ **(b)** $E(X^2)$ **(c)** $V(X)$
(d) $E(2^X)$ **(e)** $E(2^{-X})$ **(f)** $E(\cos(\pi X))$

3.24 Compute $E(W)$ and $V(W)$ for the random variable W of Prob. 3.7.

3.25 Let X have the pf given by:

$$f(x) = \frac{2x}{n(n + 1)}, \quad x = 1, 2, \ldots, n; \quad f(x) = 0 \quad \text{elsewhere}$$

Compute $E(X)$. [*Hint:* Use Eq. (3.19).]

3.26 The random variable X has the pf defined by

$$f(x) = \frac{1}{x(x + 1)}, \quad x = 1, 2, \ldots, n, \ldots$$

Show that this probability function has the property that $\sum_{1 \le x < \infty} x f(x) = \infty$, and therefore it does not satisfy the finiteness condition (3.9).

3.27 Let X be a Bernoulli random variable. Show that

$$V(X) \le \frac{1}{4}$$

(*Hint:* The variance of a Bernoulli random variable is given by

$$V(X) = p(1 - p)$$

[see Eq. (3.24)]. Then show that $p(1 - p) \le 1/4$.)

3.28 The random variable X has the discrete uniform distribution [see Eq. (3.17)]. Show that

$$V(X) = \frac{(N - 1)(N + 1)}{12}$$

[*Hint:* Use the shortcut formula and Eq. (3.16) for the moments of the discrete uniform distribution.]

3.4   THE HYPERGEOMETRIC DISTRIBUTION

The *hypergeometric distribution* arises when one takes a random sample of size n, without replacement, from a population of size N divided into two classes consisting of D elements of the first kind and $N - D$ elements of the second kind. Such a population is called *dichotomous*.

---

example 3.14   We give two examples of dichotomous populations.

1.   Consider a lot of 100 steel bolts, of which 5 are defective. Here the population is split into *defective* and *nondefective* items, with $N = 100$ and $D = 5$.

2. Consider a population of 1000 school children, 900 of whom have been vaccinated against the measles; here $N = 1000$ and $D = 900$. In this case, the dichotomy is *vaccinated* and *nonvaccinated*. ∎

The experiment of drawing a random sample of size n from a dichotomous population is best explained in the context of an *urn model*.

---

example 3.15 We represent the dichotomous population from which the sample is being drawn as an urn filled with D beads colored red and $N - D$ beads colored white. We denote the number of red beads in a sample of size n by X. Compute the probability function of X.

Solution. We begin by noting that the number of red beads x in the sample satisfies the following inequalities:

$$\max(0, n - (N - D)) \leq x \leq \min(n, D). \tag{3.26}$$

To see this, we note that the number of red beads in the sample cannot exceed the sample size n nor the total number of red beads D; consequently, $x \leq \min(n, D)$. Similarly, $x \geq 0$ and $n - x \leq N - D$, since this inequality simply expresses the fact that the number of white beads in the sample cannot exceed the total number of white beads. Solving this inequality for x yields $x \geq n - (N - D)$, and therefore $\max(0, n - (N - D)) \leq x$. Combining these two inequalities yields Eq. (3.26). The formula for the probability function is given in Eq. (3.27).

■ THEOREM 3.2

Suppose an urn is filled with D beads colored red and $N - D$ beads colored white. A random sample of size n, without replacement, is drawn from the urn and the number of red beads in the sample is denoted by X. Then the pf $h(x) = P(X = x)$ is given by

$$h(x) = \frac{\binom{D}{x}\binom{N - D}{n - x}}{\binom{N}{n}} \tag{3.27}$$

where $\max(0, n - (N - D)) \leq x \leq \min(n, D)$. We define $h(x) = 0$ elsewhere. ∎

The pf $h(x)$ defined by Eq. (3.27) is called the *hypergeometric distribution* with parameters n, N, D.

Derivation of Eq. (3.27). We know from Theorem 2.2 that there are $\binom{N}{n}$ ways of selecting a sample of size n from a lot of size N. To each of these outcomes we assign the probability $\binom{N}{n}^{-1}$. There are $\binom{D}{x} \times \binom{N-D}{n-x}$ ways of choosing x red beads from the D red ones and $n - x$ beads from the remaining $N - D$ white ones. The ratio of these two quantities is the pf $h(x)$ defined in Eq. (3.27). ∎

Applications of the Hypergeometric Distribution to Acceptance Sampling

Acceptance sampling is a set of methods for accepting or rejecting a lot of N items on the basis of inspecting a sample of size n items selected at random from the lot. In detail, acceptance sampling requires that you specify three numbers:

- N = the number of items from which the sample is to be taken
- n = the number of items to be sampled
- c = maximum allowable number of defective items in a sample of size n, called the *acceptance number*

The significance of the acceptance number c is that the lot is rejected if the sample contains more than c defective items.

---

example 3.16 Steel bolts are shipped in lots of 100. Ten bolts are selected at random and inspected. If one or more of these bolts is found to be defective, the entire lot is rejected; if none is defective, the lot is accepted. Suppose there are five defective bolts in the lot; what is the probability that the lot will be accepted?

Solution. In this context the defective bolts correspond to the red beads and the nondefective bolts correspond to the white beads. The number of defective bolts in the sample, denoted X, has a hypergeometric distribution with parameters $N = 100, n = 10, D = 5$. The lot will be accepted only if there are no defectives in the sample; this means the acceptance number c equals 0. Consequently, the probability of acceptance is $P(X = 0)$. Therefore,

$$P(\text{acceptance}) = P(X = 0) = \frac{\binom{95}{10}}{\binom{100}{10}} = 0.5838 \qquad \blacksquare$$

Is This a Good Acceptance Sampling Plan? The answer is that it depends on the percentage of defective items that is acceptable. Suppose the maximum allowable percentage of defective items in the lot is specified to be 1 percent. Since the percentage of defective items in this case equals 5 percent > 1 percent, it follows that this lot should be rejected with high probability. However, the preceding calculation shows that the probability of accepting this lot equals 0.5838, which is uncomfortably high. One way to reduce the probability of accepting a defective lot is to double the sample size to $n = 20$, say. In this case the acceptance probability equals

$$P(\text{acceptance}) = P(X = 0) = \frac{\binom{95}{20}}{\binom{100}{20}} = 0.3198$$

This probability is still too high; at this point one should consider looking for an alternative supplier.

Computing the Probability of Acceptance

An important criterion for evaluating the performance of a sampling plan is to compute the probability of accepting the lot. Suppose we have a sampling plan defined by the two numbers (n, c), and the true number of defective items in the lot is denoted by D; what is the probability of acceptance? Since the number of defective items X in a sample of size n taken from a lot of size N containing D defective items has the hypergeometric distribution $h(x)$, it follows that probability of acceptance is given by

$$P(\text{acceptance}) = P(X \le c) = \sum_{0 \le x \le c} h(x) \tag{3.28}$$

The Mean and Variance of the Hypergeometric Distribution

Proposition 3.2 gives the formulas for the mean and variance of the hypergeometric distribution.

■ PROPOSITION 3.2

The random variable X has the hypergeometric distribution with parameters n, N, D. Its mean and variance are given by

$$E(X) = n \times \frac{D}{N}, \qquad V(X) = \frac{N-n}{N-1} \times n \times \frac{D}{N}\left(1 - \frac{D}{N}\right) \tag{3.29}$$

■

The complete proof of Eq. (3.29) is somewhat tedious; refer to Sec. 5.7 at the end of Chap. 5 for the details. Intuitively, the quantity D/N is the probability that a bead drawn at random is red. The quantity $n(D/N)$ represents the expected number of red beads in a sample of size n and

$$n \times \frac{D}{N}\left(1 - \frac{D}{N}\right)$$

represents the variance of X if the sampling were done with replacement. In this case X has a binomial distribution; see Example 3.20 in Sec. 3.5. The quantity

$$\frac{N-n}{N-1}$$

is called the *finite-population correction factor*.

---

example 3.17   Suppose X has a hypergeometric distribution with parameters $n = 10$, $N = 10{,}000$, $D = 500$. Compute $E(X)$ and $V(X)$.

Solution. This is a straightforward application of Eq. (3.29).

$$E(X) = 10 \times \frac{500}{10{,}000} = 0.5$$

$$V(X) = \frac{10{,}000 - 10}{10{,}000 - 1} \times 10 \times 0.05 \times (1 - 0.05)$$

$$= 0.4703 \qquad \blacksquare$$

A Recurrence Relation for the Hypergeometric Probabilities

A recurrence relation for the binomial coefficients was derived in Sec. 2.3 [Eq. (2.19)], so it is not surprising that we can also derive a recurrence relation for the hypergeometric probabilities. The recurrence relation is given by

$$h(x + 1) = \frac{(n - x)(D - x)}{(x + 1)(N - D - n + x + 1)} \times h(x) \qquad (3.30)$$

where $\max(0, n - (N - D)) \le x \le \min(n, D)$.

Derivation of Eq. (3.30). The recurrence relation is a consequence of the following formula for the ratio $R_h(x)$:

$$R_h(x) = \frac{h(x + 1)}{h(x)}$$

$$= \frac{(n - x)(D - x)}{(x + 1)(N - D - n + x + 1)}, \qquad (3.31)$$

where $\max(0, n - (N - D)) \le x \le \min(n, D)$.
    We omit the derivation of Eq. (3.31) since it is just a straightforward algebraic manipulation of the ratio of two binomial coefficients. Note that

$$R_h(x) = 0 \quad \text{for } x \ge \min(n, D)$$

which means that the recurrence relation 3.30 is valid for all integers $x \ge \min(n, D)$.

---

example **3.18** Compute the probability function for the number of hearts in a five-card poker hand using the recurrence relation (3.30).

Solution. In this case we have

$$n = 5, \qquad N = 52, \qquad D = 13, \qquad x = 0, 1, 2, 3, 4, 5$$

Consequently,

$$h(0) = \frac{\binom{13}{0}\binom{39}{5}}{\binom{52}{5}} = 0.2215$$

and

$$R_h(x) = \frac{(5 - x)(13 - x)}{(x + 1)(35 + x)}, \qquad x = 0, 1, \ldots$$

Therefore,

$$h(1) = \frac{5 \times 13}{35} \times 0.2215 = 0.4114$$

$$h(2) = \frac{4 \times 12}{2 \times 36} \times 0.4114 = 0.2743$$

$$h(3) = \frac{3 \times 11}{3 \times 37} \times 0.2743 = 0.0815$$

$$h(4) = \frac{2 \times 10}{4 \times 38} \times 0.0815 = 0.0107$$

$$h(5) = \frac{1 \times 9}{5 \times 39} \times 0.0107 = 0.0005 \qquad \blacksquare$$

3.5 THE BINOMIAL DISTRIBUTION

A surprising variety of statistical experiments can be reduced to the analysis of the mathematical model of tossing a coin n times with probability p of tossing a head. The essential features of this model are that the results of the successive coin tosses are mutually independent and the probability of getting a head remains constant. It is the simplest example of n independent repetitions of an experiment under identical conditions. Our main interest is in obtaining the probability function for the number of heads in n tosses of a coin.

▨ PROPOSITION 3.3

Let X denote the number of heads in n tosses of a coin with probability p for heads. Then the probability function of X is given by

$$P(X = x) = \binom{n}{x} p^x (1 - p)^{n-x}, \quad x = 0, 1, \ldots, n \qquad (3.32)$$

$\blacksquare$

Proof. The sample space consists of 2^n sample points that can be represented as a sequence of H's and T's like so:

$$\omega = (T, H, H, T, \ldots, H, T)$$

To each sample point ω that consists of x H's and of $n - x$ T's we assign the probability $p^x(1 - p)^{n-x}$. To justify this assignment of probabilities, we reason as follows. We assume that (1) the probability of tossing a head is the same for each toss, and (2) the outcomes of the coin tosses are mutually independent. Therefore, the probability of getting x H's and $n - x$ T's equals

$$P(T, H, \ldots, H, T) = P(T)P(H) \cdots P(H)P(T)$$
$$= (1 - p)p \cdots p(1 - p)$$
$$= p^x(1 - p)^{n-x}$$

The event $\{X = x\}$ contains all sample points ω that consist of x H's and $n - x$ T's, each of which is assigned the same probability $p^x(1-p)^{n-x}$. In addition, there are $\binom{n}{x}$ such sample points, since each one corresponds to a labeling of the n trials using x H's and $n - x$ T's. Summing the probabilities of the sample points in the event $\{X = x\}$ yields Eq. (3.32).

The probability function defined by Eq. (3.32) is called the *binomial distribution*; it is one of the three most important discrete distributions, the other two being the hypergeometric and the Poisson distributions. To see why it is called the binomial distribution, apply the binomial expansion [Eq. (2.3)] to the expression $(p + (1 - p))^n$, where $0 \le p \le 1$. We obtain the following expansion:

$$1 = 1^n = (p + (1 - p))^n = \sum_{0 \le x \le n} \binom{n}{x} p^x (1 - p)^{n-x} \tag{3.33}$$

This shows that

$$f_X(x) = \binom{n}{x} p^x (1 - p)^{n-x}, \quad x = 0, 1, \ldots, n$$

defines a pf in the sense of Eq. (3.4). Notice that the term that appears on the right-hand side of Eq. (3.32) also appears as the xth term (counting from 0) of the binomial expansion; this is the reason Eq. (3.32) is called the binomial distribution. The individual terms, denoted by $b(x; n, p)$ and defined by

$$b(x; n, p) = \binom{n}{x} p^x (1 - p)^{n-x}, \quad x = 0, 1, \ldots, n \tag{3.34}$$

are called *binomial probabilities*.

A Recurrence Relation for the Binomial Probabilities

We now derive a recurrence relation for the binomial probabilities that is useful for both computational and theoretical purposes.

$$b(x + 1; n, p) = b(x; n, p) \times \frac{(n - x)}{(x + 1)} \frac{p}{1 - p} \tag{3.35}$$

$$b(0; n, p) = (1 - p)^n \tag{3.36}$$

The recurrence relation is a consequence of the following formula for the ratio $R_b(x)$:

$$R_b(x) = \frac{b(x + 1; n, p)}{b(x; n, p)} = \frac{(n - x)}{(x + 1)} \frac{p}{1 - p} \quad \text{for } 0 \le x \le n \tag{3.37}$$

We denote the distribution function of the binomially distributed random variable X by $B(x; n, p)$; the formula for $B(x; n, p)$ is

$$B(x; n, p) = \sum_{0 \le j \le x} b(j; n, p) \tag{3.38}$$

Table A.1 in the appendix lists the values of $B(x; n, p)$ for various values of the parameters n, p. The values of the binomial probabilities can be computed from the formula

$$b(x; n, p) = B(x; n, p) - B(x - 1; n, p) \qquad (3.39)$$

How to Use the Binomial Tables

---

example 3.19   Use Table A.1 to compute (1) $B(6; 10, 0.4)$ and (2) $b(6; 10, 0.4)$.

Solution. Table A.1 gives the values of $B(x; n, p)$ for $n = 5, 10, 15, 20, 25$ and $p = 0.05, 0.1, 0.15, 0.2, \ldots, 0.45, 0.50$.

1. To find $B(6; 10, 0.4)$ first look for the number 10 in the first column, labeled n, and then look for the number 6 in the second column, labeled x. We then proceed to the right until we find the entry 0.945 in the column headed by 0.4. Thus $B(6; 10, 0.4) = 0.945$.
2. To compute $b(6; 10; 0.4)$ we use Eq. (3.39):

$$b(6; 10, 0.4) = B(6; 10, 0.4) - B(5; 10, 0.4) = 0.945 - 0.834 = 0.111 \qquad \blacksquare$$

To compute $B(x; n, p)$, $0.5 < p < 1$, we use the formula

$$B(x; n, p) = 1 - B(n - 1 - x; n, 1 - p) \qquad (3.40)$$

The effect of the parameter p on the shape of the binomial distribution can be understood by looking at the probability histograms shown in Figs. 3.3, 3.4, and 3.5.

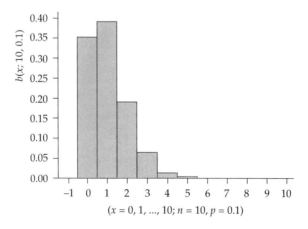

$(x = 0, 1, \ldots, 10; n = 10, p = 0.1)$

Figure 3.3 Probability histogram of binomial probability function ($x = 0, 1, \ldots, 10$; $n = 10$, $p = 0.1$)

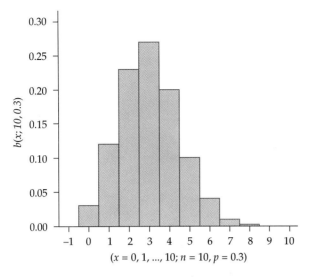

Figure 3.4 Probability histogram of binomial probability function ($x = 0, 1, \ldots, 10$; $n = 10$, $p = 0.3$)

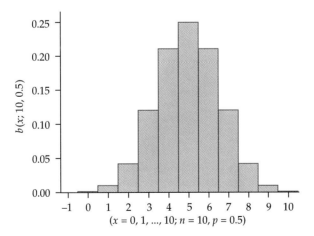

Figure 3.5 Probability histogram of binomial probability function ($x = 0, 1, \ldots, 10$; $n = 10$, $p = 0.5$)

The Mean and Variance of the Binomial Distribution

Proposition 3.4 gives the formulas for the mean and variance of the binomial distribution.

▨ PROPOSITION 3.4

For a random variable X having a binomial distribution with parameters $n, p,$

$$E(X) = np \tag{3.41}$$
$$V(X) = np(1 - p) \tag{3.42}$$

■

These formulas will be derived as corollaries of the addition formula for the mean and variance of sums of independent random variables (see Sec. 5.2.3).

---

example 3.20   Consider again the urn model of the previous section: An urn contains D beads colored red and $N - D$ beads colored white. This time we take a random sample of size n, *with replacement,* from the urn and let X denote the number of red beads in the sample. Show that X has a binomial distribution with parameters $n, p = D/N$.

Solution. The sample space consists of 2^n sequences of n letters, each letter an R or a W. A sample point is simply a sequence of R's and W's denoted by WWRW...R. Since we are sampling with replacement, the numbers of red and white beads remain constant at each draw, and therefore the probability of drawing a red bead at each trial equals $p = D/N$; similarly, the probability of drawing a white bead equals $1 - p = (N - D)/N$. To each sample point we assign the probability $p^x(1 - p)^{n-x}$, where x and $n - x$ equal the numbers of R's and W's, respectively, that appear in the sample. Reasoning just as we did in Proposition 3.3, we see that X has binomial distribution (3.32) with parameters n, p. ■

The Binomial Approximation to the Hypergeometric Distribution

Suppose we have an urn filled with 10,000 beads, of which 500 are red and the rest white. If we take a sample, without replacement, of size 10 from the urn, then the number of red beads in the sample is a random variable X with hypergeometric distribution with parameters

$$n = 10, \qquad N = 10,000, \qquad D = 500$$

Notice that each time a bead is drawn the proportion p' of red beads that remain satisfies the inequality $490/9990 = 0.049 \leq p' \leq 500/9990 = 0.0501$. The lower bound follows from the fact that the maximum number of red beads that can be drawn is 10, and the upper bound follows from the observation that the minimum number of red beads that can be drawn is 0. Consequently, the proportion of red beads that remain in the urn is close to $0.05 = 500/10,000 = D/N = p$. If we now take a sample of size 10 with replacement, then the number of red beads in the sample has a binomial distribution with parameters $n = 10$ and $p = D/N = 500/10{,}000 = 0.05$. This suggests that there is very little difference in this case between sampling without replacement and sampling with replacement. More precisely, we have the following theorem, called the *binomial approximation to the hypergeometric distribution:*

■ THEOREM 3.3

The random variable X has the hypergeometric distribution $h(x)$. Suppose N and D become large while their ratio remains fixed with $D/N = p$. Then

$$\lim_{D,N \to \infty, D/N = p} h(x) = \binom{n}{x} p^x (1-p)^{n-x}, \quad x = 0, 1, \ldots, n \qquad (3.43)$$

■

Proof. We defer the proof to Sec. 3.7 at the end of this chapter.

A Rule of Thumb for Approximating the Hypergeometric Distribution. It has been found that the binomial approximation is quite accurate when the sample size is less than 5 percent of the population size, that is, when the ratio $n/N < 0.05$.

---

example 3.21

Suppose $n = 10$, $N = 10{,}000$, $D = 500$; so $p = 500/10{,}000 = 0.05$. Compute the binomial approximations of (1) $h(0)$ and (2) $h(1)$, and compare them with the exact values.

Solution

1. The exact value of $h(0) = 0.5986$. The binomial approximation of $h(0)$ is $b(0; 10, 0.05)$, so

$$h(0) \approx (1 - 0.05)^{10} = 0.5987$$

2. The exact value of $h(1) = 0.3153$. The binomial approximation of $h(1)$ is $b(1; 10, 0.05)$, so

$$h(1) \approx 10 \times 0.05 \times (1 - 0.05)^9 = 0.3151 \qquad ■$$

The exact hypergeometric probabilities and their binomial approximations are displayed in Table 3.11.

TABLE 3.11

x	$b(x; 10, 0.05)$	$h(x)$	x	$b(x; 10, 0.05)$	$h(x)$
0	0.5987	0.5986	3	0.0105	0.0104
1	0.3151	0.3153	4	0.0009	0.0007
2	0.0746	0.0746	5	0.0001	0.0000

---

example 3.22

A sampling plan uses a sample of size $n = 10$ taken from a lot of size $N = 10{,}000$, acceptance number $c = 1$, and $D = 500$. Compute the probability of acceptance using (1) the hypergeometric distribution and (2) the binomial approximation to the hypergeometric distribution.

Solution

1. Using the hypergeometric distribution, the probability of acceptance is given by

$$P(\text{acceptance}) = h(0) + h(1)$$
$$= 0.5986 + 0.3153 = 0.9139$$

2. Using the binomial approximation to the hypergeometric distribution, the probability of acceptance is given by

$$P(\text{acceptance}) \approx b(0; 10, 0.05) + b(1; 10, 0.05)$$
$$= 0.5987 + 0.3151 = 0.9138 \qquad \blacksquare$$

3.5.1 Chebyshev's Inequality

The variance of a random variable controls the spread of its distribution about the expected value; in particular, a small variance implies that large deviations from the expected value are improbable. The precise version of this statement is called *Chebyshev's inequality*. This inequality is remarkable in that no knowledge of the pf $f(x)$ is required; it is necessary only to know μ and σ^2.

■ THEOREM 3.4

Let X be a random variable with expected value μ and variance σ^2. Then

$$P(|X - \mu| \geq d) \leq \frac{\sigma^2}{d^2} \qquad (3.44)$$
$$\blacksquare$$

Proof. The details of the proof appear in Sec. 3.7 at the end of this chapter.

---

example 3.23 Let X denote the number of heads that appear in $n = 1000$ tosses of a fair coin. Estimate the probability $P(|X - 500| \geq 50)$.

Solution. The expected number of heads and the variance of the number of heads are given by

$$\mu = E(X) = 1000 \times 0.5 = 500, \quad V(X) = 1000 \times \tfrac{1}{2} \times \tfrac{1}{2} = 250$$

Inserting these values and $d = 50$ into Eq. (3.44) yields

$$P(|X - 500| \geq 50) \leq \frac{250}{2500} = 0.10$$

The complement of the event is

$$(|X - 500| \geq 50)' = (|X - 500| < 50) \subset (|X - 500| \leq 50)$$

Therefore,

$$P(|X - 500| \leq 50) \geq 0.90 \qquad \blacksquare$$

Sometimes it is more illuminating to measure deviations from the mean μ in terms of units of the standard deviation. More precisely, set $d = k\sigma$ in

Eq. (3.44); the resulting inequality, which is also called Chebyshev's inequality, states that

$$P(|X - \mu| \geq k\sigma) \leq \frac{1}{k^2} \qquad (3.45)$$

For instance, the probability that the random variable X deviates from its mean μ by more than three units of standard deviation is less than $1/9$.

PROBLEMS

3.29 The random variable X has a hypergeometric distribution with parameters $n = 4$, $N = 20$, $D = 6$.
(a) Compute $h(x)$, $x = 0, 1, 2, 3, 4$, using Eq. (3.27).
(b) Compute $h(x)$, $x = 0, 1, 2, 3, 4$, using the recurrence formula (3.30). Which method is more efficient for computing the hypergeometric probabilities?

3.30 A lot of 10 components contains 2 that are defective. A random sample of size 4 is drawn. Let X denote the number of defective components drawn.
(a) Show that X has a hypergeometric distribution and identify the parameters N, D, n.
Use the formula from part (a) to compute the following probabilities:
(b) $P(X = 0)$ (c) $P(X = 1)$ (d) $P(X > 1)$

3.31 Suppose the sampling plan is to sample 10 percent of a lot of N items with the acceptance number $c = 0$. Compute the acceptance probability in each of the following cases:
(a) $N = 50$, $D = 1$ (b) $N = 100$, $D = 2$ (c) $N = 200$, $D = 4$

3.32 Suppose a sample of size 15 is taken from a lot of 50 items, and the lot is accepted if the number X of defective items in the sample is less than or equal to 1 (i.e., we accept the lot if $X \leq 1$). Compute P(acceptance) for the following values of D:
(a) $D = 1$ (b) $D = 3$ (c) $D = 6$

3.33 A group of 10 items contains 2 that are defective. Articles are taken one at a time from the lot and tested. Let Y denote the number of articles tested in order to find the first defective item. Compute the pf of Y.

3.34 An urn contains two nickels and three dimes. We select two coins at random, without replacement. Let $X =$ monetary value of the coins in the sample; e.g., if the sample consists of a nickel and dime, $X = 15$ cents.
(a) Compute the pf of X.   (b) Compute $E(X)$.   (c) Compute $V(X)$.

3.35 An urn contains two white balls and four red balls. Balls are drawn one by one without replacement until two white balls have appeared. Denote by U the number of balls that must be drawn until both white balls have appeared. Compute the pf of U.

3.36 Compute each of the following binomial probabilities in two ways: (1) by using Eq. (3.32) and (2) by using Eq. (3.39) together with Table A.1 in the appendix.

(a) $b(0; 20, 0.1)$, $b(1; 20, 0.1)$, $b(5; 20, 0.1)$

(b) $b(2; 10, 0.4)$, $b(4; 10, 0.4)$, $b(7; 10, 0.4)$

3.37 The random variable X has the binomial distribution $B(x; n, p)$. Compute the following probabilities using Table A.1.

(a) $P(X \le 4) = B(4; 10, p)$ for $p = 0.1, 0.4, 0.6, 0.8$

(b) $P(X > 9) = 1 - B(8; 20, p)$ for $p = 0.2, 0.5, 0.7, 0.9$

3.38 For $n = 20$ and $p = 0.5$:

(a) Estimate $P(|X - 10| \ge 5)$ using Chebyshev's inequality.

(b) Compute $P(|X - 10| \ge 5)$ using Table A.1.

3.39 Show that $b(x; n, p) = b(n - x; n, 1 - p)$ and give an intuitive explanation for this identity.

3.40 (a) Compute

$$b(x; 15, 0.3) \quad \text{for } x = 0, 1, 2, 3$$

directly using Eq. (3.32).

(b) Compute

$$b(x; 15, 0.3) \quad \text{for } x = 0, 1, 2, 3$$

using the recurrence formula (3.35).

3.41 A multiple-choice quiz has 10 questions each with four alternatives. A passing score is 6 or more correct. If a student attempts to guess the answer to each question, what is the probability that she passes?

3.42 Let Y denote the number of times a 1 appears in 720 throws of a die. Compute $E(Y)$ and $V(Y)$.

3.43 The lifetime (measured in hours) of a randomly selected battery is a random variable denoted by T. Assume that $P(T \le 3.5) = 0.10$. This means that 90 percent of the batteries have lifetimes exceeding 3.5 hours. A random sample of 20 batteries is selected for testing.

(a) Let $Y =$ the number of batteries in the sample with a lifetime less than or equal to 3.5 hours. Show that the pf of Y is binomial and identify the parameters n, p.

(b) Compute $E(Y)$ and $V(Y)$.

(c) Compute $P(Y > 4)$.

(d) Suppose 6 batteries (out of the 20 tested) had lifetimes that were less than or equal to 3.5 hours. Is this consistent with the assumption that $P(T \le 3.5) = 0.10$? Justify your conclusions by computing $P(Y \ge 6)$ and then interpreting this probability.

3.44 A sampling plan has $n = 20$ and the lot size $N = 1000$.

(a) Compute the acceptance probabilities for $D = 50$ and acceptance number $c = 2$. (*Hint:* Use the binomial approximation to the hypergeometric distribution.)

(b) Compute the acceptance probabilities for $D = 50$ and acceptance number $c = 1$.

(c) Compute the acceptance probabilities for $D = 50$ and acceptance number $c = 0$.

3.45 It is known that 40 percent of patients naturally recover from a certain disease. A drug is claimed to cure 80 percent of all patients with the disease. An experimental group of 20 patients are given the drug. What is the probability that at least 12 recover if:

(a) The drug is worthless and hence the true recovery rate 40 percent?

(b) The drug is really 80 percent effective?

3.46 (Continuation of Prob. 3.45) Suppose 200 patients are given the drug. Let R denote the number of patients who recover. Compute $E(R)$ and $V(R)$ for each of the following cases:

(a) The drug is worthless.

(b) The drug is really 80 percent effective as claimed.

3.47 Use Chebyshev's inequality to estimate the probability that the sample proportion obtained from a poll of voters with a sample size of $n = 2500$ differs from the true proportion by an amount less than or equal to 0.04.

3.48 Let X denote the number of heads in n tosses of a fair coin. Use Chebyshev's inequality to obtain a lower bound on the following probabilities:

(a) $P(225 \leq X \leq 275), n = 500$

(b) $P(80 \leq X \leq 120), n = 200$

(c) $P(45 \leq X \leq 55), n = 100$

3.49 R. Wolf (1882) threw a die 20,000 times and recorded the number of times each of the six faces appeared. The results follow.

Face	1	2	3	4	5	6
Frequency	3407	3631	3176	2916	3448	3422

Source: D. J. Hand et al., *Small Data Sets*, London, Chapman & Hall, 1994.

(a) Compute an upper bound on the probability that the number of times the number 5 appeared is greater than or equal to the observed value, 3448. Assume that the dice are fair, and use Chebyshev's inequality to obtain the upper bound.

(b) Compute an upper bound on the probability that the number of times the number 4 appeared is less than or equal to the observed value, 2916.

3.50 Can Chebyshev's inequality be improved? Consider the random variable X with the following probability function:

$$P(X = \pm a) = p, \qquad P(X = 0) = 1 - 2p, \quad 0 < p < 0.5$$

Show that

$$P(|X| \geq a) = \frac{V(X)}{a^2}$$

3.51 The probability that a disk drive lasts 6000 or more hours equals 0.8. In a batch of 20,000 disk drives, how many would be expected to last 6000 or more hours?

3.6   THE POISSON DISTRIBUTION

Table A.1 in the appendix lists the values of $B(x; n, p)$ for selected values of n and p in the ranges $n = 5, 10, 15, 20, 25$ and $p = 0.1, 0.2, \ldots, 0.5$. There are many instances, however, where one must compute $B(x; n, p)$ for values of $n \gg 25$ and $p \ll 0.1$. (The notation $n \gg 25$ means that n is very much larger than 25, and $p \ll 0.1$ means that p is very much smaller than 0.1.) As an example, we consider the problem of determining whether or not the occurrence of eight cases of leukemia among 7076 children less than 15 years of age in a small city is so much larger than the expected rate as to warrant further study by public health officials. (The following account has been adapted from R. J. Larsen and M. L. Marx, *An Introduction to Mathematical Statistics and its Applications*, Englewood Cliffs, N.J., Prentice-Hall, 1986.) An examination of public health records in the neighboring towns showed that in the previous 5 years there had been 286 diagnosed cases of leukemia out of 1,152,695 children. From this we estimate the probability that a child less than 15 years old will be stricken with leukemia to be $p = 286/1,152,695 = 0.000248$, a prevalence rate of 248 per 100,000. With this information we can compute the probability of the occurrence of 8 or more cases of leukemia in a population of 7076 children. In a population of size n, the number of occurrences of leukemia is a binomially distributed random variable X with parameters $n = 7076$, $p = 0.000248$. Now $P(X \geq 8) = 1 - P(X \leq 7)$; consequently,

$$P(X \geq 8) = 1 - \sum_{0 \leq j \leq 7} \binom{7076}{x}(0.000248)^x(0.999752)^{7076-x} \tag{3.46}$$

Although the summands on the right-hand side of the preceding equation are not to be found in any table, and are not easily computable, it is possible to *approximate* the probability

$$\binom{7076}{x}(0.000248)^x(0.999752)^{7076-x}$$

by means of a technique due to the French mathematician S. Poisson, called the *Poisson approximation to the binomial distribution*. The Poisson approximation may be used whenever n is very large, p is very small, and $np = E(X)$ is *moderate*. In the case $n = 7076$, $p = 0.000248$, we have $7076 \times 0.000248 = 1.75$. In other words, in the population of 7076 children less than 15 years old, we would expect to see 1.75 cases of leukemia; instead there were 8. Intuitively, this seems to be a very rare event and we would therefore like to compute its probability.

▧ THEOREM 3.5

The Poisson approximation to the binomial. Assume $np_n = \lambda > 0$ for $0 < n < \infty$; then

$$\lim_{n \to \infty} b(x; n, p_n) = p(x; \lambda) = e^{-\lambda} \frac{\lambda^x}{x!}, \quad x = 0, 1, \ldots, \quad \lambda > 0 \qquad (3.47)$$

■

We defer the proof of the Poisson approximation to the binomial to Sec. 3.7. In the meantime we give some applications.

example 3.24

1. Use the Poisson approximation to estimate the probability of eight or more leukemia cases in a population of size $n = 7076$ when the probability of getting leukemia is $p = 0.000248$.

    Solution. In this case

    $$np = 7076 \times 0.000248 = 1.75 = \lambda$$

    Therefore,

    $$P(X \le 7) \approx \sum_{0 \le x \le 7} e^{-1.75} \frac{1.75^x}{x!} = 0.999518$$

    Consequently,

    $$P(X \ge 8) \approx 1 - 0.999518 = 0.000482$$

    What do we conclude from the data? Is this a statistical fluke, or are there environmental factors responsible for this unusual cluster of leukemia cases? Clearly, there is cause for concern, and further scientific studies will be necessary to resolve the mystery.

2. Suppose 1.5 percent of the ball bearings made by a machine are defective, and the ball bearings are packed 200 to a box. Compute the pf of the number of defective ball bearings in the box and compare it to the Poisson approximation.

    Solution. Denoting the number of defective ball bearings by X, we see that X has a binomial distribution with parameters $n = 200$, $p = 0.015$, i.e.,

    $$P(X = x) = b(x; 200, 0.015)$$

    The Poisson approximation, with $\lambda = 200 \times 0.015 = 3$, yields

    $$P(X = x) \approx \frac{e^{-3} 3^x}{x!}, \quad x = 0, 1, \ldots, 100$$

    Table 3.12 gives an idea of the accuracy of the approximation.

    TABLE 3.12

x	$b(x; 200, 0.015)$	$p(x; 3)$
0	0.0487	0.0498
1	0.1482	0.1494
2	0.2246	0.2240
3	0.2257	0.2240

■

We now show that the quantities

$$p(x; \lambda) = e^{-\lambda}\frac{\lambda^x}{x!}, \quad x = 0, 1, \ldots$$

define a probability function, a result that is not too surprising since $p(x; \lambda)$ is itself the limit of a sequence of probability functions.

▦ DEFINITION 3.6

The random variable X is said to have a *Poisson distribution* with parameter $\lambda > 0$ if

$$P(X = x) = e^{-\lambda}\frac{\lambda^x}{x!}, \quad x = 0, 1, \ldots \tag{3.48}$$

∎

That $p(x; \lambda)$ defines a pf follows from the calculation

$$\sum_{0 \le x < \infty} p(x; \lambda) = \sum_{0 \le x < \infty} e^{-\lambda}\frac{\lambda^x}{x!} = e^{-\lambda}\sum_{0 \le x < \infty}\left(\frac{\lambda^x}{x!}\right) = e^{-\lambda}e^{\lambda} = 1$$

A Recurrence Relation for the Poisson Probabilities

We now derive a recurrence formula for the Poisson probabilities that has useful computational and theoretical applications.
Let $R_p(x) = p(x + 1; \lambda)/p(x; \lambda)$, $x = 0, 1, \ldots$; then

$$R_p(x) = \frac{p(x + 1; \lambda)}{p(x; \lambda)} = \frac{\lambda}{x + 1}, \quad x = 0, 1, \ldots \tag{3.49}$$

We leave the derivation of Eq. (3.49) as Prob. 3.53. From Eq. (3.49) we obtain the following recurrence relation for the Poisson probabilities:

$$p(x + 1; \lambda) = \frac{\lambda}{x + 1}p(x; \lambda), \quad x = 0, 1, \ldots \tag{3.50}$$

The Expected Value and Variance of the Poisson Distribution

The expected value and variance of the random variable X with Poisson distribution (3.47) are given by

$$E(X) = \lambda, \quad V(X) = \lambda \tag{3.51}$$

We derive these expressions for the mean and variance from the recurrence relation (3.50). Specifically, the recurrence formula (3.50) implies that $(x + 1)$ $p(x + 1; \lambda) = \lambda p(x; \lambda)$, so

$$\sum_{0 \le x < \infty} (x + 1)p(x + 1; \lambda) = \sum_{0 \le x < \infty} \lambda p(x; \lambda)$$

$$= \lambda \sum_{0 \le x < \infty} p(x; \lambda)$$

$$= \lambda$$

On the other hand,

$$\sum_{0 \le x < \infty} (x + 1)p(x + 1; \lambda) = \sum_{0 \le x < \infty} xp(x; \lambda)$$

$$= E(X)$$

Therefore, $E(X) = \lambda$.

To prove that $V(X) = \lambda$, it suffices to show that $E(X^2) = \lambda^2 + \lambda$. We multiply both sides of the recurrence relation by $(x + 1)^2$ and obtain

$$(x + 1)^2 p(x + 1; \lambda) = \lambda(x + 1)p(x; \lambda)$$

Summing both sides of this equation with respect to x yields

$$\sum_{0 \le x < \infty} (x + 1)^2 p(x + 1; \lambda) = \sum_{0 \le x < \infty} \lambda(x + 1)p(x; \lambda)$$

$$= \lambda \sum_{0 \le x < \infty} xp(x; \lambda) + \lambda \sum_{0 \le x < \infty} p(x; \lambda)$$

$$= \lambda E(X) + \lambda$$

$$= \lambda^2 + \lambda$$

On the other hand,

$$\sum_{0 \le x < \infty} (x + 1)^2 p(x + 1; \lambda) = \sum_{0 \le x < \infty} x^2 p(x; \lambda)$$

$$= E(X^2)$$

Therefore, $E(X^2) = \lambda^2 + \lambda$.

How to Use the Poisson Tables. The (cumulative) Poisson probability distribution is defined by

$$P(x; \lambda) = \sum_{0 \le j \le x} p(j; \lambda)$$

Table A.2 in the appendix lists values of $P(x; \lambda)$ for selected values of λ and x. We compute the Poisson probabilities $p(x; \lambda)$ from the equation

$$p(x; \lambda) = P(x; \lambda) - P(x - 1; \lambda)$$

PROBLEMS

3.52 The random variable X has a binomial distribution with parameters $n = 20$ and $p = 0.01$. In each of the following cases compute (i) the exact value of $P(X = x)$ and (ii) its Poisson approximation.
(a) $x = 0$ (b) $x = 1$ (c) $x = 2$

3.53 Verify Eq. (3.49).

3.54 (a) Compute $p(x; 2)$, $x = 0, 1, 2, 3$, using Eq. (3.47).
(b) Compute $p(x; 2)$, $x = 0, 1, 2, 3$, using the recurrence relation (3.50).

3.55 (a) Compute the Poisson probabilities $p(x; 1.5)$, $x = 0, 1, 2, 3$, using Eq. (3.47).

(b) Compute the Poisson probabilities $p(x; 1.5)$, $x = 0, 1, 2, 3$, using the formula

$$p(x; \lambda) = P(x; \lambda) - P(x - 1; \lambda)$$

where the values of the distribution function

$$P(x; \lambda) = \sum_{0 \leq y \leq x} p(y; \lambda)$$

are provided in Table A.2 in the appendix.

3.56 A lot of 10,000 articles has 150 defective items. A random sample of 100 articles is selected. Let X denote the number of defective items in the sample. Give the formula for:
(a) The *exact* pf $f_X(x)$ of X.
(b) The binomial approximation of $f_X(x)$.
(c) The Poisson approximation of the binomial distribution of part (b).

3.57 If 99 percent of the students entering a junior high school are vaccinated against the measles, what is the probability that a group of 50 students includes:
(a) No unvaccinated students?
(b) Exactly one unvaccinated student?
(c) Two or more unvaccinated students?

3.58 A batch of dough yields 500 cookies. How many chocolate chips should be mixed into the dough so that the probability of a cookie having no chocolate chips is ≤ 0.01?

3.59 A taxi company has a limousine with a seating capacity of N. The cost of each seat is $5, and the price per seat is $10. The number of reservations X is Poisson distributed with parameter $\lambda = 5$, i.e., $P(X = x) = e^{-5}5^x/x!$, $x = 0, 1, \ldots$. Compute the expected net revenue when:
(a) $N = 4$
(b) $N = 5$
(c) Which is more profitable to operate, a four- or five-seat limousine?

3.60 A sampling plan has $n = 100$, $c = 2$, and the lot size $N = 10,000$. Compute the approximate acceptance probabilities for the following values of D:
(a) $D = 50$ **(b)** $D = 100$ **(c)** $D = 200$
(*Hint:* Use the Poisson approximation to the binomial.)

3.7 MATHEMATICAL DETAILS AND DERIVATIONS

Derivation of the Binomial Approximation to the Hypergeometric Distribution (Theorem 3.3)

Our proof uses the fact that the ratio $R_h(x)$ of the hypergeometric probabilities [Eq. (3.31)] converges to the the corresponding ratio $R_b(x)$ [Eq. (3.37)] for the binomial probabilities. In detail,

$$\lim_{D,N \to \infty, D/N = p} R_h(x) = \lim_{D,N \to \infty, D/N = p} \frac{(n-x)(D-x)}{(x+1)(N-D-n+x+1)}$$

$$= \frac{(n-x)}{(x+1)} \frac{p}{1-p}$$

$$= R_b(x)$$

The last step in the preceding sequence of calculations is justified by noting that

$$\lim_{D,N \to \infty, D/N = p} \frac{(D-x)}{(N-D-n+x+1)} = \frac{p}{1-p}$$

which follows by dividing the numerator and denominator by N, using the fact that $p = D/N$, and noting that both x/N and n/N go to zero as N goes to infinity.

The proof will be completed by showing that

$$\lim_{D,N \to \infty, D/N = p} h(0) = (1-p)^n = b(0; n, p) \tag{3.52}$$

Proof of Eq. (3.52)

$$\lim_{D,N \to \infty, D/N = p} h(0) = \lim_{D,N \to \infty, D/N = p} \frac{(N-D)(N-D-1)\cdots(N-D-n+1)}{N(N-1)\cdots(N-n+1)}$$

$$= \lim_{D,N \to \infty, D/N = p} \frac{(1-p)(1-p-1/N)\cdots[1-p-(n-1)/N]}{1(1-1/N)\cdots[1-(n-1)/N]}$$

$$= (1-p)^n \tag{3.53}$$

The next to last line in the preceding sequence of equations is obtained by dividing the numerator and denominator by N; the limit follows from the fact that all terms of the form j/N go to zero as N tends to infinity. This completes the derivation of Eq. (3.52).

For instance, to show that

$$\lim_{D,N \to \infty, D/N = p} h(1) = b(1; n, p)$$

we reason as follows:

$$\lim_{D,N \to \infty, D/N = p} h(1) = \lim_{D,N \to \infty, D/N = p} R_h(0)h(0)$$

$$= R_b(0)b(0; n, p) = \frac{np}{1-p}(1-p)^n$$

$$= np(1-p)^{n-1} = b(1; n, p)$$

Derivation of Chebyshev's Inequality

Chebyshev's inequality is a consequence of the observation that the sum of a set of nonnegative numbers is greater than the sum taken over a subset. In detail,

$$\sigma^2 = \sum_x (x - \mu)^2 f(x)$$

$$\geq \sum_{(x:|x-\mu|\geq d)} (x - \mu)^2 f(x)$$

$$\geq \sum_{(x:|x-\mu|\geq d)} d^2 f(x)$$

$$= d^2 \sum_{(x:|x-\mu|\geq d)} f(x) = d^2 P(|X - \mu| \geq d)$$

We have thus shown that $d^2 P(|X - \mu| \geq d) \leq \sigma^2$. Dividing both sides of this inequality by d^2 yields Chebyshev's inequality (3.44).

Derivation of the Poisson Approximation to the Binomial

The derivation of the Poisson approximation to the binomial is similar to that used to obtain the binomial approximation to the hypergeometric. That is, we begin with a recurrence relation for the Poisson probabilities and then show that it is the limit of the recurrence relation for the binomial. In detail, $p_n = \lambda/n$ implies that $\lim_{n\to\infty} p_n = 0$, and therefore,

$$\lim_{n\to\infty} R_b(x) = \lim_{n\to\infty} \frac{(n - x)p_n}{(x + 1)(1 - p_n)}$$

$$= \lim_{n\to\infty} \frac{np_n - xp_n}{(x + 1)(1 - p_n)}$$

$$= \frac{\lambda}{x + 1} = R_p(x), \ x = 0, 1, \ldots$$

The next step is to use the following limit, which can be found in any standard calculus text:

$$\lim_{n\to\infty} \left(1 - \frac{\lambda}{n}\right)^n = e^{-\lambda} \tag{3.54}$$

Setting $p_n = \lambda/n$ in Eq. (3.54) and noting that $b(0; n, p_n) = (1 - p_n)^n$, we obtain the Poisson approximation for $b(0; n, p_n)$:

$$\lim_{n\to\infty, np_n=\lambda} b(0; n, p_n) = e^{-\lambda} = p(0; \lambda)$$

To obtain the Poisson approximation for $x = 1$, for example, we proceed as follows:

$$\lim_{n\to\infty, np_n=\lambda} b(1; n, p_n) = \lim_{n\to\infty, np_n=\lambda} R_b(0)b(0; n, p_n)$$

$$= R_p(0)p(0; \lambda) = \lambda e^{-\lambda} = p(1; \lambda)$$

A similar argument is used to derive the Poisson approximation for all integers x. This completes the proof of Eq. (3.47).

3.8 CHAPTER SUMMARY

In this chapter we studied the fundamental concept of the distribution of a random variable X defined on a sample space. It is analogous to the concept of a variable defined on a population S studied in Chap. 1. Similarly, the expected value and variance of X are analogues of the sample mean and sample variance. Random variables are classified according to their distribution functions, among the most important of which are the hypergeometric, binomial, and Poisson.

To Probe Further. The pioneering contributions of the mathematicians J. Bernoulli (1654–1705), A. De Moivre (1667–1754), P. S. Laplace (1749–1827), and S. D. Poisson (1781–1840) to the theory of probability are described in S. M. Stigler, *History of Statistics: The Measurement of Uncertainty before 1900*, Cambridge, Mass., Belknap Press of Harvard University Press, 1986.

4 CHAPTER

Continuous Random Variables and Their Distribution Functions

The principal element of the analysis of probabilities is an exponential integral that has presented itself in several very different mathematical theories.... This same function is connected with general physics.... We have discovered in recent years that it also represents the diffusion of heat in the interior of solid substances. Finally, it determines the probability of errors and mean results of numerous observations; it reappears in the questions of insurance and in all difficult applications of the science of probabilities.

J. B. J. Fourier (1768–1830), French mathematician

4.1 ORIENTATION

In this chapter we study random variables whose range is not a discrete subset but is a subinterval of the real numbers. To compute probabilities, the distribution function, expected values, moments, etc., we use the concepts and methods of calculus in an essential way. We then turn our attention to the study of the *normal distribution,* considered to be the most important probability distribution of all.

Organization of Chapter

4.2 DEFINITION AND EXAMPLES OF CONTINUOUS RANDOM VARIABLES

Random variables whose range is an interval of real numbers arise naturally in a variety of applications. Consider, for example, the lifetime X of an electronic component. Theoretically, the component can function forever, or it can be defective and not function at all. The range of X in this case is the set of nonnegative real numbers, denoted by $R_+ = [0, \infty)$.

Informally, we say that a random variable X is *continuous* if its df $F_X(x)$ is a continuous function defined by a definite integral in the sense of the following definition. This is in sharp contrast to the discrete random variables studied in Chap. 3, whose df's are discontinuous functions, as shown, for example, in Fig. 3.2.

■ DEFINITION 4.1

A random variable X is said to be *continuous* if its df is a continuous function of the following type:

$$F_X(x) = \int_{-\infty}^{x} f_X(t)\, dt \qquad (4.1)$$

where

$$f_X(t) \geq 0 \qquad (4.2)$$

and

$$\int_{-\infty}^{\infty} f_X(t)\, dt = 1 \qquad (4.3)$$

■

The function $f_X(x)$ is called the *probability density function* (pdf) of the random variable X. Figure 4.1 displays the graph of a pdf $f_X(x)$. Its distribution

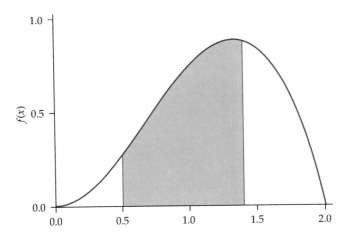

Figure 4.1 The area of the shaded part equals the probability that X lies between 0.5 and 1.4

function has a simple geometric interpretation: It is the area under the curve $y = f_X(t)$ from $-\infty$ to x. Similarly, $P(a \le X \le b)$ is the area under the curve $y = f_X(x)$ from a to b. When there is little or no danger of confusion, we drop the subscript X and simply write $F(x)$ and $f(x)$.

We can compute the probability $P(a \le X \le b)$ in terms of the distribution function as follows:

$$P(a \le X \le b) = F_X(b) - F_X(a)$$

One consequence of our definition is that

$$P(X = a) = \int_a^a f(x)\,dx = 0$$

This is because the area under the curve from a to a is zero. From this it follows that when X is a continuous random variable, the following probabilities are equal:

$$
\begin{aligned}
P(a < X \le b) &= P(a < X < b) \\
&= P(a \le X < b) \\
&= P(a \le X \le b) \\
&= F_X(b) - F_X(a)
\end{aligned}
$$

Using the fundamental theorem of calculus, we can derive the following important relation between the df F_X and the pdf $f(x)$:

$$F_X'(x) = f(x) \quad \text{for all } x \text{ at which } f \text{ is continuous} \tag{4.4}$$

It is useful to restate the equation $F_X'(x) = f(x)$ in differential form as follows:

$$
\begin{aligned}
P(x < X \le x + \Delta x) &= F_X(x + \Delta x) - F_X(x) \\
&\approx F_X'(x)\Delta x = f(x)\Delta x
\end{aligned}
\tag{4.5}
$$

The Exponential Distribution

The exponential distribution is widely used in the engineering sciences to model a variety of quantities, such as the length of time X (in microseconds) required to run a program on an interactive computer system or the lifetime of a bearing. Its pdf is given by

$$f(x) = \theta e^{-\theta x}, \quad 0 < x < \infty, \quad \theta > 0 \tag{4.6}$$

$$f(x) = 0 \quad \text{elsewhere} \tag{4.7}$$

We will see later that the expected value of a random variable with an exponential distribution equals $1/\theta$. We obtain the distribution function by integrating the probability density function. Consequently,

$$F(x) = 0, \quad x \leq 0$$

$$F(x) = \int_0^x \theta\, e^{-\theta t}\, dt = 1 - e^{-\theta x}, \quad 0 \leq x < \infty$$

The graphs of the exponential pdf and its df, for $\theta = 2$, are displayed in Fig. 4.2.

---

example 4.1 Let X have an exponential distribution with $\theta = 4$. Compute (1) $P(0.1 \leq X \leq 0.3)$ and (2) $P(X \geq 0.5)$.

Solution

$$P(0.1 \leq X \leq 0.3) = e^{-4 \times 0.1} - e^{-4 \times 0.3}$$

$$= 0.3691$$

$$P(X \geq 0.5) = e^{-4 \times 0.5}$$

$$= e^{-2} = 0.1353 \qquad \blacksquare$$

The Uniform Distribution

We say that the random variable U is *uniformly distributed* on the interval $[0, 1]$ if its pdf is given by

$$f(x) = 1, \quad 0 < x < 1 \tag{4.8}$$

$$f(x) = 0 \quad \text{elsewhere} \tag{4.9}$$

The graph of $f(x)$ is displayed in Fig. 4.3.

Notice that $f(x)$ is another example of a function that is defined *piecewise;* that is, the formula that produces the output $f(x)$ depends on the interval where the input x is located. Consequently, its df $F(x)$ is also defined piecewise.

$$F(x) = 0, \quad x \leq 0$$

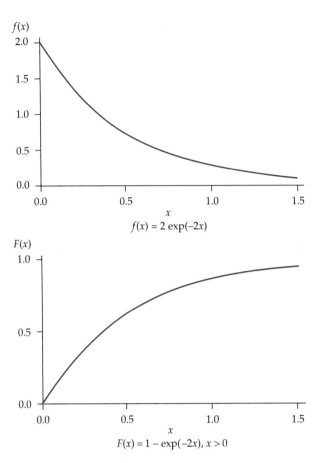

Figure 4.2 Graphs of the probability density function $f(x) = 2e^{-2x}$ and its distribution function

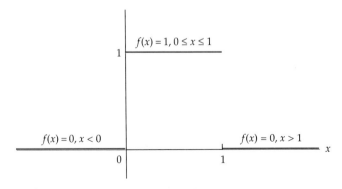

Figure 4.3 Graph of the pdf of the uniform distribution

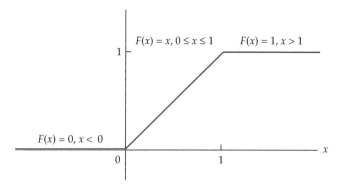

Figure 4.4 Graph of the df of the uniform distribution

$$F(x) = \int_0^x 1\,dt = x, \quad 0 \le x \le 1$$

$$F(x) = 1, \quad x \ge 1$$

The graph of $F(x)$ is displayed in Fig. 4.4.

The random variable U is the mathematical model of a *random number generator*, which is a computer program that produces a random number lying in the interval $[0, 1]$. In this instance randomness means that the number produced lies in the interval $[\alpha, \beta]$ with probability $\beta - \alpha$, where $0 \le \alpha < \beta \le 1$. The uniform distribution is widely used as a tool to simulate the df's of other random variables; see Sec. 4.6 for additional details.

---

example 4.2   Let X be uniformly distributed on $[0, 1]$. Compute (1) $P(0.25 \le X \le 0.6)$ and (2) $P(X \ge 0.7)$.

Solution

$$P(0.25 \le X \le 0.6) = 0.6 - 0.25 = 0.35$$
$$P(X \ge 0.7) = 1 - 0.7 = 0.3 \qquad\blacksquare$$

The Uniform Distribution on the Interval $[a, b]$

A random variable X is uniformly distributed over the interval $[a, b]$ if the probability that X lies in a subinterval $[c, d] \subset [a, b]$ is proportional to the length of the subinterval. This means that its pdf is given by

$$f(x) = \frac{1}{b-a}, \quad a < x < b$$

$$f(x) = 0 \quad \text{elsewhere}$$

The df of a random variable uniformly distributed over the interval $[a, b]$ is given by

$$F(x) = 0, \quad x \le a$$

$$F(x) = \int_a^x \frac{1}{b-a}\, dt = \frac{x-a}{b-a}, \quad a \le x \le b$$

$$F(x) = 1, \quad x \ge b$$

A random variable X that is uniformly distributed on $[0, L]$ is the mathematical model for the experiment of choosing a random point on an interval of length L.

The uniform distribution can also be used as a mathematical model for choosing a random direction in the plane, as will now be explained. A line through the origin is specified by the angle $\theta, 0 \le \theta < 2\pi$ (measured in radians), that it makes with the positive x axis. Let Θ denote a random variable that is uniformly distributed over $[0, 2\pi)$. We say that $(\cos \Theta, \sin \Theta)$ is a *random point on the unit circle*; the line connecting $(0, 0)$ to $(\cos \Theta, \sin \Theta)$ defines a random direction, or random vector of unit length, in the plane.

---

example 4.3   In this example we discuss a common error that occurs when students evaluate indefinite integrals instead of definite integrals. Consider the pdf defined by

$$f(x) = 2(1 - x), \quad 0 < x < 1$$
$$f(x) = 0 \quad \text{elsewhere}$$

$$(4.10)$$

Using the formula $F'(x) = f(x)$, students compute $F(x)$ via the recipe

$$F(x) = \int 2(1 - x)\, dx = 2x - x^2$$

Unfortunately, this answer cannot be correct because $F(3) = -3$, and this is absurd since $0 \le F(x) \le 1$. The error, which is quite common, is to forget that $F_X(x)$ is obtained by computing a *definite integral*. In addition, one should not forget that the function $f(x)$ is defined piecewise. The correct calculation is

$$F(x) = 0, \quad x \le 0$$

$$F(x) = \int_0^x 2(1 - t)\, dt = 2x - x^2, \quad 0 \le x \le 1$$

$$F(x) = 1, \quad 1 \le x < \infty$$ ∎

Medians and Percentiles of a Continuous df

A median of a continuous df $F(x)$ is defined to be a solution to the equation $F(x) = 0.5$. Similarly, the $100p$th percentile of $F(x)$ is defined to be a solution to the equation $F(x) = p$. Although in all the examples considered in this text the solution is unique, this is not so in general. Little or no harm is done, however, by ignoring these possibilities. Since $F(x)$ is continuous and nondecreasing on every interval $[a, b]$, it follows from the *intermediate value theorem* of calculus that, given any p satisfying $0 < p < 1$, there exists at least one solution to the equation

$$F(x) = p, \quad 0 < p < 1$$

The solution is denoted by η_p.

---

example 4.4 Compute the median of (1) the exponential distribution and (2) the df defined by Eq. (4.10).

Solution

1. The median of the exponential distribution satisfies the equation

$$1 - e^{-\theta \eta_{0.5}} = 0.5 \quad \text{or} \quad \eta_{0.5} = -\frac{\ln 0.5}{\theta} = \frac{\ln 2}{\theta}$$

2. The median satisfies the quadratic equation

$$F(x) = 2x - x^2 = 0.5$$

which has two solutions, $x = 1 \pm \sqrt{0.5}$, only one of which is the median. The solution $x = 1 + \sqrt{0.5} = 1.7071$ cannot possibly be the median since $F(1.7071) = 1$. Consequently, $\eta_{0.5} = 1 - \sqrt{0.5} = 0.293$.   ■

PROBLEMS

4.1 The random variable Y has the exponential distribution

$$F_Y(y) = 1 - \exp(-0.75y), \quad 0 \le y < \infty$$
$$F_Y(y) = 0, \quad y \le 0$$

(a) Compute $P(1 < Y < 2)$.
(b) Compute $P(Y > 3)$.
(c) Compute the 95th percentile, $\eta_{0.95}$.

4.2 The random variable X has the pdf given by

$$f(x) = 2x, \quad 0 < x < 1$$
$$f(x) = 0 \quad \text{elsewhere}$$

(a) Sketch the graph of $f(x)$ for all values of x.
(b) Compute the df $F(x)$ for all values of x and sketch its graph.
(c) Shade the area corresponding to $P(0.25 < X < 0.5)$ and then compute its value via the formula $F(0.5) - F(0.25)$.
(d) Compute the median and 90*th* percentile of the distribution.

4.3 Let Y denote a random variable with continuous df $F(y)$ defined by

$$F(y) = 0, \quad y \le 0$$
$$F(y) = y^3/8, \quad 0 < y \le 2$$
$$F(y) = 1, \quad y > 2$$

(a) Plot $F(y)$ for all y.

(b) Compute $P(0.2 < Y < 0.5)$ and $P(Y > 0.6)$.

(c) Compute the pdf $f(y)$ for all y and sketch its graph.

4.4 The distribution of the lifetime (in hours) of a battery for a laptop computer is modeled as an exponential random variable with parameter $\theta = 1/3.5$. Compute the safe life of the battery, which is defined to be the 10th percentile of the distribution.

4.5 The percentage of alcohol in a certain compound is given by $100X$, where X is a random variable with pdf given by

$$f(x) = k(1 - x)^2, \quad 0 < x < 1$$
$$f(x) = 0 \quad \text{elsewhere}$$

(a) Find the value of k.

(b) Sketch the graph of $f(x)$ for all values of x.

(c) Compute the df $F(x)$ for all values of x.

(d) Sketch the graph of $F(x)$.

(e) What is the probability that the percentage of alcohol is less than 25 percent? More than 75 percent?

(f) Compute the median of the distribution.

4.6 An archer shoots an arrow at a circular target. Suppose the distance between the center and the impact point of the arrow is a continuous random variable X (in meters) with pdf given by

$$f(x) = 6x(1 - x), \quad 0 < x < 1$$
$$f(x) = 0 \quad \text{otherwise}$$

(a) Sketch the graph of $f(x)$.

(b) Compute the df $F(x)$ for all values of x and sketch its graph.

(c) The bull's-eye of the target is a circle of radius 10 cm. What is the probability that the archer hits the bull's-eye?

4.7 The random variable X has pdf given by

$$f(x) = (1 + x), \quad -1 \le x \le 0$$
$$f(x) = (1 - x), \quad 0 \le x \le 1$$
$$f(x) = 0 \quad \text{otherwise}$$

(a) Sketch the graph of $f(x)$.

(b) Compute the df $F(x)$ for all values of x and sketch its graph.

(c) Compute $P(-0.4 < X < 0.6)$.

(d) Compute the median of the distribution.

(e) Compute the 10th and 90th percentiles of the distribution.

4.8 The lifetime T (in hours) of a component has a df given by

$$F(t) = 0, \quad t \le 1$$

$$F(t) = 1 - \frac{1}{t^2}, \quad t \ge 1$$

(a) Sketch the graph of $F(t)$.
(b) Compute the pdf $f(t)$ of T and sketch its graph.
(c) Find $P(T > 100)$.
(d) Compute $P(T > 150 \,|\, T > 100)$; i.e., given that the component is still functioning after 100 hours, determine the probability that it will continue to function for another 50 hours.

4.9 The tensile strength S (measured in grams per square centimeter) of a fiber thread has a pdf given by

$$f(x) = kx^2(2 - x), \quad 0 < x < 2$$
$$f(x) = 0 \quad \text{elsewhere}$$

(a) Find the value of k.
(b) Compute $P(0.5 < S < 1.5)$.
(c) Compute $P(S > 1)$.

4.10 Suppose the lifetime of a fan on a diesel engine has an exponential distribution with $1/\theta = 28{,}700$ hours. Compute the probability of a fan failing on:
(a) a 5,000-hour warranty.
(b) an 8,000-hour warranty.

4.11 Suppose the number of days to failure of a diesel locomotive has an exponential distribution with parameter $1/\theta = 43.3$ days. Compute the probability that the locomotive operates one entire day without failure.

4.12 Let X denote a random variable with continuous df $F(x)$ defined as follows:

$$F(x) = 0, \quad x \le -1$$
$$F(x) = (x^3/2) + 1/2, \quad -1 \le x \le 1$$
$$F(x) = 1, \quad x \ge 1$$

(a) Plot $F(x)$ for all x.
(b) Compute the pdf $f(x)$ for all x and sketch its graph.
(c) Compute the 10th and 90th percentiles of the distribution.
(d) Compute $P(X > 0.4)$.

4.13 Which of the following functions $F(x)$ are continuous df's of the form in Definition 4.1? Calculate the corresponding pdf $f(x)$ where appropriate.
(a) $F(x) = 0, \quad x \le 0$

$$F(x) = \frac{x}{1 + x}, \quad x > 0$$

(b) $F(x) = 0, \quad x \le 2$

$$F(x) = 1 - \frac{4}{x^2}, \quad x > 2$$

(c) $F(x) = 0, \quad x < -\pi/2$
$$F(x) = \sin x, \quad -\pi/2 < x < \pi/2$$
$$F(x) = 1, \quad x \ge \pi/2$$

4.3   EXPECTED VALUE, MOMENTS, AND VARIANCE OF A CONTINUOUS RANDOM VARIABLE

The definitions of $E(X)$, $E(\phi(X))$, $E(X^k)$, and variance for continuous random variables are similar to those already given for discrete random variables, except that sums are replaced by integrals.

■ **DEFINITION 4.2**

Let X be a continuous random variable with pdf $f(x)$. Then $E(X)$ and $E(\phi(X))$ are defined by the equations

$$E(X) = \mu_X = \int_{-\infty}^{\infty} xf(x)\,dx \tag{4.11}$$

provided that

$$\int_{-\infty}^{\infty} |x|f(x)\,dx < \infty \tag{4.12}$$

$$E\big(\phi(X)\big) = \int_{-\infty}^{\infty} \phi(x)f(x)\,dx \tag{4.13}$$

provided that

$$\int_{-\infty}^{\infty} |\phi(x)|f(x)\,dx < \infty \tag{4.14}$$

■

The kth moment μ_k, $k = 1, 2, \ldots$, of a continuous random variable X is defined by the equation

$$\mu_k = E(X^k), \quad \text{where } k = 1, 2, \ldots$$

The kth moment is computed by evaluating the integral

$$E(X^k) = \int_{-\infty}^{\infty} x^k f(x)\,dx \tag{4.15}$$

The variance of a random variable X is defined by

$$V(X) = E((X - \mu_X)^2) = \int_{-\infty}^{\infty} (x - \mu_X)^2 f(x)\,dx \tag{4.16}$$

provided that

$$\int_{-\infty}^{\infty} (x - \mu_X)^2 f(x)\,dx < \infty$$

Notation. The standard deviation is also denoted by

$$\sigma(X) = \sqrt{V(X)}$$

The shortcut formula (3.23) for computing $V(X)$ remains valid:

$$V(X) = E(X^2) - \mu_X^2$$

▬▬ **REMARK 4.1**

The interpretation of the mean and variance as the center of mass and spread of the distribution remains valid in the continuous case.   ■

---

example **4.5**

Compute the first two moments and the variance for the following random variables.

1. The pdf of X is given by

$$f(x) = 2(1 - x), \quad 0 < x < 1$$
$$f(x) = 0 \quad \text{elsewhere}$$

Solution. $E(X)$, $E(X^2)$, and $V(X)$ are computed by evaluating the following definite integrals and using the shortcut formula for computing the variance.

$$\mu_X = \int_{-\infty}^{\infty} x f(x)\, dx = \int_0^1 x 2(1 - x)\, dx = \frac{1}{3}$$

$$E(X^2) = \int_{-\infty}^{\infty} x^2 f(x)\, dx = \int_0^1 x^2 2(1 - x)\, dx = \frac{1}{6}$$

$$V(X) = \frac{1}{6} - \left(\frac{1}{3}\right)^2 = \frac{1}{18}$$

2. X has the exponential distribution [Eq. (4.6)].

Solution. Using an appropriate integration by parts, one can show that

$$\mu_X = \int_{-\infty}^{\infty} x f(x)\, dx = \int_0^{\infty} x \theta e^{-\theta x}\, dx = \frac{1}{\theta} \tag{4.17}$$

$$E(X^2) = \int_{-\infty}^{\infty} x^2 f(x)\, dx = \int_0^{\infty} x^2 \theta e^{-\theta x}\, dx = \frac{2}{\theta^2} \tag{4.18}$$

$$V(X) = \frac{1}{\theta^2} \tag{4.19}$$

■

---

example **4.6**

Suppose the lifetime T of an electronic component has pdf $f(t)$. The manufacturing cost and selling price of each component are C_1 and C_2 dollars, respectively. The component is guaranteed for K hours; if $T \leq K$, then the manufacturer replaces the faulty component without charge. What is the manufacturer's expected profit per component?

Solution. The first step is to express the profit R as a function of the lifetime T:

$$R = C_2 - C_1 \quad \text{if } T > K \tag{4.20}$$
$$R = C_2 - 2C_1 \quad \text{if } T \leq K \tag{4.21}$$

Equation (4.20) asserts that if the component's lifetime exceeds the warranty period, then the net profit is $C_2 - C_1$. Equation (4.21) asserts that if the component fails during the warranty, then the manufacturer's profit is decreased by the cost of producing the defective component plus the cost of replacement, which equals $2C_1$. Thus R is a *discrete* random variable whose expectation is easily computed by the methods of Sec. 3.3. This yields the following formula for $E(R)$:

$$E(R) = (C_2 - 2C_1) + C_1 P(T > K) \qquad (4.22)$$

∎

---

example 4.7 Suppose the lifetime T of the component in the preceding example is exponentially distributed with $\theta^{-1} = 1000$, which implies that $E(T) = 1000$ hours. The manufacturing cost and selling price of each component are \$6 and \$10, respectively. If the item fails before 500 hours, the manufacturer replaces the faulty component without charge. Compute the expected profit per customer.

Solution. The assumption that T has an exponential distribution means that

$$P(T > K) = e^{-K/1000}$$

The length of the warranty period is 500 hours, so $K = 500$; the manufacturing cost and selling price are $C_1 = \$6$, $C_2 = \$10$. Inserting these values into Eq. (4.22) yields

$$E(R) = (10 - 12) + 6e^{-500/1000}$$

$$= \$1.64$$

∎

Chebyshev's Inequality

Chebyshev's inequality (3.44), which we derived in the previous chapter, remains valid for continuous random variables.

■■ **PROPOSITION 4.1**

Let X be a random variable with expected value μ and variance σ^2. Then

$$P(|X - \mu| \geq d) \leq \frac{\sigma^2}{d^2} \qquad (4.23)$$

Equivalently,

$$P(|X - \mu| \leq d) \geq 1 - \frac{\sigma^2}{d^2} \qquad (4.24)$$

∎

We omit the proof, which is similar to the one already given except that sums are replaced by integrals.

PROBLEMS

4.14 Compute $E(X)$, $E(X^2)$, and $V(X)$ for the random variable X with pdf given by:

(a) $f(x) = \frac{1}{2} - \frac{x}{4}$, $-1 \leq x \leq 1$; $f(x) = 0$ elsewhere.

(b) $f(x) = (1/2) \sin x, \ 0 \le x \le \pi; \ f(x) = 0$ elsewhere.
(c) $f(x) = 3(1 - x)^2, 0 < x < 1; \ f(x) = 0$ elsewhere.
(d) $f(x) = x, \ 0 \le x \le 1; \ f(x) = 2 - x, \ 1 \le x \le 2; \ f(x) = 0$ elsewhere.
(e) $f(x) = (1 + x), -1 \le x \le 0; f(x) = (1 - x), 0 \le x \le 1; f(x) = 0$ elsewhere.
(f) $f(x) = (1/2) e^{-|x|}, \ -\infty < x < \infty.$

4.15 Let S denote the random variable described in Prob. 4.9. Compute:
  (a) $E(S)$ **(b)** $V(S)$

4.16 If X is uniformly distributed over the interval $[a, b]$, show that:
  (a) $E(X) = (a + b)/2$
  (b) $V(X) = (b - a)^2/12$

4.17 Let Θ be uniformly distributed over the interval $[0, 2\pi]$. Compute:
  (a) $E(\cos \Theta), \ V(\cos \Theta)$
  (b) $E(\sin \Theta), \ V(\sin \Theta)$
  (c) $E(|\cos \Theta|)$
  (d) Does $E(\cos \Theta) = \cos E(\Theta)$?

4.18 Show that if U is uniformly distributed over $[0, 1]$, then $X = a + (b - a)U$ is uniformly distributed over $[a, b]$.

4.19 Verify Eqs. (4.17) and (4.18) by performing integration by parts.

4.20 Let X denote a random variable with continuous df $F(x)$ defined as follows:

$$F(x) = 0, \quad x \le -1$$

$$F(x) = (x^3/2) + 1/2, \quad -1 \le x \le 1$$

$$F(x) = 1, \quad x \ge 1$$

Compute $E(X)$, $E(X^2)$, and $V(X)$.

4.21 Suppose the lifetime T of an electronic component is exponentially distributed with $\theta^{-1} = 1000$, and the warranty period is 700 hours. The manufacturing cost and selling price of each component are $6 and $10, respectively. If the item fails before K hours, the manufacturer replaces the faulty component without charge.
  (a) Compute the expected profit per customer if the warranty period is (*i*) 700 hours, (*ii*) 900 hours.
  (b) How long can the warranty period be extended before the expected profit turns into a loss?
  (*Hint:* Review Example 4.6.)

4.4   THE NORMAL DISTRIBUTION

The probability density function of the *normal distribution* depends on two parameters, μ and σ, and is defined by the equation

$$f(x; \mu, \sigma) = \frac{1}{\sigma \sqrt{2\pi}} e^{-(x-\mu)^2/2\sigma^2}, \quad -\infty < x < \infty \qquad (4.25)$$

where

$$-\infty < \mu < \infty, \, 0 < \sigma < \infty$$

The mean and variance of the normal distribution equal μ and σ^2; we will derive this result later in the chapter. The graph of the normal pdf is a bell-shaped curve called the *normal curve*. Its statistical importance rests on the fact that the histograms of many probability functions and the relative frequency histograms of many data sets can be approximated by a normal curve. In the physics and engineering literature, the normal distribution is also called the *Gaussian distribution*.

In Fig. 4.5 we display the graphs of two normal curves with the same mean ($\mu_1 = \mu_2 = 3$) but different variances ($\sigma_1 = 2$ and $\sigma_2 = 0.5$).

In Fig. 4.6 we display the graphs of two normal curves with different means ($\mu_1 = 1.5 < 2.5 = \mu_2$) but the same variance ($\sigma_1 = \sigma_2 = 0.5$).

Figures 4.5 and 4.6 give us some insight into the significance of the constants μ and σ.

1. The maximum value of $f(x; \mu, \sigma)$ occurs at $x = \mu$, and its graph is *symmetric* about μ.
2. Looking at Fig. 4.5, we see that the *spread* of the distribution is determined by σ. A small value of σ produces a sharp peak at $x = \mu$; consequently, most of the area under this normal curve is close to μ. A large value of σ, on the other hand, produces a smaller, more rounded bulge at $x = \mu$. The area under this normal curve is less concentrated about μ.

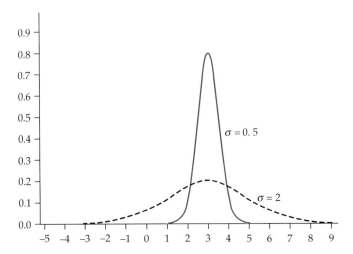

Figure 4.5 Two normal probability density functions: mean = 3, standard deviations = 2, 0.5

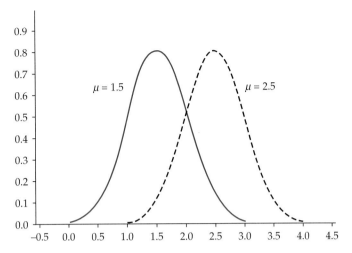

Figure 4.6 Two normal probability density functions: means = 1.5, 2.5, standard deviation = 0.5

3. Table 4.1 gives the value of the probability $P(\mu - k\sigma < X < \mu + k\sigma)$ for $k = 1, 2, 3$.

TABLE 4.1

k	Interval	$P(\mu - k\sigma < X < \mu + k\sigma)$
1	$\mu \pm \sigma$	0.6826
2	$\mu \pm 2\sigma$	0.9544
3	$\mu \pm 3\sigma$	0.9974

Note that most of the area under the normal curve is concentrated in the interval $[\mu - 3\sigma, \mu + 3\sigma]$; that is, 99 percent of the area under the normal curve lies within three units of standard deviation of the mean. Consequently, if σ is small, then most of the area is concentrated in a small interval centered at μ; if σ is large, the area is spread out over a larger interval.

Using techniques from advanced calculus, one can show that

$$\int_{-\infty}^{\infty} \frac{1}{\sigma \sqrt{2\pi}} e^{-(x-\mu)^2/2\sigma^2} \, dx = 1$$

Thus $f(x; \mu, \sigma)$ defines a pdf in the sense of Definition 4.1.

DEFINITION 4.3

A random variable X with distribution function given by

$$F(x) = \int_{-\infty}^{x} \frac{1}{\sigma \sqrt{2\pi}} e^{-(t-\mu)^2/2\sigma^2} \, dt$$

is said to have a *normal distribution* with mean μ and variance σ^2. We express this by writing $X \sim N(\mu, \sigma^2)$. ∎

The fact that $E(X) = \mu$, $V(X) = \sigma^2$ will be demonstrated later in this section.

▓ DEFINITION 4.4

A normally distributed random variable with $\mu = 0$ and $\sigma = 1$ is said to have the *standard normal distribution*. We denote it by the letter Z. ∎

The pdf of Z is given by:

$$f(x) = \frac{1}{\sqrt{2\pi}} e^{-x^2/2}; \quad -\infty < x < \infty \tag{4.26}$$

Its df $\Phi(z)$ is defined by the definite integral

$$\Phi(z) = \int_{-\infty}^{z} \frac{1}{\sqrt{2\pi}} e^{-x^2/2} \, dx \tag{4.27}$$

We compute $P(Z < z) = \Phi(z)$ by using Table A.3 in the appendix.

---

example 4.8   Use Table A.3 in the appendix to compute (1) $P(Z < 1.5)$, (2) $P(Z < -1.5)$, (3) $P(-1.5 < Z < 1.5)$.

Solution

1. $P(Z < 1.5) = 0.9332$
2. $P(Z < -1.5) = 0.0668$
3. $P(-1.5 < Z < 1.5) = \Phi(1.5) - \Phi(-1.5) = 0.9332 - 0.0668 = 0.8664$ ∎

We now show that the mean $E(Z) = 0$ and variance $V(Z) = 1$.

▓ PROPOSITION 4.2

Let Z be a standard normal random variable; then

$$E(Z) = 0 \quad \text{and} \quad V(Z) = 1$$ ∎

Proof. The proposition is proved by evaluating the following convergent improper integrals:

$$E(Z) = \int_{-\infty}^{\infty} \frac{1}{\sqrt{2\pi}} x e^{-x^2/2} \, dx = 0 \tag{4.28}$$

$$E(Z^2) = \int_{-\infty}^{\infty} \frac{1}{\sqrt{2\pi}} x^2 e^{-x^2/2} \, dx = \int_{-\infty}^{\infty} \frac{1}{\sqrt{2\pi}} x d(-e^{-x^2/2})$$

$$= \int_{-\infty}^{\infty} \frac{1}{\sqrt{2\pi}} e^{-x^2/2} \, dx = 1 \tag{4.29}$$

Remarks on the Calculations. Equation (4.28) follows at once from the fact that the integrand is an *odd* function; Eq. (4.29) follows from a routine integration by parts, together with the fact that

$$\lim_{x \to \pm\infty} xe^{-x^2/2} = 0$$

In many statistical applications it is necessary to compute $P(Z \geq z) = 1 - \Phi(z)$. We call this function the *tail* of the normal distribution. The *critical values* $z(\alpha)$ of the normal distribution are given by

$$P(Z \geq z(\alpha)) = \alpha \qquad (4.30)$$

---

example 4.9   Compute the following critical values: (1) $z(0.1)$, (2) $z(0.05)$, (3) $z(0.025)$.

Solution

1. Refer to Table A.3 and look for the entry that is closest to 0.9, which appears to be 0.8997. Consequently, $z(0.1) = 1.28$.
2. Similarly, $z(0.05) = 1.645$.
3. $z(0.025) = 1.96$ ∎

Standardizing a Normal Random Variable

To compute the distribution function of a normal random variable X with mean μ and variance σ^2, we rescale it into a standard normal random variable via the transformation $Z = (X - \mu)/\sigma$. We can also use the transformation $X = \mu + \sigma Z$ $(\sigma > 0)$ to obtain a normal random variable with mean μ and variance σ^2 from the standard one. [The proof that $X = \mu + \sigma Z \sim N(\mu, \sigma^2)$ is given in Proposition 4.3.] This yields the following formula for the distribution function of X.

$$P(X \leq x) = \Phi\left(\frac{x - \mu}{\sigma}\right) \qquad (4.31)$$

where $\Phi(z)$ is the standard normal distribution whose values are listed in Table A.3.

Before deriving Eq. (4.31), we illustrate its use in the following example.

---

example 4.10   Given that X is $N(25, 16)$, compute the following probabilities: (1) $P(X < 20)$, (2) $P(X > 33)$.

Solution

$$P(X < 20) = \Phi\left(\frac{20 - 25}{4}\right) = \Phi(-1.25) = 0.1056$$

$$P(X > 33) = 1 - P(X \leq 33) = 1 - \Phi\left(\frac{33 - 25}{4}\right) = 1 - \Phi(2.0) = 0.0228 \quad \blacksquare$$

Equation (4.31) for the distribution function of a normal random variable is a consequence of the next proposition, which is of independent interest.

■ PROPOSITION 4.3

The random variable $X = \mu + \sigma Z$ ($\sigma > 0$), where $Z \sim N(0, 1)$, has a normal distribution with mean value μ and variance σ^2.   ■

Proof. We begin with the observation that the event $(X \le x)$ can be expressed in terms of an equivalent event involving Z; more precisely, we have the following relationship between X and Z:

$$(X \le x) = (\mu + \sigma Z \le x)$$

$$= \left(Z \le \frac{x - \mu}{\sigma} \right)$$

Consequently,

$$P(X \le x) = P(\mu + \sigma Z \le x)$$

$$= P\left(Z \le \frac{x - \mu}{\sigma} \right)$$

$$= \Phi\left(\frac{x - \mu}{\sigma} \right)$$

We leave it as an exercise to verify that

$$\frac{d}{dx} \Phi\left(\frac{x - \mu}{\sigma} \right) = \frac{1}{\sigma \sqrt{2\pi}} e^{-(x-\mu)^2 / 2\sigma^2}$$

Conversely, the standardized random variable defined by $(X - \mu)/\sigma$ has a standard normal distribution; that is,

$$\frac{X - \mu}{\sigma} \sim N(0, 1)$$

We omit the proof, which is similar to that just given for Proposition 4.3. The representation $X = \mu + \sigma Z$ also implies that the parameters μ and σ^2 are the mean and variance of X. The details follow.

■ COROLLARY 4.1

If X is $N(\mu, \sigma^2)$ distributed, then

$$E(X) = \mu \quad \text{and} \quad V(X) = \sigma^2$$ ■

Proof. Using the representation $X = \mu + \sigma Z$, we see that

$$E(X) = E(\mu + \sigma Z) = E(\mu) + E(\sigma Z)$$

$$= \mu + \sigma E(Z) = \mu$$

Similarly, $$V(X) = V(\mu + \sigma Z)$$

$$= V(\sigma Z) = \sigma^2 V(Z) = \sigma^2$$

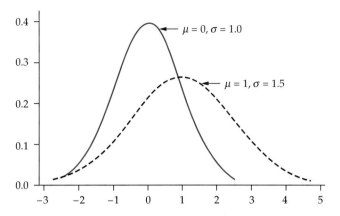

Figure 4.7 Standardizing a normal probability density: mean = 1 and standard deviation = 1.5

---

example 4.11 Graph the pdf of (1) $N(1, (1.5)^2)$ and (2) $(X - 1)/1.5$.

Solution. The pdf of $N(1, (1.5)^2)$ is displayed in Fig. 4.7. The pdf of $(X - 1)/1.5$ is $N(0, 1)$ and is also displayed in Fig. 4.7. ∎

The *critical value* $x(\alpha)$ of a normal distribution with mean μ and variance σ^2 is defined by the condition that the area under the normal curve to the right of $x(\alpha)$ equals α. That is, $P(X \geq x(\alpha)) = \alpha$. We compute $x(\alpha)$ using the formula

$$x(\alpha) = \mu + \sigma z(\alpha) \qquad (4.32)$$

where $z(\alpha)$ is the critical value of the standard normal distribution.

The derivation of Eq. (4.32) is left as an exercise; see Prob. 4.31.

4.4.1 The Normal Approximation to the Binomial Distribution

The problem of computing probabilities associated with the binomial distribution $b(x; n, p)$ becomes tedious, if not practically impossible, when n is large, say, $n > 25$. (Table A.1, for instance, lists only the binomial probabilities for $n = 5, 10, 15, 20, 25$.) How do we compute the distribution function of a binomial distribution when $n = 50$ or $n = 1000$? To compute these probabilities we turn to the most basic theorem in probability theory: the normal approximation to the binomial distribution, which is itself a special case of a more remarkable result known as the central limit theorem, which is discussed later in this book (Sec. 6.2.1).

We denote by S_n the binomially distributed random variable with parameters n, p whose distribution function we wish to approximate. This is to emphasize the fact that its distribution depends on n. When n is large, it is not

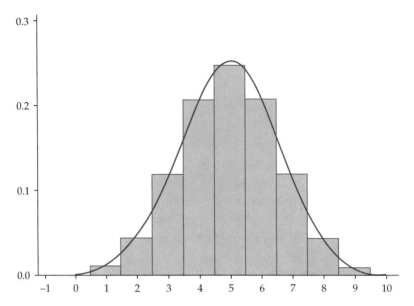

Figure 4.8 Normal approximation to the binomial: $n = 10, p = 0.5$

an easy matter to compute accurately and quickly a probability of the form

$$P(\alpha \le S_n \le \beta) = \sum_{x=\alpha}^{\beta} b(x; n, p)$$

The normal approximation to the binomial states that for large n the probability histogram of the binomial is well approximated by the normal curve with $\mu = np$ and $\sigma^2 = np(1 - p)$. Figure 4.8 is the graph of the normal pdf with $\mu = 5, \sigma^2 = 2.5$ superimposed on the probability histogram of the binomial distribution with parameters $n = 10, p = 0.5$ [so $np = 5, np(1 - p) = 2.5$].

The most striking feature of the histogram is its similarity to the graph of a normal pdf $(1/\sigma \sqrt{2\pi}) e^{-(x-\mu)^2/2\sigma^2}$.

Informal Version of the Normal Approximation to the Binomial

When the random variable S_n has a binomial distribution, Fig. 4.8 suggests that the probability $P(\alpha \le S_n \le \beta)$ can be approximated by computing $P(\alpha \le X \le \beta)$, where X is normally distributed with the same mean np and the same variance $np(1 - p)$. More precisely, the normal approximation to the binomial asserts that for large n [one rule of thumb requires that $np \ge 5$ and $n(1-p) \ge 5$], the distribution of S_n is well approximated by a normal random variable with $\mu = np$ and $\sigma^2 = np(1 - p)$.

This is the content of a famous theorem, due to the French mathematicians De Moivre and Laplace, called the *normal approximation to the binomial distribution.* The formula for the normal approximation is

$$P(\alpha \le S_n \le \beta) \approx P(\alpha \le X \le \beta)$$

$$= P\left(\frac{\alpha - np}{\sqrt{np(1-p)}} \le \frac{X - np}{\sqrt{np(1-p)}} \le \frac{\beta - np}{\sqrt{np(1-p)}}\right)$$

Consequently,

$$P(\alpha \le S_n \le \beta) \approx \Phi\left(\frac{\beta - np}{\sqrt{np(1-p)}}\right) - \Phi\left(\frac{\alpha - np}{\sqrt{np(1-p)}}\right) \qquad (4.33)$$

The significance of this theorem is that it enables us to reduce the tedious task of calculating binomial sums of the form

$$P(\alpha \le S_n \le \beta) = \sum_{x=\alpha}^{\beta} b(x; n, p)$$

to evaluating a corresponding area under the standard normal curve.

---

example 4.12   The random variable X has a binomial distribution with parameters $n = 10$, $p = 0.5$. Compute (1) the exact value of $P(4 \le X \le 8)$ and (2) the normal approximation to $P(4 \le X \le 8)$.

Solution

1. The exact answer is obtained from table A.1 in the appendix:
$$P(4 \le X \le 8) = 0.989 - 0.172 = 0.817$$

2. We obtain the normal approximation by computing the area under the approximating normal curve from $4 \le x \le 8$ (see Fig. 4.8):

$$P(4 \le X \le 8) = P\left(\frac{4-5}{\sqrt{2.5}} \le \frac{X-5}{\sqrt{2.5}} \le \frac{8-5}{\sqrt{2.5}}\right)$$

$$\approx P(-0.63 \le Z \le 1.90)$$

$$= 0.9713 - 0.2643 = 0.7070 \quad \text{(normal approximation)} \quad \blacksquare$$

REMARK 4.2

In Example 4.15 we show how to obtain a better approximation using the *continuity correction.* $\blacksquare$

---

example 4.13   Let X denote the number of heads that appear when a fair coin is tossed 100 times. Use the normal approximation to estimate $P(X > 60)$ and $P(45 < X < 60)$.

Solution. X has a binomial distribution with parameters $n = 100$, $p = 0.5$. In particular, $np = 50$, $\sqrt{np(1-p)} = 5$. Using the normal approximation, we obtain

$$P(X > 60) = P(60 < X \le 100) \approx \Phi\left(\frac{100 - 50}{5}\right) - \Phi\left(\frac{60 - 50}{5}\right)$$

$$= \Phi(5) - \Phi(2) = 1 - 0.9772 = 0.0228$$

We estimate $P(45 < X < 60)$ in a similar way:

$$P(45 \le X \le 60) \approx \Phi\left(\frac{60 - 50}{5}\right) - \Phi\left(\frac{45 - 50}{5}\right)$$

$$= \Phi(2) - \Phi(-1) = 0.9972 - 0.1587 = 0.8385 \qquad \blacksquare$$

---

example 4.14   Consider a clinical trial of a new drug. The natural recovery rate for a certain disease is 30 percent. A new drug is developed, and the manufacturer claims that it boosts the recovery rate to 50 percent. Two hundred patients are given the drug. What is the probability that at least 80 patients recover if:

1. The drug is worthless?
2. It is really 50% effective as claimed?

Solution

1. If the drug is worthless, then the the number of recoveries X is binomially distributed with parameters $n = 200$ and $p = 0.3$. Thus $np = 60$ and $\sqrt{np(1 - p)} = \sqrt{60 \times 0.7} = \sqrt{42} = 6.48$. Using the normal approximation to the binomial, we see that

   $$P(80 \le X \le 200) \approx \Phi(21.6) - \Phi(3.09) = 1 - 0.999 = 0.001$$

   In other words, if we gave the drug to 200 patients and more than 80 recovered, it would be most unlikely that the drug was completely ineffective.
2. If the drug is really as effective as claimed, then X is binomially distributed with parameters $n = 200$ and $p = 0.5$. Thus $np = 100$ and $\sqrt{np(1 - p)} = \sqrt{50} = 7.07$. Using the normal approximation to the binomial once again, we see that

$$P(80 \le X \le 200) \approx \Phi\left(\frac{100}{7.07}\right) - \Phi\left(\frac{-20}{7.07}\right)$$

$$= \Phi(14.14) - \Phi(-2.83)$$

$$= 1 - 0.0023 = 0.9977 \qquad \blacksquare$$

Formal Version of the Normal Approximation to the Binomial

Mathematical statisticians prefer to state the normal approximation to the binomial in the form of a limit theorem for the corresponding distributions as follows.

■ THEOREM 4.1

Let S_n denote the binomially distributed random variable with parameters n, p. Then

$$\lim_{n \to \infty} P\left(\frac{S_n - np}{\sqrt{np(1-p)}} \le z\right) = \int_{-\infty}^{z} \frac{1}{\sqrt{2\pi}} e^{-x^2/2}\, dx \qquad (4.34)$$

∎

Although the proof of this important result is beyond the scope of this book, its intuitive content and implications, as we have seen, are not.

The Continuity Correction

The accuracy of the normal approximation is improved by using the *continuity correction*:

$$P(\alpha \le S_n \le \beta) \approx \Phi\left(\frac{\beta - np + 0.5}{\sqrt{np(1-p)}}\right) - \Phi\left(\frac{\alpha - np - 0.5}{\sqrt{np(1-p)}}\right) \qquad (4.35)$$

---

example **4.15**

The random variable X has a binomial distribution with $n = 10$, $p = 0.5$, $np = 5$, $np(1-p) = 2.5$. Compute the value of $P(4 \le X \le 8)$ using (1) Table A.1, (2) the normal approximation with the continuity correction, and (3) the normal approximation without the continuity correction.

Solution

$$P(4 \le X \le 8) = 0.989 - 0.172 = 0.817 \quad \text{(exact answer)}$$

$$P(4 \le X \le 8) \approx P\left(\frac{-1.5}{\sqrt{2.5}} \le Z \le \frac{3.5}{\sqrt{2.5}}\right)$$

$$= P(-0.95 \le Z \le 2.21) = 0.9864 - 0.1711$$

$$= 0.8153 \quad \text{(with continuity correction)}$$

$$P(4 \le X \le 8) \approx P(-0.63 \le Z \le 1.90)$$

$$= 0.7070 \quad \text{(without continuity correction)} \qquad ∎$$

Note that for large n the effect of the correction term $0.5/\sqrt{n}$ becomes so small that it can be safely ignored; in the present case, however, the continuity correction significantly improves the accuracy of the normal approximation. The continuity correction is graphically displayed in Fig. 4.9.

---

example **4.16**

We now apply the normal approximation to analyze and interpret the results of Deming's red bead experiment described in Chap. 1 (Example 1.3).

Following Deming, we calculate how much of the observed variation is attributable to the system. Let X denote the number of red beads produced by a worker. The exact distribution of X is hypergeometric with

$$n = 50, \qquad N = 4000, \qquad D = 800, \qquad p = \frac{800}{4000} = 0.20$$

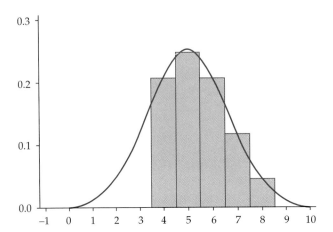

Figure 4.9 Normal approximation to Y = binomial ($n =$ 10, $p = 0.5$), $P(4 \leq Y \leq 8) \approx P(3.5 \leq X \leq 8.5)$ (continuity correction)

Note that

$$\frac{n}{N} = \frac{50}{4000} = 0.0125 < 0.05$$

Consequently, we can apply the binomial approximation to the hypergeometric and assume that X is $B(x; 50, p)$ distributed. Applying the normal approximation to the binomial, with $n = 50$, yields

$$P(50p - 3\sqrt{50p(1-p)} \leq X \leq 50p + 3\sqrt{50p(1-p)}) \approx \Phi(3) - \Phi(-3) = 0.9974$$

Deming calls the quantities U and L, defined by

$$L = 50p - 3\sqrt{50p(1-p)}, \qquad U = 50p + 3\sqrt{50p(1-p)}$$

the *limits of variation attributable to the system*. Since management does not know the true proportion of red beads in the mixture, it uses the sample proportion, denoted $\hat{p}$, which equals $51/300 = 0.17$. The estimated limits of variation attributed to the system are given by

$$L = 50\hat{p} - 3\sqrt{50\hat{p}(1-\hat{p})} = 8.5 - 7.97 = 0.53$$
$$U = 50\hat{p} + 3\sqrt{50\hat{p}(1-\hat{p})} = 8.5 + 7.97 = 16.47$$

Let us now apply these theoretical results to the actual data from Deming's red bead experiment. Looking at Table 1.2, we see that the observed variation in the workers' performance can be accounted for by chance alone; that is, the number of red beads produced by each worker lies within the interval [0.53, 16.47]. ∎

---

example 4.17 Overbooking occurs when the number of tickets sold by an airline exceeds the seating capacity of an aircraft. This is done to protect the airline from "no shows." This practice is, unfortunately, not limited to airline reservations. Assume that a college can admit at most 820 freshmen. Assume also that it sends out 1600 acceptances and that each student comes to the college with probability 0.5.

1. What is the probability that the school ends up with more students than it can accommodate?

   Solution. Let T denote the total number of freshmen who decide to enroll. Clearly, T is binomially distributed with parameters $n = 1600$ and $p = 0.5$. Consequently,

   $$E(T) = np = 800 \quad \text{and} \quad \sqrt{V(T)} = \sqrt{np(1-p)} = \sqrt{400} = 20$$

   The college ends up with more students than it can accommodate if $T \geq 821$; but, according to the normal approximation, $T \approx N(800, 400)$, and therefore

   $$P(821 \leq T < \infty) = P\left(\frac{821 - 800}{20} \leq \frac{T - E(T)}{\sqrt{V(T)}} < \infty\right)$$

   $$= P\left(\frac{21}{20} \leq Z < \infty\right) = 0.15$$

2. Find the largest number n that the college could accept if it wants the probability of getting more students than it can accommodate to be less than 0.05.

   Solution. The condition that $P(T \geq 821) = 0.05$ yields a quadratic equation for $\sqrt{n}$ that is derived by using the normal approximation. In detail,

   $$0.05 = P(T \geq 821) = P\left(\frac{T - (n/2)}{\sqrt{n}/2} \geq \frac{821 - (n/2)}{\sqrt{n}/2}\right)$$

   Consequently,

   $$\frac{821 - (n/2)}{\sqrt{n}/2} = 1.645$$

   Setting $x = \sqrt{n}$ yields the following equation for x:

   $$\frac{821 - (x^2/2)}{x/2} = 1.645$$

   Rearranging terms yields the quadratic equation

   $$\frac{x^2}{2} + 0.8225x - 821 = 0$$

   There are two solutions, only one of which is meaningful:

   $$x = 39.7074$$

   which implies that $n = x^2 = 1576$. ∎

4.4.2 Distribution of the Sample Proportion $\hat{p}$

For the purposes of statistical inference it is more convenient to apply the normal approximation to the *sample proportion* $\hat{p} = S_n/n$. For example, a public opinion survey, such as the Gallup poll, usually reports the approval rating for a politician as a sample proportion. In this context, the normal approximation is a very useful tool for studying the fluctuations of $\hat{p}$ about its true value p; it is particularly useful for constructing a *confidence interval* for the unknown proportion p. A confidence interval is a measure of the accuracy, or reliability, of $\hat{p}$ as an estimator of p; it is discussed in detail in Chap. 7.

To apply the normal approximation to $\hat{p} = S_n/n$, we begin by dividing the numerator and denominator of $(S_n - np)/\sqrt{np(1 - p)}$ by n. This yields the normal approximation for the sample proportion $\hat{p}$:

$$\frac{S_n - np}{\sqrt{np(1 - p)}} = \frac{S_n/n - p}{\sqrt{p(1 - p)/n}}$$

$$= \frac{\hat{p} - p}{\sqrt{p(1 - p)/n}} \approx N(0, 1) \tag{4.36}$$

---

example 4.18   Suppose the current approval rating of a politician is 52 percent. How likely is it that a poll based on a sample of size $n = 2000$ would show that her approval rating is less than 50 percent?

Solution. We are interested in computing the probability $P(\hat{p} < 0.5)$ under the assumption that $p = 0.52$. Using the normal approximation as expressed in Eq. (4.36), we proceed as follows:

$$P(\hat{p} < 0.5) = P\left(\frac{\hat{p} - 0.52}{\sqrt{0.52 \times 0.48/2000}} < \frac{0.5 - 0.52}{0.0112}\right)$$

$$\approx P(Z < -1.79) = 0.0367 \qquad \blacksquare$$

PROBLEMS

4.22 Let Z be a standard normal random variable. Use the table of the normal distribution (Table A.3) to compute the following probabilities.
(a) $P(Z < 1)$ (b) $P(Z < -1)$ (c) $P(|Z| < 1)$
(d) $P(Z < -1.64)$ (e) $P(Z > 1.64)$ (f) $P(|Z| < 1.64)$
(g) $P(Z > 2)$ (h) $P(|Z| > 2)$ (i) $P(|Z| < 1.96)$

4.23 Estimate each of the following probabilities using Chebyshev's inequality, and compare the estimates with the exact answers obtained from the table of the normal distribution.
(a) $P(|Z| > 1)$ (b) $P(|Z| > 1.5)$ (c) $P(|Z| > 2)$

4.24 Find c such that:
(a) $P(Z < c) = 0.25$ (b) $P(Z < c) = 0.75$
(c) $P(|Z| < c) = 0.5$ (d) $P(Z < c) = 0.85$
(e) $P(Z > c) = 0.025$ (f) $P(|Z| < c) = 0.95$

4.25 Miles per gallon (mpg) is a measure of the fuel efficiency of an automobile. The mpg of a compact automobile selected at random is normally distributed with $\mu = 29.5$ and $\sigma = 3$.

(a) What proportion of the automobiles have a fuel efficiency greater than 24.5 mpg?

(b) What proportion of the automobiles have a fuel efficiency less than 34.0 mpg?

4.26 The results of a statistics exam are assumed to be normally distributed with $\mu = 65$ and $\sigma = 15$. In order to get an A, a student must rank in the 90th percentile or above. Similarly, in order to get at least a B, a student must rank in the 80th percentile or above.

(a) What is the minimum score required to get an A?

(b) What is the minimum score required to get at least a B?

4.27 The distribution of the lifetime (in hours) of a battery for a laptop computer is modeled as a normal random variable with mean $\mu = 3.5$ and standard deviation $\sigma = 0.4$. Compute the safe life of the battery, which is defined as the 10th percentile of the distribution.

4.28 Suppose that a manufacturing process produces resistors whose resistance (measured in ohms) is normally distributed with $\mu = 0.151$ and $\sigma = 0.003$. The specifications for the resistors require that the resistance equal 0.15 ± 0.005 ohms. What proportion of the resistors fail to meet the specifications?

4.29 The random variable X has a normal distribution with mean and standard deviation given by $\mu = 12, \sigma = 6$.

(a) Compute $P(X < 0)$, $P(X > 20)$.

(b) Find c such that $P(|X - 12| < c) = 0.9, 0.95, 0.99$.

4.30 The random variable X has a normal distribution with mean and standard deviation given by $\mu = 60, \sigma = 5$. Compute:

(a) $P(X < 50)$ (b) $P(X > 65)$

(c) The value of c such that $P(|X - 60| < c) = 0.95$.

(d) The value of c such that $P(X < c) = 0.01$.

4.31 Derive Eq. (4.32) for the critical value of a normal distribution.

4.32 Assume that math SAT scores are normally distributed with $\mu = 500$ and $\sigma = 100$.

(a) What proportion of the students have a math SAT score greater than 650?

(b) What proportion of the students have a math SAT score less than 550?

(c) Ten students are selected at random. What is the probability that at least one scored 650 or better?

(d) [Part (c) continued] What is the probability that all 10 scored at most 550?

4.33 Suppose the combined SAT scores of students admitted to a college are normally distributed with $\mu = 1200$ and $\sigma = 150$. Compute the value of c for which $P(|X - 1200| < c) = 0.5$. The interval $[1200 - c, 1200 + c]$ is the interquartile range.

4.34 The random variable X has a binomial distribution with parameters $n = 25$ and $p = 0.5$.
(a) Compute the exact value of $P(9 \le X \le 16)$ using Table A.1.
(b) Compute $P(9 \le X \le 16)$ using the normal approximation to the binomial.
(c) Compute the same probability in part (b) using the continuity correction. Does the continuity correction improve the accuracy of the approximation?

4.35 Let X denote the number of heads that appear when a fair coin is tossed 300 times. Use the normal approximation to the binomial to estimate the following probabilities:
(a) $P(X \ge 160)$ **(b)** $P(X \le 140)$
(c) $P(|X - 150| \ge 20)$ **(d)** $P(135 < X < 165)$

4.36 Let Y denote the number of times a 1 appears in 720 throws of a die. Using the normal approximation to the binomial, approximate the following probabilities:
(a) $P(Y \ge 130)$ **(b)** $P(Y \ge 140)$
(c) $P(|Y - 120| \ge 20)$ **(d)** $P(105 < Y < 120)$

4.37 A certain type of seed has probability 0.7 of germinating. Out of a package of 100 seeds, let Y denote the number of seeds that germinate.
(a) Give the formula for the exact distribution of Y.
(b) Calculate the probability that at least 75 percent of the seeds germinate.
(c) Calculate the probability that at least 80 percent germinate.
(d) Calculate the probability that no more than 65 percent germinate.

4.38 The natural recovery rate for a certain disease is 40 percent. A new drug is developed, and the manufacturer claims that it boosts the recovery rate to 80 percent. Two hundred patients are given the drug. What is the probability that at least 100 patients recover if:
(a) The drug is worthless?
(b) It is really 80 percent effective as claimed?

4.39 Suppose that the velocity V of a particle of mass m is $N(0, 1)$ distributed. Denote its kinetic energy by $K = mv^2/2$. Show that $E(K) = m/2$ and $V(K) = m^2/2$.

4.40 Suppose the probability that a person cancels a hotel reservation is 0.1 and the hotel's capacity is 1000 rooms. To minimize the number of empty rooms, the hotel routinely overbooks. How many reservations above 1000 should the hotel accept if it wants the probability of overbooking to be less than 0.01?

4.5   OTHER IMPORTANT CONTINUOUS DISTRIBUTIONS

Although the normal distribution is the most important one, there remain a wide variety of phenomena in science and engineering for which it is inappropriate. For example, the distribution of the lifetime of a TV monitor is not symmetric, but skewed to the right. The proportion of TV monitors that fail during the first year of service is a variable whose range is the unit interval [0, 1]. In this section we present the *Weibull* and *beta* distributions, which are particularly useful for modeling lifetimes and proportions. The gamma distribution, which includes the exponential, chi-square, and exponential distributions as special cases, is discussed in Chap. 6.

4.5.1   The Weibull Distribution

The *Weibull distribution* is widely used by engineers for modeling the lifetimes of a component. The nonnegative random variable T with distribution function

$$F(t) = 1 - e^{-(t/\beta)^\alpha}, \quad t \geq 0$$

$$F(t) = 0, \quad t \leq 0 \tag{4.37}$$

is said to have a Weibull distribution with *shape parameter* $\alpha > 0$ and *scale parameter* $\beta > 0$. The pdf of the Weibull distribution is

$$F'(t) = f(t) = \frac{\alpha}{\beta^\alpha} t^{\alpha-1} e^{-(t/\beta)^\alpha}, \quad t > 0$$

$$f(t) = 0, \quad t \leq 0$$

Observe that the exponential distribution is obtained as a special case by setting $\alpha = 1, \beta = 1/\theta$.

---

example **4.19**   The lifetime, measured in years, of a brand of TV picture tubes has a Weibull distribution with parameters $\alpha = 2$, $\beta = 13$. Compute the probability that a TV tube fails before the expiration of a two-year warranty.

Solution. To compute $P(T \leq 2)$, we use Eq. (4.37):

$$P(T \leq 2) = 1 - \exp(-(2/13)^2) = 0.0234 \qquad ∎$$

The first and second moments of the Weibull distribution are computed using Eq. (4.15). Thus,

$$E(T) = \int_0^\infty \frac{\alpha}{\beta^\alpha} t^\alpha e^{-(t/\beta)^\alpha} \, dt$$

$$E(T^2) = \int_0^\infty \frac{\alpha}{\beta^\alpha} t^{\alpha+1} e^{-(t/\beta)^\alpha} \, dt$$

These integrals are evaluated numerically by making the following change of variable:

$$y = \left(\frac{t}{\beta}\right)^\alpha, \qquad dt = \frac{\beta}{\alpha} y^{(1/\alpha)-1} \, dy$$

After some careful algebraic manipulations, we obtain the following equations for the first two moments:

$$E(T) = \beta \int_0^\infty y^{1/\alpha} e^{-y} \, dy \tag{4.38}$$

$$E(T^2) = \beta^2 \int_0^\infty y^{2/\alpha} e^{-y}\, dy \qquad (4.39)$$

A formula for these integrals in terms of the *gamma function* is given in Chap. 6.

4.5.2   The Beta Distribution

The burden of computing the exact sum of binomial probabilities, which can be quite laborious for large n, is made much easier by representing the sum as a definite integral. The integrand turns out to be a *beta distribution*, which has many other statistical applications—Bayesian estimation, for example. The *tail probability* of the binomial distribution, with n fixed, is a function $\pi(p)$ of the parameter p and is defined as follows:

$$\pi(p) = \sum_{k \le x \le n} b(x; n, p) \qquad (4.40)$$

The function $\pi(p)$ plays a useful role in testing hypotheses about a population proportion, and in that context it is called a *power function*. We now derive a formula for the tail probability of the binomial distribution that reveals an unexpected connection between the binomial distribution, which is discrete, and the beta distribution, which is continuous.

■■■ **THEOREM 4.2**

The tail probability function of a binomial random variable X with parameters n, p is given by the following definite integral:

$$\pi(p) = P(X \ge k \mid p) = \int_0^p n \binom{n-1}{k-1} t^{k-1} (1-t)^{n-k}\, dt \qquad (4.41)$$
■

Refer to Sec. 4.7 for the proof. The following consequence of Eq. (4.41) is useful in hypothesis testing.

■■■ **PROPOSITION 4.4**

The function $\pi(p)$ is an *increasing* function of p; in particular,

$$\text{if } p_1 \le p_2, \text{ then } \pi(p_1) \le \pi(p_2)$$
■

Proof. The integrand appearing in the right-hand side of Eq. (4.41) is a nonnegative function; therefore, the definite integral, which is just the area under the integrand from 0 to p, increases as p increases.

The integrand appearing on the right hand side of Eq. (4.41) is a special case of the beta distribution, which we now define. The *beta function* $B(\alpha, \beta)$ is defined by the integral

$$B(\alpha, \beta) = \int_0^1 x^{\alpha-1} (1-x)^{\beta-1}\, dx \qquad (4.42)$$

A random variable Y is said to have a *beta distribution* when its density function $f(x)$ has the form

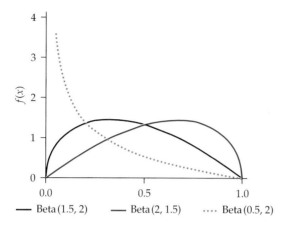

Figure 4.10 Probability density functions of three beta distributions

$$f(x) = \frac{1}{B(\alpha, \beta)} \times x^{\alpha-1}(1-x)^{\beta-1}, \quad 0 < x < 1, \quad \alpha > 0, \quad \beta > 0 \quad (4.43)$$

$f(x) = 0$ elsewhere

It is easy to see that the integrand in Eq. (4.41) is of this form with $\alpha = k$ and $\beta = n - k + 1$. We also use the notation Beta(α, β) to denote a beta distribution with parameters α, β. Figure 4.10 displays the probability density functions of three beta distributions: Beta(1.5, 2), Beta(2, 1.5), and Beta(0.5, 2). Note that the density of Beta(0.5, 2) defines a convergent improper integral at $x = 0$.

■ **PROPOSITION 4.5**

The mean and variance of the beta distribution with parameters α and β are given by

$$E(X) = \frac{\alpha}{\alpha + \beta}, \quad V(X) = \frac{\alpha\beta}{(\alpha + \beta + 1)(\alpha + \beta)^2} \quad (4.44)$$

■

These expressions for the mean and variance are a consequence of general results concerning the gamma function that are discussed in Chap. 6.

PROBLEMS

4.41 The lifetime, measured in years, of a TV picture tube has a Weibull distribution with parameters $\alpha = 2$, $\beta = 3$.

(a) Compute the probability that a tube will fail on a (1) one-year, (2) two-year, and (3) three-year warranty.

(b) Compute the expected profit per TV tube if the warranty is for one year and the cost and selling price are $50 and $100, respectively. [*Hint:* Use Eq. (4.22).]

4.42 The life of an engine fan, measured in hours, has a Weibull distribution with $\alpha = 1.053$, $\beta = 26,710$. Compute:
   (a) The median life.
   (b) The probability that it fails on an 8000-hour warranty.

4.43 Compute $B(\alpha, \beta)$ for:
   (a) $\alpha = 1$, $\beta = 1$
   (b) $\alpha = 1$, $\beta = 2$
   (c) $\alpha = 2$, $\beta = 3$
   (d) $\alpha = 3$, $\beta = 2$

4.44 Suppose that X has a beta distribution with parameters $\alpha = 3$, $\beta = 2$. Compute $E(X)$ and $V(X)$.

4.45 The proportion X of computer disk drives that fail during their first year of service has a Beta(3, 2) distribution.
   (a) Compute the df $F(x) = P(X \le x)$.
   (b) The warranty period is 90 days. What proportion of the disk drives will require warranty service?

4.46 Show that the function $O(p)$ defined by

$$O(p) = P(X \le c) = \sum_{x=0}^{c} b(x; n, p)$$

is a decreasing function of p.

4.6 FUNCTIONS OF A RANDOM VARIABLE

We can obtain a normal random variable with mean μ and variance σ^2 from the standard normal random variable via the formula $X = \mu + \sigma Z$; that is the content of Proposition 4.3. The relationship between Z and X can be expressed in the form $X = g(Z)$, where $g(t) = \mu + \sigma t$. We say that X is a *function of the random variable* Z. Conversely, if X is $N(\mu, \sigma^2)$, then the random variable defined by the function $Z = h(X)$, where $h(x) = (x - \mu)/\sigma$, has a standard normal distribution. It is this relationship between Z and X that we used to compute the df of X in terms of the df of Z. It is a remarkable fact that most random variables, including many of those discussed in this text, can be represented as a function $g(U)$ of a uniformly distributed random variable U. This is very useful because most computers, including some pocket calculators, have a *random number generator* whose output u can be viewed as the value $U(\omega) = u$, where U denotes a random variable that is uniformly distributed over the interval $[0, 1]$.

example 4.20   Let U be uniformly distributed over the interval $[0, 1]$ and set $X = -\ln U$. Show that X is exponentially distributed with parameter $\theta = 1$.

Solution. In this case $X = g(U)$, where $g(u) = -\ln u$. Since $0 < U < 1$, it follows that $\ln 0 < \ln U < \ln 1$, so $-\infty < \ln U < 0$; therefore, $0 < -\ln U < \infty$. Thus, the range of $-\ln U$ is $(0, \infty)$. Moreover,

$$(X \le x) = (-\ln U \le x)$$
$$= (e^{-\ln U} \le e^x)$$
$$= \left(\frac{1}{U} \le e^x \right)$$
$$= (U \ge e^{-x})$$

Therefore,

$$P(X \le x) = P(U \ge e^{-x}) = 1 - e^{-x} \qquad \blacksquare$$

---

example **4.21**   Compute the distribution of $X = U^2$, where U is uniformly distributed over the interval $[0, 1]$.

Solution. The problem is to compute $F_X(x)$ in terms of $F_U(u)$; we proceed as follows.

$$(X \le x) \equiv (U^2 \le x) \equiv (-\sqrt{x} \le U \le \sqrt{x})$$

Consequently,

$$F_X(x) = F_U(\sqrt{x}) - F_U(-\sqrt{x}) = F_U(\sqrt{x})$$

since $F_U(u) = 0$, $u \le 0$. Thus

$$F_X(x) = 0, \quad x \le 0$$
$$F_X(x) = \sqrt{x}, \quad 0 \le x \le 1$$
$$F_X(x) = 1, \quad 1 \le x \qquad \blacksquare$$

---

example **4.22**   Compute the distribution of $X = Z^2$, where Z is $N(0, 1)$.

Solution. Using exactly the same reasoning as in the preceding example, we obtain

$$F_X(x) = \Phi(\sqrt{x}) - \Phi(-\sqrt{x}) = 2\Phi(\sqrt{x}) - 1, \quad x \ge 0$$

where $\Phi(z)$ is the standard normal distribution function. We obtain the density function of X by computing $F_X'(x)$ via the chain rule of differential calculus:

$$F_X'(x) = 2\Phi'(\sqrt{x}) \frac{1}{2\sqrt{x}}$$
$$= 2 \frac{1}{\sqrt{2\pi}} e^{-(\sqrt{x})^2/2} \frac{1}{2\sqrt{x}} \qquad (4.45)$$
$$= \frac{1}{\sqrt{2\pi}} x^{-1/2} e^{-x/2}, \quad x > 0 \qquad \blacksquare$$

▮ REMARK 4.3

In Chap. 6 we show that this is equivalent to the assertion that Z^2 has a chi-square distribution with one degree of freedom.   ▮

We have thus proved the following proposition:

PROPOSITION 4.6

If Z is $N(0,1)$, then Z^2 has a *chi-square* distribution with one degree of freedom. ∎

---

example 4.23   Compute the distribution function of $Y = U^{-1}$, where U is uniformly distributed over the interval $[0,1]$.

Solution

$$(Y \le y) \equiv (U^{-1} \le y) \equiv (U \ge y^{-1})$$

Thus,

$$F_Y(y) = P(U \ge y^{-1}) = 1 - F_U(y^{-1})$$

Therefore,

$$F_Y(y) = 0, \quad y \le 1$$
$$F_Y(y) = 1 - y^{-1}, \quad 1 \le y$$
∎

PROBLEMS

4.47 Let U be a uniformly distributed random variable on the interval $[-1,1]$. Let $X = U^2$.
(a) Compute the df $F(x) = P(X \le x)$ for all x.
(b) Compute the pdf $f(x)$ for all x.

4.48 Let U be a uniformly distributed random variable on the interval $[-1,1]$. Let $Y = 4 - U^2$.
(a) Compute $G(y) = P(Y \le y)$ for all y.
(b) Compute the pdf $g(y)$ for all y.

4.49 The random variable U is uniformly distributed on $[0,1]$. Show that $Y = -\theta^{-1}\ln U$ has an exponential distribution with parameter θ, i.e., $P(Y < y) = 1 - e^{-\theta y}$.

4.50 Let $R = \sqrt{X}$, where X has an exponential distribution with parameter $\theta = 1/2$. Show that R has a Rayleigh distribution:

$$F_R(r) = \int_0^r t e^{-t^2/2}\, dt, \quad r > 0$$

$$F_R(r) = 0 \quad \text{otherwise}$$

4.51 Let Y have pdf $g(y) = (1+y)^{-2}, y > 0$, and $g(y) = 0$ otherwise.
(a) Compute the df $G(y)$ for all y.
(b) Let $X = 1/Y$. Verify that X and Y have the same df.

4.52 Let $X = \tan U$, where U is uniformly distributed on the interval $[-\pi/2, \pi/2]$.
(a) Show that

$$F_X(x) = \frac{1}{2} + \frac{1}{\pi}\arctan x$$

(b) Show that

$$f_X(x) = \frac{1}{\pi}\frac{1}{1+x^2}$$

4.53 The random variable T has df F given by

$$F(t) = 0, \quad t \le 1$$

$$F(t) = 1 - \frac{1}{t^2}, \quad t \ge 1$$

Compute the df and pdf of $X = T^{-1}$.

4.54 The random variable X has df F given by

$$f(x) = (1+x), \quad -1 \le x \le 0$$
$$f(x) = (1-x), \quad 0 \le x \le 1$$
$$f(x) = 0 \quad \text{otherwise}$$

Compute the df and pdf of $Y = X^2$.

4.55 The random variable X has pdf f given by

$$f(x) = 2(1-x), \quad 0 < x < 1$$
$$f(x) = 0 \quad \text{elsewhere}$$

Compute the df and pdf of $Y = \sqrt{X}$.

4.56 Suppose the random variables X and Y have the same distribution function $F(x) = P(X \le x) = P(Y \le x)$. Then $X = Y$.

Is this statement true or false? If true, give a proof; if false, produce a counterexample.

4.57 State whether each of the following statements is true or false. If false, give a counterexample.

(a) $E(X^2) = \left(E(X)\right)^2$
(b) $E(1/X) = 1/E(X)$
(c) $E(X) = 0 \Rightarrow X = 0$

4.7   MATHEMATICAL DETAILS AND DERIVATIONS

Proof of Theorem 4.2

1. Differentiate both sides of Eq. (4.40) with respect to p:

$$\frac{d}{dp}\pi(p) = \frac{d}{dp}\sum_{k \le x \le n} b(x; n, p) = \sum_{k \le x \le n}\frac{d}{dp}b(x; n, p)$$

2. The derivative of $b(x; n, p)$ is given by

$$\frac{d}{dp}b(x; n, p) = n\binom{n-1}{x-1}p^{x-1}(1-p)^{n-x} - n\binom{n-1}{x}p^x(1-p)^{n-x-1}$$

where we set

$$\binom{n-1}{n} = 0$$

3. Inserting the formula for $(d/dp)b(x; n, p)$ obtained in step 2 into the right-hand side of the expression for $(d/dp)\pi(p)$ in step 1 yields a "telescoping sum" with the result that

$$\frac{d}{dp}\pi(p) = n\binom{n-1}{k-1}p^{k-1}(1-p)^{n-k} > 0, \quad 0 < p < 1 \tag{4.46}$$

The fact that the derivative $\pi'(p) > 0$ proves that $\pi(p)$ is an increasing function of p. We obtain Eq. (4.41) for the tail of the binomial distribution by integrating both sides of Eq. (4.46) and using the fundamental theorem of calculus, which states that

$$\pi(p) - \pi(0) = \int_0^p \pi'(t)\,dt$$

4.8 CHAPTER SUMMARY

Engineers use random variables of the continuous type to model the lifetimes of manufactured components, the compressive strength of concrete slabs, etc. The distribution function of a continuous random variable is given by

$$F_X(x) = \int_{-\infty}^x f_X(t)\,dt$$

where
$$f_X(t) \geq 0$$

$$\int_{-\infty}^{\infty} f_X(t)\,dt = 1$$

The function $f_X(x)$ is called the probability density function (pdf) of the random variable X. Of particular importance are the uniform, exponential, Weibull, and beta distributions. The most important continuous distribution, however, is the normal distribution, which is used to model a wide variety of experimental data as well as to approximate the binomial distribution.

To Probe Further. The first statement of the normal approximation to the binomial in English appears in A. De Moivre's *The Doctrine of Chances* (1738). Motivating his research was the difficulty of computing the binomial probabilities for large values of n. "Although the solution of problems of chance," wrote De Moivre, "often require that several terms of the binomial $(a + b)^n$ be added together, nevertheless in very high powers [that is, when n is very large] the thing appears so laborious, and of so great a difficulty that few people have undertaken that task." (*Source:* S. M. Stigler, *History of Statistics: The Measurement of Uncertainty before 1900*, Cambridge, Mass., Belknap Press of Harvard University Press, 1986, p. 74.)

The historical origins and the scientific and philosophical implications of statistics are explored in Theodore M. Porter, *The Rise of Statistical Thinking, 1820–1900*, Princeton, N. J., Princeton University Press, 1986. The quotation at the beginning of this chapter is also taken from this source (p. 99).

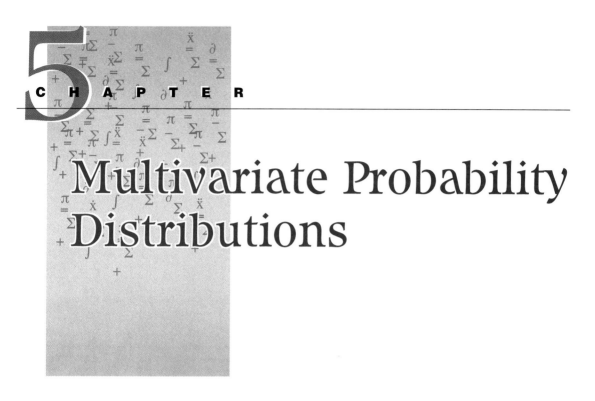

5

Multivariate Probability Distributions

Physicians have nothing to do with what is called the law of large numbers,
a law which, according to a great mathematician's expression, is always true
in general and false in particular.

Claude Bernard (1813–1878), French physiologist

5.1 ORIENTATION

We frequently have the occasion to study how one or more variables influence another. For instance, what is the effect of doubling the amount of
fertilizer on crop yield? How useful are a student's college grade point average and score on the law school admissions test as predictor variables for
future performance in law school? In environmental studies the air quality of
a metropolitan region is determined by simultaneously measuring several pol-
lutants, such as ozone, nitrogen oxide, and hydrocarbons; before undertaking
a plan to reduce the pollution levels, the environmental scientist must study
their interactions because reducing the level of just one pollutant could actually
increase the combined harmful effects of the others. Statisticians use the *joint
probability function* to study the interactions among such variables. We begin
with the discrete case and follow with the continuous case; the only difference
between the two is that the latter uses some basic tools from multivariate
calculus.

173

Organization of Chapter

5.2 THE JOINT PROBABILITY FUNCTION

The concept of a joint probability function is best introduced in the context of a simple example. Consider the sample space corresponding to the experiment of rolling two dice (Example 3.2), which consists of the ordered pairs (i, j) $(i = 1, \ldots, 6; \ j = 1, \ldots, 6)$. Assume that the two dice have different colors, white and red, say. Denote by X the number appearing on the white die and Y the number appearing on the red die. Thus $X(i, j) = i$ and $Y(i, j) = j$. The ordered pair of random variables (X, Y) is called a *random vector*. The event (i, j) can be expressed as follows:

$$(i, j) = \{X = i\} \cap \{Y = j\}$$

More generally, let X and Y be two discrete random variables defined on the same sample space Ω with discrete ranges

$$R_X = \{x_1, x_2, \ldots\} \quad \text{and} \quad R_Y = \{y_1, y_2, \ldots\}$$

The intersection of the events $\{X = x\}$ and $\{Y = y\}$ is the event denoted by

$$\{X = x, Y = y\} = \{X = x\} \cap \{Y = y\}$$

■ **DEFINITION 5.1**

The function $f(x, y)$ defined by

$$f(x, y) = P(X = x, Y = y) \tag{5.1}$$

is called the *joint probability function (jpf)* of the random vector (X, Y). ■

---

example 5.1   Compute the joint probability function for the experiment of throwing two dice.

Solution. Let X, Y denote the numbers on the faces of the two dice. Their joint probability function is

$$f(x, y) = \frac{1}{36}, \quad x = 1, \ldots, 6, \quad y = 1, \ldots, 6$$

$$f(x, y) = 0 \quad \text{otherwise.}$$ ■

The quantity $f(x, y) = P(X = x, Y = y)$ is the probability of an event, and the events $\{X = x, Y = y\}$ form a partition of the sample space. Consequently, a joint probability function has the following two properties:

$$0 \leq f(x, y) \leq 1 \tag{5.2}$$

$$\sum_{x} \sum_{y} f(x, y) = 1 \tag{5.3}$$

---

example **5.2**   A computer store sells three computer models priced at \$1600, \$2000, and \$2400 and two brands of monitors priced at \$400 and \$800, respectively. Table 5.1 gives the jpf of X, the computer price, and Y, the monitor price. *Interpretation:* $P(X = 1600, Y = 400) = 0.30$, e.g., means that 30 percent of the computer systems sold consisted of the \$1600 computer and the \$400 monitor.

TABLE 5.1 JOINT PROBABILITY
FUNCTION OF (X, Y)

	Y	
X	**400**	**800**
1600	0.30	0.25
2000	0.20	0.10
2400	0.10	0.05

∎

Computing $P(X \in A, Y \in B)$

We use the joint probability function to compute $P(X \in A, Y \in B)$ via the formula

$$P(X \in A, Y \in B) = \sum_{x \in A} \sum_{y \in B} f(x, y) \tag{5.4}$$

The joint distribution function $F(x, y)$ is an important special case defined by

$$F(x, y) = P(X \leq x, Y \leq y) \tag{5.5}$$

For instance, with reference to the jpf in Table 5.1, we see that

$$F(2000, 400) = P(X \leq 1600, Y \leq 400) = f(1600, 400) + f(2000, 400)$$
$$= 0.30 + 0.20 = 0.50$$

Marginal Distribution Functions

The probability function $f_X(x)$ is obtained by summing the jpf $f(x, y)$ over the variable y. Similarly, $f_Y(y)$ is obtained by summing the jpf over the variable x.

$$f_X(x) = \sum_y f(x,y) \tag{5.6}$$

$$f_Y(y) = \sum_x f(x,y) \tag{5.7}$$

The probability functions f_X and f_Y are called the *marginal distributions*.

Derivation of Eqs. (5.6) and (5.7). The events $(Y = y), y \in R_Y$, form a partition of the sample space, and therefore

$$(X = x) = \cup_y(X = x) \cap (Y = y)$$

Consequently,

$$f_X(x) = P(X = x) = P(\cup_y(X = x) \cap (Y = y))$$
$$= \sum_y P(X = x, Y = y)$$
$$= \sum_y f(x,y)$$

The derivation of Eq. (5.7) is similar and is therefore omitted.

---

example 5.3

Refer to the joint probability function in Table 5.1. Compute the marginal probability functions.

Solution. To compute $f_X(1600)$, for instance, we sum the entries across the first row and obtain the value $f_X(1600) = 0.55$; to compute $f_Y(400)$ we sum the entries down the first column and obtain the value $f_Y(400) = 0.60$. The complete solution is displayed in Table 5.2.

TABLE 5.2   JOINT PROBABILITY FUNCTION OF (X, Y)

	Y		
X	400	800	$f_X(x)$
1600	0.30	0.25	0.55
2000	0.20	0.10	0.30
2400	0.10	0.05	0.15
$f_Y(y)$	0.60	0.40	

---

example 5.4

(This is a continuation of Example 5.2.) The total cost of a computer system (computer and monitor) is given by the random variable $T = X + Y$. Compute the probability function and expected value of the total cost T.

Solution. The range of T is the set of all possible values of $X + Y$; these values are listed in the first row of the Table 5.3. We compute $f_T(2400)$, for example, by noting that

$$(X + Y = 2400) = (X = 2000, Y = 400) \cup (X = 1600, Y = 800)$$

Therefore,

$$P(X + Y = 2400) = P(X = 2000, Y = 400) + P(X = 1600, Y = 800)$$
$$= 0.20 + 0.25 = 0.45$$

We compute the remaining entries in the same way.

TABLE 5.3

t	2000	2400	2800	3200
$f_T(t)$	0.30	0.45	0.20	0.05

We compute $E(T)$ directly from Table 5.3.

$$E(T) = 2000 \times 0.30 + 2400 \times 0.45 + 2800 \times 0.2 + 3200 \times 0.05 = 2400 \qquad \blacksquare$$

Conditional Distribution of Y Given $X = x$

The *conditional probability* of the event $Y = y$ given that $X = x$, which we denote by $P(Y = y \mid X = x)$, is defined in exactly the same way as we defined the conditional probability $P(B \mid A) = P(A \cap B)/P(A)$; it is given by

$$P(Y = y \mid X = x) = \frac{P(Y = y, X = x)}{P(X = x)}$$
$$= \frac{f(x, y)}{f_X(x)}, \quad f_X(x) > 0 \tag{5.8}$$

---

example 5.5   (This is a continuation of Example 5.2.) A customer buys a \$400 monitor. What is the probability that she also bought a \$1600 computer?

Solution. With reference to the joint probability function in Table 5.1, we see that

$$P(X = 1600 \mid Y = 400) = \frac{f(1600, 400)}{f_Y(400)} = \frac{0.30}{0.60} = 0.50 \qquad \blacksquare$$

The *conditional distribution of Y given $X = x$* (where $f_X(x) > 0$) is the function $f(y \mid x)$ defined by

$$f(y \mid x) = P(Y = y \mid X = x) = \frac{f(x, y)}{f_X(x)} \quad f_X(x) > 0 \tag{5.9}$$

The condition $f_X(x) > 0$ is intuitively clear, since the only x values that we observe are those for which $f_X(x) > 0$. We now verify that for each $x \in R_X$ the function $y \mapsto f(y \mid x)$ satisfies Eq. (3.4) and is therefore a probability function. In detail,

$$\sum_y f(y \mid x) = \sum_y \frac{f(x, y)}{f_X(x)} = \frac{f_X(x)}{f_X(x)} = 1$$

The Product Rule for Computing the Joint Probability Function

The analogue of the product rule for events [Eq. (2.25)] is the following product rule for computing joint probabilities:

$$P(X = x, Y = y) = P(Y = y \mid X = x) \times P(X = x) \tag{5.10}$$

In the general case we display the joint probability function as in Table 5.4.

TABLE 5.4   JOINT PROBABILITY FUNCTION OF (X, Y)

			Y		
X	y_1	$\cdots$	y_j	$\cdots$	$f_X(x)$
x_1	$f(x_1, y_1)$	$\cdots$	$f(x_1, y_j)$	$\cdots$	$\sum_y f(x_1, y)$
$\vdots$	$\vdots$	$\vdots$	$\vdots$	$\vdots$	$\vdots$
x_i	$f(x_i, y_1)$	$\cdots$	$f(x_i, y_j)$	$\cdots$	$\sum_y f(x_i, y)$
$\vdots$	$\vdots$	$\vdots$	$\vdots$	$\vdots$	$\vdots$
$f_Y(y)$	$\sum_x f(x, y_1)$	$\cdots$	$\sum_x f(x, y_j)$	$\cdots$	

The marginal distribution functions f_X and f_Y are obtained by summing the entries in Table 5.4 across the rows and columns, with the row and column sums placed in the corresponding rightmost column and last row, respectively.

---

example 5.6

A random sample of size 2 is taken without replacement from an urn containing two nickels, three dimes, and three quarters. The numbers of nickels, dimes, and quarters in the sample are denoted by X, Y, Z, respectively. Compute the joint probability function of X and Y.

Solution. The solution is displayed in Table 5.5.

TABLE 5.5   JOINT PROBABILITY FUNCTION OF (X, Y)

		Y		
X	0	1	2	$f_X(x)$
0	3/28	9/28	3/28	15/28
1	6/28	6/28	0	12/28
2	1/28	0	0	1/28
$f_Y(y)$	10/28	15/28	3/28	

Details of the Computations. The denominator 28 in each of the nonzero terms is the total number of ways of choosing two coins from eight, which is $\binom{8}{2} = 28$. The event $(X = 0, Y = 0)$ describes a sample consisting of two quarters only; the numerator is the number of ways of choosing two quarters from three, which equals $\binom{3}{2} = 3$. The other entries are computed in a similar way. For instance, the joint probabilities $f(0, 1)$ and $f(1, 0)$ are computed as follows:

$$f(0, 1) = \frac{\binom{3}{1}\binom{3}{1}}{\binom{8}{2}} = \frac{9}{28}$$

$$f(1, 0) = \frac{\binom{2}{1}\binom{3}{1}}{\binom{8}{2}} = \frac{6}{28}$$

Note that $f(x, y) = 0$ whenever $x + y > 2$; this is because the total number of coins in the sample cannot exceed the sample size, which is 2. ∎

The Joint Probability Function of n Random Variables

The concept of a joint probability function is easily extended to three or more discrete random variables.

▇ DEFINITION 5.2

The joint probability function of n discrete random variables $X_1, \ldots, X_n$ is defined to be

$$f(x_1, \ldots, x_n) = P(X_1 = x_1, \ldots, X_n = x_n) \tag{5.11}$$

∎

5.2.1 Independent Random Variables

The concept of independence for events developed in Sec. 2.4.3 can be extended to two or more random variables. We recall that the events A and B are said to be independent if

$$P(A \cap B) = P(A)P(B)$$

▇ DEFINITION 5.3

We say that the random variables X and Y are mutually independent if

$$P(X = x, Y = y) = P(X = x)P(Y = y) \tag{5.12}$$

for every x and y. We say that the random variables $X_1, \ldots, X_n$ are mutually independent if

$$P(X_1 = x_1, \ldots, X_n = x_n) = P(X_1 = x_1) \cdots P(X_n = x_n) \tag{5.13}$$

for every $x_i, i = 1, \ldots, n$. ∎

We can express condition (5.12) for independence in terms of the marginal distribution functions f_X and f_Y as follows:

■ **PROPOSITION 5.1**

The random variables X and Y are independent if and only if their joint probability function is the product of their marginal distribution functions, i.e.,

$$f(x,y) = f_X(x)f_Y(y), \quad \text{for all } x, y \tag{5.14}$$

More generally, the random variables $X_1, \ldots, X_n$ are mutually independent if and only if their joint probability function is the product of their marginal distribution functions, i.e.,

$$f(x_1, \ldots, x_n) = f_{X_1}(x_1) \cdots f_{X_n}(x_n) \tag{5.15}$$

■

Bernoulli Trials Process

A *Bernoulli trials process* is a sequence of independent and identically distributed Bernoulli random variables $X_1, \ldots, X_n$. It is the mathematical model of n repetitions of an experiment under identical conditions, with each experiment producing only two outcomes called *success/failure, heads/tails,* etc.

Two important examples will now be described.

1. **Quality control:** As items come off a production line, they are inspected for defects. When the ith item inspected is defective, we record $X_i = 1$ and set $X_i = 0$ otherwise.
2. **Clinical trials:** Patients with a disease are given a drug. If the ith patient recovers, we set $X_i = 1$ and set $X_i = 0$ otherwise.

The Sample Space of a Bernoulli Sequence of Trials

example 5.7 The simplest mathematical model of a Bernoulli sequence of trials is tossing a coin n times with probability p for heads. Let $X_i = 1$ if the ith toss comes up heads, and set $X_i = 0$ if the ith toss comes up tails. The sample space consists of 2^n sequences of n digits, each digit being a 1 or a 0, so each sample point ω can be represented as a sequence of 1s and 0s in the following format:

$$\omega = (0, 1, 1, 0, \ldots, 1, 0)$$

which corresponds to the event (tail, head, head, tail, ..., head, tail).

The random variables X_i are identically distributed with

$$P(X_i = 0) = 1 - p, \qquad P(X_i = 1) = p$$

since the probability of tossing a head is the same for each toss. The coin is said to be fair when $p = 0.5$.

It is reasonable to assume that the outcomes of the ith and jth tosses are independent events (the coin has no memory), and therefore X_i and X_j are mutually independent random variables; indeed, the same reasoning shows that the random variables $X_1, \ldots, X_n$ are mutually independent, and therefore by Eq. (5.13) we have

$$P(0, 1, \ldots, 1, 0) = P(X_1 = 0)P(X_2 = 1) \cdots P(X_{n-1} = 1)P(X_n = 0)$$
$$= qp \cdots pq$$
$$= p^x q^{n-x}$$

In other words, we assign to each sample point ω, consisting of x 1s and $n - x$ 0s, the probability $p^x q^{n-x}$. ∎

The formal mathematical definition follows.

■ DEFINITION 5.4

A *Bernoulli trials process* is a sequence of independent and identically distributed (iid) random variables $X_1, \ldots, X_n$, where each random variable takes on only one of two values, 0 or 1. The number $p = P(X_i = 1)$ is called the probability of success, and the number $1 - p = P(X_i = 0)$ is called the probability of failure.

The sum T defined by

$$T = \sum_{1 \le i \le n} X_i \qquad (5.16)$$

is called the *number of successes in n Bernoulli trials*. It has a binomial distribution. This is a consequence of exactly the same reasoning we used earlier to derive the binomial distribution (see Proposition 3.3). This is why a Bernoulli trials process is also called a *binomial experiment*. ∎

PROBLEMS

5.1 Let (X, Y) have the joint probability function specified in the following table:

	Y			
X	2	3	4	5
0	1/24	3/24	1/24	1/24
1	1/12	1/12	3/12	1/12
2	1/12	1/24	1/12	1/24

(a) Compute the marginal probability functions.
(b) Are X and Y independent?

5.2 Let (X, Y) have the joint probability function displayed in Prob 5.1. Compute the following probabilities.
(a) $P(X \le 1, Y \le 3)$
(b) $P(X > 1, Y > 3)$
(c) $P(X = 2, Y > 2)$
(d) $P(Y = y \mid X = 0), y = 2, 3, 4, 5$
(e) $P(Y = y \mid X = 1), y = 2, 3, 4, 5$
(f) $P(Y = y \mid X = 2), y = 2, 3, 4, 5$

5.3 Let X and Y have the joint probability function given in Table 5.5. Compute:
(a) $P(Y \leq 1 \mid X = x), x = 0, 1, 2$
(b) $P(Y = y \mid X = 0), y = 0, 1, 2$

5.4 Two numbers are selected at random and without replacement from the set
$$\{1, 1, 1, 1, 0, 0, 0, 0, 0, 0\}$$
Let $X_i(i = 1, 2)$ denote the ith number that is drawn.
(a) Compute the joint probability function of (X_1, X_2). Are (X_1, X_2) mutually independent?
(b) Compute the probability function of $X_1 + X_2$.

5.5 Consider the experiment of drawing in succession and without replacement two coins from an urn containing two nickels and three dimes. Let X_1 and X_2 denote the monetary values (in cents) of the first and second coins drawn, respectively. Compute:
(a) The joint probability function of the random variables X_1, X_2.
(b) The probability function of the random variable $T = X_1 + X_2$, which represents the total monetary value of the coins drawn.

5.6 A bridge hand is a set of 13 cards selected at random from a deck of 52. Compute the probability of:
(a) Getting x hearts.
(b) Getting y clubs.
(c) Getting x hearts and y clubs.

5.7 *Nontransitive dice:* Consider three six-sided dice whose faces are labeled as follows:
$$\text{Die } 1 = \{5, 7, 8, 9, 10, 18\}$$
$$\text{Die } 2 = \{2, 3, 4, 15, 16, 17\}$$
$$\text{Die } 3 = \{1, 6, 11, 12, 13, 14\}$$
Let Y_i denote the number that turns up when the ith die is rolled. Assume that the three random variables Y_1, Y_2, Y_3 are mutually independent. Show that $P(Y_1 > Y_2) = P(Y_2 > Y_3) = P(Y_3 > Y_1) = 21/36$.

5.8 A fair die is thrown n times and the successive outcomes are denoted by $X_1, \ldots, X_n$. Let
$$M_n = \max(X_1, \ldots, X_n)$$
Compute the conditional probabilities p_{ij} defined by
$$p_{ij} = P(M_n = j \mid M_{n-1} = i), \quad i = 1, \ldots, 6, \quad j = 1, \ldots, 6$$
Display your answers in the format of a 6×6 matrix, with first row given by
$$p_{11}, p_{12}, \ldots, p_{16}$$
and ith row given by
$$p_{i1}, p_{i2}, \ldots, p_{i6}$$

5.9 The random variable Y has a geometric distribution
$$P(Y = j) = pq^{j-1}, \quad j = 1, \ldots, \quad 0 < p < 1$$

(a) Show that $P(Y > a) = q^a$.

(b) Show that $P(Y > a + b \mid Y > b) = P(Y > a)$.

(c) Show that $P(Y \text{ is odd }) = 1/(1 + q)$. [*Hint:* Use Eq. (3.7) with a suitable choice for r.]

(d) Compute the probability function of $(-1)^Y$.

5.2.2 Mean and Variance of a Sum of Random Variables

The main results of this section are some useful formulas for the mean and variance of sums of random variables expressed in terms of the means and variances of the individual summands. We begin with the simplest case of two summands. All of our results are consequences of the following exceedingly useful theorem, whose proof is omitted.

■ **THEOREM 5.1**

Let (X, Y) denote a random vector with joint probability function $f(x, y)$, and let $\phi(x, y)$ denote a function of two real variables x, y. Then $\phi(X, Y)$ is a random variable and

$$E(\phi(X, Y)) = \sum_x \sum_y \phi(x, y)f(x, y) \quad \text{(discrete case)} \qquad (5.17)$$

■

Examples and Applications

1. **The addition rule for computing** $E(aX + bY)$**.** For random variables X and Y and arbitrary constants a and b, we have

$$E(aX + bY) = aE(X) + bE(Y) \qquad (5.18)$$

Derivation of Eq. (5.18). Choosing $\phi(x, y) = ax + by$ in Eq. (5.17) and computing the summations yields

$$E(aX + bY) = \sum_x \sum_y (ax + by)f(x, y)$$

$$= \sum_x \sum_y axf(x, y) + \sum_y \sum_x byf(x, y)$$

$$= \sum_x ax \sum_y f(x, y) + \sum_y by \sum_x f(x, y)$$

$$= a \sum_x xf_X(x) + b \sum_y yf_Y(y)$$

$$= aE(X) + bE(Y)$$

2. **The addition rule for** $E(\sum_{1 \le i \le n} a_i X_i)$**.** We now extend Eq. (5.18) to the expectation of a finite sum of random variables. Let $X_1, \ldots, X_n$ be a sequence of random variables and $a_1, \ldots, a_n$ be a sequence of arbitrary constants; then

$$E\left(\sum_{1 \le i \le n} a_i X_i\right) = \sum_{1 \le i \le n} a_i E(X_i) \tag{5.19}$$

The straightforward proof of Eq. (5.19) is omitted.

---

example 5.8   The addition rule (5.19) will now be used to compute the expected value of a binomial random variable given earlier [Eq. (3.41)]. A Bernoulli random variable X_i has expected value p. The random variable $T = \sum_{1 \le i \le n} X_i$ has a binomial distribution. Therefore,

$$E(T) = E\left(\sum_{1 \le i \le n} X_i\right) = \sum_{1 \le i \le n} E(X_i)$$

$$= p + \cdots + p = np$$ ∎

The Covariance Cov(X, Y)

Let X and Y be two random variables with means μ_X, μ_Y and variances σ_X^2, σ_Y^2. We seek a formula for the variance of the sum, denoted by $V(X + Y)$. The computation of $V(X + Y)$ requires the new concept of the *covariance* of X and Y, denoted Cov(X, Y) and defined by

$$\text{Cov}(X, Y) = E((X - \mu_X)(Y - \mu_Y)) \tag{5.20}$$

It corresponds to choosing

$$\phi(x, y) = (x - \mu_X)(y - \mu_Y)$$

in Eq. (5.17). The covariance measures how the two variables vary together. Suppose, for example, that (X, Y) represents the heights of a mother and her son. We would expect that a mother taller than average (so $X > \mu_X$) is more likely to have a son taller than average (so $Y > \mu_Y$), with the inequalities reversed when the mother is shorter than average. This suggests that the product $(X - \mu_X)(Y - \mu_Y)$ is likely to be positive since both factors are likely to have the same sign. In other words, in this case we would expect that $\text{Cov}(X, Y) = E((X - \mu_X)(Y - \mu_Y)) \ge 0$.

Intuitively, a positive value of the covariance indicates that X and Y tend to increase together whereas a negative value indicates that an increase in X is accompanied by a decrease in Y. A classic example of the latter is the drop in stock market averages that accompanies an increase in interest rates.

As in the case of the variance, there is a shortcut formula for the computation of the covariance:

$$\text{Cov}(X, Y) = E(XY) - \mu_X \mu_Y \tag{5.21}$$

Derivation of Eq. (5.21). Expand the product

$$(X - \mu_X)(Y - \mu_Y) = XY - \mu_X Y - \mu_Y X + \mu_X \mu_Y$$

Then take expected values of both sides:

$$\begin{aligned}
E((X - \mu_X)(Y - \mu_Y)) &= E(XY - \mu_X Y - \mu_Y X + \mu_X \mu_Y) \\
&= E(XY) - \mu_X E(Y) - \mu_Y E(X) + \mu_X \mu_Y \\
&= E(XY) - 2\mu_X \mu_Y + \mu_X \mu_Y \\
&= E(XY) - \mu_X \mu_Y
\end{aligned}$$

Note that

$$\text{Cov}(X, X) = V(X)$$

---

example **5.9** Compute $\text{Cov}(X, Y)$ for the random vector with joint probability function given in Table 5.5.

Solution. This problem highlights the computational efficiency of the shortcut formula. For instance, there is only one nonzero summand in the computation of $E(XY) = 1 \times 1 \times (6/28)$. A routine computation using the marginal distribution functions yields $\mu_X = 0.5$, $\mu_Y = 0.75$. Consequently,

$$\text{Cov}(X, Y) = 0.214 - 0.5 \times 0.75 = -0.1607 \qquad \blacksquare$$

Proposition 5.2 tells us that the covariance of independent random variables equals zero.

▨ PROPOSITION 5.2

Let X and Y be independent random variables. Then

$$E(XY) = E(X)E(Y) \tag{5.22}$$

and therefore

$$\text{Cov}(X, Y) = 0 \tag{5.23}$$

$\blacksquare$

Proof. The hypothesis of independence means that $f(x, y) = f_X(x)f_Y(y)$; consequently,

$$\begin{aligned}
E(XY) &= \sum_x \sum_y xy f(x, y) \\
&= \sum_x \sum_y xy f_X(x) f_Y(y) \\
&= \sum_x x f_X(x) \sum_y y f_Y(y) \\
&= E(X)E(Y)
\end{aligned}$$

Derivation of Eq. (5.23). If X and Y are independent, then so are $X - \mu_X$ and $Y - \mu_Y$. Moreover, $E(X - \mu_X) = E(Y - \mu_Y) = 0$; consequently,

$$\begin{aligned}
\text{Cov}(X, Y) &= E((X - \mu_X)(Y - \mu_Y)) \\
&= E(X - \mu_X)E(Y - \mu_Y) = 0
\end{aligned}$$

Addition Rule for Computing the Variance of a Sum of Random Variables

The following addition rule is useful for computing the variance of the sum in terms of the variances and covariances of the summands.

$$V(X + Y) = V(X) + V(Y) + 2\operatorname{Cov}(X, Y) \tag{5.24}$$

$$V\left(\sum_{1 \leq i \leq n} X_i\right) = \sum_{1 \leq i \leq n} V(X_i) + 2\sum_{i < j} \operatorname{Cov}(X_i, X_j) \tag{5.25}$$

Derivation of Eq. (5.24). We proceed in the usual way by first writing down the definition of $V(X + Y)$ and then expanding and rearranging the terms:

$$\begin{aligned}
V(X + Y) &= E([X + Y - \mu_{X+Y}]^2) \\
&= E([(X - \mu_X) + (Y - \mu_Y)]^2) \\
&= E((X - \mu_X)^2) + E(Y - \mu_Y)^2) + 2E((X - \mu_X)(Y - \mu_Y)) \\
&= V(X) + V(Y) + 2\operatorname{Cov}(X, Y)
\end{aligned}$$

The derivation of Eq. (5.25) is similar and is omitted.

The Addition Rule for Computing the Variance of a Sum of Independent Random Variables

When X and Y are independent [so $\operatorname{Cov}(X, Y) = 0$], Eq. (5.25) implies that the variance of the sum of two independent random variables equals the sum of the variances:

$$V(X + Y) = V(X) + V(Y) \tag{5.26}$$

Similarly, Eq. (5.25) implies that the variance of a sum of mutually independent independent random variables $X_1, \dots, X_n$ [so $\operatorname{Cov}(X_i, X_j) = 0, i \neq j$] equals the sum of the variances:

$$V(X_1 + \cdots + X_n) = V(X_1) + \cdots + V(X_n) \tag{5.27}$$

---

example **5.10**   Equation (5.27) will now be used to compute the variance of a binomial random variable given earlier [Eq. (3.42)]. The distribution of $\sum_{1 \leq i \leq n} X_i$, where the summands are mutually independent Bernoulli random variables, is binomial with parameters n, p. A Bernoulli random variable X_i has variance $p(1 - p)$. The sum of the variances appearing on the right-hand side of Eq. (5.27) equals $np(1 - p)$. Consequently, the variance of a binomial random variable with parameters n, p equals $np(1 - p)$.   ∎

More generally, the variance of the linear combination

$$a_1 X_1 + \cdots + a_n X_n$$

where $X_1, \dots, X_n$ is a sequence of independent random variables and $a_1, \dots, a_n$ is a sequence of constants, is given by

$$\begin{aligned}
V(a_1 X_1 + \cdots + a_n X_n) &= V(a_1 X_1) + \cdots + V(a_n X_n) \\
&= a_1^2 V(X_1) + \cdots + a_n^2 V(X_n)
\end{aligned} \tag{5.28}$$

Computing the Covariance under a Change of Scale

Recall that changing the system of units in which X is measured (from Fahrenheit to Celsius, for example) means replacing X by $aX + b$; this is called a change of scale. Unfortunately, the usefulness of the covariance as a measure of association is marred by the fact that its value depends on the system of units used. The following proposition states how $\text{Cov}(X, Y)$ behaves under a simultaneous change of scale for X and Y.

▨ **PROPOSITION 5.3**

Let the random variables $aX + b, cY + d$ be obtained from X, Y via a linear change of scale. Then

$$\text{Cov}(aX + b, cY + d) = ac\,\text{Cov}(X, Y) \qquad (5.29)$$

■

Proof. The details of the proof are left as an exercise (Prob. 5.17).

The Correlation Coefficient

We now introduce the concept of the *correlation coefficient* $\rho(X, Y)$, defined by

$$\rho(X, Y) = \frac{\text{Cov}(X, Y)}{\sigma_X \sigma_Y} \qquad (5.30)$$

We say that the random variables are *uncorrelated* when $\rho = 0$. Independent random variables are always uncorrelated because their covariance $\text{Cov}(X, Y) = 0$, which implies that $\rho(X, Y) = 0$ as well. The converse is not true; that is, it is easy to construct a pair of uncorrelated random variables that are dependent (see Prob. 5.36).

The correlation coefficient measures the strength of the linear relationship between the random variables X and Y. For instance, it can be shown that $\rho = \pm 1$ implies that constants a, b exist such that $Y = aX + b$; if $\rho = 1$, then $a > 0$, whereas $\rho = -1$ implies $a < 0$. The correlation coefficient plays an important role in regression analysis, covered in (Chap. 10), and so we defer a more detailed discussion of it until then.

The correlation coefficient has the following properties:

$$-1 \le \rho(X, Y) \le 1 \qquad (5.31)$$
$$\rho(aX + b, cY + d) = \rho(X, Y) \quad \text{if } ac > 0 \qquad (5.32)$$
$$\rho(aX + b, cY + d) = -\rho(X, Y) \quad \text{if } ac < 0 \qquad (5.33)$$

The derivations of Eqs. (5.32) and (5.33) are left as an exercise (Prob. 5.18). The derivation of Eq. (5.31) requires the use of Schwarz's inequality and is therefore omitted.

Notation. When there is no chance of confusion, we drop the explicit dependence on X, Y and write

$$\rho = \rho(X, Y)$$

example **5.11**   Refer to the joint probability function of Table 5.5. Compute the correlation coefficient.

Solution. The essential steps in the computation of ρ follow:

$$E(X) = 0.5, \qquad E(Y) = 0.75$$
$$V(X) = 0.32, \qquad V(Y) = 0.40$$
$$E(XY) = \frac{6}{28}$$

and therefore

$$\text{Cov}(X, Y) = 0.214 - 0.5 \times 0.75 = -0.1607$$
$$\rho(X, Y) = \frac{-0.1607}{\sqrt{0.32 \times 0.40}} = -0.449$$

Because $\text{Cov}(X, Y) \neq 0$, it follows that X and Y are not independent. ∎

example **5.12**   Suppose X and Y are random variables with

$$E(X) = 4, \qquad E(Y) = -1, \qquad E(X^2) = 41, \qquad E(Y^2) = 10, \qquad E(XY) = 6$$

Using only this information:

1. Compute $V(X + Y)$.

   Solution. The basic idea of the computation is to use Eq. (5.24), which means we first compute $V(X)$, $V(Y)$, and $\text{Cov}(X, Y)$. Now

   $$V(X) = E(X^2) - E(X)^2 = 41 - 16 = 25$$
   $$V(Y) = E(Y^2) - E(Y)^2 = 10 - 1 = 9$$
   $$\text{Cov}(X, Y) = E(XY) - E(X)E(Y) = 6 - (4 \times -1) = 10$$
   $$V(X + Y) = V(X) + V(Y) + 2\,\text{Cov}(X, Y) = 25 + 9 + 2 \times 10 = 54$$

2. Compute the correlation coefficient.

   Solution

   $$\rho = \frac{\text{Cov}(X, Y)}{\sigma_X \sigma_Y} = \frac{10}{5 \times 3} = \frac{10}{15} = 0.667$$

3. Are X and Y independent?

   Solution. No, because $\rho = 0.667 \neq 0$. ∎

5.2.3   The Law of Large Numbers for Sums of iid Random Variables

It has been observed that the relative frequency of heads in a large number of tosses of a fair coin approaches $1/2$. The *law of large numbers* is a precise statement of the proposition that the long-run relative frequency of an event approaches a limit that equals its probability. In its most general form it is an assertion about the limiting behavior of the sample mean of an iid sequence of random variables. The sample mean of a sequence of iid random variables is denoted by $\overline{X}$ and is defined by

$$\overline{X} = \frac{\sum_{1 \le i \le n} X_i}{n}$$

The Mean and Variance of the Sample Mean $\overline{X}$

We compute the mean and variance of $\overline{X}$ by using the addition rules (5.19) and (5.27) for a sum of independent random variables with

$$a_i = \frac{1}{n}, \qquad E(X_i) = \mu, \qquad V(X_i) = \sigma^2$$

Consequently, the mean and variance of $\overline{X}$ are given by

$$E(\overline{X}) = \mu \quad \text{and} \quad V(\overline{X}) = \frac{\sigma^2}{n} \tag{5.34}$$

▨ **THEOREM 5.2**

The law of large numbers. Let $X_1, \ldots, X_n$ denote a sequence of independent random variables with $E(X_i) = \mu$ and $V(X_i) = \sigma^2$. Then for every $d > 0$,

$$\lim_{n \to \infty} P(|\overline{X} - \mu| > d) = 0 \tag{5.35}$$

Proof. Applying Chebyshev's inequality to $\overline{X}$, we see that

$$P(|\overline{X} - \mu| > d) \le \frac{\sigma^2}{nd^2}$$

Consequently,

$$\lim_{n \to \infty} P(|\overline{X}_n - \mu| > d) \le \lim_{n \to \infty} \frac{\sigma^2}{nd^2} = 0 \qquad ∎$$

The Law of Large Numbers for the Sample Proportion

The proportion of successes in a Bernoulli sequence of trials is called the *sample proportion* and is denoted by $\hat{p}$, where

$$\hat{p} = \frac{\sum_{1 \le i \le n} X_i}{n} - \overline{X}$$

so

$$E(\hat{p}) = p \quad \text{and} \quad V(\hat{p}) \le \frac{1}{4n} \qquad \text{(Prob. 3.27)}$$

Using Chebyshev's inequality, we obtain the following upper bound on the probability that the sample proportion deviates from the true proportion by an amount greater than or equal to the quantity d:

$$P(|\hat{p} - p| \geq d) \leq \frac{p(1-p)}{nd^2}$$

$$\leq \frac{1}{4nd^2}$$

(5.36)

Applying the law of large numbers to the sample proportion $\hat{p}$ yields the following result:

$$\lim_{n \to \infty} P(|\hat{p} - p| \geq d) = 0 \tag{5.37}$$

In other words, as the sample size increases, the distribution of $\hat{p}$ becomes more concentrated around the true proportion p. For example, a Gallup poll, using a sample size of $n = 3500$, predicted that Ronald Reagan would win the 1980 election with 55.3 percent of the popular vote. In this case the sample proportion $\hat{p}$ equals 0.553, which is just the proportion of voters in the sample who said they would vote for Reagan. On election day, the final results were somewhat different, as Reagan's actual percentage of the vote was 51.6 percent; the margin of error (predicted $-$ actual) equals 3.7 percent.

example 5.13   Use Chebyshev's inequality to estimate the probability that the sample proportion obtained from a poll of voters based on a sample of size $n = 3500$ differs from the true proportion by an amount less than or equal to 0.03.

Solution. We have to compute the probability

$$P(|\hat{p} - p| < 0.03) = 1 - P(|\hat{p} - p| \geq 0.03)$$

Using inequality (5.36) with $n = 3500$, we obtain the upper bound

$$P(|\hat{p} - p| \geq 0.03) \leq \frac{1}{4 \times 3500 \times (0.03)^2} = 0.0794$$

Therefore,

$$P(|\hat{p} - p| < 0.03) = 1 - P(|\hat{p} - p| \geq 0.03)$$
$$\geq 1 - 0.0794 = 0.9206 \quad \blacksquare$$

The next example illustrates an important link between the binomial distribution and the empirical distribution function.

example 5.14   Let $X_1, \ldots, X_n$ denote a sequence of mutually independent random variables with the same distribution $F(x)$. We call such a sequence a *random sample of size* n from the df $F(x)$; another frequently used term is *iid sequence*. Show that the random variable $n\hat{F}_n(x)$ has a binomial distribution with parameters n, $p = F(x)$.

Solution. This result plays an important role in linking probability theory to statistical inference, so we state it separately as a proposition.

▓▓ PROPOSITION 5.4

Let $\hat{F}_n(x)$ be the empirical distribution of a random sample $X_1, \ldots, X_n$ taken from the df $F(x)$. Then $n\hat{F}_n(x)$ has a binomial distribution with parameters $n, p = F(x)$.   ∎

Proof. Define the related sequence Y_i as follows:

$$Y_i = 1, \quad \text{if } X_i \le x$$
$$Y_i = 0, \quad \text{if } X_i > x$$

In other words, $Y_i = 1$ if the ith element of the sample $X_i \le x$, and $Y_i = 0$ otherwise. Thus, the sum $\sum_{1 \le i \le n} Y_i$ counts the number of observations in the sample that are less than or equal to x. Consequently,

$$\#\{X_i \le x\} = \sum_{1 \le i \le n} Y_i$$

and therefore

$$\hat{F}_n(x) = \frac{\#\{X_i \le x\}}{n} = \frac{\sum_{1 \le i \le n} Y_i}{n}$$

Note that the Y_i are mutually independent since the X_i are; in addition, they take on only the two values 0 and 1; therefore, the sequence $Y_1, \ldots, Y_n$ forms a Bernoulli sequence of trials with

$$P(Y_i = 1) = P(X_i \le x) = F(x)$$

We have thus shown that

$$n\hat{F}_n(x) = \sum_{1 \le i \le n} Y_i \tag{5.38}$$

Consequently, $n\hat{F}_n(x)$ has a binomial distribution with parameters $n, p = F(x)$, as claimed.   ∎

The Fundamental Theorem of Mathematical Statistics

Let $\hat{F}_n(x)$ denote the empirical distribution function corresponding to the iid sequence $X_1, \ldots, X_n$ with the distribution $F(x)$. The *fundamental theorem of mathematical statistics,* in its simplest form, states that the empirical distribution function $\hat{F}_n(x)$ resembles the distribution function $F(x)$ from which the random sample is drawn. In detail, according to Proposition 5.4, $n\hat{F}_n(x)$ has a binomial distribution with parameters $n, p = F(x)$. Applying Theorem 5.2 in this context yields the following version of the fundamental theorem of mathematical statistics:

$$\text{For every } d > 0, \quad \lim_{n \to \infty} P(|\hat{F}_n(x) - F(x)| > d) = 0 \tag{5.39}$$

In other words, the empirical distribution function $\hat{F}_n(x)$ converges to the (unknown) df $F(x)$; that is,

$$\lim_{n \to \infty} \hat{F}_n(x) = F(x) \tag{5.40}$$

PROBLEMS

5.10 The random vector (X, Y) has the joint probability function specified in the following table:

	Y			
X	2	3	4	5
0	1/24	3/24	1/24	1/24
1	1/12	1/12	3/12	1/12
2	1/12	1/24	1/12	1/24

Compute $\text{Cov}(X, Y)$ and $\rho(X, Y)$.

5.11 Refer to the joint probability function in Table 5.2. Compute $\text{Cov}(X, Y)$.

5.12 Suppose X and Y are random variables with

$$E(X) = 2, \qquad E(Y) = 3, \qquad E(X^2) = 53, \qquad E(Y^2) = 45, \qquad E(XY) = -21$$

Using only this information, compute:
(a) $V(X + Y)$
(b) $V(2X - 3Y)$
(c) The correlation coefficient.
(d) Are X and Y independent?

5.13 Machines 1 and 2 have i breakdowns per day ($i = 0, 1, 2, 3, 4$). The probabilities for i for each machine are listed in the following table:

	Number of breakdowns				
	0	1	2	3	4
Machine 1	0.3	0.2	0.2	0.2	0.1
Machine 2	0.4	0.3	0.1	0.1	0.1

(a) Assuming that the two machines operate independently of one another, compute the jpf of X_1, the number of machine 1 breakdowns, and X_2, the number of machine 2 breakdowns.
(b) Compute the probability that machine 1 has more breakdowns than machine 2.
(c) Compute the probability that machines 1 and 2 have the same number of breakdowns.
(d) Compute the probability function of the total number of breakdowns $T = X_1 + X_2$.

5.14 *Weird dice:* Consider two six-sided dice whose faces are labeled as follows:
$$\text{Die 1} = \{1, 2, 2, 3, 3, 4\}$$
$$\text{Die 2} = \{1, 3, 4, 5, 6, 8\}$$

Let Y_i denote the number that turns up when the ith die is rolled. Assume that the two random variables Y_1 and Y_2 are independent. Compute the probability function of $Y = Y_1 + Y_2$ and display your results in the format of a horizontal bar chart as in Table 3.3.

5.15 The random variables X and Y are independent with $E(X) = 2$, $V(X) = 9$, $E(Y) = -3$, $V(Y) = 16$. Compute:
(a) $E(3X - 2Y)$
(b) $V(3X)$
(c) $V(-2Y)$
(d) $V(3X - 2Y)$

5.16 Two numbers are selected at random and without replacement from the set

$$\{1,1,1,1,0,0,0,0,0,0\}$$

Denote the ith number that is drawn by X_i $(i = 1, 2)$.
(a) Compute the probability function of the sample mean $\overline{X}$.
(b) Compute $E(\overline{X})$ and $V(\overline{X})$.

5.17 (a) Show that $\mathrm{Cov}(aX, cY) = ac\,\mathrm{Cov}(X, Y)$.
(b) Let X, Y, W be random variables. Show that

$$\mathrm{Cov}(X + Y, W) = \mathrm{Cov}(X, W) + \mathrm{Cov}(Y, W)$$

(c) Give the details of the derivation of Eq. (5.29).

5.18 Derive the properties (5.32) and (5.33) of the correlation coefficient.

5.19 Let $(X_1, \ldots, X_n)$ be mutually independent random variables having the discrete uniform distribution [see Eq. (3.17)].
(a) Compute the probability function for the random variable M_n defined by

$$M_n = \max(X_1, \ldots, X_n)$$

[*Hint:* Compute $P(M_n \le x)$.]
(b) Compute the probability function for the random variable L_n defined by

$$L_n = \min(X_1, \ldots, X_n)$$

[*Hint:* Compute $P(L_n \ge x)$.]

5.3   THE MULTINOMIAL DISTRIBUTION

Recall that the mathematical model of n independent repetitions of an experiment with two mutually exclusive and exhaustive outcomes is called a *Bernoulli trials process.* We now generalize this model to the case where each experiment results in one of k mutually exclusive and exhaustive outcomes denoted by $C_1, C_2, \ldots, C_k$; these outcomes are sometimes called *cells* or *categories.* We denote their probabilities by

$$P(C_i) = p_i \quad \text{where} \sum_{1 \le i \le k} p_i = 1$$

The values $p_i = P(C_i)$ are called *cell probabilities*. We use the term *multinomial experiment* to describe this model.

---

example 5.15   Rolling a six-sided die n times is a simple example of a multinomial experiment. Here, of course, $k = 6$ and C_i denotes the event that an i is thrown. The assertion that the die is fair is equivalent to the assertion that $P(C_i) = 1/6$. ■

Let $Y_1, Y_2, \ldots, Y_k$ denote the respective frequencies of the outcomes $C_1, C_2, \ldots, C_k$, and note that $\sum_{1 \le i \le k} Y_i = n$, so these random variables are *not* independent. A formula for the joint distribution of these variables, called the *multinomial distribution*, is

$$P(Y_1 = y_1, \ldots, Y_k = y_k) = \binom{n}{y_1, \ldots, y_k} p_1^{y_1} \cdots p_k^{y_k}$$

$$= \frac{n!}{y_1! \cdots y_k!} p_1^{y_1} \cdots p_k^{y_k}$$

(5.41)

We do not derive this formula in detail, except to point out that it is similar to the counting argument used to derive the expression for the multinomial coefficients in Eq. (2.22). Observe that each Y_i has a binomial distribution with parameters n, p_i (why?); consequently, $E(Y_i) = np_i$. The observed value y_i is called the *observed frequency*, and the value np_i is called the *expected frequency* or *theoretical frequency*.

---

example 5.16   Consider the experiment of tossing three fair coins and repeating this experiment eight times. Let C_i ($i = 0, 1, 2, 3$) denote the event that i heads were thrown. Let Y_i equal the number of times i heads occurred. Compute (1) $E(Y_i)(i = 0, 1, 2, 3)$ and (2) $P(Y_0 = 1, Y_1 = 3, Y_2 = 3, Y_3 = 1)$.

Solution. The random vector (Y_0, Y_1, Y_2, Y_3) has a multinomial distribution with probabilities p_i given by

$$p_i = P(C_i) = \binom{3}{i} 2^{-3}, \quad i = 0, 1, 2, 3$$

and $n = 8$ repetitions of the experiment.

1. Each Y_i has the binomial distribution with parameters $n = 8$ and p_i as just defined. Consequently,

$$E(Y_i) = 8p_i = \binom{3}{i}, \quad i = 0, 1, 2, 3$$

2. We compute $P(Y_0 = 1, Y_1 = 3, Y_2 = 3, Y_3 = 1)$ by inserting the values $y_0 = 1, y_1 = 3, y_2 = 3, y_3 = 1, n = 8$ (and the values for p_i just computed) into Eq. (5.41). Thus,

$$P(Y_0 = 1, Y_1 = 3, Y_2 = 3, Y_3 = 1) = \frac{8!}{1!3!3!1!} \frac{1}{8} \left(\frac{3}{8}\right)^3 \left(\frac{3}{8}\right)^3 \frac{1}{8}$$

$$= 0.0487 \qquad \blacksquare$$

The multinomial distribution plays an important role in connection with the chi-square test, discussed in Sec. 9.3.

5.4   THE POISSON PROCESS

The Poisson distribution is more than just an approximation to the binomial; it, rightly so, is now regarded as one of the three fundamental distributions, together with the binomial and the normal, in probability theory. It plays an important role in operations research, particularly in the analysis of *queuing systems.*

---

example **5.17**   A *queuing system* denotes any service facility to which customers or jobs arrive, receive service, and then depart. We use the word *customer* in a general sense to refer to a telephone call arriving at a telephone exchange, an order for a component stocked in a warehouse, a broken component brought to a repair shop, or a packet of digital data arriving at some node in a complex computer network. The *service time S* denotes the amount of time required to service the customer. The length of a telephone call and the time to repair a broken component are examples of service times. In most applications both the customer arrivals and their service times are random variables. Queuing systems are classified according to the

- *Input process,* which denotes the probability distribution of the customer arrivals
- *Service distribution,* which denotes the probability distribution of the service time
- *Queuing discipline,* which refers to the order of service [e.g., first come first served (FCFS)] $\blacksquare$

---

example **5.18**   We now show that, under intuitively reasonable hypotheses, the input process to a queuing system has a Poisson distribution. To fix our ideas, let $X(t)$ denote the number of phone calls, say, arriving at an exchange during the time interval $[0, t]$. In the general case we speak of an "event" occurring in the time interval $[0, t]$. Notice that the number of telephone calls received during the time interval $[t_1, t_2]$ equals $X(t_2) - X(t_1)$. We are interested in computing the probability function $P(X(t) = k)$. Under certain reasonable conditions, to be explained, we show that $X(t)$ has a Poisson distribution with parameter λt, i.e.,

$$P(X(t) = k) = e^{-\lambda t} \frac{(\lambda t)^k}{k!}, \quad k = 0, 1, \dots \qquad (5.42)$$

Equation (5.42) has some interesting consequences that are worth noting before we proceed to its derivation. Since $X(t)$ has a Poisson distribution with parameter λt, it follows at once that $E(X(t)) = \lambda t$. Consequently, the expected number of phone calls in an interval of time of length t is proportional to the length of the interval. Notice that the constant $\lambda = E(X(t))/t$, so λ is the expected (or average) number of phone calls per unit time; for this reason λ is called the *intensity* of the Poisson process.

An Intuitive Derivation of Eq. (5.42). Partition the interval $[0, t]$ into n subintervals of equal length t/n. With each subinterval we associate the Bernoulli random variable X_i defined as follows:

$X_i = 1$ if a phone call was received during the ith time interval

$X_i = 0$ if no phone call was received during the ith time interval

We observe that this model is mathematically equivalent to tossing a coin n times, with the role of heads or tails now played by the arrival or nonarrival of a phone call during the ith time interval. To complete the analogy, we make the following reasonable assumptions:

- The probability p_n that a phone call arrives is the same for each subinterval, and the number of arrivals in disjoint time intervals are mutually independent.
- The probability p_n is approximately proportional to the length of the interval, that is,

$$p_n \approx \lambda \left(\frac{t}{n} \right)$$

Consequently, $S_n = \sum_{1 \le i \le n} X_i$ has a binomial distribution with parameters n, p_n, and it is reasonable to hope that S_n is a good approximation to $X(t)$, especially as $n \to \infty$. Since $p_n \approx \lambda t/n$, it follows at once that

$$\lim_{n \to \infty} np_n = \lambda t$$

Therefore, by the Poisson approximation to the binomial, we get

$$P(X(t) = k) = \lim_{n \to \infty} P(S_n = k) = e^{-\lambda t} \frac{(\lambda t)^k}{k!}, \quad k = 0, 1, \ldots \quad (5.43)$$

■

---

example 5.19   Consider a warehouse in which auto parts are stocked to satisfy consumer demands. Suppose orders for parts arrive at the warehouse according to a Poisson process at the rate $\lambda = 1.5$ per hour. The warehouse is open 8 hours per day. The warehouse begins the day with an initial inventory of 15 parts.

1. What is the probability that the warehouse exhausts its inventory after 4 hours of operation?
2. What is the probability that the inventory is exhausted after 8 hours of operation?
3. What is the probability that at the end of 8 hours the warehouse has exactly 5 parts in inventory? At least 5 parts in inventory?

Solution. Let $X(t)$ denote the number of orders for the component during the time period $[0, t]$, where t is measured in hours. $X(t)$ has a Poisson distribution with parameter $1.5t$; consequently,

$$P(X(t) = k) = p(k; 1.5t), \quad k = 0, 1, \dots$$

1. Here $t = 4$ and the problem is reduced to computing $P(X(4) \geq 15)$. But $X(4)$ has a Poisson distribution with parameter $1.5 \times 4 = 6$. Using appendix Table A.2, we see that $P(X(4) \geq 15) = 1 - 0.9995 = 0.0005$.

2. Here $t = 8$, and therefore $X(8)$ has a Poisson distribution with parameter $1.5 \times 8 = 12$. Proceeding exactly as in part 1, we obtain $P(X(8) \geq 15) = 1 - 0.772 = 0.228$.

3. The warehouse will have 5 components in inventory provided $X(8) = 10$; therefore, the answer is given by $P(X(8) = 10) = p(10; 12) = 0.105$. The warehouse will have at least 5 components in inventory provided $X(8) \leq 10$; the problem is solved by computing $P(X(8) \leq 10) = 0.347$. ∎

The input process $X(t)$ has several additional properties that are worth considering in some detail. For example, the same reasoning that led us to conclude that the number of phone calls arriving during the time interval of length t is Poisson distributed with parameter λt can also be used to show that the number of phone calls arriving during any subinterval $[s, s + t]$ of length t is Poisson distributed with parameter λt. Moreover, it is intuitively clear that the number of phone calls in non-overlapping time intervals are mutually independent random variables in the sense of Definition 5.3. We summarize this discussion in the following definition.

▩ **DEFINITION 5.5**

$X(t), t \geq 0$, is a *Poisson process* with intensity $\lambda > 0$ if

1. For $s \geq 0$ and $t > 0$ the random variable $X(s + t) - X(s)$ has the Poisson distribution with parameter λt, i.e.,

$$P(X(t + s) - X(s) = k) = e^{-\lambda t}\frac{(\lambda t)^k}{k!}, \quad k = 0, 1, \dots$$

and

2. For any time points $0 = t_0 < t_1 < \cdots < t_n$, the random variables
$$X(t_1) - X(t_0), X(t_2) - X(t_1), \dots, X(t_n) - X(t_{n-1})$$
are mutually independent. ∎

The Poisson process is an example of a *stochastic process,* a collection of random variables indexed by the time parameter t.

5.5 APPLICATIONS OF BERNOULLI RANDOM VARIABLES TO RELIABILITY THEORY (OPTIONAL)

Recall that a Bernoulli random variable X assumes only the two values 0 and 1. In spite of their deceptively simple structure, Bernoulli random variables are exceedingly useful in a wide variety of problems, some of which have already been discussed. Additional applications are now given.

example 5.20   Consider a complex system (e.g., a space telescope or nuclear reactor) consisting of n components, each of which has only two states: a functioning state and a failed state. The state of the ith component can be represented as a Bernoulli random variable X_i, where $X_i = 1$ means the ith component is functioning and $X_i = 0$ means the ith component has failed. The *reliability* p_i of the ith component is defined by

$$p_i = P(X_i = 1) = E(X_i)$$

Similarly, the state of the system as a whole is a Bernoulli random variable denoted by X, where $X = 1$ means that the system is functioning and $X = 0$ means that the system has failed. The reliability r of the system is defined to be

$$r = P(X = 1) = E(X) \tag{5.44}$$

The state of the system X is called the *system function*. It is a function, possibly a complicated one, of the states of its components. Reliability engineers classify an n-component system into one of three classes: a *series* system, a *parallel* system, or a *k-out-of-n* system.

1. **Series system.** A system of components connected in series functions if and only if each of its components functions (Fig. 5.1). Consequently, the system function X is given by

$$X = X_1 \times X_2 \times \cdots \times X_n \tag{5.45}$$

Equation (5.45) simply states that $X = 1$ if and only if each factor on the right equals 1.

Figure 5.1 Series system

2. **Parallel system.** A parallel system functions if and only if *at least* one of its components functions (Fig. 5.2). Consequently, the system function is given by

$$X = 1 - (1 - X_1)(1 - X_2) \cdots (1 - X_n) \tag{5.46}$$

Suppose, for example, that $X_2 = 1$, so $1 - X_2 = 0$; this implies that the product $(1 - X_1)(1 - X_2) \cdots (1 - X_n) = 0$ and thus $X = 1$. In words, if the second component is functioning, then so is the system. The same argument clearly shows that $X = 1$ if $X_i = 1$ for at least one i.

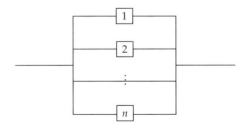

Figure 5.2 Parallel system

3. **k-out-of-n system.** A k-out-of-n system functions if and only if at least k of the n components function. An airplane that functions if and only if *at least* two of its three engines function is an example of a 2-out-of-3 system. The function for a 2-out-of-3 system is given by

$$X = X_1 X_2 X_3 + X_1 X_2 (1 - X_3) + X_1 (1 - X_2) X_3 + (1 - X_1) X_2 X_3 \quad (5.47)$$

∎

Computing the Reliability of a System

Avionic systems are designed so that the failure of one component does not lead to a chain reaction of failures of the other components. In particular, the three jet engines are installed in such a way that that their systems are *independent* of one another, so that the failure of a fuel pump or oil filter in one engine does not affect the operation of similar components in the other two engines. This notion of independence is expressed mathematically as follows:

▨ **DEFINITION 5.6**

Two components are independent if and only if their corresponding system functions X_1 and X_2 are mutually independent random variables. More generally, n components are mutually independent if their corresponding system functions $X_1, \ldots, X_n$ are mutually independent random variables:

$$P(X_1 = x_1, \ldots, X_n = x_n) = P(X_1 = x_1) \cdots P(X_n = x_n) \quad (5.48)$$

∎

If we make the assumption that the components are independent, then the reliability of the system is easily computed directly from Eqs. (5.45)–(5.47).

1. **Reliability of a series system.** According to Eq. (5.45), the system function is $X = X_1 \times \cdots \times X_n$, and the reliability of the system is given by

$$r = E(X) = E(X_1 \times \cdots \times X_n)$$
$$= E(X_1) \times \cdots \times E(X_n)$$
$$= p_1 \cdots p_n$$

2. **Reliability of a parallel system.** The function for a parallel system is $X = 1 - (1 - X_1)(1 - X_2) \cdots (1 - X_n)$, and the reliability of a parallel system is

$$r = E(X) = E(1 - (1 - X_1) \cdots (1 - X_n))$$
$$= 1 - E(1 - X_1) \cdots E(1 - X_n)$$
$$= 1 - (1 - p_1) \cdots (1 - p_n)$$

3. **Reliability of a 2-out-of-3 system.** Recall that the system function for a 2-out-of-3 system is given by

$$X = X_1 X_2 X_3 + X_1 X_2 (1 - X_3) + X_1 (1 - X_2) X_3 + (1 - X_1) X_2 X_3$$

Therefore, the reliability of this system is given by

$$r = E(X) = p_1 p_2 p_3 + p_1 p_2 (1 - p_3) + p_1 (1 - p_2) p_3 + (1 - p_1) p_2 p_3$$

If, in addition to being independent, the components have the same reliability $p = P(X_i = 1)$, then the previous formulas simplify to the following:

$$r = p^n \quad \text{(series system)} \tag{5.49}$$
$$r = 1 - (1 - p)^n \quad \text{(parallel system)} \tag{5.50}$$
$$r = p^3 + 3p^2(1 - p) \quad \text{(2-out-of-3 system)} \tag{5.51}$$

---

example 5.21   Let us now apply the general theory to some specific examples.

1. Suppose $p = 0.95$. How many components must be linked in parallel so that
   the system reliability $r = 0.99$?

   Solution. It is clear that n satisfies the equation
   $$1 - (1 - 0.95)^n = 0.99$$

   Equivalently,
   $$0.05^n = 0.01$$

   Taking logarithms of both sides yields
   $$n = \frac{\ln 0.01}{\ln 0.05} = 1.5372$$

   Since n must be the *smallest* integer satisfying the inequality $n \geq 1.5372$, we
   must choose $n = 2$.

2. Suppose the system consists of $n = 3$ components linked in parallel. What
   must the value of p be so that $r = 0.99$?

   Solution. In this case we have to solve the following equation for p:
   $$1 - (1 - p)^3 = 0.99$$

   which yields
   $$p = 1 - 0.01^{1/3} = 0.7846$$

3. *Calculating the reliability of a k-out-of-n system.* Assume that the system consists
   of n independent components, each of which has the same reliability p. The
   system functions if and only if at least k out of the n components function.
   It is easy to see that the total number of functioning components X has a
   binomial distribution with parameters n, p. Consequently, the reliability of
   the system is given by

   $$r = P(X \geq k) = \sum_{k \leq x \leq n} \binom{n}{x} p^x (1 - p)^{n-x} \qquad \blacksquare$$

PROBLEMS

5.20 A die is thrown six times. Let $Y_i = $ the number of times that i appears.
   (a) Compute $P(Y_1 = 1, \ldots, Y_6 = 1)$.
   (b) Compute $P(Y_1 = 2, Y_3 = 2, Y_5 = 2)$.
   (c) Compute $E(Y_i), i = 1, \ldots, 6$.

5.21 The output of a steel plate manufacturing plant is classified into one of three categories: no defects, minor defects, and major defects. Suppose that $P(\text{no defects}) = 0.75, P(\text{minor defects}) = 0.20$, and $P(\text{major defects}) = 0.05$. A random sample of 20 steel plates is inspected. Let Y_1, Y_2, Y_3 denote the respective numbers of steel plates with no defects, minor defects, and major defects.
   (a) Compute $E(Y_i), i = 1, 2, 3$.
   (b) Compute $P(Y_1 = 15, Y_2 = 4, Y_3 = 1)$.

5.22 The "placebo effect" refers to treatment where a substantial proportion of the patients report a significant improvement even though the treatment consists of nothing more than giving the patient a sugar pill or some other harmless inert substance. Ten patients are given a new treatment for symptoms of a cold. The patients' reactions fall into one of three categories: helped, harmed, and no effect. From previous experience with placebos, it is known that $P(\text{helped}) = 0.50, P(\text{harmed}) = 0.30$, and $P(\text{no effect}) = 0.20$. Let Y_1, Y_2, Y_3 denote the respective numbers of patients in the three categories (helped, harmed, no effect). Assuming that the new treatment is no more effective than a placebo, compute:
   (a) $E(Y_i), i = 1, 2, 3$
   (b) $P(Y_1 = 5, Y_2 = 3, Y_3 = 2)$

5.23 Suppose that math SAT scores are $N(500, 100^2)$ distributed. Six students are selected at random. What is the probability that of their math SAT scores, two are less than or equal to 450, two are between 450 and 600, and two are greater than or equal to 600?

5.24 Telephone calls arrive according to a Poisson process $X(t)$ with intensity $\lambda = 2$ per minute. Compute:
   (a) $E(X(1)), E(X(5))$
   (b) $E(X(1)^2), E(X(5)^2)$
   (c) What is the probability that no phone calls arrive during the interval 9:10 to 9:12?

5.25 $X(t)$ is a Poisson process with intensity $\lambda = 1.5$. Compute:
   (a) $P(X(2) \le 3)$
   (b) $P(X(2) = 2, X(4) = 6)$. [*Hint:* $X(2)$ and $X(4) - X(2)$ are independent random variables.]

5.26 By considering the various possibilities, verify that the 2-out-of-3 system function is given by Eq. (5.47).

5.27 For the following system compute:
   (a) The system function X in terms of $X_1, \ldots, X_n$.
   (b) The reliability of the system, assuming the components are independent with $P(X_i = 1) = p$.

5.28 For the following system compute:

(a) The system function X in terms of $X_1, \ldots, X_n$.

(b) The reliability of the system, assuming the components are independent with $P(X_i = 1) = p$.

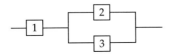

5.29 *High-priority freight service* (*Source:* Wayne Nelson, *Applied Life Data Analysis*, New York, John Wiley & Sons, 1982.) A high-priority freight train requires three locomotives for a one-day run. If any one of these locomotives fails, the train is delayed and the railroad must pay a large penalty. Analysis of statistical data indicates that the locomotives fail independently and their failure times can be approximated by an exponential distribution with expected failure time $\lambda^{-1} = 43.3$ days. Thus, this is a series system.

(a) Compute the reliability of the system.

(b) To reduce the chance of delay, suppose the railroad adds a fourth locomotive; thus, the train is delayed only if two or more locomotives fail. This is an example of a 3-out-of-4 system. Determine the reliability provided by the addition of a fourth locomotive.

5.6 THE JOINT DENSITY FUNCTION (CONTINUOUS CASE)

Intuitively, a density function of a continuous random variable X is a non-negative function $f(x)$ that assigns to each subinterval of the line $[a, b]$ a probability (or a mass) defined by the definite integral

$$P(a \leq X \leq b) = \int_a^b f(x)\, dx$$

Thus, $P(a \leq X \leq b)$ can be represented as the area under the curve $y = f(x)$, $a \leq x \leq b$.

In a similar way, the joint density function of a continuous random vector (X, Y) is a nonnegative function of two variables $f(x, y)$ that assigns to each rectangle in the plane

$$[a, b] \times [c, d] = \{(x, y) : a \leq x \leq b, c \leq y \leq d\}$$

a probability (or mass) defined by the double integral

$$P(a \leq X \leq b, c \leq Y \leq d) = \int_a^b \int_c^d f(x, y)\, dy\, dx \qquad (5.52)$$

The joint density function of a continuous random vector is sketched in Fig. 5.3.

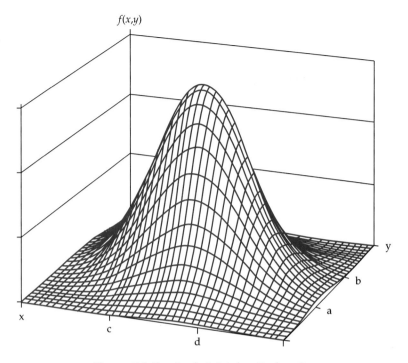

$f(x,y)$

y

b

a

x

c

d

Figure 5.3 Graph of a joint density function

Thus, $P(a \leq X \leq b, c \leq Y \leq d)$ can be represented as the volume underneath the surface $z = f(x, y)$ with rectangular base $[a, b] \times [c, d]$.

▬ **DEFINITION 5.7**

$f(x, y)$ is a joint density function (jdf) if it satisfies the following two conditions:

$$f(x, y) \geq 0 \tag{5.53}$$

$$\int_{-\infty}^{\infty} \int_{-\infty}^{\infty} f(x, y) \, dy \, dx = 1 \tag{5.54}$$

∎

---

example 5.22   Let

$$f(x, y) = \tfrac{1}{3}(x + y), \quad 0 < x < 1, \quad 0 < y < 2$$

$$f(x, y) = 0 \quad \text{elsewhere}$$

To verify that $f(x, y)$ is a joint density function, we must show that

$$\int_{-\infty}^{\infty} \int_{-\infty}^{\infty} f(x, y) \, dy \, dx = \tfrac{1}{3} \int_0^1 \int_0^2 (x + y) \, dy \, dx = 1$$

It suffices to show that

$$\int_0^1 \int_0^2 (x + y)\, dy\, dx = 3$$

We evaluate this double integral in the usual way, by first evaluating

$$\int_0^2 (x + y)\, dy = xy + \frac{y^2}{2}\bigg|_{y=0}^{y=2} = 2x + 2$$

and then evaluating

$$\int_0^1 (2x + 2)\, dx = x^2 + 2x\bigg|_0^1 = 3$$

This proves that $f(x, y)$ is a joint density function. ∎

The Marginal Density Functions

The probability density functions of the random variables X and Y are called the *marginal probability density functions*. Proposition 5.5 gives a formula for computing them.

▇ PROPOSITION 5.5

If (X, Y) have the joint density function $f(x, y)$, then the *marginal probability density functions* $f_X(x)$ and $f_Y(y)$ are given by

$$f_X(x) = \int_{-\infty}^{\infty} f(x, y)\, dy$$

and

$$f_Y(y) = \int_{-\infty}^{\infty} f(x, y)\, dx$$ ∎

---

example 5.23   Compute the marginal pdf's and expected values for the random variables X, Y whose joint density function is given in Example 5.22.

Solution. The marginal pdf's are obtained by evaluating the following integrals:

$$f_X(x) = \int_{-\infty}^{\infty} f(x, y)\, dy$$

$$= \int_0^2 \tfrac{1}{3}(x + y)\, dy = \tfrac{1}{3}(2x + 2), \quad 0 < x < 1$$

$$f_X(x) = 0 \quad \text{elsewhere}$$

Notice that the answer $f_X(x) = \frac{1}{3}(2x + 2)$, $-\infty < x < \infty$, cannot possibly be correct, since $f_X(-3) = -4/3 < 0$, and a marginal density function can never assume a negative value. Similarly,

$$f_Y(y) = \int_{-\infty}^{\infty} f(x, y)\, dx$$

$$= \int_0^1 \tfrac{1}{3}(x + y)\, dx = \tfrac{1}{3}(\tfrac{1}{2} + y),\ 0 < y < 2$$

$$f_Y(y) = 0 \quad \text{elsewhere}$$

We compute $E(X)$ and $E(Y)$ by evaluating the following integrals:

$$E(X) = \int_0^1 x\tfrac{1}{3}(2x + 2)\, dx = \tfrac{5}{9}$$

$$E(Y) = \int_0^2 y\tfrac{1}{3}(\tfrac{1}{2} + y)\, dy = \tfrac{11}{9} \qquad \blacksquare$$

The concept of a joint density function is easily extended to three or more random variables.

DEFINITION 5.8

$f(x_1, x_2, \ldots, x_n)$ is the joint density function of n continuous random variables $X_1, X_2, \ldots, X_n$ if

$$P(a_1 \le X_1 \le b_1, \ldots, a_n \le X_n \le b_n) = \int_{a_1}^{b_1} \cdots \int_{a_n}^{b_n} f(x_1, \ldots, x_n)\, dx_n \cdots dx_1 \qquad \blacksquare$$

example 5.24 In this example we consider the concept of a uniformly distributed random vector. We give an informal definition first. We say that (X, Y) is uniformly distributed over a region $\mathcal{R}$ of finite area $A(\mathcal{R})$ if the probability that the random vector is in $\mathcal{C} \subset \mathcal{R}$ is proportional to the area of $\mathcal{C}$. The formal definition follows.

DEFINITION 5.9

Let $\mathcal{R}$ denote a region of the plane with finite area $A(\mathcal{R})$, and let $\mathcal{C} \subset \mathcal{R}$. The random vector (X, Y) is *uniformly distributed over the region* $\mathcal{R}$ if

$$P((X, Y) \in \mathcal{C}) = \frac{A(\mathcal{C})}{A(\mathcal{R})} \qquad \blacksquare$$

Let us now derive the formula for the joint density function $f(x, y)$. Recall that the area $A(\mathcal{R})$ can be expressed as the double integral of the function $f(x, y) \equiv 1$ computed over the region $\mathcal{R}$:

$$A(\mathcal{R}) = \int_{\mathcal{R}} \int 1\, dy\, dx$$

The joint density function in this case is the (piecewise) defined function

$$f(x,y) = \frac{1}{A(\mathcal{R})} \quad \text{for } (x,y) \in \mathcal{R}$$

$$f(x,y) = 0 \quad \text{otherwise} \qquad \blacksquare$$

---

example 5.25 Let (X, Y) have the joint density given by

$$f(x,y) = 1, \quad 0 < x < 1, \quad 0 < y < 1$$

$$f(x,y) = 0 \quad \text{elsewhere}$$

Here $\mathcal{R}$ is the unit rectangle, so $A(\mathcal{R}) = 1$.

1. To compute $P(X < 0.4, Y < 0.6) = P(0 < X < 0.4, \ 0 < Y < 0.6)$, we only have to compute the area of the rectangle $\mathcal{C} = \{(x,y) : 0 < x < 0.4, \ 0 < y < 0.6\}$, which equals $0.4 \times 0.6 = 0.24$.
2. To compute $P(X < Y)$, we first sketch the region

$$\mathcal{C} = \{(x,y) : x < y \quad \text{and} \quad 0 \le x \le 1, 0 \le y \le 1\}$$

and then compute its area. It is easy to verify that $\mathcal{C}$ is the triangle with vertices $(0,0), (0,1), (1,1)$, so its area equals $1/2$; consequently, $P(X < Y) = 1/2$. $\blacksquare$

Independent Random Variables

The definition of independent random variables in the continuous case is similar to the discrete-case definition [Eq. (5.13)].

▬ PROPOSITION 5.6

The continuous random variables X and Y are independent if and only if their joint density function is the product of their marginal density functions, i.e.,

$$f(x,y) = f_X(x)f_Y(y), \quad \text{for all } x, y \tag{5.55}$$

More generally, the random variables $X_1, \ldots, X_n$ are *mutually independent* if and only if their joint density function is the product of their marginal densities, i.e.,

$$f(x_1, \ldots, x_n) = f_{X_1}(x_1) \times \cdots \times f_{X_n}(x_n) \tag{5.56}$$

$\blacksquare$

5.6.1   Functions of Random Vectors

The concepts of independence, covariance, and correlation are easily extended to continuous random vectors; the only difference is that sums are replaced by integrals. In particular, all the results concerning the expected value and variance of a sum of discrete random variables obtained in Sec. 5.2.3 carry over without change. For this reason we shall content ourselves with a brief outline of the basic results.

A Formula for $E(\phi(X, Y))$

To compute $E(\phi(X, Y))$ in the continuous case, we use the following theorem, which is the analogue of Eq. (5.17).

▬ THEOREM 5.3

Let (X, Y) denote a continuous random vector with joint density function $f(x, y)$, and let $\phi(x, y)$ denote a function of two variables. Then $\phi(X, Y)$ is a random variable and

$$E\left(\phi(X, Y)\right) = \int_{-\infty}^{\infty} \int_{-\infty}^{\infty} \phi(x, y) f(x, y)\, dx\, dy \tag{5.57}$$

■

Examples and Applications

1. **The addition rule for computing** $E(aX + bY)$**.** For random variables X and Y and arbitrary constants a and b, we have

$$E(aX + bY) = aE(X) + bE(Y)$$

Proof. Let $\phi(x, y) = ax + by$ in Eq. (5.57); then use the fact that the double integral of a sum of two functions is the sum of the double integrals:

$$E(aX + bY) = \int_{-\infty}^{\infty} \int_{-\infty}^{\infty} (ax + by) f(x, y)\, dx\, dy$$

$$= \int_{-\infty}^{\infty} \int_{-\infty}^{\infty} axf(x, y)\, dx\, dy + \int_{-\infty}^{\infty} \int_{-\infty}^{\infty} byf(x, y)\, dx\, dy$$

$$= \int_{-\infty}^{\infty} ax \left[\int_{-\infty}^{\infty} f(x, y)\, dy \right] dx + \int_{-\infty}^{\infty} by \left[\int_{-\infty}^{\infty} f(x, y)\, dx \right] dy$$

$$= a \int_{-\infty}^{\infty} xf_X(x)\, dx + b \int_{-\infty}^{\infty} yf_Y(y)\, dy$$

$$= aE(X) + bE(Y)$$

2. Using this result, we can extend the addition rule [Eq. (5.19)] to sums of continuous random variables:

$$E\left(\sum_{1 \leq i \leq n} a_i X_i \right) = \sum_{1 \leq i \leq n} a_i E(X_i)$$

3. **Computing** $E(XY)$. Let $\phi(x, y) = xy$ in Eq. (5.57); then evaluate the double integral

$$E(XY) = \int_{-\infty}^{\infty} \int_{-\infty}^{\infty} xyf(x, y)\, dx\, dy$$

---

example 5.26

Compute $E(XY)$ for the random vector of Example 5.22.

Solution. We compute the double integral

$$\int_0^1 \int_0^2 xy\frac{1}{3}(x + y)\, dy\, dx = E(XY)$$

in the usual way, by first integrating with respect to y,

$$\int_0^2 xy\frac{1}{3}(x + y)\, dy = \frac{1}{3}\left(\frac{x^2y^2}{2} + \frac{xy^3}{3}\right)\Bigg|_{y=0}^{y=2} = \frac{2}{3}x^2 + \frac{8}{9}x$$

and then integrating with respect to x:

$$\int_0^1 \left(\frac{2}{3}x^2 + \frac{8}{9}x\right) dx = \frac{2}{3}$$

Thus, $E(XY) = 2/3$. ∎

4. **The expected value of the product of independent random variables.** Proposition 5.2 states that the expected value of the product of two independent (discrete) random variables equals the product of their expected values. In the next proposition we extend this result to the case where the random variables are continuous.

■ **PROPOSITION 5.7**

Let X and Y be independent random variables. Then

$$E(XY) = E(X)E(Y) \tag{5.58}$$

■

Proof. If X and Y are independent, then their joint density function $f(x, y) = f_X(x)f_Y(y)$; therefore,

$$E(XY) = \int_{-\infty}^{\infty} \int_{-\infty}^{\infty} xyf_X(x)f_Y(y)\, dx\, dy$$

$$= \int_{-\infty}^{\infty} xf_X(x)\, dx \cdot \int_{-\infty}^{\infty} yf_Y(y)\, dy$$

$$= E(X)E(Y)$$

5. **A formula for Cov**(X, Y)**.** We define the covariance as in the discrete case and compute it by evaluating the double integral:

$$\text{Cov}(X, Y) = E((X - \mu_X)(Y - \mu_Y))$$
$$= \int_{-\infty}^{\infty} \int_{-\infty}^{\infty} (x - \mu_X)(y - \mu_Y) f(x, y) \, dy \, dx$$

6. The shortcut formula for the covariance of X and Y still holds:
$$\text{Cov}(X, Y) = E(XY) - \mu_X \mu_Y$$

▓ PROPOSITION 5.8

Let X and Y be independent random variables. Then
$$\text{Cov}(X, Y) = 0 \tag{5.59}$$

■

Proof. We omit the proof since it is the same as the one given in the discrete case [see the derivation of Eq. (5.23)].

---

example 5.27 Show that the random variables of Example 5.22 are not independent.

Solution. In our previous discussion we showed that $E(X) = 5/9$, $E(Y) = 11/9$, and $E(XY) = 2/3$; consequently,
$$\text{Cov}(X, Y) = \tfrac{2}{3} - \tfrac{55}{81} = -0.0123 \neq 0$$
Since $\text{Cov}(X, Y) \neq 0$, it follows that X and Y are not independent. ■

7. **The correlation coefficient.** The correlation coefficient is defined by
$$\rho(X, Y) = \frac{\text{Cov}(X, Y)}{\sigma_X \sigma_Y}$$
which is the same formula as in the discrete case [see Eq. (5.30)].

---

example 5.28 Compute the correlation coefficient for the random variables X and Y of Example 5.22.

Solution. In the previous example we showed that $\text{Cov}(X, Y) = -0.0123$. A straightforward computation, the details of which are omitted, yields the result
$$V(X) = 0.0802 \quad \text{and} \quad \sigma(X) = 0.2832$$
$$V(Y) = 0.2840 \quad \text{and} \quad \sigma(Y) = 0.5329$$

Thus,
$$\rho(X, Y) = \frac{-0.123}{0.2832 \times 0.5329} = -0.0818$$
■

8. **Addition rule for computing the variance of a sum of random variables.** The following addition rule for computing the variance of a sum of random variables in terms of the variances and covariances of the summands is an extension of formulas previously derived for discrete random variables. We omit the derivations.

$$V(X + Y) = V(X) + V(Y) + 2 \operatorname{Cov}(X, Y)$$

$$V\left(\sum_{1 \le i \le n} X_i\right) = \sum_{1 \le i \le n} V(X_i) + 2 \sum_{i < j} \operatorname{Cov}(X_i, X_j)$$

5.6.2   Conditional Distributions and Conditional Expectations

The conditional distribution of Y given $X = x$ defines a new probability function that we denote by the symbol $f(y \mid x)$ [see Eq. (5.9)]. In particular, it has a mean called the *conditional expectation of Y given* $X = x$, denoted $E(Y \mid X = x)$ and defined by the equation

$$E(Y \mid X = x) = \sum_y y f(y \mid x) \tag{5.60}$$

Observe that $E(Y \mid X = x)$ is a function of the random variable X that equals $E(Y \mid X = x)$ when $X = x$; this random variable is denoted $E(Y \mid X)$ and is called the conditional expectation of Y given X. We now prove the interesting and useful result that its expectation equals $E(Y)$.

▬ THEOREM 5.4

The expected value of the conditional expectation $E(Y \mid X)$ is $E(Y)$. In symbols,

$$E(E(Y \mid X)) = E(Y) \tag{5.61}$$

■

Proof

$$E(E(Y \mid X)) = \sum_x E(Y \mid X = x) f_X(x)$$

$$= \sum_x \left[\sum_y y f(y \mid x) \right] f_X(x)$$

$$= \sum_y y \left[\sum_x f(y \mid x) f_X(x) \right]$$

$$= \sum_y y \left[\sum_x f(x, y) \right]$$

$$= \sum_y y f_Y(y) = E(Y)$$

Regression of Y on X

Statisticians call the function $E(Y \mid X = x)$ the *regression of Y on X*. The following notation is also used:

$$\mu_{Y|x} = E(Y \mid X = x)$$

Because of its importance, we restate it in the form of a definition.

■■■ **DEFINITION 5.10**

The function

$$\mu_{Y|x} = E(Y \mid X = x)$$

is called the *regression of Y on X*.   ■

The Continuous Case

We now extend the concepts of conditional distribution and conditional expectation to the case where the random vector (X, Y) is of the continuous type. We emphasize that the underlying concepts are the same in both the discrete and continuous cases; the only difference is that we replace sums by integrals in some of the definitions and computations.

■■■ **DEFINITION 5.11**

Let (X, Y) be a continuous random vector with joint density function $f(x, y)$. The *conditional density of $Y = y$ given $X = x$*, denoted $f(y \mid x)$, is defined to be

$$f(y \mid x) = \frac{f(x, y)}{f_X(x)}, \quad f_X(x) > 0 \tag{5.62}$$

$$f(y \mid x) = 0 \quad \text{elsewhere} \qquad ■$$

We leave it to the reader to verify that $f(y \mid x)$ defined by Eq. (5.62) is a pdf for each x such that $f_X(x) > 0$.

The conditional density $f(y \mid x)$ has an expectation that is given by

$$E(Y \mid X = x) = \int_{-\infty}^{\infty} yf(y \mid x) \, dy \tag{5.63}$$

We call $E(Y \mid X = x)$ the *conditional expectation of Y given $X = x$*. Note that, as in the discrete case discussed previously, $E(Y \mid X = x)$ is a function of the random variable X that equals $E(Y \mid X = x)$ when $X = x$; this random variable is denoted $E(Y \mid X)$ and is also called the conditional expectation of Y given X. The analogue of Eq. (5.61) remains valid:

$$E(E(Y \mid X)) = E(Y) \tag{5.64}$$

Proof of Eq. (5.64)

$$E(E(Y \mid X)) = \int E(Y \mid X = x) f_X(x) \, dx$$

$$= \int \left[\int yf(y \mid x) \, dy \right] f_X(x) \, dx$$

$$= \int y \left[\int f(y \mid x) f_X(x) \, dx \right] dy$$

$$= \int y \left[\int f(x, y) \, dx \right] dy$$

$$= \int yf_Y(y) \, dy = E(Y)$$

The function $E(Y \mid X = x)$ is also called the *regression of Y on X* and is denoted $\mu_{Y|x}$.

---

example 5.29   We continue our study of Example 5.22. In this case

$$f(y \mid x) = \frac{(x+y)/3}{(2x+2)/3} = \frac{x+y}{2x+2}, \quad 0 < x < 1, \quad 0 < y < 2$$

so

$$E(Y \mid X = x) = \int_0^2 yf(y \mid x)\, dy = \frac{1}{2x+2}\int_0^2 y(x+y)\, dy = \frac{6x+8}{6x+6}, \quad 0 < x < 1$$

∎

5.6.3   The Bivariate Normal Distribution

We have seen that under certain reasonable hypotheses a random variable X has an approximate normal distribution; this is the content of the normal approximation to the binomial and its generalization, the central limit theorem, which we study in Sec. 6.2.1. It is also true (but not easy to prove) that under certain reasonable conditions, too technical to state here, sums of mutually independent random vectors have an approximate *bivariate normal distribution*, defined as follows.

▆ DEFINITION 5.12

The random vector (X, Y) has a bivariate normal distribution with mean vector (μ_X, μ_Y), variances (σ_X^2, σ_Y^2), and correlation coefficient ρ if

$$f(x, y) = \frac{1}{2\pi\sigma_X\sigma_Y\sqrt{1-\rho^2}}$$

$$\times \exp\left\{ -\frac{1}{2(1-\rho^2)}\left[\left(\frac{x-\mu_X}{\sigma_X}\right)^2 - 2\rho\frac{(x-\mu_X)(y-\mu_Y)}{\sigma_X\sigma_Y} + \left(\frac{y-\mu_Y}{\sigma_Y}\right)^2 \right] \right\}$$

$$(5.65)$$

∎

It can be shown by a straightforward but tedious calculation that if (X, Y) has the bivariate normal density function (5.65), then

$$E(X) = \mu_X, \qquad E(Y) = \mu_Y$$

$$V(X) = \sigma_X^2, \qquad V(Y) = \sigma_Y^2, \qquad \rho(X, Y) = \rho$$

In addition, the marginal and conditional distributions are also normally distributed. More precisely, it can be shown that

$$f_X(x) = N(\mu_X, \sigma_X^2) \tag{5.66}$$

$$f_Y(y) = N(\mu_Y, \sigma_Y^2) \tag{5.67}$$

$$f(y \mid x) = N\left(\mu_Y + \rho\frac{\sigma_Y}{\sigma_X}(x - \mu_X), \sigma_Y^2(1 - \rho^2)\right) \tag{5.68}$$

$$f(x \mid y) = N\left(\mu_X + \rho\frac{\sigma_X}{\sigma_Y}(y - \mu_Y), \sigma_X^2(1 - \rho^2)\right) \qquad (5.69)$$

Consequently,

$$\mu_{Y|x} = \mu_Y + \rho\frac{\sigma_Y}{\sigma_X}(x - \mu_X) \qquad (5.70)$$

$$\mu_{X|y} = \mu_X + \rho\frac{\sigma_X}{\sigma_Y}(y - \mu_Y) \qquad (5.71)$$

---

example 5.30   Suppose the heights of sons (Y) and mothers (X) (in inches) is a bivariate normal distribution with

$$\mu_X = 62.48, \qquad \mu_Y = 68.65, \qquad \sigma_X = 5.712, \qquad \sigma_Y = 7.344, \qquad \rho = 0.5$$

Then

$$\mu_{Y|X=x} = 68.65 + 0.5 \times \frac{7.344}{5.712} \times (x - 62.48)$$

For a mother with height 65 inches, this formula "predicts" that the son's height will be

$$\mu_{Y|X=65} = 68.65 + 0.5 \times \frac{7.344}{5.712} \times (65 - 62.48) = 70.27 \text{ inches.} \qquad \blacksquare$$

PROBLEMS

5.30 Let (X, Y) have the joint density given by

$$f(x, y) = 1/4, \quad -1 < x < 1, \quad -1 < y < 1$$
$$f(x, y) = 0 \quad \text{elsewhere}$$

Compute:

(a) $P(X < 0.4, Y < 0.6)$ (d) $P(Y > 2X)$

(b) $P(X < 0.4), P(Y < 0.6)$ (e) $P(Y = X)$

(c) $P(X < Y)$ (f) $P(X + Y < 0.5), P(X + Y < 1.5)$

(*Hint:* No integrations are required. Just sketch the regions and compute the areas using elementary geometry.)

5.31 Given

$$f(x, y) = c(x + 2y), \quad 0 < x < 1, \quad 0 < y < 1$$
$$f(x, y) = 0 \quad \text{elsewhere}$$

compute:

(a) c (c) μ_X, μ_Y

(b) $f_X(x), f_Y(y)$ (d) σ_X, σ_Y

5.32 Let the random vector (X, Y) have the joint density function
$$f(x, y) = xe^{-xy-x}, \quad x > 0, \quad y > 0$$
$$f(x, y) = 0 \quad \text{elsewhere}$$

Compute:
(a) $f_X(x)$, μ_X, σ_X
(b) $f_Y(y)$, μ_Y, σ_Y
(c) Are X and Y independent?

5.33 Let (X, Y) be uniformly distributed over the unit circle $\{(x, y) : x^2 + y^2 \leq 1\}$. Its joint density function is given by

$$f(x, y) = \frac{1}{\pi}, \quad x^2 + y^2 \leq 1$$

$$f(x, y) = 0 \quad \text{elsewhere}$$

Compute:
(a) $P(X^2 + Y^2 \leq 1/4)$ **(d)** $P(Y < 2X)$
(b) $P(X > Y)$ **(e)** Let $R = X^2 + Y^2$.
(c) $P(X = Y)$ Compute $F_R(r) = P(R \leq r)$.

(*Hint:* No calculations involving double integrals are necessary. Just sketch the region of integration, and evaluate the probability by first computing the corresponding area.)

5.34 Let X, Y have the joint pdf
$$f(x, y) = c(x + xy), \quad 0 \leq x \leq 1, \quad 0 \leq y \leq 1$$
$$f(x, y) = 0 \quad \text{elsewhere}$$

(a) Compute the value of c.
(b) Compute the marginal pdf's $f_X(x)$ and $f_Y(y)$.
(c) Are the random variables X and Y independent?
(d) Compute $P(X + Y \leq 1)$ by first writing it as a double integral of the joint density function $f(x, y)$ over an appropriate region of the plane. Then evaluate the double integral.

5.35 Let X, Y have the joint pdf
$$f(x, y) = \tfrac{6}{7}(x + y)^2, \quad 0 \leq x \leq 1, \quad 0 \leq y \leq 1$$
$$f(x, y) = 0 \quad \text{elsewhere}$$

By integrating over an appropriate region, compute the following probabilities:
(a) $P(X + Y \leq 1)$
(b) $P(Y > X)$
(c) Compute the marginal pdf's $f_X(x)$ and $f_Y(y)$.

5.36 Let (X, Y) be uniformly distributed over the unit circle $\{(x, y) : x^2 + y^2 \leq 1\}$. Its joint density function is given by

$$f(x, y) = \frac{1}{\pi}, \quad x^2 + y^2 \leq 1$$

$$f(x, y) = 0 \quad \text{elsewhere}$$

(a) Compute the marginal distributions.

(b) Show that $\text{Cov}(X, Y) = 0$ but that X and Y are not independent.

5.37 Let X, Y have the joint density function given by

$$f(x, y) = \tfrac{3}{5}(xy + y^2), \quad 0 \le x \le 2, \quad 0 \le y \le 1$$
$$f(x, y) = 0 \quad \text{elsewhere}$$

Compute:

(a) $P(Y < \tfrac{1}{2} \mid X < \tfrac{1}{2})$ (b) $f(y \mid x)$ (c) $E(Y \mid X = x)$

5.38 Let the random vector (X, Y) have the joint density function

$$f(x, y) = xe^{-xy-x}, \quad x > 0, \quad y > 0$$
$$f(x, y) = 0 \quad \text{elsewhere}$$

Compute:

(a) $f(y \mid x)$ (b) $\mu_{Y|x}$

5.39 Let X and Y have a bivariate normal distribution with parameters

$$\mu_X = 2, \qquad \mu_Y = 1, \qquad \sigma_X^2 = 9, \qquad \sigma_Y^2 = 16, \qquad \rho = 3/4$$

Compute:

(a) $P(Y < 1)$ (c) $E(Y \mid X = 0)$

(b) $P(Y < 1 \mid X = 0)$ (d) $V(X + Y)$

5.40 Let X and Y have a bivariate normal distribution with parameters

$$\mu_X = 2, \qquad \mu_Y = 1, \qquad \sigma_X^2 = 9, \qquad \sigma_Y^2 = 16, \qquad \rho = -3/4$$

Compute:

(a) $P(Y < 3)$ (c) $E(Y \mid X = 2)$

(b) $P(Y < 3 \mid X = 2)$ (d) $V(X + Y)$

5.41 Let X and Y have a bivariate normal distribution with parameters

$$\mu_X = 2, \qquad \mu_Y = 1, \qquad \sigma_X^2 = 9, \qquad \sigma_Y^2 = 16, \qquad \rho = -2/3$$

Compute:

(a) $P(X < 4)$ (c) $E(X \mid Y = 1)$

(b) $P(X < 4 \mid Y = 1)$ (d) $V(X + Y)$

5.7 MATHEMATICAL DETAILS AND DERIVATIONS

Computing the Mean and Variance of the Hypergeometric Distribution

Our purpose here is to give the details of the proof of Proposition 3.2. Although it is possible to compute the mean directly via the formula

$$E(X) = \sum_{0 \le j \le n} jh(j)$$

it is actually easier to first represent X as a sum of Bernoulli random variables and then use Eq. (5.19). Let $X_i = 1$ if the ith bead drawn is red, and let

$X_i = 0$ otherwise; clearly, $X = X_1 + X_2 + \cdots + X_n$ and therefore $E(X) = E(X_1) + E(X_2) + \cdots + E(X_n)$. Because each bead has an equal chance of being selected on the ith draw, it follows that $P(X_i = 1) = D/N$, since there are N beads and D of them are red. Moreover, the random variables X_i are Bernoulli random variables; consequently,

$$E(X_i) = P(X_i = 1) = \frac{D}{N}, \quad i = 1, \ldots, n$$

$$V(X_i) = \frac{D}{N}\left(1 - \frac{D}{N}\right), \quad i = 1, \ldots, n$$

Therefore,

$$E(X) = E(X_1) + E(X_2) + \cdots + E(X_n) = n \times \frac{D}{N}$$

$$V(X) = \sum_{1 \le i \le n} V(X_i) + 2\sum_{i<j} \mathrm{Cov}(X_i, X_j)$$

$$= n \times \frac{D}{N}\left(1 - \frac{D}{N}\right) + 2\sum_{i<j} \mathrm{Cov}(X_i, X_j)$$

We have thus reduced the computation of the variance to the task of computing $\mathrm{Cov}(X_i, X_j)$. The details from here on are somewhat more complicated, so we content ourselves with a sketch, omitting some tedious algebraic manipulations. First observe that

$$\mathrm{Cov}(X_i, X_j) = E(X_i X_j) - E(X_i)E(X_j)$$

$$= \frac{D}{N}\frac{D-1}{N-1} - \left(\frac{D}{N}\right)^2$$

$$= -\frac{D(N-D)}{N^2(N-1)}$$

Therefore,

$$2\sum_{i<j} \mathrm{Cov}(X_i, X_j) = -2\binom{n}{2}\frac{D(N-D)}{N^2(N-1)}$$

Putting these results together yields

$$V(X) = \sum_{1 \le i \le n} V(X_i) + 2\sum_{i<j} \mathrm{Cov}(X_i, X_j)$$

$$= n \times \frac{D}{N}\left(1 - \frac{D}{N}\right) - 2\binom{n}{2}\frac{D(N-D)}{N^2(N-1)}$$

$$= \frac{N-n}{N-1} \times n \times \frac{D}{N}\left(1 - \frac{D}{N}\right)$$

5.8   CHAPTER SUMMARY

The joint distribution function tells us how one variable influences another. It is also a useful tool for deriving formulas for the mean and variance of a sum of random variables. Statisticians measure the interactions between variables by computing conditional distributions, correlations, and conditional expectations. Stochastic processes, such as the Poisson process, are characterized by their joint distribution functions.

To Probe Further. Several texts offer an elementary introduction to stochastic processes and reliability theory. Among the better ones are:

1. H. M. Taylor and S. Karlin, *An Introduction to Stochastic Modeling*, Orlando, Fla., Academic Press, 1984.
2. S. M. Ross, *Introduction to Probability Models*, 4th ed., San Diego, Academic Press, 1989.
3. R. Barlow and F. Proschan, *Theory of Reliability and Life Testing*, New York, Holt, Rinehart & Winston, 1975.

The story of how Galton discovered the bivariate normal distribution is presented in S. M. Stigler's *History of Statistics: The Measurement of Uncertainty before 1900* (Cambridge, Mass., Belknap Press of Harvard University Press, 1986).

The quotation at the head of this chapter is found in T. M. Porter, *The Rise of Statistical Thinking* (Princeton, N. J., Princeton University Press, 1986, p. 161).

The nontransitive dice problem (5.7) appears in C. Wang, *Sense and Nonsense of Statistical Inference* (New York, Marcel Dekker, 1993), and the weird dice problem (5.14) appears in J. Gallian, *Math Horizons*, Feb. 1995, pp. 30–31.

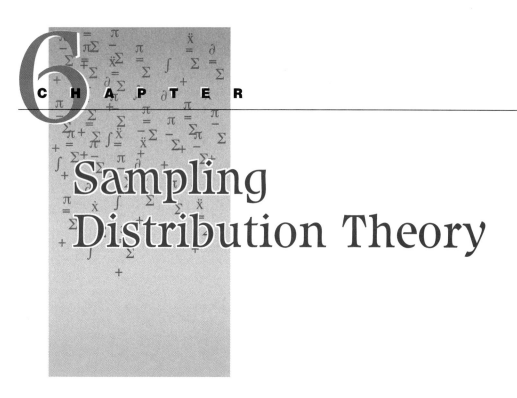

6 CHAPTER

Sampling Distribution Theory

Everybody believes in the normal approximation, the experimenters because they think it is a mathematical theorem, the mathematicians because they think it is an experimental fact.

G. Lippman (1845–1921), French physicist

6.1 ORIENTATION

In this chapter we study the distribution of the sample mean. The exact distribution of the sample mean of a random sample taken from a normal distribution is itself normal. The *central limit theorem* asserts that for large values of the sample size n the distribution of the sample mean is approximately normal with mean μ and standard deviation $\sigma / \sqrt{n}$. We also introduce the *gamma, chi-square* (denoted χ^2), t, and F distributions. These distributions are related to one another as well as to the normal distribution itself. For instance, the chi-square distribution is a special case of the gamma distribution, and it can be shown that t^2 has an F distribution (see Sec. 6.4.2). The exponential and *Erlang* distributions, which are omnipresent in applied probability theory, are both special cases of the gamma distribution.

Organization of Chapter

Section 6.4   The Distribution of the Sample Variance
Section 6.5   Mathematical Details and Derivations

6.2 SAMPLING FROM A NORMAL DISTRIBUTION

Many problems in probability theory and statistical inference can be reduced to the study of the distribution of the *sample total* $T = \sum X_i$ and the *sample mean* $\overline{X} = \sum \overline{X}_i/n$ of a sequence of independent and identically distributed random variables $X_1, \ldots, X_n$. The common distribution function is called the *parent distribution*. The mathematical model of a *random sample* taken from a distribution is a sequence of independent and identically distributed (iid) random variables. The formal definition follows.

▒ DEFINITION 6.1

A *random sample* of size n is a sequence of n mutually independent random variables $X_1, \ldots, X_n$ with a common distribution function $F(x) = P(X_i \leq x)$, $i = 1, 2, \ldots, n$.   ■

The term *sampling from a normal population* refers to the important special case when the parent df is a normal distribution. Snedecor and Cochran (*Statistical Methods*, 7th ed., Ames, Iowa State University Press, 1980) list several situations in which it is reasonable to assume that the sample comes from a normal population.

1. The distributions of many variables such as heights, weights, and SAT scores are approximately normal.
2. Even if the distribution of X_i is not normal, the distribution of the sample mean $\overline{X}$ is approximately normal, provided the sample size is sufficiently large.
3. Sometimes a suitable transformation of the data such as $Y_i = g(X_i)$, $i = 1, 2, \ldots, n$, produces a data set that is approximately normal.
4. Last, but not least, the normal distribution is mathematically convenient to work with.

Since the random variables X_i are identically distributed, they have a common mean $E(X_i) = \mu$ and common variance $V(X_i) = \sigma^2$. We call μ and σ^2 the *population mean* and *population variance*, respectively. We then speak of a random sample of size n from a distribution (or population) with mean μ and variance σ^2.

Computing the Distribution Function of the Sample Total T and Sample Mean $\overline{X}$ from a Normal Distribution

The calculation of the df of the sample total T and sample mean $\overline{X}$ is not an easy matter except for certain special cases, such as when the X_i have a Bernoulli distribution; in this case, as discussed earlier, T has a binomial distribution with parameters n, p.

We can also compute the exact distribution of T and $\overline{X}$ when the parent distribution is normal. These results are a consequence of the following theorem, which states that sums of mutually independent, normally distributed random variables are themselves normally distributed.

THEOREM 6.1

Let the n random variables X_i, $i = 1, 2, \ldots, n$, be mutually independent and normally distributed, with $X_i \sim N(\mu_i, \sigma_i^2)$. Then

$$X = \sum_{1 \le i \le n} a_i X_i \sim N(\mu, \sigma^2)$$

where

$$\mu = \sum_{1 \le i \le n} a_i \mu_i \quad \text{and} \quad \sigma^2 = \sum_{1 \le i \le n} a_i^2 \sigma_i^2 \qquad \blacksquare$$

Proof. The proof that X is itself normally distributed is omitted since it requires sophisticated mathematical techniques that are beyond the scope of this course. The assertion that $E(X) = \sum_{1 \le i \le n} a_i \mu_i$ is a consequence of the addition rule for the expected value of a sum:

$$E\left(\sum_{1 \le i \le n} a_i X_i \right) = \sum_{1 \le i \le n} a_i \mu_i$$

Similarly, the assertion

$$V\left(\sum_{1 \le i \le n} a_i X_i \right) = \sum_{1 \le i \le n} a_i^2 \sigma_i^2$$

is a consequence of the rule for calculating the variance of a sum of independent random variables. The significance of the result is this: The distribution of $\sum_{1 \le i \le n} a_i X_i$ is also normal. The following corollary is a consequence of Theorem 6.1.

COROLLARY 6.1

Let $X_1, \ldots, X_n$ be a random sample from a normal population with mean μ and variance σ^2. Then the distribution of the sample total and sample mean are both normal with parameters given by

$$T \sim N(n\mu, n\sigma^2) \quad \text{and} \quad \overline{X} \sim N\left(\mu, \frac{\sigma^2}{n} \right) \qquad \blacksquare$$

Proof. We emphasize the following points: $\overline{X}$ is itself normally distributed with the same mean as each individual X_i but with variance reduced by a factor of $1/n$. To show that the distribution of T is normal, we apply Theorem 6.1 with $a_i = 1$. Similarly, the distribution of $\overline{X}$ is also normal because $\overline{X} = T/n$.

Figure 6.1 shows how the shape of the sampling distribution changes as the sample size n increases. Here the parent distribution is $N(0, 1)$.

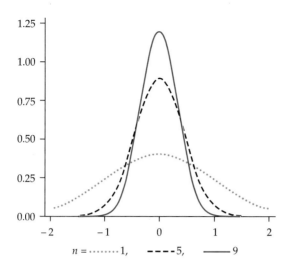

Figure 6.1 Probability density function of sample mean—standard normal parent ($n = 1, 5, 9$)

Looking at Fig. 6.1, notice that the variability of the sample mean $\overline{X}$ decreases as the sample size increases. This is a consequence of the fact that its variance $V(\overline{X}) = \sigma^2/n$ goes to zero as $n \to \infty$. Chebyshev's inequality tells us that the sample standard deviation $\sigma(\overline{X}) = \sigma/\sqrt{n}$ is a useful measure of how much the sample mean $\overline{X}$ deviates from the population mean μ. For this reason the quantity $\sigma(\overline{X})$ is called the *standard error of the mean*.

The following examples and the problems illustrate a variety of applications of Theorem 6.1.

---

example **6.1**

The time for a worker to assemble a component is normally distributed with mean $\mu = 15$ minutes and standard deviation $\sigma = 2$. Denote the mean assembly times of 16 day-shift workers and 9 night-shift workers by $\overline{X}$ and $\overline{Y}$, respectively. Assume that the assembly times of the workers are mutually independent. (1) Describe the distribution of $\overline{X} - \overline{Y}$. (2) Suppose the observed mean assembly times of the day-shift and night-shift workers was $\bar{x} = 14.5$ and $\bar{y} = 16$ minutes, respectively. (Thus $\bar{x} - \bar{y} = -1.5$.) Compute $P(\overline{X} - \overline{Y} < -1.5)$.

Solution

1. $\overline{X}$ is normally distributed with $\mu = 15$ and $\sigma^2(\overline{X}) = 4/16$, and $\overline{Y}$ is normally distributed with $\mu = 15$ and $\sigma^2(\overline{Y}) = 4/9$. Consequently, $\overline{X} - \overline{Y}$ is normally distributed with mean 0 and standard deviation

$$\sigma(\overline{X} - \overline{Y}) = \sqrt{\sigma^2(\overline{X}) + \sigma^2(\overline{Y})} = \frac{5}{6}$$

2.
$$P(\overline{X} - \overline{Y} < -1.5) = P\left(\frac{\overline{X} - \overline{Y}}{5/6}\right) = P(Z < -1.8) = 0.0359 \quad \blacksquare$$

example **6.2**   A cable car has a load capacity of 5000 lb. Assume that the weight of a randomly selected person is given by a normally distributed random variable W with mean 175 lb and standard deviation 20. Determine the maximum number of passengers allowed on the cable car so that their total weight exceeds the 5000-lb weight limit with a probability less than 0.05.

Solution. Let n denote the number of passengers allowed on the cable car. The first step is to describe the distribution of $T = \sum_{1 \le i \le n} W_i$, where W_i is the weight of the ith person. Each $W_i \sim N(175, 400)$; consequently, T is $N(n \times 175, n \times 400)$, by Corollary 6.1. Therefore,

$$\frac{T - n \times 175}{20 \sqrt{n}} \sim N(0, 1)$$

An overload occurs if $T > 5000$, i.e., if

$$\frac{T - n \times 175}{20 \sqrt{n}} > \frac{5000 - n \times 175}{20 \sqrt{n}}$$

This event will have probability 0.05 provided

$$\frac{5000 - n \times 175}{20 \sqrt{n}} = 1.645$$

which, after some algebra, is transformed into the equation

$$175n + 32.9 \sqrt{n} - 5000 = 0$$

Substituting $x = \sqrt{n}$ results in the quadratic equation

$$175x^2 + 32.9x - 5000 = 0$$

which has only one nonnegative solution, $x = 5.25$; therefore, $n = x^2 = 27.58$. Because the solution has to be an integer, $n = 27$ is the maximum number of passengers allowed on the cable car.   ∎

6.2.1   The Central Limit Theorem

In this section we continue our study of the distribution of the sample mean, but we drop the assumption of normality and suppose instead that our random sample comes from a general distribution $F(x)$ with (population) mean μ and (population) variance σ^2.

▰ PROPOSITION 6.1

Let $X_1, \ldots, X_n$ denote a random sample of size n, with common mean μ and variance σ^2. Then

$$E(\overline{X}) = \mu \tag{6.1}$$

$$\sigma(\overline{X}) = \frac{\sigma}{\sqrt{n}} \tag{6.2}$$

$$E(T) = n\mu \tag{6.3}$$

$$\sigma(T) = \sigma \sqrt{n} \tag{6.4}$$

■

Proof. Since $\overline{X} = (1/n)\sum_{1\leq i\leq n} X_i$, it follows from the addition rule [with $a_i = 1/n$ and $E(X_i) = \mu$], that

$$E(\overline{X}) = \sum_{1\leq i\leq n} \frac{1}{n}E(X_i) = n \times \frac{\mu}{n} = \mu$$

Similarly, it follows from the addition rule for the variance of a sum of independent random variables (with $a_i = 1/n$) that

$$\sigma^2(\overline{X}) = V\left(\sum_{1\leq i\leq n} \frac{X_i}{n}\right) = n \times \frac{\sigma^2}{n^2} = \frac{\sigma^2}{n}$$

and therefore

$$\sigma(\overline{X}) = \frac{\sigma}{\sqrt{n}}$$

The computations of $E(T)$ and $\sigma(T)$ proceed along the same lines.

Figure 6.2 displays the pdf's of the sample mean $\overline{X}$ when the parent distribution is the exponential distribution with $\theta = 1$ and $n = 1, 5, 9$.

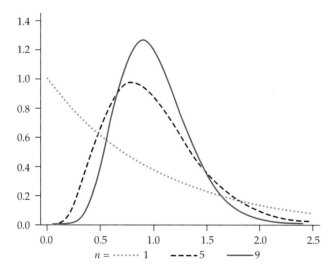

Figure 6.2 Probability density function of sample mean— exponential parent ($n = 1, 5, 9$)

The most noteworthy feature is the fact that as the sample size increases, the shape of the distribution of $\overline{X}$ appears to be very close to a normal distribution. That is the content of the next theorem, called the *central limit theorem* (CLT) because of its central importance in both probability and statistical inference.

■ THEOREM 6.2

The Central Limit Theorem (CLT). Let $X_1, \ldots, X_n$ be a random sample from a distribution with mean μ and variance σ^2. Then, for n sufficiently large, the distribution of $\overline{X}$ is approximately normal with mean μ and standard deviation $\sigma(\overline{X}) = \sigma/\sqrt{n}$. In particular,

$$P\left(a \leq \frac{\overline{X} - \mu}{\sigma/\sqrt{n}} \leq b\right) \approx \Phi(b) - \Phi(a).$$

The distribution of the sample total T is also approximately normal with $\mu_T = n\mu$ and $\sigma(T) = \sigma\sqrt{n}$. ■

The CLT is particularly useful because it yields a computable approximation to the distribution of the sample mean in terms of the normal distribution. It is a far-reaching and remarkable generalization of the normal approximation to the binomial as given in Eq. (4.33). The only conditions imposed are (1) the random variables $X_1, \ldots, X_i, \ldots$ are iid and (2) their means and variances are finite. A sequence of iid Bernoulli random variables is just an important special case.

---

example 6.3   Suppose a machine set for filling 1-lb boxes of sugar yields a weight W with $E(W) = 16.0$ oz and $\sigma(W) = 0.2$. A carton of sugar contains 48 boxes. (1) Describe the distribution of the weight T of the carton. (2) Compute the probability that the total weight of the carton exceeds 48.2 lb.

Solution

1. The total weight of the carton $T = \sum_{i=1}^{48} W_i$, with $E(W_i) = 16.0$ oz and $\sigma(W_i) = 0.2$. It follows from the central limit theorem that T is approximately normally distributed with $\mu = 48 \times 16 = 768$ oz and $\sigma(T) = 0.2 \times \sqrt{48} = 1.39$.

2. We first convert 48.2 lb to ounces. Thus,

$$P(T > 48.2 \text{ lb}) = P(T > 771.2 \text{ oz})$$

$$= P\left(\frac{T - 768}{1.39} > \frac{771.2 - 768}{1.39}\right)$$

$$= P(Z > 2.30) = 0.0107 \qquad ■$$

PROBLEMS

6.1 Let X_1, X_2, X_3 be three independent, identically distributed normal random variables with $\mu = 50, \sigma^2 = 20$. Let $X = X_1 - 2X_2 + 2X_3$. Compute:
(a) $E(X)$ (c) $P(|X - 50| \leq 25)$
(b) $V(X)$ (d) The 90th percentile of the distribution of X.

6.2 A toy consists of three parts whose respective weights (measured in grams) are denoted by X_1, X_2, X_3. Assume that the weights are independent and normally distributed with

$$X_1 \sim N(150, 36), \qquad X_2 \sim N(100, 25), \qquad X_3 \sim N(250, 60)$$

(a) Give a formula for the distribution of total weight of the toy $T = X_1 + X_2 + X_3$; name it and identify the parameters.

(b) What proportion of the toys have weight greater than 520 g?

(c) Suppose the toys are shipped 24 to a box. The box weighs 200 g. Calculate the expected value and variance of the total weight of the box.

(d) Calculate the distribution of the total weight W of the box. What is the probability that the total weight of the box is greater than 12,300 g?

6.3 Suppose that the time (measured in minutes) to repair a component is normally distributed with $\mu = 65, \sigma = 10$.

(a) What is the proportion of components that are repaired in less than one hour?

(b) Give a formula for the distribution of the total time required to repair eight components. What is the probability that eight components are repaired in an eight-hour working day?

6.4 Let $X_1, \ldots, X_n$ be iid $N(2, 4)$-distributed random variables.

(a) If $n = 100$, compute $P(1.9 < \overline{X} < 2.1)$.

(b) How large must n be so that $P(1.9 < \overline{X} < 2.1) = 0.9$?

6.5 Let $X_1, \ldots, X_9$ be iid $N(2, 4)$-distributed and $Y_1, \ldots, Y_4$ be iid $N(1, 1)$-distributed. The Y_i are assumed to be independent of the X_i.

(a) Describe the distribution of $\overline{X} - \overline{Y}$.

(b) Compute $P(\overline{X} > \overline{Y})$.

6.6 Let $\overline{X}$ denote the sample mean of a random sample of size $n_1 = 16$ taken from a normal distribution $N(\mu, 36)$, and let $\overline{Y}$ denote the sample mean of a random sample of size $n_2 = 25$ taken from a different normal distribution $N(\mu, 9)$.

(a) Describe the distribution of $\overline{X} - \overline{Y}$. Name it and identify the parameters.

(b) Compute $P(\overline{X} - \overline{Y} > 5)$.

(c) Compute $P(|\overline{X} - \overline{Y}| > 5)$.

6.7 To determine the difference, if any, between two brands of radial tires, 12 tires of each brand are tested. Assume that the lifetimes of both brands of tires come from the same normal distribution with $\sigma = 3300$.

(a) Describe the distribution of the difference of the sample means $\overline{X} - \overline{Y}$. Name it and identify the parameters.

(b) The engineers recorded the following values for the mean lifetime of each brand:

$$\overline{x} = 38,500, \qquad \overline{y} = 41,000 \qquad \text{so} \quad \overline{x} - \overline{y} = -2500$$

If in fact the lifetimes of both brands have the same distribution, what is the probability that the difference between the sample means is at least as large as the observed difference of 2500? That is, compute $P(|\overline{X} - \overline{Y}| > 2500)$.

(c) [Part (*b*) continued] Compute $P(\overline{X} - \overline{Y} < -2500)$.

6.8 Assume that the miles-per-gallon (mpg) of two brands of gasoline come from the same normal distribution with $\sigma = 2.0$. Four cars are driven with brand A gasoline, and six cars are driven with brand B gasoline. Denote the average mpg rates obtained with brand A and brand B gasolines by $\overline{X}$ and $\overline{Y}$, respectively.

(a) Describe the distribution of the difference of the sample means $\overline{X} - \overline{Y}$. Name it and identify the parameters.

(b) The engineers recorded the following mpg rates for each brand:

$$\overline{x} = 32.5 \quad \text{and} \quad \overline{y} = 31, \quad \text{so} \quad \overline{x} - \overline{y} = 1.5$$

If, in fact, the mpg rates of the two brands have the same distribution, what is the probability that the difference between the sample means is at least as large as the observed difference of 1.5? That is, compute $P(|\overline{X} - \overline{Y}| > 1.5)$.

(c) [Part (*b*) continued] Compute $P(\overline{X} - \overline{Y} > 1.5)$.

6.9 An elevator has weight capacity 3000 lb. Assume that the weight X of a randomly selected person is $N(175, 400)$ distributed, i.e., $\mu = 175, \sigma = 20$. Let W_n denote the total weight of n passengers.

(a) Give the formulas for $E(W_n)$ and $V(W_n)$.

(b) Describe the distribution of W_n.

(c) If 18 people get on the elevator, what is the probability that the total weight of the passengers exceeds the weight capacity of the elevator?

(d) What is the maximum number N of people allowed on the elevator such that the probability of their combined weight exceeding the 3000-lb load limit is less than 0.05?

6.10 Boxes of cereal are filled by machine, and the net weight of the filled box is a random variable with mean $\mu = 10$ oz and variance $\sigma^2 = 0.5$ oz^2. A carton of cereal contains 48 boxes. Let T be the total weight of a carton.

(a) Give a formula for the approximate distribution of T using the central limit theorem.

(b) Use the central limit theorem to estimate $P(T > 31 \text{ lb})$.

6.11 The random variables $X_1, \ldots, X_{40}$ are independent with the same probability function given by

x	-1	0	1
$f(x)$	0.2	0.2	0.6

(a) Compute the mean and variance of the sample total $T = X_1 + \cdots + X_{40}$.

(b) What is the largest value that T can assume? The smallest?

(c) What does the central limit theorem say about the distribution of T? Use it to compute $P(T > 25)$ and $P(T < 0)$.

6.3   THE GAMMA DISTRIBUTION

The *gamma distribution* is used to model all kinds of phenomena in statistics, engineering, and science. For example, the exponential, chi-square, and Erlang distributions are themselves gamma distributions (we will treat these special cases in more detail later). The definition of the gamma distribution requires the *gamma function*, a topic of independent interest.

■ **DEFINITION 6.2**

The *gamma function* $\Gamma(\alpha)$ is defined by the convergent improper integral

$$\Gamma(\alpha) = \int_0^\infty y^{\alpha-1} e^{-y}\, dy, \quad \alpha > 0 \tag{6.5}$$

■

Properties of the Gamma Function

1. A recurrence formula for the gamma function:

$$\Gamma(1) = \int_0^\infty e^{-y}\, dy = 1 \tag{6.6}$$

Recurrence relation: $\Gamma(\alpha) = (\alpha - 1)\Gamma(\alpha - 1)$ \tag{6.7}

The recurrence relation (6.7) is obtained by an integration by parts. The details are left as an exercise (Prob. 6.12). An interesting consequence is the following explicit formula for $\Gamma(n)$ for n a positive integer greater than 1:

$$\Gamma(2) = 1\Gamma(1) = 1$$
$$\Gamma(3) = 2\Gamma(2) = 2 = 2!$$
$$\Gamma(4) = 3\Gamma(3) = 3 \times 2 = 3!$$

and, in general,

$$\Gamma(n) = (n - 1)\Gamma(n - 1) = (n - 1)! \tag{6.8}$$

■ **DEFINITION 6.3**

A random variable X with pdf given by

$$g(x; \alpha, \beta) = \left(\frac{1}{\Gamma(\alpha)\beta^\alpha}\right) x^{\alpha-1} e^{-x/\beta}, \quad x > 0 \tag{6.9}$$

$$g(x; \alpha, \beta) = 0, x \le 0$$

where $0 < \alpha < \infty$, $0 < \beta < \infty$, is said to have a *gamma distribution* with parameters α and β.   ■

We prove that Eq. (6.9) defines a probability density function by manipulating the integral defining the gamma function. Make the change of variable $y = x/\beta$, $\beta > 0$, in the integral (6.5), which is thereby transformed into

$$\Gamma(\alpha) = \int_0^\infty \left(\frac{x}{\beta}\right)^{\alpha-1} e^{-x/\beta} \left(\frac{1}{\beta}\right) dx$$

Combining the terms involving β in the denominator and dividing through by $\Gamma(\alpha)$, we obtain

$$\int_0^\infty \left(\frac{1}{\Gamma(\alpha)\beta^\alpha}\right) x^{\alpha-1} e^{-x/\beta} \, dx = 1$$

It follows that the function $g(x;\alpha,\beta)$ defined in Eq. (6.9) is a probability density function. Thus, the constant $\Gamma(\alpha)\beta^\alpha$ is used to make $g(x;\alpha,\beta)$ a probability density function. The symbol $\Gamma(\alpha,\beta)$ denotes the gamma distribution with parameters α and β. The following additional points should be noted.

2. In some important special cases the distribution of the sum of mutually independent gamma random variables is itself a gamma distribution. We omit the proof since it requires techniques beyond the scope of this text.

■ **THEOREM 6.3**

Suppose X_i are independent random variables with gamma distributions $\Gamma(\alpha_i, \beta)$. Then

$$\sum_i X_i \sim \Gamma\left(\sum_i \alpha_i, \beta\right)$$ ■

3. Finally, we note that a gamma random variable X is always nonnegative, that is, $P(X \geq 0) = 1$.

Except for certain special cases, such as $\alpha = 1$, we need a computer or a table to compute the values of the gamma distribution defined by

$$G(x;\alpha,\beta) = \int_0^x g(t;\alpha,\beta)\,dt$$

■ **PROPOSITION 6.2**

The mean and variance of the gamma distribution with parameters α and β are given by

$$E(X) = \alpha\beta \quad \text{and} \quad V(X) = \alpha\beta^2 \qquad (6.10)$$ ■

The proof is via a straightforward integration that we leave as an exercise (Prob. 6.14).

The Mean and Variance of the Weibull Distribution

Equations for the first two moments of the Weibull distribution were derived in Sec. 4.5.1 [Eqs. (4.38) and (4.39)]. These equations, and the variance, are conveniently expressed in terms of the gamma function:

$$E(T) = \beta \int_0^\infty y^{1/\alpha} e^{-y}\, dy$$

$$= \beta \Gamma\left(1 + \frac{1}{\alpha}\right)$$

$$E(T^2) = \beta^2 \int_0^\infty y^{2/\alpha} e^{-y}\, dy$$

$$= \beta^2 \Gamma\left(1 + \frac{2}{\alpha}\right)$$

$$V(T) = \beta^2 \left[\Gamma\left(1 + \frac{2}{\alpha}\right) - \Gamma\left(1 + \frac{1}{\alpha}\right)^2\right]$$

A Formula for the Beta Function

The beta function, defined by the definite integral of Eq. (4.42), can be expressed in terms of the gamma function as follows:

$$B(\alpha, \beta) = \frac{\Gamma(\alpha)\Gamma(\beta)}{\Gamma(\alpha + \beta)}$$

The derivation is omitted because it involves a change of variables for a double integral.

By choosing special values for the parameters α and β one obtains many distributions that are widely used in statistics and probability theory. Here are some examples:

1. The *exponential distribution* with parameter λ is obtained by setting $\alpha = 1$ and $\beta = \lambda^{-1}$.
2. The *Erlang distribution* is obtained by setting $\alpha = n$, where n is a positive integer.
3. The *chi-square distribution* with n degrees of freedom is a special case of the gamma distribution with $\alpha = n/2$, where n is a positive integer and $\beta = 2$. A random variable with a chi-square distribution with n degrees of freedom is denoted by χ_n^2. The chi-square distribution plays an important role in statistical inference, particularly in K. Pearson's chi-square goodness-of-fit test. The graphs of χ_n^2 for $n = 2, 4, 7$ are displayed in Fig. 6.3.

Some Applications of the Chi-Square Distribution

The distribution of a sum of squares of n independent standard normal random variables has a χ_n^2 distribution. The normalized sample variance of a random sample taken from a normal distribution also has a chi-square distribution, but with $n - 1$ degrees of freedom. A precise statement of this result is given later in Theorem 6.9.

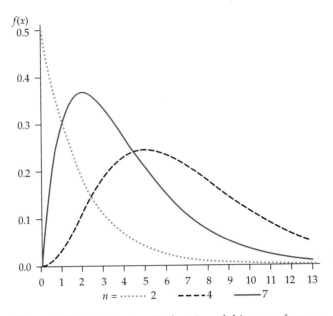

Figure 6.3 Probability density function of chi-square for $n = 2, 4, 7$

THEOREM 6.4

Let $Z_1, \ldots, Z_n$ be iid standard normal random variables. Then the sum of squares

$$\chi^2 = \sum_{1 \leq i \leq n} Z_i^2$$

has a chi-square distribution with n degrees of freedom.   ■

Proof. We give the proof only for the special case $n = 1$ (the case $n \geq 2$ requires more advanced techniques and is therefore omitted). In Example 4.22 we showed that the pdf $f(x)$ of Z^2 is given by the following formula [Eq. (4.45)]:

$$f(x) = \frac{1}{\sqrt{2\pi}} x^{-1/2} e^{-x/2}, \quad x > 0$$

$$f(x) = 0, \quad x \leq 0$$

This is a gamma distribution with parameters $\alpha = 1/2$, $\beta = 2$.

Percentiles of the Chi-Square Distribution

Table A.5 lists the values of the $100p$th percentile $Q_n(p)$, where

$$P(\chi^2 \leq Q_n(p)) = p \tag{6.11}$$

Figure 6.4 displays the relation between p and $Q_n(p)$.

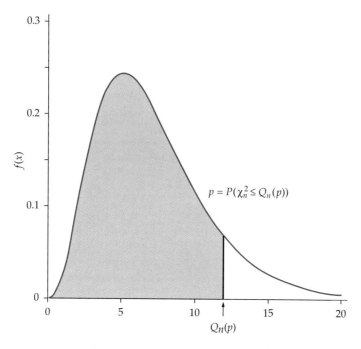

Figure 6.4 $100p$th percentiles of chi-square distribution

The critical value of the chi-square distribution is the point $\chi_n^2(\alpha)$ on the x axis such that

$$P(\chi^2 > \chi_n^2(\alpha)) = \alpha \qquad (6.12)$$

It follows that the critical value is given by

$$\chi_n^2(\alpha) = Q_n(1 - \alpha)$$

since

$$P(\chi^2 > \chi_n^2(\alpha)) = 1 - P(\chi^2 \le \chi_n^2(\alpha)) = 1 - \alpha$$

---

example **6.4**   Use Table A.5 to compute the following critical values of the chi-square distribution: (1) $\chi_{10}^2(0.95)$, (2) $\chi_{10}^2(0.05)$.

Solution

1. For $\alpha = 0.95$, $n = 10$, the critical value $\chi_{10}^2(0.95) = Q_{10}(0.05) = 3.940$.
2. For $\alpha = 0.05$, $n = 10$, the critical value $\chi_{10}^2(0.05) = Q_{10}(0.95) = 18.307$. ∎

Warning. The symbol α that appears in Eq. (6.12) is not to be confused with the α that appears in the definition of the gamma distribution [Eq. (6.5)].

It is an interesting and useful result that sums of mutually independent χ^2 random variables are also χ^2 distributed. A precise statement of this result follows; the proof uses advanced methods and is therefore omitted.

▨ **THEOREM 6.5**

Let $Y_1, \ldots, Y_k$ be independent chi-square random variables with $n_1, \ldots, n_k$ degrees of freedom, respectively. Then their sum $\sum_{i=1}^{k} Y_i$ has a chi-square distribution with $n = n_1 + \cdots + n_k$ degrees of freedom.   ■

The χ^2 distribution has other useful applications, as we illustrate in the next example.

example 6.5   *The distribution of the error R of a range finder device.* Suppose we have a device that records the position of an object located at coordinates (a, b) in the plane as

$$(X, Y) = (a + X_1, b + X_2)$$

where the random variables $X_i, i = 1, 2$ are iid and $X_i \sim N(0, \sigma^2)$. In other words, the X, Y coordinates recorded by the device are subject to normally distributed random errors as specified. The random variable R defined by

$$R = \sqrt{(X - a)^2 + (Y - b)^2} = \sqrt{X_1^2 + X_2^2}$$

represents the distance between the true position of the object and the recorded position. Show that

$$P(R \leq r) = P\left(\chi_2^2 \leq \left(\frac{r}{\sigma}\right)^2\right)$$

Solution. We begin with the observation that $X_i/\sigma \sim N(0, 1)$, $i = 1, 2$, and therefore Theorem 6.4 implies

$$\frac{R^2}{\sigma^2} = \frac{X_1^2}{\sigma^2} + \frac{X_2^2}{\sigma^2} \sim \chi_2^2$$

Finally,

$$P(R \leq r) = P\left(\frac{R^2}{\sigma^2} \leq \left(\frac{r}{\sigma}\right)^2\right)$$

$$= P\left(\chi_2^2 \leq \left(\frac{r}{\sigma}\right)^2\right)$$ ■

The Poisson Process and the Gamma Distribution

The gamma distribution, and the exponential distribution in particular, are associated in a natural way with the Poisson process, as will now be shown. Let messages arrive at a computer node according to a Poisson process $X(t)$ with intensity $\lambda > 0$. (It might be helpful to review Example 5.17.) Denote by W_n the waiting time until the nth message arrival; we define $W_0 = 0$. Thus, W_1 is the waiting time to the arrival of the first message, W_2 is the waiting time to the arrival of the second message, etc. We now prove that the waiting time until the arrival of the nth message has an Erlang distribution.

■ **THEOREM 6.6**

Suppose messages arrive to a computer node according to a Poisson process with arrival rate λ; then the distribution of W_n, the waiting time until the nth message arrival, is Erlang with parameters $\alpha = n$, $\beta = \lambda^{-1}$. That is, we have the following formula for the distribution of W_n:

$$P(W_n \leq t) = \int_0^t \frac{\lambda^n x^{n-1} e^{-\lambda x}}{(n-1)!} \, dx \qquad (6.13)$$

$$= 0, \quad t < 0 \qquad ■$$

Proof. We begin with the observation that the event $(W_n \leq t) = (X(t) \geq n)$. This is because the right-hand side is the event that at least n messages arrived during the time interval $[0, t]$, which is equivalent to the assertion that the waiting time until the nth message arrival is less than or equal to t. Now,

$$P(W_n \leq t) = P(X(t) \geq n) = \sum_{k=n}^{\infty} p(k; \lambda t)$$

$$= 1 - \sum_{k=0}^{n-1} p(k; \lambda t)$$

$$= 1 - \sum_{k=0}^{n-1} e^{-\lambda t} \frac{(\lambda t)^k}{k!}$$

We have thus shown that

$$F_{W_n}(t) = P(W_n \leq t) = 1 - \sum_{k=0}^{n-1} e^{-\lambda t} \frac{(\lambda t)^k}{k!} \qquad (6.14)$$

It is left as an exercise to show that

$$F'_{W_n}(t) = \frac{\lambda^n t^{n-1} e^{-\lambda t}}{(n-1)!}$$

The random variable $T_i = W_i - W_{i-1}$ is called the ith *interarrival time*. Notice that W_n can be written as the "telescoping sum"

$$W_n = W_1 + (W_2 - W_1) + \cdots + (W_n - W_{n-1}) = T_1 + \cdots + T_n \qquad (6.15)$$

■ **THEOREM 6.7**

The random variables $T_1, \ldots, T_n$ are mutually independent and exponentially distributed with parameter λ. ■

Proof. We consider only the case $n = 1$ (the general case requires a more complex and lengthy argument that we omit). It is easy to see that $T_1 = W_1$, so it has an Erlang distribution with $n = 1$. Substituting $n = 1$ into the right-hand side of Eq. (6.14), we see that $P(W_1 \leq t) = 1 - e^{-\lambda t}$, which is an exponential distribution with parameter λ. This yields the following alternative version of Theorem 6.6, which is frequently useful.

▬ THEOREM 6.8

Let T_i, $i = 1, \ldots, n$, denote n exponential random variables that are independent and identically distributed (with parameter λ); then $\sum_{1 \leq i \leq n} T_i$ has a gamma distribution with parameters $\alpha = n$, $\beta = \lambda^{-1}$.   ∎

Proof. There is not much to prove; the theorem is a consequence of the representation $W_n = \sum_{1 \leq i \leq n} T_i$ [Eq. (6.15)], Theorem 6.7, and Theorem 6.3 (with $\alpha_i = 1$, $\beta = 1/\lambda$).

Another Interpretation of $W_n = \sum_{1 \leq i \leq n} T_i$

Suppose we have n electronic components with iid exponentially distributed lifetimes denoted $T_1, \ldots, T_n$. If the remaining $n - 1$ components are to be used as spares, then W_n represents the total lifetime of all the components. The following example is a typical application of the previous theorems.

example 6.6   How many spare parts are needed for the space shuttle? Suppose a critical component of an experiment to be performed on the space shuttle has an expected lifetime $E(T) = 10$ days and is exponentially distributed with parameter $\lambda = 0.1$. As soon as a component fails, it is replaced by a new one whose lifetime T has the same exponential distribution. If the mission lasts 10 days and two spare components are on board, what is the probability that these three components last long enough for the space shuttle to accomplish its mission? More generally, suppose the space shuttle is equipped with n such components, including $n - 1$ spares, and denote the probability that these n components last long enough for the shuttle to complete its mission by $r(n)$. Our problem is to find a formula for $r(n)$.

Solution. Denote the lifetime of the ith component by T_i and note that the total lifetime has the same distribution as W_n since

$$W_n = \sum_{1 \leq i \leq n} T_i$$

and therefore

$$r(n) = P(W_n > 10) = \sum_{k=0}^{n-1} e^{-\lambda t} \frac{(\lambda t)^k}{k!} \tag{6.16}$$

where $\lambda = 0.1$, $t = 10$. Since $\lambda t = 1$, $r(n)$ is easily computed via the formula

$$r(n) = e^{-1} \left(1 + 1 + \frac{1}{2!} + \cdots + \frac{1}{(n-1)!} \right)$$

The values of $r(n)$ for $n = 2, 3, 4, 5$ are

$$r(2) = e^{-1}(1 + 1) = 0.74$$
$$r(3) = e^{-1}(1 + 1 + 1/2) = 0.92$$
$$r(4) = e^{-1}(1 + 1 + 1/2 + 1/6) = 0.981$$
$$r(5) = e^{-1}(1 + 1 + 1/2 + 1/6 + 1/24) = 0.9963$$

If we use two spares, so that the total number of components is three, then the probability that the space shuttle completes its mission is given by $r(3) = 0.92$. Since each launch costs tens of millions of dollars, it is clear that the probability of failure is uncomfortably high. How many components are needed to increase the probability of success to 0.99? Since $r(4) = 0.981 < 0.99 < 0.9963 = r(5)$, it is now clear that at least five components are needed. ■

6.4 THE DISTRIBUTION OF THE SAMPLE VARIANCE

In Sec. 6.2 we derived some results concerning the df of the sample mean $\overline{X}$ taken from (1) a normal distribution and (2) an arbitrary distribution. In this section we study the distribution of the *sample variance* s^2, which is defined by the equation

$$s^2 = \frac{1}{n - 1} \sum_{1 \le i \le n} (X_i - \overline{X})^2 \tag{6.17}$$

The *sample standard deviation* s is the square root of s^2.

The following proposition will be used in the next chapter to show that s^2 is an *unbiased estimator* of the population variance σ^2.

▨ **PROPOSITION 6.3**

Let $X_1, \ldots, X_n$ denote a random sample of size n taken from an arbitrary distribution function F, with mean μ and variance σ^2; then

$$E(s^2) = \sigma^2 \tag{6.18}$$

■

Proof. To prove Eq. (6.18), we use the algebraic identity

$$\sum_{1 \le i \le n} (x_i - c)^2 = \sum_{1 \le i \le n} (x_i - \overline{x})^2 + n(\overline{x} - c)^2 \tag{6.19}$$

We give the derivation of this equation in Sec. 6.5 at the end of the chapter.

We return to the proof of Proposition 6.3. We first substitute X_i and $\overline{X}$ for x_i and $\overline{x}$ in Eq. (6.19) and get

$$\sum_{1 \le i \le n} (X_i - c)^2 = \sum_{1 \le i \le n} (X_i - \overline{X})^2 + n(\overline{X} - c)^2$$

Put $c = \mu$ and take the expectation of both sides to obtain

$$E\left(\sum_{1 \le i \le n} (X_i - \mu)^2 \right) = E\left(\sum_{1 \le i \le n} (X_i - \overline{X})^2 \right) + nE(\overline{X} - \mu)^2$$

Next, observe that

$$E(X_i - \mu)^2 = V(X_i) = \sigma^2 \quad \text{and} \quad nE(\overline{X} - \mu)^2 = nV(\overline{X}) = \sigma^2$$

Consequently, $E(\sum_{1 \le i \le n}(X_i - \mu)^2) = n\sigma^2$, and therefore

$$n\sigma^2 = E\left(\sum_{1 \le i \le n}(X_i - \overline{X})^2\right) + \sigma^2$$

Solving this equation for σ^2 yields Eq. (6.18).

If in Proposition 6.3 we assume that the random sample comes from a normal distribution, then it is possible to compute the exact distribution of the sample variance s^2; this is the content of the next theorem, whose proof lies outside the scope of this text.

◼ THEOREM 6.9

Let s^2 be the sample variance of a random sample taken from a normal distribution. Then

$$\frac{(n-1)s^2}{\sigma^2} \quad \text{is } \chi^2_{n-1} \text{ distributed} \qquad ◼$$

---

example 6.7   Suppose a random sample of size $n = 19$ is taken from a normal distribution with $\sigma^2 = 9$. Compute the probability that the sample standard deviation s lies between 2 and 4.

Solution. Our problem is to compute the probability $P(a \le s \le b)$, where $a = 2$, $b = 4$. We sketch the general procedure first and then apply it to the particular case:

$$P(a \le s \le b) = P\left(\frac{(n-1)a^2}{\sigma^2} \le \frac{(n-1)s^2}{\sigma^2} \le \frac{(n-1)b^2}{\sigma^2}\right)$$

$$= P\left(\frac{(n-1)a^2}{\sigma^2} \le \chi^2_{n-1} \le \frac{(n-1)b^2}{\sigma^2}\right)$$

Substituting

$$n - 1 = 18, \qquad \sigma^2 = 9, \qquad a^2 = 4, \qquad b^2 = 16$$

into the last equation and using the chi-square tables, we compute the following approximation:

$$P(2 \le s \le 4) = P(8 \le \chi^2_{18} \le 32) \approx 0.98 - 0.02 = 0.96$$

6.4.1   Student's t Distribution

The sample mean taken from a normal distribution with mean μ and variance σ^2 is also normally distributed with mean μ and variance σ^2/n. That is the content of Corollary 6.1; consequently,

$$\frac{\sqrt{n}(\overline{X} - \mu)}{\sigma} \sim N(0, 1)$$

In practice, the population variance σ^2 will not be known, and it is therefore reasonable to study the distribution of $\sqrt{n}(\overline{X} - \mu)/s$, where s is the sample standard deviation.

▩ **THEOREM 6.10**

Let s be the sample standard deviation of a random sample of size n taken from a normal distribution. Then

1. $\overline{X}$ and s are mutually independent random variables and

2. $\qquad\qquad\qquad \dfrac{\sqrt{n}(\overline{X} - \mu)}{s}$ is t_{n-1} distributed $\qquad\qquad$ (6.20)

where t_ν denotes Student's t distribution with ν degrees of freedom, as described in Definition 6.4. ∎

▩ **DEFINITION 6.4**

We say that the random variable t_ν, where $\nu = 1, 2, \ldots$, has Student's t distribution with ν degrees of freedom if its pdf is given by

$$f(t|\nu) = c(\nu)\left(1 + \frac{t^2}{\nu}\right)^{-(\nu+1)/2}, \quad -\infty < t < \infty$$

$$c(\nu) = \frac{\Gamma([\nu + 1]/2)}{\sqrt{\pi\nu}\,\Gamma(\nu/2)}$$ ∎

▩ **REMARK 6.1**

It can be shown that

$$\lim_{\nu \to \infty} f(t|\nu) = \frac{1}{\sqrt{2\pi}} \, e^{-t^2/2}$$

In particular, $f(t|\nu)$ resembles a normal curve, and the standard normal curve itself serves as an approximation for large values of ν, say for $\nu > 30$. ∎

The graphs of Student's t distribution for $\nu = 1, 4, 9$ are displayed in Fig. 6.5.

The proof of Theorem 6.10, which is too advanced to be included here, is based on the following result, whose proof is also outside the scope of this text.

▩ **THEOREM 6.11**

Let $Z \sim N(0, 1)$ and $V \sim \chi_n^2$ be mutually independent. Then

$$\frac{Z}{\sqrt{V/n}} \sim t_n \qquad\qquad (6.21)$$ ∎

The t distribution plays an important role in *estimation* and *hypothesis testing*, as will be explained in detail elsewhere.

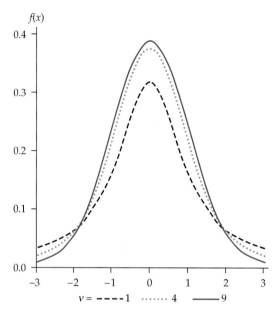

Figure 6.5 Probability density function of Student's t for $\nu = 1, 4, 9$

The critical values $t_\nu(\alpha)$ are defined in the usual way by the condition

$$P(t > t_\nu(\alpha)) = \alpha$$

and are tabulated in Table A.4 of the appendix.

---

example 6.8   Compute (1) $t_{12}(0.05)$; (2) $t_{12}(0.01)$; (3) $t_{15}(0.025)$; (4) c, where $P(t_6 \geq c) = 0.05$; (5) c, where $P(|t_6| \geq c) = 0.05$.

Solution

1. $t_{12}(0.05) = 1.782$
2. $t_{12}(0.01) = 2.681$
3. $t_{15}(0.025) = 2.131$
4. If $P(t_6 \geq c) = 0.05$, then $c = 1.943$.
5. If $P(|t_6| \geq c) = 0.05$, then $c = 2.447$.   ∎

6.4.2 The *F* Distribution

The *F* distribution is widely used in statistical inference, notably in the analysis of variance (ANOVA). For this reason we shall postpone discussing its applications until these topics are presented and content ourselves here with its definition.

■ **THEOREM 6.12**

Let U_1 and U_2 be independent χ^2 random variables, with ν_1 and ν_2 degrees of freedom, respectively. Then the random variable F defined by

$$F = \frac{U_1/\nu_1}{U_2/\nu_2}$$

has the *F* distribution with ν_1 and ν_2 degrees of freedom. ■

The graph of a typical *F* density function is displayed in Fig. 6.6.

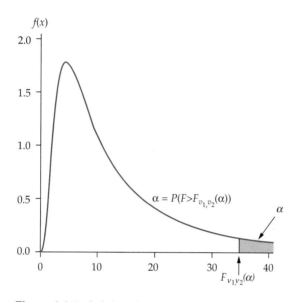

Figure 6.6 Probability density function of *F* distribution

The critical value of the *F* distribution is the point $F_{\nu_1,\nu_2}(\alpha)$ on the *x* axis such that

$$P(F > F_{\nu_1,\nu_2}(\alpha)) = \alpha$$

(see Fig. 6.6). Table A.6 displays the critical values for $\alpha = 0.05$ and 0.01 and for selected values of ν_1 and ν_2.

---

example **6.9**   Compute (1) $F_{2,27}(0.05)$ and (2) $F_{4,40}(0.01)$.

Solution

$$F_{2,27}(0.05) = 3.35$$
$$F_{4,40}(0.01) = 3.83$$

■

■■ **COROLLARY 6.2**

If t_n has a t distribution with n degrees of freedom, then t_n^2 has an F distribution with $\nu_1 = 1$, $\nu_2 = n$ degrees of freedom. ■

Proof. Using Eq. (6.21), we see that

$$t_n^2 = \left(\frac{Z}{\sqrt{V/n}}\right)^2 = \frac{Z^2}{V/n}$$

Observe that the term Z^2 in the numerator is χ_1^2 (why?) and the term in the denominator is χ_n^2. The corollary, then, is a special case of Theorem 6.12.

PROBLEMS

6.12 Derive Eq. (6.8) by performing a suitable integration by parts.

6.13 (a) Show that $\Gamma(1/2) = \sqrt{\pi}$ by making the change of variables $x = u^2/2$ in the integral

$$\Gamma(1/2) = \int_0^\infty x^{-1/2} e^{-x}\, dx$$

(b) Using the result of part (a), compute $\Gamma(3/2)$, $\Gamma(5/2)$.

6.14 Derive the formulas for the mean and variance of the gamma distribution given in Eq. (6.10).

6.15 (a) If X has a gamma distribution with $E(X) = \mu_1$ and $E(X^2) = \mu_2$, show that

$$\alpha = \frac{\mu_1^2}{\mu_2 - \mu_1^2} \quad \text{and} \quad \beta = \frac{\mu_2 - \mu_1^2}{\mu_1}$$

(b) Use the result of part (a) to find the parameters of the gamma distribution when $\mu_1 = 2$, $\mu_2 = 5$.

6.16 Use Table A.5 to compute $\chi_n^2(\alpha)$ for:
(a) $n = 10$, $\alpha = 0.1$ **(c)** $n = 15$, $\alpha = 0.1$
(b) $n = 10$, $\alpha = 0.05$ **(d)** $n = 20$, $\alpha = 0.9$

6.17 Use Table A.5 to compute numbers $a < b$ such that
$$P(a < \chi_n^2 < b) = p$$
and satisfying the conditions
$$P(\chi_n^2 < a) = P(\chi_n^2 > b)$$

(a) $n = 10$, $p = 0.9$ **(c)** $n = 10$, $p = 0.95$
(b) $n = 15$, $p = 0.9$ **(d)** $n = 20$, $p = 0.95$

6.18 An artillery shell is fired at a target. The distance (in meters) from the point of impact to the target is modeled by the random variable $7.22\chi_2^2$. If the shell hits within 10 meters of the target, the target is destroyed.
(a) What is the probability that the target is destroyed by the shell?

(b) If shells are fired repeatedly, how many rounds are required to ensure a 90 percent probability of destroying the target? (Note: The decision of how many shells to fire must be made ahead of time, since we cannot see whether we have destroyed the target.)

6.19 The speed V of a molecule in a gas at equilibrium has the pdf

$$f(v) = av^2 e^{-bv^2}, \quad v \geq 0$$
$$f(v) = 0 \quad \text{otherwise}$$

Here $b = m/2kT$, where m is the mass of the molecule, T is the absolute temperature, and k is Boltzmann's constant.

(a) Compute the constant a in terms of b.

Let $Y = (m/2) \times V^2$ denote the kinetic energy of the molecule. Compute:

(b) The df $G(y) = P(Y \leq y)$ of Y.

(c) The pdf $g(y) = G'(y)$ of Y.

6.20 Suppose a random sample of size $n = 16$ is taken from a normal distribution with $\sigma^2 = 5$. Compute the probability that the sample standard deviation s lies between 1.5 and 2.9.

6.21 Look up the following critical values for the t distribution in the tables.

(a) $t_9(0.05)$ **(c)** $t_{18}(0.025)$

(b) $t_9(0.01)$ **(d)** $t_{18}(0.01)$

6.22 Find c such that:

(a) $P(t_{14} \geq c) = 0.05$ **(c)** $P(t_{14} \geq c) = 0.01$

(b) $P(|t_{14}| \geq c) = 0.05$ **(d)** $P(|t_{14}| \geq c) = 0.01$

6.23 Look up the following critical values for the F distribution in the tables.

(a) $F_{3,20}(0.05)$ **(c)** $F_{4,30}(0.05)$

(b) $F_{3,20}(0.01)$ **(d)** $F_{4,30}(0.01)$

6.5   MATHEMATICAL DETAILS AND DERIVATIONS

We now derive the algebraic identity (6.19), restated here for convenience.

$$\sum_{1 \leq i \leq n} (x_i - c)^2 = \sum_{1 \leq i \leq n} (x_i - \bar{x})^2 + n(\bar{x} - c)^2$$

We begin with the algebraic identity

$$(x_i - c)^2 = [(x_i - \bar{x}) + (\bar{x} - c)]^2$$
$$= (x_i - \bar{x})^2 + 2(x_i - \bar{x})(\bar{x} - c) + (\bar{x} - c)^2$$

Therefore,

$$\sum_{1 \leq i \leq n} (x_i - c)^2 = \sum_{1 \leq i \leq n} (x_i - \bar{x})^2 + 2(x_i - \bar{x})(\bar{x} - c) + (\bar{x} - c)^2$$

$$= \sum_{1 \leq i \leq n} (x_i - \bar{x})^2 + \sum_{1 \leq i \leq n} 2(x_i - \bar{x})(\bar{x} - c) + \sum_{1 \leq i \leq n} (\bar{x} - c)^2$$

The middle summand equals zero because $\sum_{1 \le i \le n} (x_i - \bar{x}) = 0$ implies

$$\sum_{1 \le i \le n} 2(x_i - \bar{x})(\bar{x} - c) = 2(\bar{x} - c) \sum_{1 \le i \le n} (x_i - \bar{x}) = 0$$

The last summand is the constant term $(\bar{x} - c)^2$ added n times, so

$$\sum_{1 \le i \le n} (\bar{x} - c)^2 = n(\bar{x} - c)^2$$

Consequently,

$$\sum_{1 \le i \le n} (x_i - c)^2 = \sum_{1 \le i \le n} (x_i - \bar{x})^2 + n(\bar{x} - c)^2$$

6.6 CHAPTER SUMMARY

Many problems in probability theory and statistical inference can be reduced to the study of the distribution of the *sample mean* $\bar{X}$ of a sequence of iid random variables $X_1, \ldots, X_n$. The exact distribution of $\bar{X}$ is not easy to compute except in certain special cases, such as sampling from the Bernoulli or normal distributions. The central limit theorem tells us that for large values of the sample size n, the distribution of the sample mean is approximately normal with mean μ and standard deviation $\sigma/\sqrt{n}$. It is a remarkable fact that the CLT remains valid if it is assumed only that the parent distribution has a finite mean and finite variance. We introduced several new families of distributions, including the *gamma, chi-square,* the t, and F distributions. These distributions are related to one another as well as to the normal distribution itself. For instance, the chi-square distribution is a special case of the gamma distribution, and t^2 has an F distribution. Other special cases of the gamma distribution include the *exponential* and *Erlang* distributions, both of which play important roles in queueing theory.

To Probe Further. The long road to the central limit theorem is traced in S. M. Stigler's *History of Statistics: The Measurement of Uncertainty Before 1900* (Cambridge, Mass., Belknap Press of Harvard University Press, 1986).

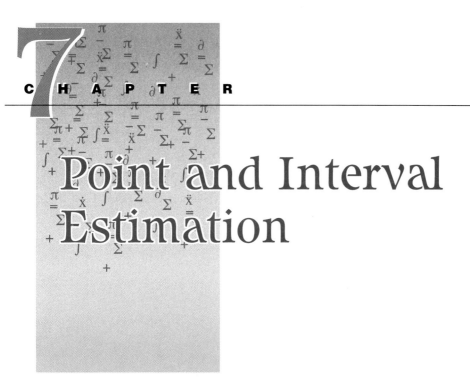

7 CHAPTER

Point and Interval Estimation

In astronomy, no experiment should be believed until confirmed by theory.

Sir Arthur Eddington (1882–1944), British astronomer

7.1 ORIENTATION

In this chapter we explore several methods for *estimating* population parameters such as the mean μ and variance σ^2 of a distribution using the sample data. We use the generic symbol θ to denote the value of the unknown parameter. The set of allowable values θ is called the *parameter space* and is denoted by Θ. In many cases the distribution function depends explicitly on the unknown parameter θ. For instance, reliability engineers often assume that their life-test data come from the exponential distribution $F(x; \theta) = 1 - \exp(-\theta x)$. The parameter space in this case is the set $\Theta = \{\theta : 0 < \theta < \infty\}$. In most cases (but not all) that we consider, the unknown parameter θ is the mean. The parameter p of a Bernoulli distribution is called a *population proportion*. The parameter space in this case is the set $\Theta = \{p : 0 < p < 1\}$.

Organization of Chapter

Section 7.2 Point Estimation of the Mean and Variance
Section 7.3 Confidence Intervals for the Mean and Variance
Section 7.4 Point and Interval Estimation for the Difference of Two Means

7.2 POINT ESTIMATION OF THE MEAN AND VARIANCE

The purpose of point estimation is to use a function of the sample data called a *point estimator* to accurately estimate the unknown parameter θ. We denote the point estimator by $\hat{\theta} = \hat{\theta}(X_1, \ldots, X_n)$. Among the most important examples of point estimators are the following:

- The *sample mean* $\overline{X}$, which is used to estimate the population mean μ, is defined by

$$\overline{X} = \frac{\sum_{1 \le i \le n} X_i}{n} = \hat{\theta}(X_1, \ldots, X_n)$$

- The *sample variance* s^2, which is used to estimate the population variance σ^2, is defined by

$$s^2 = \frac{1}{n-1} \sum_{1 \le i \le n} (X_i - \overline{X})^2 = \hat{\theta}(X_1, \ldots, X_n)$$

- The *sample standard deviation* s, which is used to estimate the standard deviation σ, is defined by

$$s = \sqrt{\frac{1}{n-1} \sum_{1 \le i \le n} (X_i - \overline{X})^2} = \hat{\theta}(X_1, \ldots, X_n)$$

Generalizing from these three examples, we see that a point estimator $\hat{\theta}$ consists of the following two elements:

1. A random sample $X_1, \ldots, X_n$, whose common distribution (also called the parent distribution) depends on the unknown values of one or more parameters.
2. A *statistic* $\hat{\theta} = \hat{\theta}(X_1, \ldots, X_n)$, which is a function of the random sample that *does not depend on any of the unknown parameters*. A statistic is a random variable, since it is a function of the random sample. We call the statistic $\hat{\theta}$ a *point estimator* of θ.

   A *point estimate* is just the value of $\hat{\theta}$ calculated from the observed sample values $X_1 = x_1, \ldots, X_n = x_n$ and is denoted by $\hat{\theta}(x_1, \ldots, x_n)$, or simply $\hat{\theta}$.

The Sampling Distribution of the Sample Mean

We compute the mean, variance, and standard deviation of the sample mean by applying Proposition 6.1:

$$E(\overline{X}) = \mu \tag{7.1}$$

$$V(\overline{X}) = \frac{\sigma^2}{n} \tag{7.2}$$

$$\sigma(\overline{X}) \,=\, \frac{\sigma}{\sqrt{n}} \qquad\qquad (7.3)$$

It follows from Eq. (7.1) that the expected value of the sample mean equals the population mean. We say that the sample mean is an *unbiased estimator* of the population mean. The pdf of the sampling distribution of an unbiased estimator is displayed in Fig. 7.1. More generally, we say that an estimator $\hat{\theta}$ of θ is *unbiased* if

$$E(\hat{\theta}) \,=\, \theta \qquad\qquad (7.4)$$

Otherwise it is is said to be *biased,* and the quantity

$$b(\hat{\theta}) \,=\, E(\hat{\theta}) - \theta$$

is called the *bias.* The bottom graph of Fig. 7.1 displays the pdf of the distribution of a biased estimator. Comparing the two sampling distributions in Fig. 7.1, we see that of the two estimators with the same variance the unbiased estimator is,

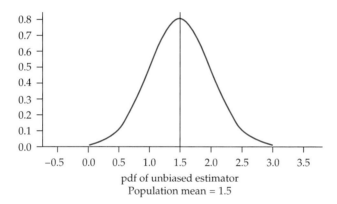

pdf of unbiased estimator
Population mean = 1.5

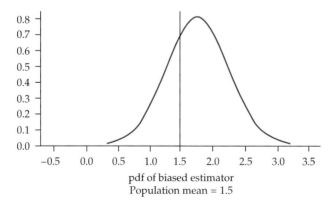

pdf of biased estimator
Population mean = 1.5

Figure 7.1 Probability density functions of sampling distributions of biased and unbiased estimators

on the average, closer to the population mean than is the biased one; thus it is the more accurate estimator.

The sample variance is an unbiased estimator of the population variance because we have previously shown (Proposition 6.3) that $E(s^2) = \sigma^2$.

Accuracy of the Sample Mean

Recall that the spread of the distribution of $\overline{X}$ is controlled by the magnitude of its standard deviation $\sigma(\overline{X})$ (see the discussion of Chebyshev's inequality, Theorem 3.4). When $\sigma(\overline{X})$ is small, most $\overline{X}$'s will be concentrated near μ with high probability. The standard deviation $\sigma(\overline{X})$ is called the *standard error of the sample mean*. Of course, when σ is unknown, which is the typical situation, $\sigma(\overline{X})$ is not computable. We therefore replace σ by its estimate s (the sample standard deviation) and obtain the *estimated standard error*, denoted $s(\overline{X})$ and defined by the equation

$$s(\overline{X}) = \frac{s}{\sqrt{n}} \tag{7.5}$$

Keep in mind that knowing only the value of $\overline{X}$ tells us nothing about how accurately it estimates μ; in particular, we would like to study the magnitude of the difference $|\overline{X} - \mu|$. Statisticians determine the accuracy of an unbiased estimator $\hat{\theta}$ of θ by computing its standard deviation $\sqrt{V(\hat{\theta})}$, which is called the *standard error of the estimator*.

▬ DEFINITION 7.1

The *standard error of an estimator* $\hat{\theta}$ of θ is denoted $\sigma(\hat{\theta})$ and is defined by

$$\sigma(\hat{\theta}) = \sqrt{V(\hat{\theta})}$$ ■

When the expression for $\sigma(\hat{\theta})$ contains unknown population parameters that can be estimated from the sample data, then we compute the *estimated standard error* $s(\hat{\theta})$ by substituting these estimates into the formula for $\sigma(\hat{\theta})$. This is exactly how we derived Eq. (7.5) for the estimated standard error of the sample mean.

The estimated standard error $s(\hat{\theta})$ plays an important role in the construction of confidence intervals for population parameters, as will be explained in Sec. 7.3.

---

example 7.1   Refer to the data set in Prob. 1.12. Compute the sample means, sample variances, sample standard deviations, and estimated standard errors for the lead levels in the blood of the exposed and control groups of children.

Solution. Denote by $\overline{X}_1$, $\overline{X}_2$ and s_1, s_2 the sample means and sample standard deviations of the lead levels in the exposed and control groups of children, respectively. Using a computer (or a scientific calculator with built-in statistical functions) we obtain the following results (rounded to two decimal places):

$$\overline{X}_1 = 31.85, \qquad s_1^2 = 207.57, \qquad s_1 = 14.41$$

$$s(\overline{X}_1) = \frac{s_1}{\sqrt{n}} = \frac{14.41}{\sqrt{33}} = 2.51$$

Similarly, for the control group, we obtain

$$\overline{X}_2 = 15.88, \qquad s_2^2 = 20.61, \qquad s_2 = 4.54$$

$$s(\overline{X}_2) = \frac{s_2}{\sqrt{n}} = \frac{4.54}{\sqrt{33}} = 0.79$$

It is interesting to note that the estimated standard error of the lead levels of the exposed group is more than three times that of the control group; in fact, $s(\overline{X}_1) = 3.174 \times s(\overline{X}_2)$. ∎

Estimating a Function $g(\theta)$ **of the Parameter** θ. To estimate the standard deviation σ, we note that $\sigma = \sqrt{\sigma^2}$; consequently, it seems reasonable to use the estimator $s = \sqrt{s^2}$ to estimate σ. This suggests the following useful principle for estimating $g(\theta)$:

> If $\hat{\theta}$ is a (good) point estimator of θ, then $g(\hat{\theta})$ is also a (good) point estimator of $g(\theta)$.

For instance, in Example 7.1 we obtained the estimate $s_1^2 = 207.57$ for the variance σ_1^2 of the lead level in the blood of the exposed group. We therefore use $\sqrt{207.57} = 14.41$ to estimate the standard deviation σ_1.

For a precise version of this principle, see Proposition 7.1 in Sec. 7.6.2 at the end of this chapter. In particular, this principle is valid for the class of *maximum-likelihood estimators* (MLEs); see Definition 7.7 for the details.

Comparing Two Unbiased Estimators. Suppose $\hat{\theta}_1$ and $\hat{\theta}_2$ are both unbiased estimators. How do we decide which one is better? The answer is to choose the estimator with the smaller variance.

To see why this is a reasonable choice, let us suppose that $V(\hat{\theta}_1) < V(\hat{\theta}_2)$. From Chebyshev's inequality (Theorem 3.4), we see that

$$P(|\hat{\theta}_i - \theta| \geq \delta) \leq \frac{V(\hat{\theta}_i)}{\delta^2} \tag{7.6}$$

It follows from Eq. (7.6) and the condition $V(\hat{\theta}_1) < V(\hat{\theta}_2)$ that $\hat{\theta}_1$ is more likely to be closer to θ than $\hat{\theta}_2$, so it is a more accurate estimator of θ.

This leads to the principle of *minimum-variance unbiased estimation*:

> When comparing two unbiased estimators, choose the one with the smaller variance. Among all unbiased estimators, choose the one with minimum variance. This estimator is called the minimum-variance unbiased estimator (MVUE).

Consistent Estimator

The sample mean is a *consistent estimator* of the population mean; that is, when the sample size n is large, then $\overline{X}$ is close to μ with high probability. More precisely, the law of large numbers (Theorem 5.2) applied to the sample mean yields the following result:

$$\lim_{n \to \infty} P(|\overline{X} - \mu| > \delta) = 0 \quad \text{for every } \delta > 0 \tag{7.7}$$

The general definition follows.

■■ **DEFINITION 7.2**

An estimator $\hat{\theta}_n$ of θ based on a random sample of size n is a *consistent estimator* of θ if

$$\lim_{n \to \infty} P(|\hat{\theta}_n - \theta| > \delta) = 0 \quad \text{for every } \delta > 0 \tag{7.8}$$

■

7.3 CONFIDENCE INTERVALS FOR THE MEAN AND VARIANCE

Suppose $\overline{X}$ is the sample mean of a random sample drawn from a normal distribution; then $\overline{X} \sim N(\mu, \sigma^2/n)$ (Corollary 6.1), and it follows that $P(\overline{X} > \mu) = 0.5 = P(\overline{X} < \mu)$. Thus, the value of $\overline{X}$ fluctuates above and below the mean μ, and since $P(\overline{X} = \mu) = 0$ (why?) it follows that $\overline{X} \neq \mu$ with probability 1. Therefore, simply reporting the value of $\overline{X}$ tells us nothing about the magnitude of the discrepancy $|\overline{X} - \mu|$. For this reason one reports an interval of values that contains the unknown parameter with high probability; such an interval is called a *confidence interval*. The following example illustrates how this is done in the special case when the data are assumed to be normally distributed with known variance; the case of unknown variance will be discussed in the next section.

We assume, then, that $X_1, \ldots, X_n$ is a random sample from a normal distribution $N(\mu, \sigma^2)$ with σ known. We begin with the fact that

$$\frac{\sqrt{n}(\overline{X} - \mu)}{\sigma} \sim N(0, 1)$$

Consequently,

$$P\left(-z(\alpha/2) \leq \frac{\sqrt{n}(\overline{X} - \mu)}{\sigma} \leq z(\alpha/2)\right) = 1 - \alpha \tag{7.9}$$

Here $z(\alpha)$ denotes the critical value of the standard normal distribution as defined in Eq. (4.30) and restated here for convenience:

$$P(Z \geq z(\alpha)) = \alpha$$

$z(\alpha)$ is, of course, the $100(1 - \alpha)$ percentile of the standard normal distribution; thus,

$$P(Z \leq z(\alpha)) = 1 - \alpha$$

Equation (7.9) is a consequence of the symmetry of the standard normal distribution, which implies that

$$P(-z(\alpha/2) \leq Z \leq z(\alpha/2)) = 1 - \alpha$$

Solving the pair of inequalities inside the parentheses of the left-hand side of Eq. (7.9) for μ yields

$$P(\overline{X} - z(\alpha/2) \times \sigma/\sqrt{n} \leq \mu \leq \overline{X} + z(\alpha/2) \times \sigma/\sqrt{n}) = 1 - \alpha \qquad (7.10)$$

It follows from Eq. (7.10) that the interval $[L, U]$, where

$$L = \overline{X} - z(\alpha/2)\sigma/\sqrt{n} \quad \text{(lower limit)} \qquad (7.11)$$

$$U = \overline{X} + z(\alpha/2)\sigma/\sqrt{n} \quad \text{(upper limit)} \qquad (7.12)$$

contains μ with probability $1 - \alpha$. And this is why it is called a $100(1 - \alpha)$ percent confidence interval for μ. It is sometimes more convenient to express the confidence interval in the following equivalent form:

$$\overline{X} \pm z(\alpha/2)\frac{\sigma}{\sqrt{n}}$$

---

example 7.2

Suppose we are given a random sample of $n = 10$ observations taken from a normal distribution $N(\mu, 4)$ with $\bar{x} = 15.1$. Construct a 95 percent confidence interval for μ.

Solution. Set $\alpha = 0.05$ and then compute

$$z(\alpha/2) = z(0.025) = 1.96; \qquad z(\alpha/2)\frac{\sigma}{\sqrt{n}} = 1.96 \times 2/\sqrt{10} = 1.24$$

The confidence interval $[L, U]$ is thus given by

$$L = \overline{X} - z(\alpha/2)\frac{\sigma}{\sqrt{n}} = 15.1 - 1.24 = 13.86$$

$$U = \overline{X} + z(\alpha/2)\frac{\sigma}{\sqrt{n}} = 15.1 + 1.24 = 16.34 \qquad \blacksquare$$

It is worth noting that the confidence interval $[L, U]$ is a (random) interval whose endpoints are the random variables displayed in Eqs. (7.11) and (7.12), respectively.

We summarize this discussion by giving a formal definition of a confidence interval.

▬ DEFINITION 7.3

A random interval of the form $[L, U]$, where L and U are statistics satisfying the condition $L \leq U$, is said to be a $100(1 - \alpha)$ percent *confidence interval* for the parameter θ if

$$P(L \leq \theta \leq U) \geq 1 - \alpha \qquad \blacksquare$$

One-Sided Confidence Intervals

There are many situations in which an engineer is interested in obtaining a *lower bound* on a population parameter. For example, an industrial engineer is never concerned about the system reliability being too high, but he is worried about it possibly being too low. We obtain a lower bound by constructing a $100(1 - \alpha)$ percent lower one-sided confidence interval, as will now be explained.

■ **DEFINITION 7.4**

A random interval of the form $[L, \infty)$ is a $100(1 - \alpha)$ percent *lower one-sided confidence interval* for the parameter θ if

$$P(L \leq \theta) \geq 1 - \alpha$$ ■

To illustrate the concept, let us construct a $100(1 - \alpha)$ percent lower one-sided confidence interval for the mean of a normal distribution based on a sample of size n drawn from a normal population $N(\mu, \sigma^2)$.
We begin with the fact that

$$P\left(\frac{\sqrt{n}(\overline{X} - \mu)}{\sigma} \leq z(\alpha)\right) = 1 - \alpha$$

Solving the inequality inside the parentheses for μ gives

$$P\left(\overline{X} - z(\alpha)\frac{\sigma}{\sqrt{n}} \leq \mu\right) = 1 - \alpha \qquad (7.13)$$

It follows from Eq. (7.13) that the random interval $[L, \infty)$, where

$$L = \overline{X} - z(\alpha)\frac{\sigma}{\sqrt{n}} \qquad (7.14)$$

contains μ with probability $1 - \alpha$.

---

example 7.3   The data set (7.15) lists nine determinations of the compressive strength of bricks measured in units of 100 pounds per square inch. Assume that the data are normally distributed with $\sigma = 5$. Obtain a lower 95 percent confidence interval for the mean compressive strength μ.

$$45.2,\ 50.4,\ 65.9,\ 52.3,\ 48.3,\ 49.6,\ 51.3,\ 62.2,\ 53.8 \qquad (7.15)$$

Solution. We use Eq. (7.14) with $\bar{x} = 53.22$, $n = 9$, and $z(0.05) = 1.645$. Consequently, the lower bound L is given by

$$L = 53.22 - 1.645 \times \frac{5}{3} = 50.48$$

So the 95 percent lower confidence interval is $[50.48, \infty)$.   ■

An Interpretation of the Confidence Interval

Each random sample $X_1, \ldots, X_n$ produces a random interval $[\overline{X} - z(\alpha/2) \times \sigma/\sqrt{n}, \overline{X} + z(\alpha/2) \times \sigma/\sqrt{n}]$. Some of these intervals will contain μ and some will not. All we can say, prior to doing the experiment, is that n repetitions of this experiment will produce n confidence intervals of which approximately $100(1-\alpha)$ percent will contain the parameter μ. After doing the experiment, the confidence interval $[L, U]$ either contains μ or does not. The student who feels uncomfortable with this explanation is in good company. This is the starting point for *Bayesian estimation*, a topic that is explored in the references listed at the end of this chapter.

Choosing the Sample Size

A frequently asked question is, how large must the sample size be in order for our point estimate $\overline{X}$ to be within d units of the unknown parameter μ? Equation (7.10) states that with probability $(1-\alpha)$ our point estimate $\overline{X}$ is within d units of the unknown parameter μ, where

$$d = z(\alpha/2) \times \frac{\sigma}{\sqrt{n}}$$

We note in passing that the width of the confidence interval equals $2d$. Thus, the accuracy of the estimator $\overline{X}$ is determined by d, the half width of the confidence interval (also called the *margin of error*). It follows that given d, σ, and α, we can solve for n and obtain the following equation for the sample size:

$$n = \left[\frac{z(\alpha/2)\sigma}{d} \right]^2 \qquad (7.16)$$

When σ is unknown, as is frequently the case, we can estimate it using data obtained from previous studies.

---

example 7.4   Refer to the data in Prob. 1.5. How large a sample size is needed for our estimate of the battery capacity to be within 1.5 units of μ?

Solution. On the basis of the data already collected, we have the estimate $s = 3.4641$. To play it safe, we round it up and assume that $\sigma = 3.5$. Setting $d = 1.5$ and $\sigma = 3.5$ in Eq. (7.16) yields

$$n = \left(\frac{1.96 \times 3.5}{1.5} \right)^2 = 20.92$$

Since n must be an integer, we round it up to $n = 21$.   ∎

Looking at Eq. (7.16), we see that the margin of error depends on α. In particular, as $\alpha \to 0$ we have $z(\alpha/2) \to \infty$. Thus, for a fixed sample size n

there is a trade-off between the accuracy of the estimator as measured by the half width of the confidence interval and the confidence level as measured by $1 - \alpha$.

7.3.1   Confidence Intervals for the Mean of a Normal Distribution: Variance Unknown

The results derived in the previous section are not very practical since it is unrealistic to assume that the variance is known but the mean is not. We now turn our attention to the problem of constructing confidence intervals for the mean of a normal distribution $N(\mu, \sigma^2)$ where σ is unknown.

In this case we use Theorem 6.10 , which states that

$$\frac{\sqrt{n}(\overline{X} - \mu)}{s} \sim t_{n-1}$$

where s is the sample standard deviation and t_{n-1} is Student's t distribution with $n - 1$ degrees of freedom. Therefore,

$$P\left(\frac{\sqrt{n}(\overline{X} - \mu)}{s} \leq t_{n-1}(\alpha)\right) = 1 - \alpha \tag{7.17}$$

From the symmetry of the t distribution, we obtain

$$P\left(-t_{n-1}(\alpha/2) \leq \frac{\sqrt{n}(\overline{X} - \mu)}{s} \leq t_{n-1}(\alpha/2)\right) = 1 - \alpha \tag{7.18}$$

Solving the inequalities for μ in Eq. (7.18) yields the $100(1-\alpha)$ percent confidence interval $[L, U]$, where

$$L = \overline{X} - t_{n-1}(\alpha/2)s/\sqrt{n} \tag{7.19}$$

$$U = \overline{X} + t_{n-1}(\alpha/2)s/\sqrt{n} \tag{7.20}$$

The confidence interval is also expressed in the form

$$\overline{X} \pm t_{n-1}(\alpha/2)\frac{s}{\sqrt{n}}$$

---

example 7.5   Refer to the data of Prob. 1.6. Construct a 95 percent confidence interval for the chamber burst strength.

Solution. The sample mean and standard deviation are $\overline{X} = 16.49$ and $s = 0.584$. (You should use your calculator or statistical software package to verify these values.) To compute a 95 percent confidence interval, we set $n = 17$, $\alpha = 0.05$ and then compute

$$t_{n-1}(\alpha/2) = t_{16}(0.025) = 2.12; \qquad t_{16}(0.025)\frac{0.584}{\sqrt{17}} = 0.30$$

Inserting these values into Eqs. (7.19) and (7.20) yields

$$L = \overline{X} - t_{16}(0.025)\frac{0.584}{\sqrt{17}} = 16.49 - 0.30 = 16.19$$

$$U = \overline{X} + t_{16}(0.025)\frac{0.584}{\sqrt{17}} = 16.49 + 0.30 = 16.79$$

The confidence interval is [16.19, 16.79].   ■

One-Sided Confidence Intervals: Unknown Variance

We conclude this section by giving the formula for constructing one-sided confidence intervals when the standard deviation is unknown.

Solving the inequality in Eq. (7.17) for μ, we obtain

$$P\left(\overline{X} - t_{n-1}(\alpha)\frac{s}{\sqrt{n}} \le \mu\right) = 1 - \alpha \tag{7.21}$$

We interpret this result in the usual way: The $100(1-\alpha)$ percent lower confidence interval is given by

$$\left[\overline{X} - t_{n-1}(\alpha)\frac{s}{\sqrt{n}}, \infty\right) \tag{7.22}$$

---

example **7.6**   Problem 1.5 lists the capacities (measured in ampere-hours) of 10 batteries; an easy calculation shows that $\overline{x} = 151$ and $s = 3.46$. From these data construct a 95 percent lower confidence interval for the battery capacity.

Solution. To construct the lower confidence interval, set $n = 10$, $s = 3.46$ in Eq. (7.22). In detail,

$$\frac{s}{\sqrt{n}} = \frac{3.46}{\sqrt{10}} = 1.0941$$

and therefore

$$t_9(0.05)\frac{s}{\sqrt{n}} = 1.833 \times 1.0941 = 2.01$$

$$\overline{X} - t_9(0.05)\frac{s}{\sqrt{n}} = 151 - 2.01 = 148.99$$

The corresponding 95 percent one-sided confidence interval is given by $[148.99, \infty)$.   ■

We note for future reference that the $100(1 - \alpha)$ percent upper confidence interval is given by

$$P\left(\mu \le \overline{X} + t_{n-1}(\alpha)\frac{s}{\sqrt{n}}\right) = 1 - \alpha \tag{7.23}$$

7.3.2 Confidence Intervals for the Mean of an Arbitrary Distribution

In this section we drop the assumption that the random sample comes from a normal distribution; instead, we assume it comes from an arbitrary distribution with population mean μ and population variance σ^2. We make the additional assumption that the sample size n is large enough ($n \geq 30$, say) so that the distribution of $\sqrt{n}(\overline{X} - \mu)/\sigma$ is close to the distribution of Z, a standard normal random variable. Thus, interval equation

$$P\left(-z(\alpha/2) \leq \frac{\sqrt{n}(\overline{X} - \mu)}{\sigma} \leq z(\alpha/2)\right) \approx P(-z(\alpha/2) \leq Z \leq z(\alpha/2)) \quad (7.24)$$

The right-hand side of Eq. (7.24) is therefore approximately equal to $1 - \alpha$. Consequently,

$$P\left(-z(\alpha/2) \leq \frac{\sqrt{n}(\overline{X} - \mu)}{\sigma} \leq z(\alpha/2)\right) \approx 1 - \alpha$$

The pair of inequalities inside the parentheses may be solved for μ, which gives

$$P(\overline{X} - z(\alpha/2) \times \sigma/\sqrt{n} \leq \mu \leq \overline{X} + z(\alpha/2) \times \sigma/\sqrt{n}) \approx 1 - \alpha \quad (7.25)$$

It follows from Eq. (7.25) that the interval $[L, U]$, where

$$L = \overline{X} - z(\alpha/2)\frac{\sigma}{\sqrt{n}} = \overline{X} - z(\alpha/2)\sigma(\overline{X})$$

$$U = \overline{X} + z(\alpha/2)\frac{\sigma}{\sqrt{n}} = \overline{X} + z(\alpha/2)\sigma(\overline{X})$$

contains μ with a probability that is approximately equal to $1 - \alpha$, so it is called an *approximate* $100(1 - \alpha)$ percent confidence interval for μ.

When σ is unknown, which is usually the case, then so is $\sigma(\overline{X})$ and we replace it with the estimated standard error $s(\overline{X})$; the approximate $100(1 - \alpha)$ percent confidence interval $[L, U]$ now takes the form

$$L = \overline{X} - z(\alpha/2)\frac{s}{\sqrt{n}} \quad (7.26)$$

$$U = \overline{X} + z(\alpha/2)\frac{s}{\sqrt{n}} \quad (7.27)$$

---

example 7.7   Refer to the 36 measurements of the heights of 10-year-old girls obtained from the StatLab data set (Table 1.5). Construct an approximate 95 percent confidence interval for the mean of the girls' heights.

Solution. We assume that these measurements constitute a random sample of size $n = 36$ from a distribution with unknown mean and unknown variance.

Using a computer, we obtained the following values for the sample mean, sample standard deviation, and estimated standard error:

$$\overline{X} = 53.34, \quad s = 2.91, \quad s(\overline{X}) = \frac{2.91}{\sqrt{36}} = 0.49$$

An approximate 95 percent confidence interval for the mean height of 10-year-old girls is given by

$$53.34 \pm 1.96 \times 0.49 = 53.34 \pm 0.95$$

The confidence interval is [52.39, 54.29]. ∎

One-Sided Confidence Intervals

In a similar way, we can construct approximate $100(1 - \alpha)$ percent upper and lower one-sided confidence intervals $(-\infty, U]$ and $[L, \infty)$, where

$$L = \overline{X} - z(\alpha)\frac{s}{\sqrt{n}} \tag{7.28}$$

$$U = \overline{X} + z(\alpha)\frac{s}{\sqrt{n}} \tag{7.29}$$

7.3.3 Confidence Intervals for the Variance of a Normal Distribution

To construct a confidence interval for the variance σ^2, we use Theorem 6.9, which asserts that

$$\frac{(n-1)s^2}{\sigma^2} \quad \text{is } \chi^2_{n-1} \text{ distributed}$$

Let $\chi^2_{n-1}(\alpha)$ denote the critical value of the chi-square distribution defined by

$$P(\chi^2 \geq \chi^2_{n-1}(\alpha)) = \alpha$$

$\chi^2_n(\alpha)$ is, of course, the $100(1 - \alpha)$ percentile of the chi-square distribution, so

$$P(\chi^2 \leq \chi^2_{n-1}(\alpha)) = 1 - \alpha \tag{7.30}$$

Therefore,

$$P\left(\chi^2_{n-1}(1 - \alpha/2) \leq \frac{(n-1)s^2}{\sigma^2} \leq \chi^2_{n-1}(\alpha/2)\right) = 1 - \alpha \tag{7.31}$$

Solving these inequalities for σ^2 in the usual way, we arrive at the following $100(1 - \alpha)$ percent confidence interval for σ^2:

$$P\left(\frac{(n-1)s^2}{\chi^2_{n-1}(\alpha/2)} \leq \sigma^2 \leq \frac{(n-1)s^2}{\chi^2_{n-1}(1 - \alpha/2)}\right) = 1 - \alpha \tag{7.32}$$

example **7.8** Refer to the 36 measurements of the heights of 10-year-old girls obtained from the StatLab data set (Table 1.5). Construct an approximate 95 percent confidence interval for the variance of the girls' heights. Assume the data come from a normal population.

Solution. Here $n - 1 = 35$ and $\alpha = 0.05$. We used the SAS software package to compute the critical values $\chi^2_{35}(0.975)$ and $\chi^2_{35}(0.025)$.

$$\chi^2_{n-1}(1 - \alpha/2) = \chi^2_{35}(0.975) = 20.57$$
$$\chi^2_{n-1}(\alpha/2) = \chi^2_{35}(0.025) = 53.20$$

Therefore,

$$\frac{(n-1)s^2}{\chi^2_{n-1}(\alpha/2)} = \frac{35 \times 8.46}{53.20} = 5.57$$

$$\frac{(n-1)s^2}{\chi^2_{n-1}(1-\alpha/2)} = \frac{35 \times 8.46}{20.57} = 14.39$$

Thus, [5.57, 14.39] is an approximate 95 percent confidence interval for σ^2. ∎

PROBLEMS

7.1 A random sample of size $n = 20$ from a normal distribution $N(\mu, 16)$ has mean $\bar{x} = 75$. Find:
(a) A 90 percent confidence interval for μ.
(b) A lower 90 percent confidence interval for μ.
(c) A 95 percent confidence interval for μ.
(d) A lower 95 percent confidence interval for μ.

7.2 Refer to the data set of Prob. 1.12. The sample mean and sample standard deviation of the $n = 33$ measurements of the lead level in the blood of the exposed group of children are $\bar{x} = 31.85$ and $s = 14.41$. Assuming the data come from a normal distribution $N(\mu, \sigma^2)$, compute:
(a) A 90 percent confidence interval for μ.
(b) A 99 percent confidence interval for μ.

7.3 Refer to the data set of Prob. 1.12. The sample mean and sample standard deviation of the $n = 33$ measurements of the lead level in the blood of the control group of children are $\bar{x} = 15.88$ and $s = 4.54$. Assuming the data come from a normal distribution $N(\mu, \sigma^2)$, compute:
(a) A 90 percent confidence interval for μ.
(b) A 99 percent confidence interval for μ.

7.4 (a) Let $X_1, \ldots, X_n$ be a random sample from the Poisson distribution

$$f(x; \lambda) = e^{-\lambda}\frac{\lambda^x}{x!}, \quad x = 0, 1, \ldots$$

Show that $\overline{X}$ is an unbiased estimator of λ.

(b) Compute the standard error $\sigma(\overline{X})$ and estimated standard error $s(\overline{X})$.

(c) Thirty-five items were inspected. The number of defects per item were recorded and are displayed in the following table:

2	3	0	1	0	0	0
0	0	2	3	3	1	3
3	1	0	0	2	0	0
2	1	2	0	1	0	3
1	4	0	0	1	0	4

Assuming the data come from a Poisson distribution with parameter λ, use the sample mean to estimate λ. Compute the standard error of your estimate.

7.5 Let $X_1, \ldots, X_n$ be a random sample from a normal distribution $N(\mu, 4)$, with μ unknown. Is

$$\hat{\theta}(X_1, \ldots, X_n) = \sum_{1 \le i \le n} (X_i - \mu)^2$$

a statistic? Justify your answer.

7.6 Eight measurements made on the viscosity of a motor oil blend yielded the values $\overline{x} = 10.23$, $s^2 = 0.64$. Assuming the data come from a normal distribution $N(\mu, \sigma^2)$, construct a 90 percent confidence interval for μ.

7.7 Suppose that $X_1, \ldots, X_{25}$ is a random sample from a normal population $N(\mu, \sigma^2)$ and that $\overline{x} = 148.30$.

(a) Assuming $\sigma = 4$, find a 95 percent confidence interval for μ.

(b) Assuming $s = 4$ (i.e., σ^2 is unknown), obtain a 95 percent confidence interval for μ.

7.8 From a sample of size 30 taken from a normal distribution, we obtain

$$\sum_{1 \le i \le 30} x_i = 675.2, \qquad \sum_{1 \le i \le 30} x_i^2 = 17,422.5$$

(a) Find a 95 percent confidence interval for μ.

(b) Find a 95 percent confidence interval for σ^2.

7.9 A sample of size $n = 6$ from a normal distribution with unknown variance produced $s^2 = 51.2$. Construct a 90 percent confidence interval for σ^2.

7.10 For a sample of 15 brand X automobiles the mean miles per gallon (mpg) was 29.3 with a sample standard deviation of 3.2. The gas tank capacity is 12 gallons.

(a) Using these data, determine a 95 percent lower one-sided confidence interval for the mpg. Assume the data come from a normal distribution.

(b) The cruising range is defined to be the distance the automobile can travel on a full tank of gas. Find a 95 percent lower one-sided confidence interval for the cruising range.

7.11 The lifetimes, measured in miles, of 100 tires were recorded and the following results were obtained:

$$\bar{x} = 38{,}000 \text{ miles}, \qquad s = 3200 \text{ miles}$$

(a) If the data cannot be assumed to come from a normal distribution, what can you say about the distribution of $\bar{X}$?

(b) Construct a lower one-sided 99 percent confidence interval for the mean lifetime.

7.12 The reverse-bias collector current (in microamperes) was measured for 20 transistors, and the data were recorded in the following table:

0.20	0.16	0.20	0.48	0.92
0.33	0.20	0.53	0.42	0.50
0.19	0.22	0.18	0.17	1.20
0.14	0.09	0.13	0.26	0.66

Source: M. G. Natrella, *Experimental Statistics, National Bureau of Standards,* vol. 91, U.S. Government Printing Office, Washington, D.C., 1963.

Assuming the data come from a normal distribution, construct a 95 percent confidence interval for μ.

7.13 An electronic parts factory produces resistors. Statistical analysis of the output suggests that the resistances can be modeled by a normal distribution $N(\mu, \sigma^2)$. Using the following data, consisting of the resistances (measured in ohms) of 10 resistors, construct:

(a) A 95 percent confidence interval for the mean resistance μ.

(b) A 99 percent confidence interval for the mean resistance μ.

0.150	0.154	0.145	0.148	0.158
0.152	0.153	0.157	0.148	0.156

7.14 (a) Compute a 95 percent confidence interval for the capacitances data in Prob. 1.14.

(b) Compute a 99 percent confidence interval.

7.15 (a) Compute a 99 percent confidence interval for the mean level of cadmium dust for the data set in Prob. 1.13. Assume that the data come from a normal distribution.

(b) Compute a 95 percent upper confidence interval.

7.16 Suppose the viscosity of a motor oil blend is normally distributed with $\sigma = 0.2$.

(a) A researcher wants to estimate the mean viscosity with a margin of error not exceeding 0.1 and a 95 percent level of confidence. How large must n be?

(b) Suppose the researcher wants to estimate the mean viscosity with the same margin of error, 0.1, but with a 99 percent level of confidence. How large must n be?

(c) Suppose the researcher wants to estimate the mean viscosity with a margin of error not exceeding 0.05 and a 95 percent level of confidence. How large must n be?

7.17 Suppose the atomic weight of a molecule is measured with two different instruments, each of which is subject to random error. More precisely, the recorded weight is $W_i = \mu + e_i$, where μ is the true atomic weight of the molecule and e_i is a random variable representing the error produced by the ith instrument. The accuracy of the instrument is determined by the variance $V(e_i) = \sigma_i^2$. Assume that $E(e_i) = 0, i = 1, 2$.

(a) Show that $E(W_i) = \mu, i = 1, 2$.

(b) Show that $W_t = tW_1 + (1-t)W_2$ is an unbiased estimator of μ for every real number t.

(c) Assuming that the error terms e_1, e_2 are mutually independent, derive a formula for $V(W_t)$ as a function of t.

(d) Find the value t^* for which $V(W_t)$ is a minimum. Show that

$$V(W_{t^*}) = \frac{\sigma_1^2 \sigma_2^2}{\sigma_1^2 + \sigma_2^2} \leq \min(\sigma_1^2, \sigma_2^2)$$

This shows that we can reduce the variance of our estimate for μ by taking a suitable weighted average of the two measurements.

7.18 Refer to the data set in Prob. 1.14. Assuming the capacitance is normally distributed, compute a 95 percent confidence interval for the variance of the capacitance.

7.19 Refer to the data set in Prob. 1.12. Assuming each variable is normally distributed:

(a) Compute a 95 percent confidence interval for the variance of lead levels in the blood of the exposed group.

(b) Compute a 95 percent confidence interval for the variance of lead levels in the blood of the control group.

7.20 Refer to the data set in Prob. 1.13. Assuming the cadmium level is normally distributed, compute a 95 percent confidence interval for the variance of the cadmium levels.

7.21 Refer to the data on the reverse-bias current in Prob. 7.12. Assuming that the data come from a normal distribution, compute a 95 percent confidence interval for the variance.

7.4 POINT AND INTERVAL ESTIMATION FOR THE DIFFERENCE OF TWO MEANS

The methods of the previous sections will now be applied to a comparative study of two populations. For example, we might be interested in the difference in weight loss obtained by following two different diets, or the difference in tensile strength between two types of steel cable. To study these and similar questions we take random samples of size n_1 and n_2 from each of the two populations, which we assume to be normally distributed with possibly different means μ_1 and μ_2, but with a common variance σ^2. We make the additional assumption that the random samples taken from the two populations are mutually independent. The sample means from the respective populations are denoted by $\overline{X}_1$ and $\overline{X}_2$, and the sample variances by s_1^2 and s_2^2. Notice that $\overline{X}_1 - \overline{X}_2$ is an unbiased estimator of the difference between the means, denoted by $\Delta = \mu_1 - \mu_2$. When the variances of the two populations are different, the problem of comparing the two means becomes somewhat more complicated and will not be discussed here.

As noted earlier, $\overline{X}_1 - \overline{X}_2$ is an unbiased estimator of Δ, and it is therefore natural to use it for the purpose of constructing confidence intervals. Before doing so, however, we need an estimator for the common variance. Of course, s_1^2 and s_2^2 are both unbiased estimators of σ^2, but it is known that the best combined estimator is the *pooled* estimator s_p^2, defined by

$$s_p^2 = \frac{(n_1 - 1)s_1^2 + (n_2 - 1)s_2^2}{n_1 + n_2 - 2} \tag{7.33}$$

Because $\overline{X}_1$ and $\overline{X}_2$ are independent, it follows that

$$V(\overline{X}_1 - \overline{X}_2) = V(\overline{X}_1) + V(\overline{X}_2)$$
$$= \sigma^2/n_1 + \sigma^2/n_2 = \sigma^2(1/n_1 + 1/n_2)$$

The estimated standard error, therefore, is given by

$$s(\overline{X}_1 - \overline{X}_2) = s_p \sqrt{\frac{1}{n_1} + \frac{1}{n_2}}$$

The confidence intervals are computed using the statistic t, defined in the next theorem (we omit the proof).

■ THEOREM 7.1

The statistic t defined by

$$t = \frac{(\overline{X}_1 - \overline{X}_2) - (\mu_1 - \mu_2)}{s_p \sqrt{\frac{1}{n_1} + \frac{1}{n_2}}} \tag{7.34}$$

has a Student t distribution with $n_1 + n_2 - 2$ degrees of freedom.   ■

We can now construct a $100(1 - \alpha)$ percent confidence interval for $\mu_1 - \mu_2$ by proceeding in exactly the same way as before; omitting the details, the final result is

$$L = \overline{X}_1 - \overline{X}_2 - t_{n_1+n_2-2}(\alpha/2)s_p \sqrt{\frac{1}{n_1} + \frac{1}{n_2}} \tag{7.35}$$

$$U = \overline{X}_1 - \overline{X}_2 + t_{n_1+n_2-2}(\alpha/2)s_p \sqrt{\frac{1}{n_1} + \frac{1}{n_2}} \tag{7.36}$$

example **7.9**   Two different weight loss programs are being compared to determine their effectiveness. Ten men were assigned to each program, so $n_1 = n_2 = 10$; their weight losses are recorded in Table 7.1. Construct a 95 percent confidence interval for the difference in weight loss between the two diets.

TABLE 7.1

Diet 1	3.4	10.9	2.8	7.8	0.9	5.2	2.5	10.5	7.1	7.5
Diet 2	11.9	13.1	11.6	6.8	6.8	8.8	12.5	8.6	17.5	10.3

Solution. A straightforward calculation yields the following results (rounded to two decimal places):

$$\bar{x}_1 = 5.86, \qquad \bar{x}_2 = 10.79, \qquad \bar{x}_1 - \bar{x}_2 = -4.93$$
$$s_1^2 = 11.85, \qquad s_2^2 = 10.67$$
$$s_p^2 = 11.26, \qquad s_p = 3.36$$

$$t_{18}(0.025)s_p\sqrt{\frac{1}{10} + \frac{1}{10}} = 3.15$$

$$L = -4.93 - 3.15 = -8.08$$
$$U = -4.93 + 3.15 = -1.78$$

The confidence interval is $[-8.08, -1.78]$, which does not include 0. The data imply that diet 2 is somewhat more effective in reducing weight. ∎

7.4.1   Paired Samples

Paired samples arise when we study the response of an experimental unit to two different treatments. Suppose, for example, we are conducting a clinical trial of a blood pressure–reducing drug. To measure its effectiveness (or ineffectiveness), we give the drug to a random sample of n patients. The data set, then, consists of the n pairs $(x_1, y_1), \ldots, (x_n, y_n)$, where x_i denotes the ith patient's blood pressure before the treatment, and y_i denotes the ith patient's blood pressure after the treatment. It is customary to call the two data sets $x_1, \ldots, x_n$ and $y_1, \ldots, y_n$ treatment 1 and treatment 2, respectively. The variable $d_i = x_i - y_i$ represents the difference in blood pressure produced by the drug on the ith patient; a positive value for d_i indicates that the drug reduced the blood pressure of the ith patient. This is an example of what is called *self-pairing*, in which a single experimental unit is given two treatments.

There are, however, many other ways in which such pairs are selected. For example, identical twins are frequently used in psychology tests, and mice of the same sex and genetic strain are used in biology experiments. After the pair has been selected, one of the two units is randomly selected for treatment 1 and the other for treatment 2. The advantages of pairing are intuitively clear: it reduces the variation in the data that is due to causes other than the treatment itself.

These considerations lead naturally to the following model for the paired sample.

▓▓ DEFINITION 7.5

Mathematical model of a paired sample. The data come from a sequence of paired random variables $(X_i, Y_i), i = 1, \ldots, n$, that denote the response of the ith experimental unit to the first and second treatments, respectively. Their difference, which is the variable we are particularly interested in, is denoted by $D_i = X_i - Y_i$. In addition, we assume that the random variables D_i $(i = 1, \ldots, n)$ are mutually independent, normally distributed random variables with a common mean and variance denoted by $E(D_i) = \Delta$ and $V(D_i) = \sigma_D^2$. The sample variance is denoted by s_D^2, where

$$s_D^2 = \frac{1}{n-1} \sum_{1 \leq i \leq n} (D_i - \overline{D})^2 \qquad \blacksquare$$

In general, the pair of random variables X_i and Y_i will not be independent, since the patient's final blood pressure, say, will be highly correlated with his or her initial blood pressure. To estimate Δ and construct confidence intervals, we begin with the fact that $D_1, \ldots, D_n$ is, by hypothesis, a random sample from the normal distribution $N(\Delta, \sigma_D^2)$. Therefore,

$$\overline{D} \sim N\left(\Delta, \frac{\sigma^2}{n}\right)$$

Consequently,

$$\frac{\sqrt{n}(\overline{D} - \Delta)}{s_D} \sim t_{n-1}$$

and therefore

$$P\left(-t_{n-1}(\alpha/2) \leq \frac{\sqrt{n}(\overline{D} - \Delta)}{s_D} \leq t_{n-1}(\alpha/2)\right) = 1 - \alpha$$

Solving for Δ in the usual way produces the $100(1 - \alpha)$ percent confidence interval $[L, U]$, where

$$L = \overline{D} - t_{n-1}(\alpha/2) \times s_D/\sqrt{n} \qquad (7.37)$$
$$U = \overline{D} + t_{n-1}(\alpha/2) \times s_D/\sqrt{n} \qquad (7.38)$$

---

example 7.10   In a study to compare two different configurations of a processor, the times to execute six different workloads (W1, ..., W6) was measured for each of the two configurations, denoted by *one cache* and *no cache*, respectively. The results, obtained from computer simulations, are displayed in Table 7.2. The difference in the execution times for each workload is displayed in the rightmost column. Compute a 95 percent confidence interval for the difference in execution times between the one-cache and no-cache configurations.

TABLE 7.2

Workload	One cache	No cache	$d_i = x_i - y_i$
W1	69	75	−6
W2	53	83	−30
W3	61	89	−28
W4	54	77	−23
W5	49	83	−34
W6	50	83	−33

Solution. This is a paired sample with workload as the experimental unit and one-cache and no-cache memories representing the two treatments. We obtain a 95 percent confidence interval for Δ by first computing

$$\overline{D} = -25.67, \qquad s_D = 10.41, \qquad t_5(0.025) = 2.571$$

The left and right endpoints of the confidence interval are given by

$$L = -25.67 - t_5(0.025) \times \frac{10.41}{\sqrt{6}} = -36.60$$

$$U = -25.67 + t_5(0.025) \times \frac{10.41}{\sqrt{6}} = -14.74$$

We can also write the confidence interval $[-36.60, -14.74]$ in the following equivalent form:

$$-25.67 \pm t_5(0.025)\frac{10.41}{\sqrt{6}} = -25.67 \pm 10.93$$

Because the difference in mean execution time between the one-cache and no-cache configurations is negative and the 95 percent confidence interval does not contain 0, we conclude that the one-cache configuration delivers significantly faster performance. We present a formal justification for this conclusion in Chap. 8. ∎

PROBLEMS

7.22 A random sample of size $n_1 = 16$ taken from a normal distribution $N(\mu_1, 36)$ has a mean $\overline{X}_1 = 30$; another random sample of size $n_2 = 25$ taken from a different normal distribution $N(\mu_2, 9)$ has a mean $\overline{X}_2 = 25$. Find:
  (a) A 99 percent confidence interval for $\mu_1 - \mu_2$.
  (b) A 95 percent confidence interval for $\mu_1 - \mu_2$.
  (c) A 90 percent confidence interval for $\mu_1 - \mu_2$.

7.23 An experiment was conducted to compare the electrical resistances (measured in ohms) of resistors supplied by two manufacturers, denoted A and B, respectively. Six resistors from each manufacturer were randomly selected and their resistances measured; the following data were obtained:

Brand A	0.140	0.138	0.143	0.142	0.144	0.137
Brand B	0.135	0.140	0.142	0.136	0.138	0.140

Assume both data sets come from normal populations. Construct a 95 percent confidence interval for the difference between the mean resistances.

7.24 The skein strengths, measured in pounds, of two types of yarn are tested, and the following results are reported:

Yarn A	99	93	99	97	90	96	93	88	89
Yarn B	93	94	75	84	91				

Construct a 95 percent confidence interval for the difference in means. Assume that both data sets come from normal populations.

7.25 To determine the difference, if any, between two brands of radial tires, 12 tires of each brand were tested and the following mean lifetimes and sample standard deviations for each of the two brands were obtained:

$$\bar{x}_1 = 38,500 \qquad s_1 = 3100$$
$$\bar{x}_2 = 41,000 \qquad s_2 = 3500 \ .$$

Assume the tire lifetimes are normally distributed. Find a 95 percent confidence interval for the difference in the mean lifetimes.

7.26 With reference to the data set on lead levels in children's blood (Prob. 1.12), find:
(a) A 90 percent confidence interval for the difference in lead level between the exposed and control groups of children.
(b) A 95 percent confidence interval for the difference in lead level between the exposed and control groups of children.
(c) A 99 percent confidence interval for the difference in lead level between the exposed and control groups of children.

7.27 The following table gives the concentration of thiol in lysate in the blood of two groups: a "normal" group and a group suffering from rheumatoid arthritis.
(a) Derive a 95 percent confidence interval for the difference in the mean concentration of thiol between the two groups.
(b) Derive a 99 percent confidence interval for the difference in the mean concentration of thiol between the two groups.

Normal	Rheumatoid
1.84	2.81
1.92	4.06
1.94	3.62
1.92	3.27
1.85	3.27
1.91	3.76
2.07	

Source: J. C. Banford, D. H. Brown, A. A. McConnell, C. J. McNeill, W. E. Smith, R. A. Hazelton and R. D. Sturrock, *Analyst*, vol. 107, 1983, p. 195.

7.28 The following data are from Charles Darwin's study of cross- and self-fertilization. Pairs of seedlings of the same age, one produced by cross-fertilization and the other by self-fertilization, were grown together so that members of each pair were grown under nearly identical conditions. The data represent the final heights of the plants after a fixed period of time.

(a) Compute a 95 percent confidence interval for the difference in the mean height. State your model assumptions.

(b) Compute a 99 percent confidence interval for the difference in the mean height.

Pair	Cross-fertilized	Self-fertilized
1	23.5	17.4
2	12.0	20.4
3	21.0	20.0
4	22.0	20.0
5	19.1	18.4
6	21.5	18.6
7	22.1	18.6
8	20.4	15.3
9	18.3	16.5
10	21.6	18.0
11	23.3	16.3
12	21.0	18.0
13	22.1	12.8
14	23.0	15.5
15	12.0	18.0

Source: D. J. Hand et al., *Small Data Sets,* London, Chapman & Hall, 1994.

7.29 The following data are the results of a study of 25 hospitalized schizophrenic patients who were treated with antipsychotic medication and after a period of time were classified as psychotic or nonpsychotic by hospital staff. Samples of cerebrospinal fluid were taken from each patient and assayed for dopamine β-hydroxylase (DBH) activity. The units of measurement are omitted. Construct a 99 percent confidence interval for the difference in DBH activity between the two groups of patients.

Judged nonpsychotic:	0.0104	0.0105	0.0112	0.0116	0.0130	0.0145	0.0154
	0.0156	0.0170	0.0180	0.0200	0.0200	0.0210	0.0230
	0.0252						
Judged psychotic:	0.0150	0.0204	0.0208	0.0222	0.0226	0.0245	0.0270
	0.0275	0.0306	0.0320				

Source: D. E. Sternberg, D. P. Van Kammen, P. Lerner, and W. E. Bunney, "Schizophrenia: Dopamine β-Hydroxylase Activity and Treatment Response," *Science*, vol. 216, 1982, pp. 1423–1425. Used with permission.

7.5   POINT AND INTERVAL ESTIMATION FOR A POPULATION PROPORTION

Estimating the proportion of defectives in a lot of manufactured items or the proportion of voters who plan to vote for a politician are examples of estimating a population proportion. The Bernoulli trials process, described in Sec. 5.2.1, is a mathematical model of this sampling procedure. We repeat, briefly, the essential details: Let $X_1, \ldots, X_n$ denote a random sample from a Bernoulli distribution with $P(X_i = 0) = 1 - p$, $P(X_i = 1) = p$; then

$$f(x; p) = p^x (1 - p)^{1-x}, \quad x = 0, 1$$

The mean, variance, standard deviation, and estimated standard error of the sample proportion $\hat{p}$ are given by

$$E(\hat{p}) = p, \qquad V(\hat{p}) = \frac{p(1 - p)}{n} \tag{7.39}$$

$$\sigma(\hat{p}) = \sqrt{\frac{p(1 - p)}{n}}, \qquad s(\hat{p}) = \sqrt{\frac{\hat{p}(1 - \hat{p})}{n}} \tag{7.40}$$

We now assume that the sample size is large enough that the normal approximation to the binomial can be brought to bear; one rule of thumb is $np > 5$ and $n(1 - p) > 5$. Of course, when p is unknown, we assume $n\hat{p} > 5$ and $n(1 - \hat{p}) > 5$. In any event, assuming the validity of the normal approximation to the binomial yields

$$\frac{\hat{p} - p}{s(\hat{p})} \approx N(0, 1)$$

Therefore,

$$P\left(-z(\alpha/2) \leq \frac{\hat{p} - p}{s(\hat{p})} \leq z(\alpha/2)\right) \approx 1 - \alpha$$

The inequalities inside the parentheses can be solved for p in exactly the same way we solved a similar pair of inequalities for μ in Sec. 7.3 [see Eq. (7.25)]. We have thus shown that

$$P(\hat{p} - z(\alpha/2) \times s(\hat{p}) \le p \le \hat{p} + z(\alpha/2) \times s(\hat{p})) \approx 1 - \alpha \qquad (7.41)$$

Putting these results together in the usual way yields an approximate $100(1 - \alpha)$ percent confidence interval for p of the form $[L, U]$, where

$$L = \hat{p} - z(\alpha/2)\sqrt{\frac{\hat{p}(1 - \hat{p})}{n}} \qquad (7.42)$$

$$U = \hat{p} + z(\alpha/2)\sqrt{\frac{\hat{p}(1 - \hat{p})}{n}} \qquad (7.43)$$

One-Sided Confidence Interval for p

An upper one-sided confidence interval for p is of importance to an engineer who is interested in obtaining an upper bound on the proportion of defectives in a lot of manufactured items. We content ourselves here with merely stating the result. Since

$$P\left(p \le \hat{p} + z(\alpha)\sqrt{\frac{\hat{p}(1 - \hat{p})}{n}}\right) \approx (1 - \alpha)$$

it follows that $[0, U]$, is a $100(1 - \alpha)$ percent upper one-sided confidence interval, where

$$U = \hat{p} + z(\alpha)\sqrt{\frac{\hat{p}(1 - \hat{p})}{n}} \qquad (7.44)$$

---

example **7.11**   The proportion of defective memory chips produced by a factory is p, $0 < p < 1$. Suppose 400 chips are tested and 10 of them are found to be defective. Compute a two-sided and a one-sided upper 95 percent confidence interval for the proportion of defective chips.

Solution. First compute the sample proportion and its estimated standard error:

$$\hat{p} = 10/400 = 0.025 \quad \text{and} \quad \sqrt{\frac{\hat{p}(1 - \hat{p})}{400}} = 0.0078$$

The choice $z_{0.025} = 1.96$ leads to an (approximate) 95 percent two-sided confidence interval $[L, U]$, where

$$L = 0.025 - 1.96 \times 0.0078 = 0.0097$$
$$U = 0.025 + 1.96 \times 0.0078 = 0.0403$$

An approximate 95 percent one-sided upper confidence interval is given by $[0, U]$, where

$$U = 0.025 + 1.645 \times 0.0078 = 0.0378 \qquad \blacksquare$$

Choosing the Sample Size

How large must the sample be in order for our point estimate $\hat{p}$ to be within d units of the unknown proportion p? The $100(1 - \alpha)$ percent confidence interval defined by Eqs. (7.42) and (7.43) indicates that with (approximate) probability $(1 - \alpha)$ our point estimate $\hat{p}$ is within d units of the unknown parameter p, where

$$d = \frac{z(\alpha/2)}{2\sqrt{n}} \geq z(\alpha/2) \times \sqrt{\frac{\hat{p}(1 - \hat{p})}{n}} \tag{7.45}$$

[To prove Eq. (7.45), observe that $p(1 - p) \leq 1/4, 0 \leq p \leq 1$.] Therefore, $p(1 - p)/n \leq 1/4n$. Consequently,

$$z(\alpha/2) \times \sqrt{\frac{\hat{p}(1 - \hat{p})}{n}} \leq \frac{z(\alpha/2)}{2\sqrt{n}}$$

no matter what the value of p is.

We now use this result to determine how large a sample is needed in order to estimate p with a given accuracy. Solving the inequality (7.45) for n yields the result

$$n \geq \left[\frac{z(\alpha/2)}{2d}\right]^2 \tag{7.46}$$

---

example 7.12   What is the minimum sample size required to estimate the proportion of voters in favor of a political party with a margin of error $d = 0.03$ and a confidence level of 90 percent?

Solution. In this case $d = 0.03$, $\alpha/2 = 0.05$, $z(0.95) = 1.645$, and therefore n satisfies the equation

$$0.03 = \frac{1.645}{2\sqrt{n}}, \quad \text{so } n = 752$$

If one is willing to live with a somewhat larger sampling error, $d = 0.05$ say, then n satisfies the equation

$$0.05 = \frac{1.645}{2\sqrt{n}}, \quad \text{so } n = 271 \qquad \blacksquare$$

7.5.1   Confidence Interval for $p_1 - p_2$

Confidence intervals for the difference between two proportions arise in a wide variety of applications. For example, we might be interested in comparing the proportion of defective items produced by two machines, M_1 and M_2. Suppose machine M_i produces n_i items, and X_i of them are defective. We denote the proportion of defective items produced by M_i by $\hat{p}_i$. The estimated standard error of the difference between the two proportions is given by

$$s(\hat{p}_1 - \hat{p}_2) = \sqrt{\frac{\hat{p}_1(1 - \hat{p}_1)}{n_1} + \frac{\hat{p}_2(1 - \hat{p}_2)}{n_2}} \qquad (7.47)$$

Assuming, as in the previous section, that both sample sizes are large enough that the normal approximation to the binomial is valid, it can be shown that the difference between the sample proportions is also approximately normally distributed with mean $p_1 - p_2$ and standard deviation $s(\hat{p}_1 - \hat{p}_2)$. Using this result, we obtain the following approximate $100(1 - \alpha)$ percent two-sided confidence interval $[L, U]$ for $p_1 - p_2$:

$$L = \hat{p}_1 - \hat{p}_2 - z(\alpha/2)s(\hat{p}_1 - \hat{p}_2) \qquad (7.48)$$
$$U = \hat{p}_1 - \hat{p}_2 + z(\alpha/2)s(\hat{p}_1 - \hat{p}_2) \qquad (7.49)$$

example 7.13   Two machines produce light bulbs. Of 400 produced by the first machine, 16 were defective; of 600 produced by the second machine, 60 were defective. Construct a 95 percent confidence interval for the difference $p_1 - p_2$.

Solution. The sample proportions are $\hat{p}_1 = 0.04$ and $\hat{p}_2 = 0.1$; therefore,

$$\hat{p}_1 - \hat{p}_2 = -0.06$$

$$s(\hat{p}_1 - \hat{p}_2) = \sqrt{\frac{0.04 \times 0.96}{400} + \frac{0.1 \times 0.9}{600}} = 0.0157$$

$$L = -0.06 - 1.96 \times 0.0157 = -0.0908$$
$$U = -0.06 + 1.96 \times 0.0157 = -0.0292 \qquad \blacksquare$$

PROBLEMS

7.30 Let $X_1, \ldots, X_n$ be a random sample from a Bernoulli distribution. Show that

$$E(\hat{p}(1 - \hat{p})) = \frac{n-1}{n} \times p(1 - p) < p(1 - p)$$

Deduce that $\hat{p}(1 - \hat{p})$ is a biased estimator of $p(1 - p)$.

7.31 From a random sample of 800 TV monitors that were tested, 10 were found to be defective. Construct a 99 percent one-sided upper confidence interval for the proportion p of defective monitors.

7.32 A random sample of 1300 voters from a large population produced the following data: 675 in favor of proposition A and 625 opposed. Find a 95 percent confidence interval for the proportion p of voters:
(a) In favor of proposition A.
(b) Opposed to proposition A.

7.33 Of a sample of 400 items 5 percent were found to be defective. Find:
(a) A 95 percent confidence interval for the proportion p of defective items.
(b) A 95 percent one-sided upper confidence interval for p.

7.34 (a) To determine the level of popular support for a political candidate, a newspaper regularly surveys 600 voters. The results of the poll are reported as a 95 percent confidence interval. Determine the margin of error.

(b) Calculate the margin of error in part (a) when the sample size is doubled to 1200.

7.35 How large a sample size is needed if we want to be 95 percent confident that the sample proportion will be within 0.02 of the true proportion of defective items and:

(a) The true proportion p is unknown?

(b) It is known that the true proportion $p \leq 0.1$?

7.36 How large a sample size is needed if we want to be 90 percent confident that the sample proportion will be within 0.02 of the true proportion of defective items and:

(a) The true proportion p is unknown?

(b) It is known that the true proportion $p \leq 0.1$?

7.37 Two machines produce computer memory chips. Of 500 chips produced by the first machine, 20 were defective; of 600 chips produced by the second machine, 40 were defective. Construct 90 percent one-sided upper confidence intervals for:

(a) p_1, the proportion of defective chips produced by machine 1.

(b) p_2, the proportion of defective chips produced by machine 2.

(c) Construct a 95 percent two-sided confidence interval for the difference $p_1 - p_2$. What conclusions can you draw if the confidence interval includes (or does not include) 0?

7.38 To compare the effectiveness of two different methods of teaching arithmetic, 60 students were divided into two groups of 30 students each. One group (the control group) was taught by traditional methods, and the other group (the treatment group), was taught by a new method. At the end of the term 20 students of the control group and 25 students of the treatment group passed the basic skills test.

(a) Find an approximate 95 percent confidence interval for $p_1 - p_2$, where p_1 and p_2 are the true proportions of students who pass the basic skills test for the control and treatment groups, respectively. What conclusions can you draw if the confidence interval includes (or does not include) 0?

(b) Find an approximate 90 percent confidence interval for $p_1 - p_2$. What conclusions can you draw if the confidence interval includes (or does not include) 0?

7.39 Samples of size $n_1 = n_2 = 150$ were taken from the weekly outputs of two factories. The numbers of defective items produced were $x_1 = 10$ and $x_2 = 15$. Construct an approximate 95 percent confidence interval for $p_1 - p_2$, the difference in the proportion of defective items produced by the two factories. What conclusions can you draw if the confidence interval includes (or does not include) 0?

7.6 SOME METHODS OF ESTIMATION (OPTIONAL)

This section introduces two methods for constructing estimates: the method of moments and the maximum-likelihood method.

7.6.1 Method of Moments

The kth moment μ_k of a random variable X is defined by $\mu_k = E(X^k)$; in terms of the distribution function, the corresponding formulas are

$$\mu_k = \sum_{i=1}^{\infty} x_i^k f(x_i) \quad \text{(discrete case)} \tag{7.50}$$

$$\mu_k = \int_{-\infty}^{\infty} x^k f(x)dx \quad \text{(continuous case)} \tag{7.51}$$

It is clear from the definition that $\mu_1 = \mu$, and it follows from Eq. (3.23) that $\sigma_X^2 = \mu_2 - \mu_1^2$.

It is not too difficult to show that if $X_1, \ldots, X_n$ is a random sample from a distribution $f(x)$, then the kth sample moment $\hat{\mu}_k$ defined by

$$\hat{\mu}_k = \frac{1}{n} \sum_{1 \leq i \leq n} X_i^k \tag{7.52}$$

is an unbiased estimate for the kth population moment μ_k. The verification of this fact is left as an exercise (Prob. 7.46). Note also that $\hat{\mu}_1 = \overline{X}$. If the distribution function of the random variable depends on one or more parameters, then so will the moments.

---

example 7.14
1. **Exponential distribution.** Suppose X is exponentially distributed with parameter λ; then

$$E(X) = \frac{1}{\lambda}$$

Consequently,

$$\lambda = \frac{1}{\mu_1}$$

The method of moments estimator (MME) is obtained by substituting $\hat{\mu}_1$ for μ_1 in the preceding equation, which yields the following estimator for λ:

$$\hat{\lambda} = \hat{\mu}_1^{-1} = \overline{X}^{-1}$$

2. **Normal distribution.** Suppose $X \sim N(\mu, \sigma^2)$. In this case

$$\mu_1 = \mu, \qquad \mu_2 = \sigma^2 + \mu^2$$

Solving these simultaneous equations for μ and σ^2 in terms of μ_1 and μ_2 yields

$$\mu = \mu_1 \quad \text{and} \quad \sigma^2 = \mu_2 - \mu_1^2$$

The MME is obtained by substituting $\hat{\mu}_i$ for μ_i in these equations, which results in the following estimators for μ and σ^2:

$$\hat{\mu} = \hat{\mu}_1 = \overline{X}$$
$$\hat{\sigma}^2 = \hat{\mu}_2 - \hat{\mu}_1^2$$

After some algebraic manipulations that we omit, it can be shown that

$$\hat{\sigma}^2 = \frac{1}{n} \sum_{1 \le i \le n} (x_i - \overline{x})^2 = \frac{n-1}{n} s^2$$

Consequently,

$$E(\hat{\sigma}^2) = \frac{n-1}{n} \sigma^2 < \sigma^2$$

Thus, the MME for σ^2 is biased.

3. **Gamma distribution.** Suppose X has a gamma distribution with parameters α, β. In this case

$$\mu_1 = \alpha\beta, \qquad \mu_2 = (\alpha + \alpha^2)\beta^2 \tag{7.53}$$

We can solve these equations for α, β in terms of μ_1, μ_2, thus obtaining the formulas

$$\alpha = \frac{\mu_1^2}{\mu_2 - \mu_1^2} \quad \text{and} \quad \beta = \frac{\mu_2 - \mu_1^2}{\mu_1}$$

The MME is obtained by substituting $\hat{\mu}_i$ for μ_i in these equations, leading to the following estimators for α and β:

$$\hat{\alpha} = \frac{\hat{\mu}_1^2}{\hat{\mu}_2 - \hat{\mu}_1^2} \quad \text{and} \quad \hat{\beta} = \frac{\hat{\mu}_2 - \hat{\mu}_1^2}{\hat{\mu}_1} \tag{7.54}$$

∎

---

example 7.15 *Estimating German Tank Production During World War II.* During World War II tanks produced by the Germans were stamped with serial numbers indicating the order in which they were produced. The problem was to estimate the number N of tanks produced, based on the serial numbers $\{x_1, \ldots, x_n\}$ of destroyed or captured German tanks. For this purpose we assume that $\{x_1, \ldots, x_n\}$ is a random sample taken from the (finite) population $\{1, 2, \ldots, N\}$. Its population mean is given by

$$\mu = \frac{1 + 2 + \cdots + N}{N} = \frac{N+1}{2}$$

Solving for N in terms of μ yields $N = 2\mu - 1$; this yields the MME

$$\hat{N} = 2\overline{X} - 1$$

To illustrate the method (and its limitations), suppose 10 tanks were captured and their serial numbers were recorded (in increasing order) as follows:

$$125, 135, 161, 173, 201, 212, 220, 240, 250, 502$$

The sample mean and corresponding estimate for N are given by

$$\overline{X} = 221.9 \quad \text{and} \quad \hat{N} = 2\overline{X} - 1 = 442.8$$

It is immediately obvious that there is something seriously wrong with this estimate since common sense tells you that the number of tanks produced must exceed the largest captured serial number, so $N \geq 502$. Indeed, the "commonsense estimator"

$$\hat{N}_1 = \max\{125, 135, 161, 173, 201, 212, 220, 240, 250, 502\} = 502$$

is clearly superior to the MME $2\overline{X} - 1$. ∎

In the next section we introduce an alternative method that generally produces estimators much better than those obtained via other methods such as the method of moments. It is called the method of maximum likelihood and the estimators it produces are called maximum-likelihood estimators (MLEs). In particular, the estimator $\hat{N}_1$ just defined is an MLE (see Example 7.16).

7.6.2 Maximum-Likelihood Estimators

The *maximum-likelihood method* is a precise formulation of the intuitively appealing criterion that a good estimate of an unknown parameter is also the most plausible. The degree of plausibility is measured by the *likelihood function*, a concept that we now explain.

▦ DEFINITION 7.6

Let $X_1, \ldots, X_n$ be a random sample of size n taken from a distribution $f(x; \theta)$. The *likelihood function* is the joint distribution function of the random sample, which is given by

$$L(x_1, \ldots, x_n; \theta) = f(x_1; \theta) \cdots f(x_n; \theta) \tag{7.55}$$

∎

When the distribution function depends on two or more parameters, then so does the likelihood function:

$$L(x_1, \ldots, x_n; \theta_1, \theta_2) = f(x_1; \theta_1, \theta_2) \cdots f(x_n; \theta_1, \theta_2)$$

To simplify the notation we write the random sample and the set of parameters using **boldface** vector notation:

$$\mathbf{x} = (x_1, \ldots, x_n) \quad \text{and} \quad \boldsymbol{\theta} = (\theta_1, \ldots, \theta_k)$$

The likelihood function is then written in the following, more compact form:

$$L(x_1, \ldots, x_n; \theta_1, \theta_2) = L(\mathbf{x}; \boldsymbol{\theta})$$

▦ DEFINITION 7.7

Let $X_1, \ldots, X_n$ be a random sample from the distribution $f(x; \theta)$; then the maximum-likelihood estimator (MLE) for θ is that value $\hat{\theta}$ that maximizes the likelihood function $L(x; \theta)$, i.e.,

$$L(\hat{\theta}) \geq L(x; \theta), \quad \theta \in \Theta$$

∎

▦ REMARK 7.1

For the purposes of computing $\max L(\mathbf{x}; \theta)$, it is sometimes more convenient to maximize the logarithm of the likelihood function instead of maximizing the likelihood function itself. This is because we compute the maximum by differentiation, and the function

$$\ln(\mathbf{x};\theta) = \sum_i \ln f(x_i;\theta)$$

is easier to differentiate than $L(\mathbf{x};\theta)$. We obtain the same maximum no matter which function we choose to maximize, since it is easily verified that

$$L(\hat{\theta}) \geq L(\mathbf{x};\theta), \quad \theta \in \Theta \quad \text{iff} \quad \ln L(\hat{\theta}) \geq \ln L(\mathbf{x};\theta), \quad \theta \in \Theta \qquad \blacksquare$$

This will be illustrated in several of the following examples as well as in some of the problems.

Examples

1. **Sampling from a Bernoulli distribution.** Recall that a Bernoulli distribution is given by the following simple formula:

    $$f(x;p) = p^x(1-p)^{1-x}, \quad x = 0, 1$$

    The likelihood function and its logarithm are given by

    $$\begin{aligned} L(\mathbf{x};p) &= p^{x_1}(1-p)^{1-x_1} \cdots p^{x_n}(1-p)^{1-x_n} \\ &= p^{\sum_{1 \leq i \leq n} x_i}(1-p)^{n-\sum_{1 \leq i \leq n} x_i} \qquad (7.56) \\ &= p^T(1-p)^{n-T} \end{aligned}$$

    where $T = \sum_{1 \leq i \leq n} x_i$, and

    $$\ln L(\mathbf{x};p) = T \ln(p) + (n-T)\ln(1-p)$$

    To compute the maximum we differentiate $\ln L(\mathbf{x};p)$ with respect to p, set it equal to zero, and solve for p, i.e.,

    $$\frac{d}{dp}\ln L(\mathbf{x};p) = \frac{T}{p} - \frac{n-T}{1-p} = 0$$

    If $T = 0$ or n, then a direct calculation shows that $\hat{p} = 0$ or n, respectively. On the other hand, for $0 < T < n$, we have

    $$\hat{p} = \frac{T}{n} = \frac{1}{n}\sum_{1 \leq i \leq n} x_i$$

    Consequently, the MLE for the binomial parameter is

    $$\hat{p} = \overline{X}$$

2. **Sampling from an exponential distribution.** Here $f(x;\mu) = \mu e^{-\mu x}, x > 0; f(x;\mu) = 0$ otherwise. The likelihood function and its logarithm are given by

    $$\begin{aligned} L(\mathbf{x};\mu) &= \mu e^{-\mu x_1} \cdots \mu e^{-\mu x_n} \qquad (7.57) \\ &= \mu^n e^{-\mu \sum_{1 \leq i \leq n} x_i}, \quad \min(x_1,\ldots,x_n) > 0 \\ &= 0 \quad \text{otherwise} \end{aligned}$$

    and $\ln L(\mathbf{x};\mu) = n\ln(\mu) - \mu T$

To compute the maximum we differentiate $\ln L(\mathbf{x}; \mu)$ with respect to μ, set it equal to zero, and solve for μ, i.e.,

$$\frac{d}{d\mu} \ln L(\mathbf{x}; \mu) = \frac{n}{\mu} - T = 0$$

Thus $\hat{\mu} = n/T = 1/\overline{X}$; since $E(1/\overline{X}) \neq 1/E(\overline{X})$, it follows that $\hat{\mu}$ is a biased estimator of μ.

3. **Sampling from a Poisson distribution.** $f(x; \lambda) = e^{-\lambda}(\lambda^x / x!)$, $x = 0, 1, \ldots$. The likelihood function and its logarithm are, respectively,

$$L(\mathbf{x}; \lambda) = \frac{e^{-n\lambda} \lambda^{\sum_{1 \leq i \leq n} x_i}}{x_1! \cdots x_n!} \tag{7.58}$$

$$\ln L(\mathbf{x}; \lambda) = -n\lambda + T \ln(\lambda) - \sum_{1 \leq i \leq n} \ln(x_i!)$$

To compute the maximum we differentiate $\ln L(\mathbf{x}; \lambda)$ with respect to λ, set it equal to zero, and solve for λ, i.e.,

$$\frac{d}{d\lambda} \ln L(\mathbf{x}; \lambda) = 0 = -n + \frac{T}{\lambda}$$

i.e.,

$$\hat{\lambda} = \frac{T}{n}$$

4. **Sampling from a normal distribution.** Let $f(x; \mu, \sigma^2)$ denote the normal density with mean μ and variance σ^2. A straightforward calculation using Eq. (4.25) gives

$$L(\mathbf{x}; \mu, \sigma^2) = \frac{1}{\sigma^n (2\pi)^{n/2}} \exp\left(-\frac{1}{2\sigma^2} \sum_{1 \leq i \leq n} (x_i - \mu)^2\right) \tag{7.59}$$

Taking logarithms of both sides yields

$$\ln L(\mathbf{x}; \mu, \sigma^2) = \ln\left(\frac{1}{\sigma^n (2\pi)^{n/2}}\right) - \frac{1}{2\sigma^2} \sum_{1 \leq i \leq n} (x_i - \mu)^2$$

$$= -n \ln(\sigma) - \frac{1}{2\sigma^2} \sum_{1 \leq i \leq n} (x_i - \mu)^2 - \frac{n}{2} \ln(2\pi)$$

Because $\ln L(\mathbf{x}; \mu, \sigma^2)$ is a function of two variables, its maximum is obtained by computing the partial derivatives with respect to μ and σ and setting them equal to zero:

$$\frac{\partial \ln L}{\partial \mu} = \frac{1}{\sigma^2} \sum_{1 \leq i \leq n} (x_i - \mu) = 0$$

$$\frac{\partial \ln L}{\partial \sigma} = -\frac{n}{\sigma} + \frac{1}{\sigma^3} \sum_{1 \leq i \leq n} (x_i - \mu)^2 = 0$$

Solving these equations for μ and σ^2, respectively, yields

$$\hat{\mu} = \bar{x} \tag{7.60}$$

$$\hat{\sigma}^2 = \frac{1}{n} \sum_{1 \leq i \leq n} (x_i - \bar{x})^2 \tag{7.61}$$

Notice that $\hat{\sigma}^2 = (n-1)/n \times s^2$, and therefore the MLE for σ^2 is biased since

$$E(\hat{\sigma}^2) = (n-1)/n \times \sigma^2 \neq \sigma^2$$

---

example 7.16   (Continuation of Example 7.15) In order to estimate the number N of German tanks produced, we assumed that the observed serial numbers $\{x_1, \ldots, x_n\}$ of the captured tanks were, in effect, a random sample taken from the (finite) population $\{1, 2, \ldots, N\}$. Consequently, the joint density function is given by

$$P(X_1 = x_1, \ldots, X_n = x_n) = \frac{1}{P_{N,n}}, \quad N \geq \max\{x_1, \ldots, x_n\}$$

$$= 0 \quad \text{otherwise}$$

This implies that the likelihood function is given by

$$L(\mathbf{x}; N) = \frac{1}{P_{N,n}}, \quad N \geq \max\{x_1, \ldots, x_n\} \tag{7.62}$$

$$= 0 \quad \text{otherwise}$$

Thus,

$$L(\mathbf{x}; N) = \frac{1}{N(N-1)\cdots(N-n+1)}, \quad N \geq \max\{x_1, \ldots, x_n\}$$

Clearly, $L(\mathbf{x}; N)$ is maximized by choosing N as small as possible consistent with the constraint $N \geq \max\{x_1, \ldots, x_n\}$; this means choosing $\hat{N} = \max\{x_1, \ldots, x_n\}$. ∎

It is frequently the case that one wants to estimate a function $g(\theta)$ of the parameter. For example, if the data come from an exponential distribution with parameter λ, then estimating the reliability at time t_0 means that one must estimate $\exp(-\lambda t_0)$, i.e., $g(\lambda) = \exp(-\lambda t_0)$. One of the nice properties of an MLE is that $g(\hat{\theta})$ is the MLE for $g(\theta)$. It is useful to restate this result more formally:

■ PROPOSITION 7.1

Let $\hat{\theta}$ be the MLE of θ, and let $g(\theta)$ denote any function of the parameter θ. Then $g(\hat{\theta})$ is the MLE of $g(\theta)$. ∎

---

example 7.17   Refer to the air conditioner failure times for the B8045 plane in Prob. 1.27. Assuming the data come from an exponential distribution with parameter λ, compute the maximum-likelihood estimates for $\hat{\lambda}$ and $\hat{R}(t)$.

Solution. It was shown that

$$\text{MLE } \hat{\lambda} = \frac{1}{\overline{X}} = \frac{1}{82} = 1.22 \times 10^{-2}$$

Consequently, the MLE for the reliability at time t is given by

$$\hat{R}(t) = \exp(-\hat{\lambda}t) = \exp(-1.22 \times 10^{-2} \times t)$$

Setting $t = 50$ hours yields the estimate $\hat{R}(50) = 0.54$. This may be interpreted as follows: The probability that the air-conditioning unit is still working after 50 hours is 0.54. ∎

Under suitable hypotheses on $f(x; \theta)$, which are too technical to be stated here, MLEs enjoy the following properties:

1. An MLE is consistent, i.e.,

$$\lim_{n \to \infty} P(|\hat{\theta} - \theta| < d) = 1 \quad \text{for } d > 0$$

2. An MLE is *asymptotically unbiased*, i.e.,

$$\lim_{n \to \infty} E(\hat{\theta}) = \theta$$

PROBLEMS

7.40 (a) Let $X_1, \ldots, X_n$ be a random sample taken from an exponential distribution with parameter λ. Show that $\overline{X}$ is an unbiased estimator of $1/\lambda$.

(b) Let $U = \min(X_1, \ldots, X_n)$. Show that U is exponentially distributed with parameter $n\lambda$. [*Hint:* Show that $P(U > x) = \exp(-n\lambda x)$.]

(c) Deduce from part (*a*) that nU is an unbiased estimator of $1/\lambda$.

(d) Compute $V(nU)$ and $V(\overline{X})$. Which estimator is better? Justify your answer by stating an appropriate criterion.

7.41 Ten measurements of the concentration (in percent) of an impurity in an ore were taken, and the following data were obtained:

$$3.8, 3.5, 3.4, 3.9, 3.7, 3.7, 3.6, 3.7, 4.0, 3.9$$

(a) Assuming these data come from a normal distribution $N(\mu, \sigma^2)$, compute the MLEs for μ, σ^2, σ.

(b) Which of the three estimators in part (*a*) are unbiased? Discuss.

7.42 The breaking strengths of 36 steel wires were measured and the following results were obtained: $\overline{x} = 9830$ psi and $s = 400$. Assume that the measurements come from a normal distribution. The safe strength is defined to be the 10th percentile of the distribution. Compute the MLE of the safe strength.

7.43 The lifetime of a component is assumed to have a gamma distribution with parameters α, β. The first two sample moments are $\hat{\mu}_1 = 96$ and $\hat{\mu}_2 = 10,368$. Use the method of moments to estimate α, β.

7.44 Let

$$f(x|\theta) = \theta x^{\theta-1}, \quad 0 < x < 1, \quad 0 < \theta$$
$$= 0 \quad \text{elsewhere}$$

(a) Show, via a suitable integration, that

$$\mu_1 = \frac{\theta}{\theta + 1}$$

(b) Use the result from part (a) to derive the method of moments estimator for θ.

7.45 In a political campaign $X = 185$ voters out of $n = 351$ voters polled are found to favor candidate A. Use these data to compute the MLE of p and $p(1 - p)$. Are both of these estimators unbiased? Discuss.

7.46 Verify that $\hat{\mu}_k$, as defined in Eq. (7.52), is an unbiased estimator for μ_k.

7.47 It is sometimes necessary (e.g., in likelihood ratio tests) to compute the maximum value of the likelihood function $L(\mathbf{x}|\hat{\theta})$. In particular, with reference to the likelihood function displayed in Eq. (7.59), show that

$$L(\mathbf{x}|\hat{\mu}, \hat{\sigma}^2) = (\hat{\sigma}\sqrt{2\pi})^{-n} e^{-n/2} \tag{7.63}$$

7.7 CHAPTER SUMMARY

In this chapter we discussed several methods for estimating population parameters such as the mean μ and variance σ^2 of a distribution. The performance of an estimator depends on several factors, including its bias, its standard error (or estimated standard error), and whether or not it is consistent. We then introduced the important concept of a confidence interval, which is an interval that contains the unknown parameter with high probability. The particular cases of the sample mean and sample proportion were then studied in some detail. We then discussed two basic methods for constructing estimators: the method of moments and the maximum-likelihood method.

To Probe Further. We did not discuss Bayesian estimation. A good introduction to this important topic is found in M. H. DeGroot, *Probability and Statistics* (Reading, Mass., Addison-Wesley, 1986). A more advanced treatment of the topics treated here is found in E.L. Lehman, *Theory of Point Estimation* (Pacific Grove, Calif., Wadsworth & Brooks/Cole, 1991). C. Howson and P. Urbach's *Scientific Reasoning: The Bayesian Approach* (LaSalle, Ill., Open Court, 1989) is a penetrating comparison of the Bayesian approach to confidence intervals with the more traditional interpretation presented here.

CHAPTER 8

Inferences about Population Means

Doubt everything or believe everything; these are two equally convenient strategies. With either we dispense with the need for reflection.

Henri Poincaré (1854–1912), French mathematician

8.1 ORIENTATION

A frequently occurring problem of no small importance is to determine whether or not a product meets a standard. The quality of the product is usually measured by a quantitative variable X defined on a certain population. Since the quality of the product is always subject to some kind of random variation, the standard is usually expressed as an assertion about the distribution of the variable X. For example, suppose X denotes the compressive strength [measured in pounds per square inch (psi)] of a brick, and the manufacturer's specifications are that the bricks have a mean strength μ greater than μ_0. The assertion $\mu > \mu_0$ is an example of a *statistical hypothesis*; it is a claim about the distribution of a variable X defined on a population. In this chapter we learn how to use statistics such as the sample mean and sample variance to test the truth or falsity of a statistical hypothesis. We then extend these methods to testing hypotheses about two distributions where the object is to study the difference between their means.

Organization of Chapter

8.2   TESTS OF STATISTICAL HYPOTHESES: BASIC CONCEPTS AND EXAMPLES

The concepts of hypothesis testing are best introduced in the context of a basic example. The following one pertains to communication engineering.

example 8.1   Consider a binary digital communications system where the transmitted signal is a pulse of amplitude μ, where $\mu = +a$ or $\mu = -a$. The positive pulse represents a 1 and the negative pulse represents a 0. As in any communications system, what is transmitted is not always the same as what is received. We assume that the received signal, denoted by Y, equals the transmitted signal plus a normal random variable ϵ that represents the *noise* in the system. That is, the received signal Y is a random variable given by

$$Y = \mu + \epsilon \tag{8.1}$$

where $\mu = -a$ or $\mu = +a$, and $\epsilon \sim N(0, \sigma^2)$, so $Y \sim N(\mu, \sigma^2)$. (*Note:* Electrical engineers use the term *Gaussian noise* to describe a communications channel of this type.) Figure 8.1 displays the graphs of the pdf of the received signal for the cases $\mu = -1$ and $\mu = +1, \sigma = 0.5$.

The function of the receiver is to decide, based on the signal received, whether the transmitted signal was $+a$ or $-a$. This is equivalent to deciding whether Y comes from a normal distribution with $\mu = -a$ or with $\mu = +a$. We let H_0 denote the hypothesis that $\mu = -a$ and H_1 the hypothesis that $\mu = +a$. We call H_0, somewhat arbitrarily, the *null hypothesis* and H_1 the *alternative hypothesis*. Each of these hypotheses completely specifies the distribution of Y since σ is assumed to be known. Such hypotheses are called *simple hypotheses*. Hypotheses that are not simple are called *composite* and will be discussed later.

Suppose the receiver uses the following *decision procedure* (also called a *test*) to determine which signal was transmitted:

$$\mu = -a \quad \text{if } Y \leq 0$$
$$\mu = +a \quad \text{if } Y > 0$$

What criteria are reasonable for evaluating the receiver's performance?

Solution. The performance of the receiver is best understood if the outcomes of the decision procedure are displayed in the format of a table. Looking at Table 8.1, we see that there are two possible types of error.

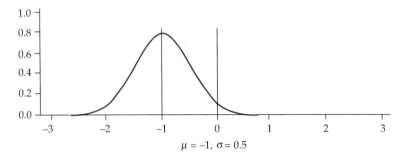

$\mu = -1, \sigma = 0.5$

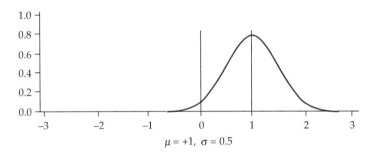

$\mu = +1, \sigma = 0.5$

Figure 8.1 Probability distribution functions of received signal Y for $\mu = -1, \mu = 1$

- A *type I error* occurs if one rejects H_0 when H_0 is true. Here this means that one decides that the received signal $\mu = +a$ even though the transmitted signal $\mu = -a$. A type I error occurs when $Y > 0$ even though $\mu = -a$.
- A *type II error* occurs if one accepts H_0 when H_0 is false (equivalently, one rejects H_1 when H_1 is true). In this case this means that one decides that the received signal $\mu = -a$ even though the transmitted signal $\mu = +a$. A type II error occurs when $Y \leq 0$ even though $\mu = +a$.

TABLE 8.1

Received signal	Decision	Transmitted signal	
		$\mu = -a$	$\mu = +a$
$Y \leq 0$	$\mu = -a$	Correct decision	Type II error
$Y > 0$	$\mu = a$	Type I error	Correct decision

Since Y is normally distributed, it follows that the type I and type II error probabilities are given by

$$P(Y > 0 \mid \mu = -a) = P\left(\frac{Y - (-a)}{\sigma} > \frac{a}{\sigma}\right)$$

$$= 1 - \Phi(a/\sigma) \quad \text{(type I error probability)}$$

$$P(Y \leq 0 \mid \mu = +a) = P\left(\frac{Y - a}{\sigma} \leq \frac{-a}{\sigma}\right)$$

$$= \Phi(-a/\sigma) \quad \text{(type II error probability)}$$

Looking at Fig. 8.1, where $a = 1$ and $\sigma = 0.5$, we see that the type I error probability is the area under the normal curve for $\mu = -1$ to the right of 0, and the type II error probability is the area under the normal curve for $\mu = +1$ to the left of 0.

The probabilities of making the four possible decisions (correct and incorrect) are listed in Table 8.2. Note that these probabilities depend on the ratio a/σ. Electrical engineers call the quantity $(a/\sigma)^2$ the *signal-to-noise ratio (SNR)*.

TABLE 8.2   DECISION PROBABILITIES

		Transmitted signal	
Received signal	**Decision**	$\mu = -a$	$\mu = +a$
$Y \leq 0$	$\mu = -a$	$\Phi(a/\sigma)$	$\Phi(-a/\sigma)$
$Y > 0$	$\mu = a$	$1 - \Phi(a/\sigma)$	$1 - \Phi(-a/\sigma)$

The performance of the receiver is measured by how well it controls the type I and type II error probabilities; a good receiver will reject H_0 with small probability when it is true and will reject H_0 with high probability when it is false. Since the error probabilities depend only the ratio a/σ, we set $a = 1$ and compute the entries in Table 8.2 for selected values of σ. The error probabilities for $a = +1, \sigma = 0.5$ (i.e., $a/\sigma = 2.0$) are displayed in Table 8.3.

TABLE 8.3   DECISION PROBABILITIES FOR $a = +1$, $\sigma = 0.5$

		Transmitted signal	
Received signal	**Decision**	$\mu = -1$	$\mu = +1$
$Y \leq 0$	$\mu = -1$	$\Phi(2.0) = 0.9772$	$\Phi(-2.0) = 0.0228$
$Y > 0$	$\mu = 1$	$1 - \Phi(2.0) = 0.0228$	$1 - \Phi(-2.0) = 0.9772$

Observe that the sum of the entries in each column of Table 8.2 equals 1. This follows from the fact that

$$P(\text{accept } H_0 \mid \mu) + P(\text{reject } H_0 \mid \mu) = 1$$

Consequently, to determine the error probabilities, it suffices to know the entries in either the first or the second row. The entries in each row are functions of the parameter μ given explicitly by

$$P(\text{accept } H_0 \mid \mu) = \Phi(-\mu/\sigma) \quad \text{(first row)}$$

$$P(\text{reject } H_0 \mid \mu) = 1 - \Phi(-\mu/\sigma) \quad \text{(second row)}$$

The function $\pi(\mu)$ defined by

$$\pi(\mu) = P(\text{reject } H_0 \mid \mu) \qquad (8.2)$$

is called the *power function* of the test; it is the probability of rejecting H_0 as a function of the parameter μ. It is this function that produces the entries in the second row. Table 8.4 displays the values of the power function for $a = 1$ and $\sigma = 0.75, 0.5, 0.43$. The entries in Table 8.4 tell us that the type I and type II error probabilities decrease as σ decreases. Note in particular that the type I and type II error probabilities for $\mu = 1, \sigma = 0.5$ are both equal to 0.0228, so the probability of making the correct decision appears to be quite high.

To pursue this example further, suppose the transmission rate is 19,200 bits per second (bps); then the expected number of type I errors per unit time equals $19,200 \times 0.0228 = 438$ (rounding to the nearest integer). This is an unacceptably high error rate in a modern communications network transmitting thousands of pulses per second.

TABLE 8.4   POWER FUNCTION $\pi(\mu)$

	$\pi(-1)$	$\pi(+1)$
$\mu = +1, \sigma = 0.75$	0.0918	0.9082
$\mu = +1, \sigma = 0.50$	0.0228	0.9772
$\mu = +1, \sigma = 0.43$	0.0099	0.9901

∎

Testing a Simple Hypothesis against a One-Sided Alternative

In most applications one or both of the hypotheses are not simple. The next example is typical.

example 8.2   Consider a manufacturer's claim that bricks produced by a new and more efficient process have a mean compressive strength $\mu > \mu_0$. Bricks that do not meet this standard are defective (the technical term is *nonconforming*) and will not be used by the customer. To test the manufacturer's claim, we select a sample of n bricks, determine their compressive strengths, and compute the sample mean. If it is large enough, we accept the claim that $\mu > \mu_0$. From past experience it is known that the compressive strength of a randomly selected brick comes from a normal distribution with known standard deviation σ. The assertion that $\mu = \mu_0$ is another example of a *null hypothesis* (denoted H_0: $\mu = \mu_0$) or *no-change hypothesis. To simplify the calculations, we will always assume that the null hypothesis is simple.* In plain English, the null hypothesis is that the new process produces bricks that are no better than those manufactured using current technology. The alternative hypothesis, that the bricks produced by the new process exceed the specifications, is denoted by H_1: $\mu > \mu_0$. In this case, the alternative hypothesis does not completely specify the distribution; it is an example of a *composite hypothesis*. An alternative hypothesis of the form $H_1: \mu > \mu_0$ (or of the form $H_1: \mu < \mu_0$) is said to be *one-sided* since H_1 consists of those values of μ that deviate from μ_0 in one direction only. The particular parameter value μ_0 is called the *null value*.

We are interested in determining whether it is plausible to conclude that the random sample comes from a normal distribution with $\mu = \mu_0$ as opposed to the possibility $\mu > \mu_0$. We use the sample mean $\overline{X}$ for our test statistic because it is intuitively clear that we would find H_1 more likely (and thus H_0 less likely) to be true when the sample mean $\overline{X}$ is very much greater than μ_0. That is, we choose a number $c > \mu_0$, called the *cutoff value*, such that an observed value of $\overline{X} > c$ is highly unlikely if, in fact, H_0 is true. In the present context the decision to accept the alternative hypothesis means that the new process produces bricks that exceed the current standards for compressive strength; the decision to accept the null hypothesis means that the new process, despite the manufacturer's claim, does not.

The statistician uses the following decision procedure to determine which of the two hypotheses is correct.

$$\text{accept (do not reject) } H_0 \text{ if } \overline{X} \le c$$

and

$$\text{reject } H_0 \text{ if } \overline{X} > c.$$

A procedure for deciding whether to accept or reject the null hypothesis is called a *test of a statistical hypothesis*. It is also a method for measuring the strength of the evidence against the null hypothesis that is inherent in the experimental data. The set of values $\mathscr{C} = \{\overline{x} : \overline{x} > c\}$ is called the *rejection region* of the test, since H_0 is rejected when $\overline{X}$ falls in $\mathscr{C}$. The distribution of the statistic $\overline{X}$ when $\mu = \mu_0$ (the null value) is called the *null distribution*.

Analyze the performance of this test by (1) computing the the type I and type II error probabilities, and (2) computing the power function of the test.

Solution. Looking at Table 8.5, we see that there are two types of error. Rejection of H_0 when H_0 is true is a type I error. A type I error occurs when $\overline{X} > c$ even though $\mu = \mu_0$. Rejection of H_1 when H_1 is true is a type II error. A type II error occurs when $\overline{X} \le c$ even though $\mu > \mu_0$.

TABLE 8.5

	H_0 is true: bricks are nonconforming ($\mu = \mu_0$)	H_1 is true: bricks are conforming ($\mu > \mu_0$)
$\overline{X} \le c$	Correct decision	Type II error
$\overline{X} > c$	Type I error	Correct decision

Using the fact that $\overline{X}$ is normally distributed with mean μ and standard deviation $\sigma/\sqrt{n}$, we see that the type I and type II error probabilities are given by

$$P(\overline{X} > c \mid \mu_0) = P\left(\frac{\sqrt{n}(\overline{X} - \mu_0)}{\sigma} > \frac{\sqrt{n}(c - \mu_0)}{\sigma}\right)$$

$$= 1 - \Phi\left(\frac{\sqrt{n}(c - \mu_0)}{\sigma}\right) \quad \text{(type I error probability)}$$

(8.3)

$$P(\overline{X} \le c \mid \mu) = P\left(\frac{\sqrt{n}(\overline{X} - \mu)}{\sigma} \le \frac{\sqrt{n}(c - \mu)}{\sigma}\right)$$

$$= \Phi\left(\frac{\sqrt{n}(c - \mu)}{\sigma}\right) \quad \text{(type II error probability)}$$

(8.4)

The probabilities of making the four possible decisions (correct and incorrect) are listed in Table 8.6. The power function $\pi(\mu)$ of the test is given by

$$\pi(\mu) = P(\text{reject } H_0 \mid \mu)$$

$$= P(\overline{X} > c \mid \mu) = 1 - \Phi\left(\frac{\sqrt{n}(c - \mu)}{\sigma}\right)$$

(8.5)

TABLE 8.6 DECISION PROBABILITIES

	H_0 is true ($\mu = \mu_0$)	H_1 is true ($\mu > \mu_0$)
$\overline{X} \le c$	$\Phi(\sqrt{n}(c - \mu_0)/\sigma)$	$\Phi(\sqrt{n}(c - \mu)/\sigma)$
$\overline{X} > c$	$1 - \Phi(\sqrt{n}(c - \mu_0)/\sigma)$	$1 - \Phi(\sqrt{n}(c - \mu)/\sigma)$

The performance of the test is measured by how well it controls the type I and type II error probabilities. As noted earlier, a good test will reject H_0 with small probability when it is true and with high probability when it is false.

Before proceeding further, we note that type I and type II errors are not of equal importance. A quality assurance engineer is much more concerned with the possibility of mistakenly using nonconforming bricks than with mistakenly rejecting bricks that in fact are conforming. For this reason we focus our attention on controlling the type I error. The *significance level of the test* is the probability α of a type I error. The type I error probability for a test with rejection region $\mathscr{C} = \{\bar{x} : \bar{x} > c\}$ is given by

$$\alpha = P(\overline{X} > c \mid H_0 \text{ is true}) = P(\overline{X} > c \mid \mu = \mu_0).$$

A *level α test* is a test whose type I error probability is α. The type I error probability α is also called the *size of the rejection region*. We choose the value c such that the probability of a type I error is less than α, where α is typically chosen to be 0.10, 0.05, or 0.01.

Constructing a Level α Test. The cutoff value c corresponding to a level α is given by

$$c = \mu_0 + z(\alpha)\frac{\sigma}{\sqrt{n}} \tag{8.6}$$

Equation (8.6) is a consequence of the following relationship between α and c:

$$\alpha = P(\overline{X} > c \mid \mu = \mu_0)$$

$$= 1 - \Phi\left(\frac{\sqrt{n}(c - \mu_0)}{\sigma}\right)$$

so

$$\frac{\sqrt{n}(c - \mu_0)}{\sigma} = z(\alpha)$$

Computing the Probability of a Type II Error at an Alternative μ. The probability of a type II error at the alternative $\mu > \mu_0$, denoted by $\beta(\mu)$, is given by

$$\beta(\mu) = P(\overline{X} \leq c \mid \mu > \mu_0) = \Phi\left(\frac{\sqrt{n}(c - \mu)}{\sigma}\right) \tag{8.7}$$

We summarize the procedure for constructing a level α test of H_0: $\mu = \mu_0$ against H_1: $\mu > \mu_0$.

1. Choose the rejection region $\mathscr{C} = \{\overline{x} : \overline{x} > c\}$ where

$$c = \mu_0 + z(\alpha)\frac{\sigma}{\sqrt{n}}$$

With this choice of c, the probability of a type I error is α; more precisely, we have

$$P\left(\overline{X} > \mu_0 + z(\alpha)\frac{\sigma}{\sqrt{n}} \mid \mu_0\right) = \alpha$$

2. Compute the sample mean $\overline{X}$ and

$$\text{accept } H_0 \text{ if } \overline{X} \leq \mu_0 + z(\alpha)\frac{\sigma}{\sqrt{n}}$$

$$\text{reject } H_0 \text{ if } \overline{X} > \mu_0 + z(\alpha)\frac{\sigma}{\sqrt{n}} \tag{8.8}$$

A test of the form (8.8) is called an *upper-tailed test*. A *lower tailed test* is defined similarly (see Example 8.4). A test of either type is called a *one-sided test*.

An Equivalent Test Based on a Standardized Test Statistic

The one-sided test given in Eq. (8.8) can be rewritten in an equivalent form using the standardized test statistic Z_0, defined by

$$Z_0 = \frac{\sqrt{n}(\overline{X} - \mu_0)}{\sigma}$$

We note that the null distribution of the test statistic Z_0 is that of a standard normal random variable; that is,

$$Z_0 = \frac{\sqrt{n}(\overline{X} - \mu_0)}{\sigma} \sim N(0, 1) \quad \text{when } \mu = \mu_0 \tag{8.9}$$

The rejection region in terms of the standardized test statistic Z_0 is the set $\{Z_0 > z(\alpha)\}$. We obtain this result by rewriting the original rejection region— refer back to Eq. (8.8)—in terms of Z_0; that is,

$$\left(\overline{X} > \mu_0 + z(\alpha)\frac{\sigma}{\sqrt{n}} \right) = \left(\frac{\sqrt{n}(\overline{X} - \mu_0)}{\sigma} > z(\alpha) \right)$$

$$= (Z_0 > z(\alpha))$$

This leads to the following equivalent test of the hypothesis H_0: $\mu = \mu_0$ against the one-sided (upper) alternative H_1: $\mu > \mu_0$:

$$\text{accept } H_0 \text{ if } \frac{\sqrt{n}(\overline{X} - \mu_0)}{\sigma} \leq z(\alpha)$$

$$\text{reject } H_0 \text{ if } \frac{\sqrt{n}(\overline{X} - \mu_0)}{\sigma} > z(\alpha) \tag{8.10}$$

---

example **8.3**

(Continuation of Example 8.2) From past experience it is known that the mean compressive strength (measured in units of 100 pounds per square inch) of bricks is normally distributed with unknown mean μ and known standard deviation $\sigma = 6$. The bricks will not be accepted unless we can be reasonably sure that $\mu > 50.0$.

1. Suppose we use the rejection region $\mathscr{C} = \{\overline{x} : \overline{x} > 53\}$ to test the null hypothesis H_0: $\mu = 50.0$ against H_1: $\mu > 50.0$. Compute the significance level of the test and the type II error probability at the alternative $\mu = 55$.
2. Find the cutoff value c such that the the corresponding test has significance level $\alpha = 0.05$. Compute the power function of this test.
3. To test this hypothesis, nine bricks are randomly selected and their compressive strengths recorded:

$$44.5, 49.7, 65.1, 51.5, 47.6, 48.9, 50.6, 61.5, 53.1$$

Use these data to test at the level $\alpha = 0.05$ the null hypothesis that the mean compressive strength of the bricks equals 50.0 against the alternative H_1: $\mu > 50.0$.

Solution

1. **Computing the significance level of the test.** We compute the probability of a type I error by substituting the values $c = 53, \sigma = 6, n = 9$, and $\mu_0 = 50.0$ into Eq. (8.3). This yields

$$\alpha = P(\overline{X} > 53 \mid \mu = 50) = 1 - \Phi\left(\frac{3 \times (53 - 50)}{6} \right) = 1 - \Phi(1.5) = 0.0668$$

Computing the type II error. In this case $c = 53, \sigma = 6, n = 9$, and $\mu = 55$. Substituting these values into Eq. (8.4) yields the following value for $\beta(55)$:

$$\beta(55) = \Phi\left(\frac{3 \times (53 - 55)}{6}\right)$$

$$= \Phi(-1.0) = 0.1587$$

The rejection regions corresponding to the sample mean $\overline{X}$ and the standardized sample mean $Z_0 = \sqrt{n}(\overline{X} - \mu_0)/\sigma$ are displayed in Fig. 8.2.

2. **Constructing a test with significance level $\alpha = 0.05$.** The rejection region for a level $\alpha = 0.05$ test is given by $\{\bar{x} : \bar{x} > 53.3\}$ since the cutoff value is

$$c = 50.0 + \frac{1.645 \times 6}{\sqrt{9}} = 53.3$$

The power function corresponding to the test with rejection region $\{\bar{x} : \bar{x} > 53.3\}$ is obtained by setting $c = 53.3, n = 9, \sigma = 6$ in Eq. (8.5). The power function is given by

$$\pi(\mu) = 1 - \Phi\left(\frac{53.3 - \mu}{2}\right)$$

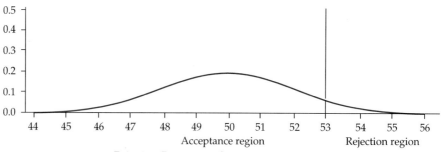

Acceptance region · Rejection region

Rejection Region Based on Sample Mean ($x > 53.0$)

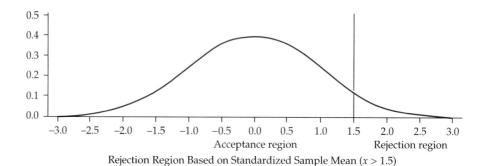

Acceptance region · Rejection region

Rejection Region Based on Standardized Sample Mean ($x > 1.5$)

Figure 8.2 Rejection regions for Example 8.3

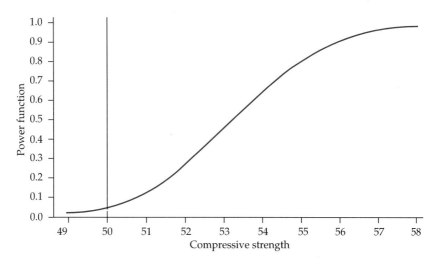

Figure 8.3 Power function of Example 8.3 for null value 50.0 and rejection region $\bar{x} > 53.3$

The graph of the power function is displayed in Fig. 8.3. Looking at this graph, we see that the probability of rejecting H_0 increases as μ increases. For example, suppose that H_1 is true. What is the probability that the test will detect this? This probability depends on μ and is in fact given by the power function evaluated at μ. To see this, we observe that if, in fact, $\mu > 50.0$ (so that H_1 is true), then $\pi(\mu) = P(\bar{X} > 53.3 \mid \mu)$ is the probability of rejecting H_0 when H_1 is true. Thus $\pi(56) = 1 - \Phi(-1.35) = 0.9115$. Notice that the type II error probability $\beta(56) = 1 - \pi(56) = 0.0885$.

3. Since $\bar{x} = 52.5 < 53.3$, we do not reject H_0.

Confidence Interval Approach. We now introduce an alternative method, based on confidence intervals, for testing hypotheses about the mean. It is increasingly popular among engineers for reasons that will soon become apparent. In this example, we are concerned that the mean compressive strength of the bricks is too low. For this purpose we observed that a one-sided $100(1 - \alpha)$ percent lower confidence interval of the form

$$\left[\bar{X} - z(\alpha)\frac{\sigma}{\sqrt{n}}, \ \infty \right)$$

is appropriate [Eq. (7.14)]. With reference to the data, we have $\bar{x} = 52.5$, $\sigma/\sqrt{n} = 2$, so the 95 percent lower confidence interval is given by

$$\left[\bar{x} - z(0.05)\frac{\sigma}{\sqrt{n}}, \ \infty \right) = 52.5 - 1.645 \times 2 = [49.2, \ \infty)$$

Conclusion. These data suggest that the true mean compressive strength could be as low as 49.2. In particular, we note that $49.2 < 50.0 = \mu_0$ implies that the null value is in the confidence interval. We interpret this to mean that the null hypothesis is plausible. This leads to the following criterion for accepting H_0. *The null hypothesis H_0: $\mu = \mu_0$ is accepted whenever μ_0 lies in the confidence interval.* Parameter values that lie outside the confidence interval are not plausible. Therefore, *we reject the null hypothesis H_0: $\mu = \mu_0$ whenever μ_0 lies outside the confidence interval;* that is, we reject the null hypothesis when

$$\mu_0 < \overline{X} - z(\alpha)\frac{\sigma}{\sqrt{n}}$$

Solving this inequality for $\overline{X}$ yields the rejection region for the level α test obtained earlier. That is, *the null hypothesis H_0: $\mu = \mu_0$ is rejected whenever*

$$\overline{X} > \mu_0 + z(\alpha)\frac{\sigma}{\sqrt{n}}$$

It is worth pointing out that when this test rejects H_0: $\mu = \mu_0$, it will also reject H_0: $\mu < \mu_0$. This is a consequence of the fact that whenever μ_0 lies outside the confidence interval, that is, whenever μ_0 satisfies the inequality

$$\mu_0 < \overline{X} - z(\alpha)\frac{\sigma}{\sqrt{n}}$$

then all parameter values $\mu < \mu_0$ also satisfy the same inequality:

$$\mu < \mu_0 < \overline{X} - z(\alpha)\frac{\sigma}{\sqrt{n}} \qquad \blacksquare$$

A Lower-Tailed Test of a Hypothesis

We construct a lower-tailed test of the null hypothesis H_0: $\mu = \mu_0$ against H_1: $\mu < \mu_0$ by modifying the reasoning previously used for an upper-tailed test [see Eq. (8.6)], so we content ourselves with a sketch. It is clear that we are very unlikely to observe a value of $\overline{X}$ that is very much smaller than the null value when H_0: $\mu = \mu_0$ is true. The appropriate rejection region in this case for a level α lower-tailed test is the set $\{\bar{x} : \bar{x} < c\}$, where

$$c = \mu_0 - z(\alpha)\frac{\sigma}{\sqrt{n}} \qquad (8.11)$$

The corresponding decision rule is

$$\text{accept } H_0 \text{ if } \overline{X} \geq \mu_0 - z(\alpha)\frac{\sigma}{\sqrt{n}}$$

$$\text{reject } H_0 \text{ if } \overline{X} < \mu_0 - z(\alpha)\frac{\sigma}{\sqrt{n}}$$

The probability $\beta(\mu)$ of a type II error at the alternative $\mu < \mu_0$ is given by

$$\beta(\mu) = P(\overline{X} \geq c \mid \mu < \mu_0) = 1 - \Phi\left(\frac{\sqrt{n}(c - \mu)}{\sigma}\right) \qquad (8.12)$$

Using similar reasoning, we see that the power function is given by

$$\pi(\mu) = \Phi\left(\frac{\sqrt{n}(c - \mu)}{\sigma}\right) \qquad (8.13)$$

The equivalent test based on the standardized statistic $Z_0 = \sqrt{n}(\overline{X} - \mu_0)/\sigma$ is given by

$$\text{accept } H_0 \text{ if } \frac{\sqrt{n}(\overline{X} - \mu_0)}{\sigma} \geq -z(\alpha)$$

$$\text{reject } H_0 \text{ if } \frac{\sqrt{n}(\overline{X} - \mu_0)}{\sigma} < -z(\alpha) \qquad (8.14)$$

The rejection region corresponding to the standardized statistic Z_0 is displayed in Fig. 8.4.

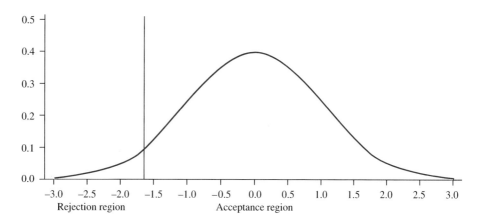

Figure 8.4 Rejection region $x < -1.645$ based on standardized sample mean: lower-tailed test, $\alpha = 0.05$

---

example 8.4   The problem of determining the concentration of arsenic in copper, measured as a percentage, is of considerable importance since the presence of even a small amount can significantly increase the copper's electrical resistance. A laboratory makes $n = 6$ measurements of the concentration of arsenic in copper, with the result $\overline{x} = 0.17$ percent. We assume that the data come from a normal distribution with known standard deviation $\sigma = 0.04$. The copper cannot be used if the concentration of arsenic in copper exceeds 0.2 percent.

1. Construct a 5 percent level test of the null hypothesis H_0: $\mu = 0.2$ against H_1: $\mu < 0.2$. Draw the graph of the power function for this test.
2. Compute the type II error probability at the alternative $\mu = 0.17$ for the lower-tailed test in part 1.
3. Test the null hypothesis at the 5 percent level by constructing a 95 percent upper confidence interval.

Solution

1. Because this is a lower-tailed test, we substitute

$$n = 6, \qquad \mu_0 = 0.2, \qquad \sigma = 0.04, \qquad z(0.05) = 1.645$$

into Eq. (8.11) and obtain the cutoff value

$$c = 0.2 - 1.645 \times \frac{0.04}{\sqrt{6}} = 0.1731$$

Since $\bar{x} = 0.17 < 0.1731$, we reject H_0; that is, we conclude that the concentration of arsenic in the copper ore is below 0.2 percent.

The same result could, of course, have been obtained using the standardized test statistic Z_0. To see this, substitute

$$\bar{x} = 0.17, \qquad \sigma = 0.04, \qquad n = 6, \qquad -z(0.05) = -1.645$$

into Eq. (8.14). This yields the result $z = -1.83 < -1.645$; the null hypothesis is again rejected.

We obtain the power function corresponding to the rejection region $\mathscr{C} = \{\bar{x} : \bar{x} < 0.1731\}$ by setting $n = 6, \sigma = 0.04, c = 0.1731$ in Eq. (8.13); the explicit formula follows:

$$\pi(\mu) = \Phi\left(\frac{\sqrt{6}(0.1731 - \mu)}{0.04}\right)$$

Its graph is displayed in Fig. 8.5.

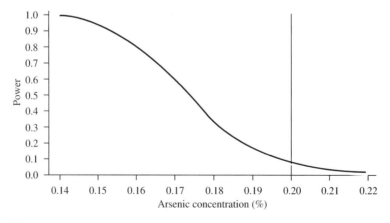

Figure 8.5 Power function of Example 8.4 for null value 0.20, rejection region $x > 0.1731$

2. We obtain the type II error probability at the alternative $\mu = 0.17 < 0.20$ from the equation $\beta(\mu) = 1 - \pi(\mu)$. Setting

$$c = 0.1731, \qquad n = 6, \qquad \sigma = 0.04, \qquad \mu = 0.17$$

in Eq. (8.13), we obtain

$$\beta(0.17) = 1 - \pi(0.17) = 1 - 0.5753 = 0.4247$$

3. A one-sided upper $100(1 - \alpha)$ percent confidence interval is given by

$$\left(-\infty, \ \overline{X} + z(\alpha)\frac{\sigma}{\sqrt{n}}\right] \tag{8.15}$$

Substituting

$$\overline{x} = 0.17, \qquad \frac{\sigma}{\sqrt{n}} = \frac{0.04}{\sqrt{6}} = 0.0163, \qquad z(0.05) = 1.645$$

into Eq. (8.15) yields the confidence interval $(-\infty, \ 0.1969]$. Since $0.1969 < 0.20$, we reject H_0: $\mu = 0.20$ and conclude that the concentration of arsenic in the ore is less than 0.20 percent. Note that the confidence interval suggests that the concentration of copper could be as large as 0.1969 percent. ■

Two-Sided Tests of a Hypothesis

We construct a two-sided test of the null hypothesis H_0: $\mu = \mu_0$ against the alternative H_1: $\mu \neq \mu_0$ by using the rejection region $\{\overline{x} : |\overline{x} - \mu_0| > c\}$. This is a reasonable choice since large values of $|\overline{x} - \mu_0|$ cast doubt on the validity of H_0. The type I error probability α is given by

$$P(|\overline{X} - \mu_0| > c \mid H_0 \text{ is true}) = P(|\overline{X} - \mu_0| > c \mid \mu = \mu_0)$$

$$= P\left(\frac{\sqrt{n}(|\overline{X} - \mu_0|)}{\sigma} > \frac{\sqrt{n}c}{\sigma}\right) = \alpha$$

This leads to the following level α test:

$$\text{accept } H_0 \text{ if } |\overline{X} - \mu_0| \leq z(\alpha/2)\sigma/\sqrt{n}$$
$$\text{reject } H_0 \text{ if } |\overline{X} - \mu_0| > z(\alpha/2)\sigma/\sqrt{n} \tag{8.16}$$

An equivalent test based on the standardized statistic Z is given by

$$\text{accept } H_0 \text{ if } \left|\frac{\sqrt{n}(\overline{X} - \mu_0)}{\sigma}\right| \leq z(\alpha/2)$$

$$\tag{8.17}$$

$$\text{reject } H_0 \text{ if } \left|\frac{\sqrt{n}(\overline{X} - \mu_0)}{\sigma}\right| > z(\alpha/2)$$

The power function is given by

$$\pi(\mu) = 1 - \Phi\left(\frac{\sqrt{n}(\mu_0 - \mu + c)}{\sigma}\right) + \Phi\left(\frac{\sqrt{n}(\mu_0 - \mu - c)}{\sigma}\right) \tag{8.18}$$

example 8.5

The resistors produced by a manufacturer are required to have a resistance of $\mu_0 = 0.150$ ohms. Statistical analysis of the output suggests that the resistances can be approximated by a normal distribution $N(\mu, \sigma^2)$ with known standard deviation $\sigma = 0.005$. A random sample of $n = 10$ resistors is drawn and the sample mean is found to be $0.152 = \bar{x}$.

1. Test, at the 5 percent level, the hypothesis that the resistors are conforming. Graph the null distribution of $\overline{X}$ and identify the rejection region of the test.
2. Compute and graph the power function of this test.
3. Test this hypothesis at the 5 percent level by constructing a 95 percent two-sided confidence interval.

Solution

1. The null hypothesis in this case is that there is no difference between the specifications and the factory output, so the null hypothesis and its alternative are given by

$$H_0: \mu = 0.150$$
$$H_1: \mu \neq 0.150$$

We construct a test that is significant at the 5 percent level by substituting the values

$$n = 10, \quad |\bar{x} - \mu_0| = |0.152 - 0.150| = 0.002,$$
$$\sigma = 0.005, \quad z(0.025) = 1.96$$

into Eq. (8.16). This yields

$$z(\alpha/2)\sigma/\sqrt{n} = 1.96 \times \frac{0.005}{\sqrt{10}} = 0.003$$

Since $|\bar{x} - 0.150| = 0.002 < 0.003$, we do not reject $H_0: \mu = 0.150$. The null distribution of $\overline{X}$ and the rejection region for the two-sided test are displayed in Fig. 8.6.

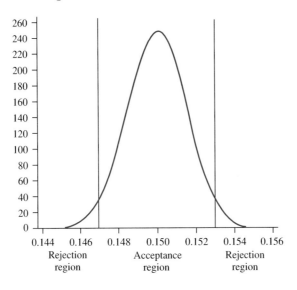

Figure 8.6 Rejection region $|x - 0.150| > 0.003$ of Example 8.5

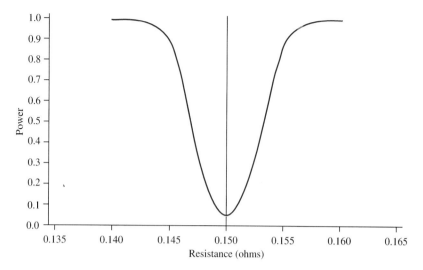

Figure 8.7 Power function of Example 8.5 for null value 0.150, rejection region $|x - 0.15| > 0.003$

2. The formula for the power function is obtained directly from Eq. (8.18) and is given by

$$\pi(\mu) = 1 - \Phi\left(\frac{\sqrt{10}(0.15 - \mu + 0.003)}{0.005}\right) + \Phi\left(\frac{\sqrt{10}(0.15 - \mu - 0.003)}{0.005}\right)$$

Figure 8.7 displays the graph.
3. The 95 percent confidence interval is $0.152 \pm 0.0031 = [0.1489, 0.1551]$. The confidence interval contains the null value 0.150, so we accept the null hypothesis. ■

Table 8.7 summarizes the various hypotheses we have studied and the corresponding rejection regions based on the standardized test statistic $\sqrt{n}(\overline{X} - \mu_0)/\sigma$.

TABLE 8.7

H_0	H_1	Rejection region
$\mu \geq \mu_0$	$\mu < \mu_0$	$\dfrac{\sqrt{n}(\overline{X} - \mu_0)}{\sigma} < -z(\alpha)$
$\mu \leq \mu_0$	$\mu > \mu_0$	$\dfrac{\sqrt{n}(\overline{X} - \mu_0)}{\sigma} > z(\alpha)$
$\mu = \mu_0$	$\mu \neq \mu_0$	$\left\|\dfrac{\sqrt{n}(\overline{X} - \mu_0)}{\sigma}\right\| > z(\alpha/2)$

8.2.1   Significance Testing

In science we have to distinguish between mathematical propositions derived from hypotheses by logical deduction and *causal inference* based on empirical data. We can never prove the Pythagorean theorem by measuring the sides of sufficiently many right triangles. Mathematical truths cannot be refuted by experiments; and experiments cannot lead to mathematical certainty. Unfortunately, the terms *reject* and *accept* used in hypothesis testing imply a certitude that cannot be justified, since no scientific hypothesis is ever finally accepted or rejected in practice. We therefore interpret *reject* as meaning that the data provide strong evidence against the null hypothesis and we interpret *accept* to mean that the data do not. For this reason statisticians now recommend that one report the *P-value* of the test, which is a measure of how much agreement there is between the data and H_0.

---

example 8.6

Return to Example 8.4. The null hypothesis is H_0: $\mu = 0.2$. We assume the data come from a normal distribution with $\sigma = 0.04$. The experimental data, based on a sample size $n = 6$, yielded a sample mean $\bar{x} = 0.17$. We are interested in the following question: If H_0 is true, so that $\mu = 0.2$ percent, what is the probability that the sample mean $\bar{X}$ would be as small as the observed value, 0.17 percent?

Solution. Using the given information ($\sigma = 0.04$ and sample size $n = 6$), we see that

$$P(\bar{X} < 0.17 \mid H_0 \text{ is true }) = P(\bar{X} < 0.17 \mid \mu = 0.2)$$

It follows that

$$P(\bar{X} < 0.17 \mid \mu = 0.2) = P\left(\frac{\sqrt{n}(\bar{X} - \mu_0)}{\sigma} < \frac{\sqrt{6}(0.17 - 0.20)}{0.04}\right)$$

$$= P(Z < -1.84) = \Phi(-1.84) = 0.033$$

That is, the probability, assuming H_0 is true, of observing a sample mean $\bar{X}$ smaller than the observed value $\bar{x} = 0.17$ is 0.033; this probability, denoted by P, is called the *P-value*. ∎

Interpreting the P-Value

A small P-value indicates that one should not accept H_0. Specifically, the test is said to be *statistically significant* at the level α if the P-value is smaller than α. Thus, instead of merely stating that H_0 was accepted (or rejected) at the significance level α, one reports the P-value of the test, and the significance of the result is left to the subjective judgment of the researcher.

In practice, then, one proceeds as follows: Specify the significance level of the test, collect the data, and then compute the P-value. A small value of P provides evidence against the null hypothesis. More precisely, the following criteria are suggested for weighing the evidence for and against the null hypothesis:

Strong evidence against H_0: $P < 0.01$

Moderately strong evidence against H_0: $0.01 < P < 0.05$

Relatively weak evidence against H_0: $P > 0.10$

In words, the P-value (for a lower-tailed test) is defined by

$$P\text{-value} = P\left(\begin{array}{c}\text{sample mean would be}\\\text{as small as the observed}\\\text{value if } H_0 \text{ is true}\end{array}\right)$$

The significance-testing approach just outlined is one that students must become familiar with since it is the one used in the current statistical software packages. Specifically, for hypothesis testing using MINITAB or SAS, the output simply lists the observed P-value and leaves it to the user to judge the significance of the result.

Some Suggestions for Computing the P-Value

Suppose that the data $\{x_1, \ldots, x_n\}$ come from a normal distribution $N(\mu, \sigma^2)$, where σ^2 is known, and suppose the test is based on the value of the statistic

$$Z_0 = \frac{\sqrt{n}(\overline{X} - \mu_0)}{\sigma}$$

When the observed value of the statistic Z_0 is z, then the P-value is

$$P\text{-value} = \begin{cases} \Phi(z) & \text{lower-tailed test} \\ 1 - \Phi(z) & \text{upper-tailed test} \\ 2[1 - \Phi(|z|)] & \text{two-tailed test} \end{cases}$$

---

example 8.7   Compute the P-value of the test of Example 8.5.

Solution. In this case

$$\overline{x} = 0.152, \qquad n = 10, \qquad \sigma = 0.005$$

Therefore,

$$z = \frac{\sqrt{n}(\overline{x} - \mu_0)}{\sigma} = \frac{\sqrt{10}(0.152 - 0.150)}{0.005} = 1.26$$

This is a two-sided test, so the P-value equals $2[1 - \Phi(1.26)] = 0.21$. This is relatively weak evidence against the the null hypothesis, so we do not reject it. ∎

8.2.2   Power Function and Sample Size

When the sample size and significance level are fixed in advance by the statistician, the type II error probability is determined. Consider, for instance, Example 8.4, where we showed that the power at the alternative $\mu = 0.17 < 0.20$ equals 0.5753. The probability of a type II error, then, equals $1 - 0.5753 = 0.4247$.

Consequently, there is a considerable risk here of incurring unnecessary costs for reprocessing the ore. This raises the following question: Is there any way of reducing the type II error probability β? When the sample size n and significance level α are fixed in advance, the answer is no, since in this case the cutoff value c must satisfy the equation $c = \mu_0 - z(\alpha)(\sigma/\sqrt{n})$ [Eq. (8.11)]. The only way, then, of increasing the power of the test at an alternative $\mu < \mu_0$ (without simultaneously increasing the type I error probability α) is to increase the sample size n. More precisely, the test will have significance level α and power $1 - \beta$ at the alternative μ if

$$\pi(\mu_0) = \alpha \tag{8.19}$$

$$\pi(\mu) = 1 - \beta \tag{8.20}$$

The sample size n that is required for the test to have power $1 - \beta$ (or type II error β) at the alternative μ is given by

$$n = \left(\frac{\sigma}{\mu_0 - \mu} \times \left[z(\alpha) + z(\beta) \right] \right)^2 \tag{8.21}$$

Derivation of Eq. (8.21)

In this case the power function is

$$\pi(\mu) = \Phi\left(\frac{c - \mu}{\sigma/\sqrt{n}} \right), \qquad -\infty < \mu < \infty$$

Inserting this expression for $\pi(\mu)$ into Eqs. (8.19) and (8.20) yields two equations for the two unknowns c and n.

$$\Phi\left(\frac{c - \mu_0}{\sigma/\sqrt{n}} \right) = \alpha \tag{8.22}$$

$$1 - \Phi\left(\frac{c - \mu}{\sigma/\sqrt{n}} \right) = \beta \tag{8.23}$$

Solving these equations for $\sqrt{n}$ (Prob. 8.17 asks you to supply the details), we obtain

$$\sqrt{n} = \frac{\sigma}{\mu_0 - \mu} \times \left[z(\alpha) + z(\beta) \right]$$

Squaring both sides of this expression yields Eq. (8.21) for the sample size. The formula (8.21) for the sample size is also valid for an upper-tailed test.

---

example 8.8   Suppose we wish to test the hypothesis H_0: $\mu = 0.2$ against H_1: $\mu < 0.2$ for the concentration of arsenic in copper at significance level $\alpha = 0.05$ with known $\sigma = 0.04$. Determine the minimum required sample size if the power of the test is to be 0.90 at the alternative $\mu = 0.17$.

Solution. We compute the minimum sample size using Eq. (8.21). That is, we set

$$\mu_0 = 0.2, \qquad \mu = 0.17, \qquad \sigma = 0.04, \qquad \alpha = 0.05, \qquad 1 - \beta = 0.9$$

This yields the result $n = 3.9^2 = 15.21$, which we round up to $n = 16$.   ∎

Power and Sample Size: Two-Sided Tests

When the alternative hypothesis is two-sided, the sample size n required to produce a test with significance level α and power $1 - \beta$ at the alternative μ is given by

$$n = \left(\frac{\sigma[z(\alpha/2) + z(\beta)]}{\mu_0 - \mu} \right)^2 \tag{8.24}$$

The derivation is similar to that for Eq. (8.21), except for the fact that an explicit solution for n is not possible. However, one can show that Eq. (8.24) is a useful approximation.

---

example 8.9   Refer to Example 8.5. In this example the resistors are required to have a resistance of $\mu_0 = 0.150$ ohms. How large must n be if you want $\alpha = 0.05$ and the type II error probability at the alternative $\mu = 0.148$ is to be 0.2?

Solution. Setting $z(\alpha/2) = z(0.05) = 1.96$, $z(\beta) = z(0.2) = 0.84$, and $\mu_0 - \mu = 0.002$ in Eq. (8.23) yields

$$\sqrt{n} = \frac{0.005(1.96 + 0.84)}{0.002} = 7, \qquad \text{so } n = 49$$ ∎

8.2.3   Large Sample Tests Concerning the Mean of an Arbitrary Distribution

When the distribution from which the sample is drawn is not known to be normal, it is still possible to make inferences concerning the mean of the distribution provided the sample size is sufficiently large so that the central limit theorem can be applied. More precisely, we drop the assumption that the random sample $X_1, \ldots, X_n$ is taken from a normal distribution; instead we assume that it comes from an arbitrary distribution with unknown mean μ and unknown variance σ^2. We make the additional assumption that the sample size n is large enough ($n \geq 30$, say) that the central limit theorem can be brought to bear; that is, we assume that the distribution of standardized variable $Z = \sqrt{n}(\overline{X} - \mu)/\sigma$ has an approximate standard normal distribution $N(0, 1)$. This is still not useful since σ is unknown; we circumvent this difficulty by replacing σ with its estimate s, the sample standard deviation, since s will be close to σ with high probability for large values of n.

Proceeding by methods similar to those used earlier (Table 8.7), we can derive approximate level α one-sided and two-sided tests displayed in Table 8.8 based on the standardized test statistic $\sqrt{n}(\overline{X} - \mu_0)/s$.

TABLE 8.8

Alternative hypothesis	Rejection region
$H_1: \mu < \mu_0$	$\dfrac{\sqrt{n}(\overline{X} - \mu_0)}{s} < -z(\alpha)$
$H_1: \mu > \mu_0$	$\dfrac{\sqrt{n}(\overline{X} - \mu_0)}{s} > z(\alpha)$
$H_1: \mu \neq \mu_0$	$\left| \dfrac{\sqrt{n}(\overline{X} - \mu_0)}{s} \right| > z(\alpha/2)$

example 8.10 A laboratory makes $n = 30$ determinations of the concentration μ of an impurity in an ore and obtains the results $\overline{x} = 0.09$ percent and $s = 0.03$ percent. If $\mu \geq 0.10$ percent, the ore cannot be used without additional processing. Consequently, we want to test $H_0: \mu = 0.1$ against $H_1: \mu < 0.1$. How plausible is the null hypothesis given that $\overline{x} = 0.09$?

Solution. We solve this problem by first computing the observed value z of the standardized statistic Z and then computing its *approximate P*-value. We say "approximate *P*-value" because the distribution of the standardized statistic is only approximately normal. Now the observed value of Z equals -1.83, since

$$z = \frac{\sqrt{n}(\overline{x} - 0.1)}{s} = \frac{\sqrt{30} \times (-0.01)}{0.03} = -1.83$$

Thus, the approximate *P*-value of the test is

$$P\text{-value} = P(Z \leq z) = P(Z \leq -1.83) = 0.03$$

Since the *P*-value $= 0.03 < 0.05$, the observed value is significant at the 5 percent level. In other words, we reject the null hypothesis $H_0: \mu = 0.1$ percent. ∎

8.2.4 Tests Concerning the Mean of a Distribution with Unknown Variance

When the variance σ^2 is unknown and the sample size n is small ($n < 30$, say) so that the applicability of the central limit theorem is questionable, we can still test hypotheses about the mean μ, provided the distribution from which the random sample is drawn is approximately normal. We begin by extending the hypothesis-testing methods of the previous sections to the important special case in which the random sample $X_1, \ldots, X_n$ comes from a normal distribution $N(\mu, \sigma^2)$ with unknown variance σ^2. The fact that $\overline{X}$ is again normally distributed with mean μ and variance σ^2/n is not very useful since σ^2 is unknown. However, we do know (Theorem 6.10) that

$$T = \frac{\sqrt{n}(\overline{X} - \mu)}{s} \quad \text{is } t_{n-1} \text{ distributed}$$

Consequently,

$$P(\overline{X} \leq c \mid \mu) = P\left(\frac{\sqrt{n}(\overline{X} - \mu)}{s} \leq \frac{\sqrt{n}(c - \mu)}{s}\right)$$

$$= P\left(t_{n-1} \leq \frac{\sqrt{n}(c - \mu)}{s}\right)$$

This result allows us to control the type I error probability using the t_{n-1} distribution instead of the normal distribution.

One-Sided Test: Unknown Variance

We now show how to construct a level α test of $H_0: \mu = \mu_0$ against $H_1: \mu < \mu_0$ using the standardized test statistic

$$T = \frac{\sqrt{n}(\overline{X} - \mu_0)}{s}$$

The method is similar to the one used to derive the lower-tailed test given in Eq. (8.14).

1. Choose a rejection region of the form

$$\mathscr{C} = \left\{\overline{x} : \frac{\sqrt{n}(\overline{x} - \mu_0)}{s} < -t_{n-1}(\alpha)\right\} \tag{8.25}$$

so that whenever the null hypothesis is true, the probability of a type I error is α; that is,

$$\text{accept } H_0 \text{ if } \frac{\sqrt{n}(\overline{X} - \mu_0)}{s} \geq -t_{n-1}(\alpha)$$

$$\text{reject } H_0 \text{ if } \frac{\sqrt{n}(\overline{X} - \mu_0)}{s} < -t_{n-1}(\alpha)$$

2. The power function, which is not easy to compute, is given by

$$\pi(\mu) = P\left(\frac{\sqrt{n}(\overline{X} - \mu_0)}{s} < t \mid \mu\right) = P(T_{n-1,\delta} < t) \tag{8.26}$$

where $T_{n-1,\delta}$ has a *noncentral t distribution* with noncentrality parameter $\delta = \sqrt{n}(\mu - \mu_0)/\sigma$. To determine the minimal sample size required for a test of power $1 - \beta$ at the alternative μ statisticians use graphs (not given here) of the power function for selected values of n and α.

example **8.11**   This is a continuation of Example 8.4, except that we now assume that the variance is unknown. Recall that the researcher was interested in determining if the concentration of arsenic in copper was greater than or equal to 0.2 percent. We change the problem slightly by assuming that the researcher made $n = 6$ measurements with the sample mean $\bar{x} = 0.17$ and sample standard deviation $s = 0.04$. Assuming that the data come from a normal distribution, test (at the 5 percent level) the null hypothesis $H_0: \mu = 0.20$ against $H_1: \mu < 0.20$.

Solution. We construct the rejection region using Eq. (8.25); that is, we set

$$\mu_0 = 0.20, \qquad n = 6, \qquad t_5(0.05) = 2.015, \qquad s = 0.04$$

and obtain the rejection region

$$\mathscr{C} = \left\{ \bar{x} : \frac{\sqrt{6}(\bar{x} - 0.20)}{0.04} < -2.015 \right\}$$

When $\bar{x} = 0.17$, the value of the standardized test statistic T is given by

$$T = \frac{\sqrt{6}(0.17 - 0.20)}{0.04} = -1.837 > -2.015$$

Consequently, we do not reject the null hypothesis and conclude that the concentration of arsenic in the copper ore is not less than 0.2 percent. Notice that our conclusion in this case differs from the one obtained in Example 8.4. That is, a value of the sample mean $\bar{X} = 0.17$ is significant at the 5 percent level when the variance is known and is not significant at the 5 percent level when the variance is unknown. Thus, knowing the value of the variance yields a test that is more sensitive to departures from the null hypothesis. ∎

Two-Sided Tests via Confidence Intervals: Variance Unknown

A $100(1 - \alpha)$ percent confidence interval for the mean of a normal distribution with unknown variance has endpoints $[L, U]$ where

$$L = \bar{X} - t_{n-1}(\alpha/2)s/\sqrt{n}$$
$$U = \bar{X} + t_{n-1}(\alpha/2)s/\sqrt{n}$$

Keeping in mind the interpretation of the confidence interval as consisting of those parameter values μ that are consistent with the null hypothesis, we arrive at the following two-sided test: Accept $H_0: \mu = \mu_0$ when

$$\bar{X} - t_{n-1}(\alpha/2)\frac{s}{\sqrt{n}} \leq \mu_0 \leq \bar{X} + t_{n-1}(\alpha/2)\frac{s}{\sqrt{n}} \tag{8.27}$$

and reject H_0 otherwise.

example **8.12**   This is a continuation of Example 8.5, except that we now assume that the variance is unknown and that $s = 0.005$. Test at the 5 percent level the null hypothesis $H_0: \mu = 0.15$ against $H_1: \mu \neq 0.15$.

Solution. We construct the 95 percent confidence interval by setting $n = 10, \mu_0 = 0.15, \alpha = 0.05, t_9(0.025) = 2.262$ in Eq. (8.27). This produces the confidence interval [0.1489, 0.1551], which contains 0.150; thus, we do not reject the null hypothesis. It is once again worth noting that the confidence interval approach tells us that μ could be as small as 0.1489 or as large as 0.1551, which is clearly more informative than merely stating that the null hypothesis is not rejected. ∎

Table 8.9 summarizes the various hypotheses we have studied and the corresponding rejection regions based on the standardized test statistic $t_{n-1} = \sqrt{n}(\overline{X} - \mu_0)/s$.

TABLE 8.9

Alternative hypothesis	Rejection region
$H_1: \mu < \mu_0$	$\dfrac{\sqrt{n}(\overline{X} - \mu_0)}{s} < -t_{n-1}(\alpha)$
$H_1: \mu > \mu_0$	$\dfrac{\sqrt{n}(\overline{X} - \mu_0)}{s} > t_{n-1}(\alpha)$
$H_1: \mu \neq \mu_0$	$\left\| \dfrac{\sqrt{n}(\overline{X} - \mu_0)}{s} \right\| > t_{n-1}(\alpha/2)$

Figure 8.8 displays the rejection regions for a two-sided test based on the standardized t distribution with $n = 16$, so $t_{n-1} = t_{15}$, and $\alpha = 0.10, 0.05$. We use the fact that $t_{15}(0.05) = 1.753$ and $t_{15}(0.025) = 2.131$.

■ **REMARK 8.1**

The one-sided and two-sided level α tests listed in Table 8.9 are still approximately correct, if one can safely assume that the distribution from which the random sample is drawn is at least approximately normal. ∎

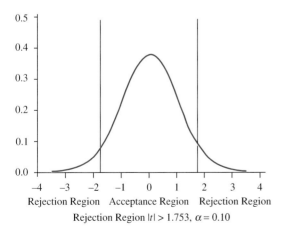

Rejection Region   Acceptance Region   Rejection Region
Rejection Region |t| > 1.753, $\alpha = 0.10$

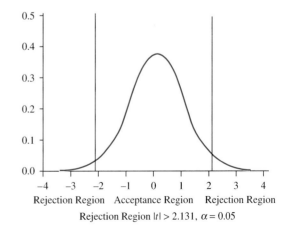

Rejection Region   Acceptance Region   Rejection Region
Rejection Region |t| > 2.131, $\alpha = 0.05$

Figure 8.8 Rejection regions for t distribution, sample size $n = 16$

PROBLEMS

8.1 The federal standard for cadmium dust in the workplace is 200 $\mu g/m^3$. To monitor the levels of cadmium dust an environmental engineer samples the air at 10-minute intervals, so each hourly average is the sample mean of 6 measurements. Assume that the level of cadmium dust is normally distributed with unknown mean μ and standard deviation $\sigma = 15 \mu g/m^3$.

   (a) Suppose management considers the air quality to be unacceptable when the hourly average (denoted by $\bar{x}$) is greater than 200 $\mu g/m^3$; that is, it suspends production whenever $\bar{x} > 200$. The environmental engineer therefore uses the rejection region $\mathcal{C} = \{\bar{x} : \bar{x} > 200\}$ for testing the hypothesis H_0: $\mu = 200$ against H_1: $\mu > 200$. What is the consequence of a type I error on production? What is the effect of a type II error on a worker's health?

   (b) [Part (a) continued] What is the significance level of this test? Compute the probability of a type I error if, in fact, $\mu = 190$. [*Hint:* Use Eq. (8.3). Compute the probability of a type II error at the alternative $\mu = 205$.]

   (c) Suppose we change the rejection region to the set $\mathcal{C} = \{\bar{x} : \bar{x} > 210\}$. Compute the significance level of this test and the probability of a type II error at the alternative $\mu = 205$. Compare the type I and type II errors of this test with those obtained in part (b). Interpret the results.

8.2 The Environmental Protection Agency (EPA) requires that when the level of lead [measured in micrograms per liter ($\mu g/L$)] reaches 15 in at least 10 percent of the samples taken, steps must be taken to reduce the levels of lead. Let $Q(p)$ denote the pth quantile of the distribution (assumed to be continuous) of the amount of lead X in a municipal water supply. Explain why the water supply meets the EPA standard when $Q(0.90) \leq 15$ and does not meet the standard when $Q(0.90) > 15$. [*Hint:* Denote the amount of lead in a water sample by X. Show that $Q(0.90) \leq 15$ implies that $P(X \leq 15) \geq P(X \leq Q(0.90)) = 0.90$.]

8.3 Suppose we use a sample of size $n = 9$ taken from a normal distribution with $\sigma = 4$ and the rejection region $\mathcal{C} = \{\bar{x} : \bar{x} < 28\}$ to test the hypothesis H_0: $\mu = 30$ against H_1: $\mu < 30$.

   (a) What is the significance level of this test? Compute the probability of a type II error at the alternative $\mu = 25$.

   (b) Suppose the cutoff value is changed to $c = 27.5$, so the rejection region is the set $\mathcal{C} = \{\bar{x} : \bar{x} < 27.5\}$. Compute the significance level of this test. Next, compute the probability of a type II error at the alternative $\mu = 25$. Compare the type I and type II errors of this test with the test of part (a). Interpret your results.

8.4 The thickness of metal wires used in the manufacture of silicon wafers is assumed to be normally distributed with mean μ (measured in microns) and standard deviation $\sigma = 1.0$. To monitor the production process the mean thickness of four wires taken from each wafer is computed. The output is considered unacceptable if the sample mean differs from the target value $\mu_0 = 10$ by more than 1.0 microns. We use the rejection region $\mathscr{C} = \{\bar{x} : |\bar{x} - 10| > 1.0\}$ to test H_0: $\mu = 10$ against H_1: $\mu \neq 10$.

(a) What is the consequence of a type I error on production? What is the effect of a type II error on the quality of the manufactured product?

(b) Compute the significance level α of this test.

(c) Compute the cutoff value c so that the significance level α equals 0.01.

8.5 Suppose we use a sample of size $n = 25$ taken from a normal distribution with $\sigma = 10$ and the rejection region $\mathscr{C} = \{\bar{x} : \bar{x} < c\}$ to test the hypothesis H_0: $\mu = 15$ against H_1: $\mu < 15$.

(a) Find the cutoff value c so that the test has significance level $\alpha = 0.1$. Compute the probability of a type II error at the alternative $\mu = 12.5$. [*Hint:* Use Eq. (8.11).]

(b) Find the cutoff value c so that the test has significance level $\alpha = 0.05$. Compute the probability of a type II error at the alternative $\mu = 12.5$. Compare the type I and type II errors of this test with the test of part (*a*). Discuss your conclusions.

8.6 The breaking strengths of steel wires used in elevator cables are assumed to come from a normal distribution with known $\sigma = 400$. Before accepting a large shipment of steel wires, an engineer wants to be confident that μ is greater than 10,000 pounds.

(a) Formulate the appropriate null and alternative hypotheses. Describe in this context, and in plain English, the consequences of a type I error and a type II error.

(b) The breaking strengths of 16 steel wires were measured, and the following result was obtained: $\bar{x} = 10,100$ pounds. Using this information, test your null hypothesis at the 5 percent significance level and state your conclusions.

(c) Use the confidence interval approach described in Example 8.3 to test your null hypothesis. What additional information concerning the value of the mean breaking strength μ does the confidence interval approach provide to the engineer? Discuss.

8.7 The concentration of an impurity in an ore, measured as a percentage, is assumed to be $N(\mu, 0.05^2)$. The ore cannot be used without additional processing if $\mu = 0.1$ percent. To determine whether or not the ore is usable, a random sample of size $n = 9$ is drawn and the ore is declared unusable if $\bar{x} > 0.08$ percent.

(a) What are the null and alternative hypotheses in this case? What is the rejection region?

(b) Compute the significance level of this test.

(c) Compute the type II error if $\mu = 0.09$ percent.

(d) Compute the P-value of the test if the observed value of the sample mean $\bar{x} = 0.09$.

8.8 Suppose we use a sample of size $n = 25$ taken from a normal distribution with $\sigma = 10$ to test the hypothesis H_0: $\mu = 50$ against H_1: $\mu \neq 50$.

(a) Construct a test of H_0 with significance level $\alpha = 0.05$ by constructing a 95 percent confidence interval for μ.

(b) Describe the rejection region of the test in part (a).

(c) What is the P-value corresponding to $\bar{x} = 53$? $\bar{x} = 46$?

8.9 Use the same test as in the previous problem, but this time assume σ is unknown and $s = 10$.

8.10 Prove that the power function $\pi(\mu)$ defined by Eq. (8.5) is a decreasing function of μ. [*Hint:* Use the fact that $\Phi(\theta)$ is an increasing function of θ.]

8.11 (a) The graph of the power function in Fig. 8.3 is an increasing function of μ. Show that this is a consequence of the fact that the power function defined by Eq. (8.5),

$$\pi(\mu) = 1 - \Phi\left(\frac{\sqrt{n}(c - \mu)}{\sigma}\right)$$

is an increasing function of μ. [*Hint:* Use the fact that $\Phi(x)$ is an increasing function of x.]

(b) The graph of the power function in Fig. 8.5 is a decreasing function of μ. Show that this is a consequence of the fact that power function defined by Eq. (8.13),

$$\pi(\mu) = \Phi\left(\frac{\sqrt{n}(c - \mu)}{\sigma}\right)$$

is a decreasing function of μ.

8.12 Twelve measurements of the ionization potential, measured in electron volts, of toluene were made, and the following data were recorded:

10.6 9.51 9.83 10.11 10.05 10.31 9.37 10.44 10.09 10.55 9.19 10.09

The published value of the ionization potential is 9.5. Assume that the data come from a normal distribution. Test the hypothesis that the new experimental data are consistent with the published value.

8.13 Seven measurements of arsenic in copper yielded a sample mean of $\bar{x} = 0.17$ percent and sample standard deviation $s = 0.03$ percent. Assuming the data come from a normal distribution $N(\mu, \sigma^2)$:

(a) Describe the test statistic and rejection region that is appropriate for testing (at the level α) H_0: $\mu = 0.2$ percent against H_1: $\mu < 0.2$ percent.

(b) Use the results of part (a) to test H_0 at the 5 percent significance level.

(c) Express the P-value in terms of the t distribution.

(d) Test the null hypothesis at the 5 percent level by constructing a suitable one-sided confidence interval. What additional information concerning the concentration of arsenic in copper does the confidence interval provide?

8.14 In a study of the effectiveness of a weight loss program, the net weight loss, in pounds, for each of 10 male subjects was recorded:

$$8.4 \ 15.9 \ 7.8 \ 12.8 \ 5.9$$

$$10.2 \ 7.5 \ 15.5 \ 12.1 \ 12.5$$

The weight loss program is judged ineffective if the mean weight loss is less than or equal to 10 pounds. Test the null hypothesis that the weight loss program is ineffective at the 5 percent significance level.

8.15 Suppose the lifetime of a tire is advertised to be 40,000 miles. The lifetimes of 100 tires are recorded, and the following results are obtained:

$$\bar{x} = 39{,}360 \text{ miles}, \qquad s = 3200 \text{ miles}$$

(a) If we do not assume that the data come from a normal distribution, what can you say about the distribution of $\bar{X}$?

(b) Describe the statistic you would choose to test H_0: $\mu = 40{,}000$ against H_1: $\mu < 40{,}000$.

(c) Compute the P-value of your test and determine whether or not it is significant at the 5 percent level.

8.16 Statistical analysis of experimental data suggests that the compressive strengths of bricks can be approximated by a normal distribution $N(\mu, \sigma^2)$. A new (and cheaper) process for manufacturing these bricks is developed, the output of which is also assumed to be normally distributed. The manufacturer claims that the new process produces bricks whose mean compressive strength exceeds the standard $\mu_0 = 2500$ psi. To test the manufacturer's claim that the new process exceeds the standard, the compressive strengths of $n = 9$ bricks, selected at random, are measured, with the following results reported: $\bar{x} = 2600$ and sample standard deviation $s = 75$.

(a) How should an engineer, skeptical of the manufacturer's claim that $\mu > 2500$, formulate the null and alternative hypotheses? Discuss, in this context, the consequences of type I and type II errors.

(b) Describe the rejection region for testing, at the 5 percent significance level, your null hypothesis of part (a). State your conclusions.

(c) Use the confidence interval approach to test this hypothesis.

(d) Compute the P-value of your test; state your conclusions.

8.17 Derive Eq. (8.21) for the sample size n by solving Eqs. (8.22) and (8.23).

8.18 Give the details of the derivation of Eq. (8.18).

8.19 The graph of the power function in Fig. 8.7 is symmetric about the null value μ_0, that is,

$$\pi(\mu_0 + d) = \pi(\mu_0 - d)$$

Show that the power function defined by Eq. (8.18) is symmetric about the null value. [*Hint:* Use the fact that $\Phi(-x) = 1 - \Phi(x)$.]

8.20 To test H_0: $\mu = 35$ against H_1: $\mu > 35$, a researcher records the observed value $\bar{x}$ of the sample mean of a random sample of size $n = 9$ taken from a normal distribution $N(\mu, 49)$. Compute the P-value of the test if the observed value of $\bar{x}$ is:

(a) $\bar{x} = 36$
(b) $\bar{x} = 37$
(c) $\bar{x} = 39$
(d) For which of the computed P-values in parts (a)–(c) would the null hypothesis be rejected at the 5 percent level of significance?

8.21 To test H_0: $\mu = 10$ against H_1: $\mu < 10$, a researcher computes the observed values $\bar{x}$ and s of a random sample of size $n = 12$ taken from a normal distribution $N(\mu, \sigma^2)$. Compute the P-value of the test if the observed values of $\bar{x}$ and s are:

(a) $\bar{x} = 9, s = 3.0$
(b) $\bar{x} = 8.5, s = 3.0$
(c) $\bar{x} = 8.5, s = 2.5$.
(d) For which of the computed P-values in parts (a)–(c), would the null hypothesis be rejected at the 5 percent level of significance?

8.22 Compute and sketch the graph of the power function of the test with rejection region $\mathscr{C} = \{\bar{x} : \bar{x} > 53\}$ in part (a) of Prob. 8.1.

8.23 Compute and sketch the graph of the power function of the test with rejection region $\mathscr{C} = \{\bar{x} : \bar{x} < 28\}$ in part (a) of Prob. 8.3.

8.24 Compute and sketch the graph of the power function of the test with rejection region $\mathscr{C} = \{\bar{x} : |\bar{x} - 10| > 1\}$ of Prob. 8.4.

8.25 Suppose you want to test H_0: $\mu = 50$ against H_1: $\mu > 50$ using a random sample of size n taken from a normal distribution $N(\mu, 81)$. Find the minimum sample size n and the cutoff value c if the probability of a type I error is to be $\alpha = 0.05$ and the probability of a type II error at the alternative $\mu = 55$ is to be $\beta(55) = 0.20$.

8.26 Suppose you want to test H_0: $\mu = 30$ against H_1: $\mu < 30$ using a random sample of size n taken from a normal distribution $N(\mu, 16)$. Find the minimum sample size n and the cutoff value c if the probability of a type I error is to be $\alpha = 0.05$ and the probability of a type II error at the alternative $\mu = 28$ is to be $\beta(28) = 0.20$.

8.3 TESTS OF HYPOTHESES ON $\mu_1 - \mu_2$

In many situations of practical interest we would like to compare two populations with respect to some numerical characteristic. For example, we might be interested in determining which of two steel cables has the greater tensile strength or which of two gasoline blends yields more miles per gallon (mpg). If we denote the tensile strengths of the two types of cable by X_1 and X_2, then the quantity of interest will be their difference, $X_1 - X_2$. The assumptions on X_1 and X_2 are the same as in Sec. 7.4, where we assumed that the variables X_1 and X_2 are both normally distributed with (possibly) different means, denoted by μ_1 and μ_2, but with the same variance, i.e., $\sigma_1^2 = \sigma_2^2 = \sigma^2$. It is worth noting that the assumption of equal variance is itself a hypothesis that can be tested; see Sec. 8.3.2. When the assumption of equal variance is found to be questionable, then the problem of testing hypotheses about the difference between the two means, $\mu_1 - \mu_2$, is considerably more complicated; we refer the interested reader to G. W. Snedecor and W. G. Cochran, *Statistical Methods,* 7th ed. (Ames, Iowa State University Press, 1980), for a discussion (with examples) of various remedies that have been proposed in the statistics literature.

We distinguish between *independent samples,* where it is assumed that the random samples selected from each of the two populations are mutually independent, and *paired samples,* as discussed in Sec. 7.4.1.

Independent Samples

Consider the problem of comparing two normal distributions, with means μ_1 and μ_2 and the same variance σ^2. In this context the null hypothesis of no difference between the means of the two populations is expressed as

$$H_0: \Delta = \mu_1 - \mu_2 = 0$$

There are various possibilities for the alternative hypothesis, such as

$$H_1: \Delta \neq 0$$

or

$$H_1: \Delta > 0, \quad \text{i.e., } \mu_1 > \mu_2$$

Consider, for example, a comparison of the effectiveness of two weight loss programs, called diet 1 and diet 2, respectively. The null hypothesis states that there is no difference between the two diets. The alternative $H_1: \Delta \neq 0$ asserts that there is a difference, whereas the alternative $H_1: \Delta > 0$ asserts that diet 1 is more effective than diet 2. To construct a level α test of $H_0: \Delta = 0$ against the two-sided alternative $\Delta \neq 0$, we proceed as follows:

1. Take random samples of sizes n_1 and n_2 from the two populations, and denote their corresponding sample means and sample variances by $\overline{X}_i$ and

$s_i^2, i = 1, 2$. Since the random samples selected from the two populations are assumed to be independent, we have

$$V(\overline{X}_1 - \overline{X}_2) = \sigma^2 \left(\frac{1}{n_1} + \frac{1}{n_2} \right)$$

The estimated standard error is

$$s(\overline{X}_1 - \overline{X}_2) = s_p \sqrt{\left(\frac{1}{n_1} + \frac{1}{n_2} \right)}$$

where s_p^2 is the pooled estimate of σ^2 [Eq. (7.33)].

2. For our test statistic we use the fact (Theorem 7.1) that when H_0 is true,

$$\frac{(\overline{X}_1 - \overline{X}_2) - (\mu_1 - \mu_2)}{s_p \sqrt{\dfrac{1}{n_1} + \dfrac{1}{n_2}}} \sim t_{n_1 + n_2 - 2} \tag{8.28}$$

has a Student t distribution with $n_1 + n_2 - 2$ degrees of freedom. Under the null hypothesis $\mu_1 = \mu_2$, it follows that

$$\frac{(\overline{X}_1 - \overline{X}_2)}{s_p \sqrt{\dfrac{1}{n_1} + \dfrac{1}{n_2}}} \sim t_{n_1 + n_2 - 2}$$

3. Choose the rejection region to be the set $\{t : |t| > t_{n_1 + n_2 - 2}(\alpha/2)\}$; with this choice of cutoff value, the probability of a type I error is α. In other words,

$$\text{accept } H_0 \text{ if } \left| \frac{(\overline{X}_1 - \overline{X}_2)}{s_p \sqrt{\dfrac{1}{n_1} + \dfrac{1}{n_2}}} \right| \leq t_{n_1 + n_2 - 2}(\alpha/2)$$

$$\text{reject } H_0 \text{ if } \left| \frac{(\overline{X}_1 - \overline{X}_2)}{s_p \sqrt{\dfrac{1}{n_1} + \dfrac{1}{n_2}}} \right| > t_{n_1 + n_2 - 2}(\alpha/2)$$

Alternatively, one can proceed by constructing a two-sided $100(1 - \alpha)$ percent confidence interval of the form $[L, U]$ [use Eqs. (7.35) and (7.36)], where

$$L = \overline{X}_1 - \overline{X}_2 - t_{n_1 + n_2 - 2}(\alpha/2) s_p \sqrt{\frac{1}{n_1} + \frac{1}{n_2}}$$

$$U = \overline{X}_1 - \overline{X}_2 + t_{n_1 + n_2 - 2}(\alpha/2) s_p \sqrt{\frac{1}{n_1} + \frac{1}{n_2}}$$

We then accept H_0 if $0 \in [L, U]$ and reject H_0 otherwise; i.e.,

$$\text{accept } H_0 \text{ if } L \leq 0 \leq U \tag{8.29}$$

$$\text{reject } H_0 \text{ if } U < 0 \text{ or } L > 0 \tag{8.30}$$

The rejection regions for lower- and upper-tailed tests are derived in a similar manner; the results are summarized in Table 8.10.

TABLE 8.10

Alternative hypothesis	Rejection region
$H_1: \mu_1 - \mu_2 < 0$	$\dfrac{(\overline{X}_1 - \overline{X}_2)}{s_p \sqrt{\dfrac{1}{n_1} + \dfrac{1}{n_2}}} < -t_{n_1 + n_2 - 2}(\alpha)$
$H_1: \mu_1 - \mu_2 > 0$	$\dfrac{(\overline{X}_1 - \overline{X}_2)}{s_p \sqrt{\dfrac{1}{n_1} + \dfrac{1}{n_2}}} > t_{n_1 + n_2 - 2}(\alpha)$
$H_1: \mu_1 - \mu_2 \neq 0$	$\left\| \dfrac{(\overline{X}_1 - \overline{X}_2)}{s_p \sqrt{\dfrac{1}{n_1} + \dfrac{1}{n_2}}} \right\| > t_{n_1 + n_2 - 2}(\alpha/2)$

A test based on the statistic defined by Eq. (8.28) is called a *two-sample t test*.

example 8.13   Refer to the weight loss data for the two diets in Table 7.1. Is there a significant difference at the 5 percent level between the two diets?

Solution. We use the confidence interval approach. Specifically, with reference to the weight loss data in Table 7.1 we obtain a 95 percent confidence interval by computing the following quantities:

$$\bar{x}_1 = 5.86, \qquad \bar{x}_2 = 10.79, \qquad \bar{x}_1 - \bar{x}_2 = -4.93$$

$$s_1^2 = 11.85, \qquad s_2^2 = 10.67$$

$$s_p^2 = 11.26, \qquad s_p = 3.36$$

$$t_{18}(0.025)s_p \sqrt{\frac{1}{10} + \frac{1}{10}} = 3.15$$

$$L = -4.93 - 3.15 = -8.08$$

$$U = -4.93 + 3.15 = -1.78$$

Since the confidence interval $[-8.08, -1.78]$ does not contain 0, we do not accept the null hypothesis.   ∎

Paired-Sample t Test

Before proceeding, it might be helpful to review the mathematical model of a paired sample as described in Definition 7.5. The key assumption made there is that the random variables $D_i = X_i - Y_i$ $(i = 1, \ldots, n)$ are mutually independent, normally distributed random variables with a common mean and variance denoted by $E(D_i) = \Delta, V(D_i) = \sigma_D^2$. As in the case of independent samples, a variety of combinations for the null and alternative hypotheses can be studied. Here we consider the two-sided alternative case only, that is, H_1: $\Delta \neq 0$. The necessary modifications for dealing with one-sided alternatives are omitted. Our problem, then, is this: Use the confidence interval approach to construct a level α test of H_0: $\Delta = 0$ against H_1: $\Delta \neq 0$.

The following test is called a *paired-sample t test*.

1. Collect the paired data (x_i, y_i), $i = 1, \ldots, n$, and compute $\overline{D}$ and s_D^2.
2. For our test statistic we use the fact that

$$\frac{(\overline{D} - \Delta)}{s_D / \sqrt{n}} \sim t_{n-1} \tag{8.31}$$

Consequently, under the null hypothesis that $\Delta = 0$, we have

$$\frac{\overline{D}}{s_D / \sqrt{n}} \sim t_{n-1}$$

3. Construct a two-sided $100(1 - \alpha)$ percent confidence interval of the form $[L, U]$ [Eqs. (7.19) and (7.20)], where

$$L = \overline{D} - t_{n-1}(\alpha/2) \times s_D / \sqrt{n}$$
$$U = \overline{D} + t_{n-1}(\alpha/2) \times s_D / \sqrt{n}$$

We then accept H_0 if $0 \in [L, U]$ and reject H_0 otherwise; that is,

$$\text{accept } H_0 \text{ if } L \leq 0 \leq U \tag{8.32}$$
$$\text{reject } H_0 \text{ if } U < 0 \text{ or } L > 0 \tag{8.33}$$

---

example 8.14   Refer to Table 7.2, which reports the results of a study comparing the execution times of six different workloads on two different configurations of a processor. The parameter of interest is Δ, the difference between the mean execution times of the two processors. Is there a significant difference at the 5 percent level between the two processor configurations?

Solution. We test the null hypothesis ($\Delta = 0$) at the 5 percent level by constructing a 95 percent confidence interval for Δ, which is given by

$$L = -25.67 - t_5(0.025) \times \frac{10.41}{\sqrt{6}} = -36.60$$

$$U = -25.67 + t_5(0.025) \times \frac{10.41}{\sqrt{6}} = -14.74$$

The confidence interval $[-36.60, -14.74]$ does not contain 0, so the null hypothesis is not accepted. Of course, the data indicate much more, that is, that a one-cache memory is significantly faster than a no-cache memory. ∎

8.3.1 The Wilcoxon Rank Sum Test for Two Independent Samples

The P-value and power of the t test change when the distribution of the data are nonnormal—highly skewed, for instance. The *Wilcoxon rank sum test*, an alternative to the t test, is a test whose P-value does not depend explicitly on the distribution of the data and whose power remains high against the alternatives even when the data are nonnormal.

1. **Assumptions on the two populations.** The independent random samples $X_1, \ldots, X_{n_1}$ and $Y_1, \ldots, Y_{n_2}$ are taken from distributions F and G with *unknown* probability density functions f and g. In particular, the distribution functions F and G are continuous.

   When F and G are both normal with the same variance σ^2 but (possibly) different means μ_1 and μ_2, then their distribution functions are given by[1]

   $$F(x) = \Phi\left(\frac{x - \mu_1}{\sigma}\right) \quad \text{and} \quad G(x) = \Phi\left(\frac{x - \mu_2}{\sigma}\right)$$

   It follows that

   $$F(x) = \Phi\left(\frac{x - \mu_1}{\sigma}\right) = \Phi\left(\frac{[x - (\mu_1 - \mu_2)] - \mu_2}{\sigma}\right)$$

   $$= G(x - (\mu_1 - \mu_2)) = G(x - \theta), \quad \theta = \mu_1 - \mu_2$$

   In other words, when $\mu_1 < \mu_2$ $(\theta < 0)$, the probability density function of the Y's is shifted to the right, and when $\mu_1 > \mu_2$ $(\theta > 0)$, the probability density function of the Y's is shifted to the left. For a graphical display of this shift, see Fig. 8.1.

2. **The null and alternative hypotheses.** The null hypothesis, as usual, is that both samples come from the same distribution, that is,

   $$H_0\colon F(x) = G(x)$$

   We consider three types of alternative hypotheses: $(\mu_1 < \mu_2)$, $(\mu_1 > \mu_2)$, and $(\mu_1 \neq \mu_2)$. Restated in terms of the parameter θ, the three alternatives are:

   $$H_1\colon F(x) = G(x - \theta), \quad \theta < 0 \ (\mu_1 < \mu_2)$$
   $$H_1\colon F(x) = G(x - \theta), \quad \theta > 0 \ (\mu_1 > \mu_2)$$
   $$H_1\colon F(x) = G(x - \theta), \quad \theta \neq 0 \ (\mu_1 \neq \mu_2)$$

---

[1]It might be helpful to review the discussion prior to Eq. (4.31).

We interpret acceptance of the alternative $H_1: \theta < 0$ or $H_1: \theta > 0$ as implying that one population has a higher mean than the other; and acceptance of $H_1: \theta \neq 0$ as implying that the two means differ.

3. **Convention.** Population 1 always denotes the smaller sample, population 2 the larger (so $n_1 \leq n_2$).

We will now illustrate the Wilcoxon rank sum test procedure, and the underlying theory, in the context of a specific example.

---

example 8.15   The data in Table 8.11 come from an experiment comparing the fuel efficiencies of two gasolines, a standard brand and a premium one. Six cars were driven with the premium gas and seven cars with the standard gas, so $n_1 = 6$, $n_2 = 7$. Test, at the 5 percent level, the null hypothesis that the premium blend is no more fuel efficient than the standard one against the one-sided alternative that the premium blend is more fuel efficient.

TABLE 8.11 MILES PER GALLON DATA
FOR TWO BRANDS OF GASOLINE

Premium, X	Standard, Y
26.6	22.7
26.0	28.1
28.4	24.0
27.5	26.8
30.9	25.5
29.0	24.7
	30.2

Solution. The null hypothesis is that the fuel efficiencies of the premium and standard gasolines are the same, and the alternative hypothesis is that the premium brand is more fuel efficient; that is, we are testing $H_0: F(x) = G(x)$ against $H_1: F(x) = G(x - \theta)$, $\theta = \mu_1 - \mu_2 > 0$.

The Wilcoxon rank sum test begins by combining the observations from both populations into a single sample of size $6 + 7 = 13$ and arranging the data in increasing order (by computing the order statistics of the combined sample), preserving the population identity as shown in Table 8.12. The first column denotes the population from which the observation is drawn (premium = 1, standard = 2), the second column displays the order statistics of the combined sample, and the last column lists the rank of each observation in the combined sample.

In this example all the observations in the combined sample are distinct, so the ranks are uniquely defined. It may happen that two or more observations are *tied*. Suppose, for example, that the recorded values for the mileages ranked 6 (26.6) and 7 (26.8) were tied with common value 26.7. Then the rank of each of these observations is the mean of the ranks that would have been assigned had they been different. In this case we would have assigned the rank 6.5 to each of them.

TABLE 8.12

Population	mpg	Rank	Population	mpg	Rank
2	22.7	1	1	27.5	8
2	24.0	2	2	28.1	9
2	24.7	3	1	28.4	10
2	25.5	4	1	29.0	11
1	26.0	5	2	30.2	12
1	26.6	6	1	30.9	13
2	26.8	7			

When H_0 is true, then the null distribution of the ranks assigned to the sample $X_1, \ldots, X_6$ is the same as if we had randomly selected 6 balls from $6 + 7 = 13$ balls numbered $1, 2, \ldots, 13$. Therefore, when the alternative $H_1 \colon \theta > 0$ is true, then the sum of the ranks assigned to the sample $X_1, \ldots, X_6$ will be larger than it would if H_0 were true. Looking at Table 8.12, we see that the four lowest-ranking observations come from population 2 (the standard brand), suggesting that the lower ranks are associated with population 2 and the higher ranks are associated with population 1. A reasonable choice for our test statistic, then, is W_1, the sum of the ranks associated with the X sample. For instance, with reference to Table 8.12, we see that the sum of the ranks associated with population 1 is

$$W_1 = 5 + 6 + 8 + 10 + 11 + 13 = 53.0$$

We then choose a number c so that an observed value of $W_1 \geq c$ implies that it is highly unlikely to have been observed if, in fact, H_0 is true.

It can be shown (we omit the proof) that under H_0 the mean and standard deviation of the rank sum W_1 (with n_1 X's and n_2 Y's) are

$$E(W_1) = \frac{n_1(n_1 + n_2 + 1)}{2}, \qquad \sigma(W_1) = \sqrt{\frac{n_1 n_2 (n_1 + n_2 + 1)}{12}} \qquad (8.34)$$

With reference to the data set in Table 8.11, the expected value and standard deviation of the rank sum are given by

$$E(W_1) = \frac{6 \times (6 + 7 + 1)}{2} = 42 \quad \text{and} \quad \sigma(W_1) = \sqrt{\frac{6 \times 7(6 + 7 + 1)}{12}} = 7.0$$

The rejection region and P-value are computed by using the result (we omit the proof) that the null distribution of W_1 is known to be approximately normal with mean $E(W_1)$ and standard deviation $\sigma(W_1)$ as given in Eq. (8.34); that is,

$$\frac{W_1 - E(W_1)}{\sigma(W_1)} \approx Z$$

where Z denotes a standard normal random variable.

The rejection region for an upper-tailed Wilcoxon rank sum test (using the continuity correction to improve the accuracy of the normal approximation) is given by

$$\text{Rejection region:} \quad \frac{W_1 - E(W_1) - 0.5}{\sigma(W_1)} > z(\alpha) \tag{8.35}$$

To illustrate the method using the fuel efficiency data (Table 8.12), we substitute

$$W_1 = 53, \qquad E(W_1) = 42, \qquad \sigma(W_1) = 7$$

into Eq. (8.35). It follows that

$$\frac{W_1 - E(W_1) - 0.5}{\sigma(W_1)} = \frac{53 - 42 - 0.5}{7} = 1.5 < 1.645 = z(0.05)$$

The (approximate) P-value is given by

$$P\text{-value} = P(Z > 1.5) = 0.0668$$

So the results are not significant at the 5 percent significance level. We do not reject H_0. ■

Table 8.13 lists the rejection regions for each of the three alternative hypotheses based on the normal approximation (with continuity correction) to the Wilcoxon rank sum statistic. Critical values for this statistic based on the exact distribution of W_1 can be found in W. J. Dixon and F. J. Massey, Jr., *Introduction to Statistical Analysis*, 4th ed. (New York, McGraw-Hill, 1983).

TABLE 8.13

Alternative hypothesis	Rejection region
$H_1: \mu_1 - \mu_2 < 0$	$\dfrac{W_1 - E(W_1) + 0.5}{\sigma(W_1)} < -z(\alpha)$
$H_1: \mu_1 - \mu_2 > 0$	$\dfrac{W_1 - E(W_1) - 0.5}{\sigma(W_1)} > z(\alpha)$
$H_1: \mu_1 - \mu_2 \neq 0$	$\dfrac{|W_1 - E(W_1)| - 0.5}{\sigma(W_1)} > z(\alpha/2)$

$$E(W_1) = \frac{n_1(n_1 + n_2 + 1)}{2} \qquad \sigma(W_1) = \sqrt{\frac{n_1 n_2(n_1 + n_2 + 1)}{12}}$$

Note: W_1 is the sum of the ranks associated with population 1 $(n_1 \leq n_2)$.

8.3.2   Tests of Hypotheses Concerning the Variance of Normal Distribution

Tests of hypotheses concerning the variance σ^2 are based on the statistic $(n-1)s^2/\sigma^2$, which is known to have a χ^2_{n-1} distribution (Theorem 6.9). Because this test is infrequently used, and this is an introductory text, we shall not pursue this matter further here. Of more importance is a test for the equality of two variances since, in our derivation of the two-sample t test, we assumed that the variances of the two normal distributions being compared were equal. In this section we present a test of this hypothesis; readers interested in how to modify the two-sample t test when $\sigma_1^2 \neq \sigma_2^2$ are advised to consult G. W. Snedecor and W. G. Cochran, *Statistical Methods*, 8th ed. (Ames, Iowa State University Press, 1989).

1. Let s_1^2 and s_2^2 denote the sample variances corresponding to random samples of sizes n_1 and n_2 taken from two independent normal distributions with the same variance σ^2; then it follows from Theorems 6.9 and 6.12 that the ratio

$$F_0 = \frac{s_1^2}{s_2^2} \sim F_{n_1-1,n_2-1} \tag{8.36}$$

has an F distribution with parameters $\nu_1 = n_1 - 1$, $\nu_2 = n_2 - 1$.

2. To test the null hypothesis $H_0: \sigma_1^2 = \sigma_2^2$ against $H_1: \sigma_1^2 \neq \sigma_2^2$ at the significance level α, choose the rejection region

$$\{F : F < F_{n_1-1,n_2-1}(1 - \alpha/2) \text{ or } F > F_{n_1-1,n_2-1}(\alpha/2)\}$$

3. The critical values $F_{\nu_1,\nu_2}(1 - (\alpha/2))$ are obtained from Table A.6 in the appendix by using the formula

$$F_{\nu_1,\nu_2}(1 - (\alpha/2)) = \frac{1}{F_{\nu_2,\nu_1}(\alpha/2)} \tag{8.37}$$

Equation (8.37) is a consequence of the relation

$$F_{\nu_1,\nu_2} \sim \frac{1}{F_{\nu_2,\nu_1}}$$

To put it another way, we accept the null hypothesis when

$$\frac{1}{F_{n_2-1,n_1-1}(\alpha/2)} < F < F_{n_1-1,n_2-1}(\alpha/2) \tag{8.38}$$

---

example 8.16

Refer to the weight loss data for the two diets in Table 7.1. Is it reasonable to assume that both the diet 1 and diet 2 data come from normal distributions with the same variance?

Solution. In this case, as previously noted, we have $n_1 = n_2 = 10$ and $s_1^2 = 11.85$, $s_2^2 = 10.67$; so the F ratio equals $11.85/10.67 = 1.11$. Since the upper-tail probabilities are given only for $\alpha = 0.05$ and $\alpha = 0.01$, we can only construct two-sided tests with significance level $\alpha = 0.10$ or $\alpha = 0.02$. Thus, for instance, $F_{9,9}(0.05) = 3.18$, and so our acceptance region is the set $\{F : 1/3.18 \leq F \leq 3.18\}$. Since $F = 1.11$, we do not reject $H_0: \sigma_1^2 = \sigma_2^2$. ∎

PROBLEMS

8.27 An experiment was conducted to compare the electrical resistances (measured in ohms) of two types of wire, and the following results were obtained:

| Wire A | 0.140 | 0.138 | 0.143 | 0.142 | 0.144 | 0.137 |
| Wire B | 0.135 | 0.140 | 0.142 | 0.136 | 0.138 | 0.140 |

(a) Test the null hypothesis that $\mu_1 = \mu_2$ against the alternative that $\mu_1 \neq \mu_2$ by constructing an appropriate two-sided 95 percent confidence interval.

(b) What is the P-value of your test?

8.28 The skein strengths, measured in pounds, of two types of yarn were tested, and the following results were reported:

| Yarn A | 99 | 93 | 99 | 97 | 90 | 96 | 93 | 88 | 89 |
| Yarn B | 93 | 94 | 75 | 84 | 91 | | | | |

(a) By constructing a suitable two-sided 95 percent confidence interval, test the hypothesis that there is no difference between the mean skein strengths of the two types of yarn.

(b) What is the P-value of your test?

(c) Use the Wilcoxon rank sum statistic to test $H_0: \mu_1 = \mu_2$ against $H_1: \mu_1 \neq \mu_2$. Use $\alpha = 0.05$. What is the P-value of the test?

8.29 To determine the difference, if any, between two brands of radial tires, 12 tires of each brand were tested and the following mean lifetimes and sample standard deviations for the two brands were obtained:

$$\bar{x}_1 = 38{,}500 \quad \text{and} \quad s_1 = 3100$$
$$\bar{x}_2 = 41{,}000 \quad \text{and} \quad s_2 = 3500$$

Assuming that the tire lifetimes are normally distributed with identical variances, test the hypothesis that there is no difference between the two lifetimes. Compute the P-value of your test.

8.30 In a code size comparison study, five determinations of the number of bytes required to code five workloads on two different processors were recorded; the data are displayed in the following table.

System 1	System 2
101	130
144	180
211	141
288	374
72	302

Source: R. Jain, *The Art of Computer Systems Performance Analysis,* New York, John Wiley & Sons, 1991. Used with permission.

Do the data indicate a significant difference between the two systems? Use $\alpha = 0.05$. (*Hint:* First decide whether the data come from two independent samples or from paired samples.)

8.31 Refer to the data in Prob. 7.27.

(a) Do the data indicate a significant difference in the mean concentration of thiol between the normal group and the rheumatoid arthritis group? Use the t test with significance level $\alpha = 0.05$. Summarize your analysis by giving the P-value.

(b) Repeat part (a) but use the Wilcoxon rank sum test.

8.32 In a comparison study of a standard versus a premium blend of gasoline, the miles per gallon (mpg) for each blend was recorded for five compact automobiles:

	mpg	
Car	Standard	Premium
1	26.9	27.7
2	22.5	22.2
3	24.5	25.3
4	26.0	26.8
5	28.5	29.7

Test, at the 5 percent significance level, the claim that the premium blend delivers more miles per gallon than the standard blend. Your null hypothesis is that "the premium blend is no more fuel efficient than the standard blend."

8.33 Refer to Charles Darwin's data in Prob. 7.28. Do the data point to a significant difference between the heights of the cross-fertilized and self-fertilized seedlings? Summarize your results by giving the P-value.

8.34 In a weight loss experiment 10 men followed diet A and 7 men followed diet B. The weight losses after one month are displayed in the following table.

Weight loss (lb)		Weight loss (lb)	
Diet A	Diet B	Diet A	Diet B
15.3	3.4	10	5.2
2.1	10.9	8.3	2.5
8.8	2.8	9.4	
5.1	7.8	12.5	
8.3	0.9	11.1	

(a) Use the two-sample t test to determine if there is any significant difference between the two diets. Assume the data are normally distributed with a common variance; use $\alpha = 0.05$. In detail: Describe the null and alternative hypotheses, and determine which diet is more effective in reducing weight.

(b) Use the Wilcoxon rank sum test to test $H_0: \mu_1 = \mu_2$ against $H_1: \mu_1 > \mu_2$. Use $\alpha = 0.05$. What is the P-value of the test? Remember that diet B is now population 1 because its sample size is the smaller one.

8.35 Refer to the data in Prob. 7.29.

(a) Do the data indicate a significant difference in the DBH activity between the nonpsychotic and psychotic groups? Use the two-sample t test, and summarize your results by stating the P-value.

(b) Repeat part (a) but use the Wilcoxon rank sum test.

8.36 The following data give the results of an experiment to study the effect of three different word-processing programs on the length of time (measured in minutes) to type a text file. A group of 24 secretaries was randomly divided into three groups of 8 secretaries each, and each group was randomly assigned to one of the three word-processing programs.

Program	Typing time (min)							
A	15	13	11	19	14	19	16	19
B	13	12	9	18	13	17	15	20
C	18	17	13	18	15	22	16	22

Use the two-sample t test to determine the effect, if any, of the word-processing program on the length of time it takes to type a text file. Assume the data are normally distributed with a common variance; use $\alpha = 0.05$. In detail:

(a) Compare the differences between word-processing programs A and B, between word-processing programs A and C, between word-processing programs B and C. Describe the null and alternative hypotheses, and state your conclusions.

(b) Discuss, on the basis of the results that you have obtained in part (a), what effect, if any, the word-processing program has on the length of time it takes to type a text file.

(c) Repeat part (a) but use the Wilcoxon rank sum test.

8.37 In a study comparing the breaking strengths (measured in psi) of two types of fiber, random samples of sizes $n_1 = n_2 = 9$ of both types of fiber were tested, and the following summary statistics were obtained: $\overline{X}_1 = 40.1, s_1^2 = 2.5; \overline{X}_2 = 41.5, s_2^2 = 2.9$. Assume both samples come from normal distributions.

Although fiber 2 is cheaper to manufacture, it will not be used unless its breaking strength is found to be greater than that of fiber 1. Thus, we want to test $H_0: \Delta = \mu_1 - \mu_2 = 0$ against $H_1: \Delta = \mu_1 - \mu_2 < 0$.

Observe that the alternative hypothesis as formulated here asserts that the breaking strength of fiber 2 is greater than that of fiber 1. Controlling the type I error here protects you from using an inferior, although cheaper, product.

(a) Use the given data to test the null hypothesis at the levels $\alpha = 0.05$, $\alpha = 0.01$.

(b) Test the null hypothesis by constructing an appropriate one-sided $100(1 - \alpha)$ percent confidence interval, where $\alpha = 0.01$.

8.38 Refer to the data in Prob. 7.28. Let σ_1^2 and σ_2^2 denote the variances of the heights of the cross-fertilized and self-fertilized seedlings, respectively. Test for the equality of the variances at the 1 percent level; that is, test the following hypothesis:

$$H_0: \frac{\sigma_1^2}{\sigma_2^2} = 1 \quad \text{against} \quad H_1: \frac{\sigma_1^2}{\sigma_2^2} \neq 1$$

8.39 In a comparison study of two independent normal populations, the sample variances were $s_1^2 = 1.1$ and $s_2^2 = 3.4$, and the corresponding sample sizes were $n_1 = 6, n_2 = 10$. Is one justified in using the two-sample t test? Justify your conclusions.

8.4 NORMAL PROBABILITY PLOTS

Many of the techniques used in statistical inference assume that the distribution $F(x)$ of the random sample $X_1, \ldots, X_n$ comes from a distribution of a particular type denoted by $F_0(x)$. In hypothesis testing, for example, we have assumed throughout that the data are normally distributed, i.e., that

$$F_0(x) = \Phi\left(\frac{x - \mu}{\sigma}\right)$$

When this assumption is violated, some of these tests perform poorly, so it is important to determine whether it is reasonable to assume that the data come from a given distribution $F_0(x)$.

As another example, consider the data in Table 8.14, which lists 24 measurements of the operating pressure (psi) of a rocket motor.

TABLE 8.14 ROCKET MOTOR OPERATING PRESSURE (PSI)

7.7401	7.7749	7.7227	7.77925	7.96195	7.4472
8.0707	7.89525	8.0736	7.4965	7.5719	7.7981
7.8764	8.1925	8.01705	7.9431	7.71835	7.87785
7.2904	7.7575	7.3196	7.6357	8.06055	7.9112

Source: I. Guttman, R. Johnson, G. Hattacharyya, and B. Reiser, "Confidence Limits for Stress–Strength Models with Explanatory Variables," *Technometrics,* vol. 30, no. 2, 1988, pp. 161–168. Used with permission.

The statistical methods used by these authors to study the reliability of the rocket motor assumed that the data came from a normal distribution. This is a special case of determining whether the data come from a given theoretical distribution $F_0(x)$ against the alternative that it does not; that is, we are interested in testing

$$H_0: F(x) = F_0(x) \quad \text{against} \quad H_1: F(x) \neq F_0(x) \qquad (8.39)$$

In this section we present two methods for studying the normality assumption:

1. The *normal probability plot*, also called *Q-Q plot*, which is an abbreviation for *quantile-quantile plot*. This is a two-dimensional graph with the property that if the random sample comes from a normal distribution, then the plotted points appear to lie in a straight line. Both SAS and MINITAB have built-in procedures for producing these plots. Our purpose here is to discuss the underlying theory, which is of independent interest.
2. The *Shapiro-Wilk test*, which is based on the fact that the correlation coefficient[2] of the normal probability plot is a measure of the strength of the linear relationship. The hypothesis of normality is rejected if the correlation falls below a critical value that depends on the sample size. In the computer printout the test statistic, denoted by W, and its P-value are printed side by side.

A Discussion of the Theory Underlying Normal Probability Plots

The basic idea behind a Q-Q plot is that when the null hypothesis $F(x) = F_0(x)$ is true, the empirical distribution function $\hat{F}_n(x)$ (see Chap. 1, Definition 1.4) of a random sample of size n drawn from the parent distribution $F_0(x)$ should approximate the null distribution $F_0(x)$. A precise statement of this result, which is known as the *fundamental theorem of mathematical statistics*, is beyond the scope of this text, so, we content ourselves with the following version:

■ THEOREM 8.1

Let $\hat{F}_n$ denote the empirical distribution function of a random sample $X_1, \ldots, X_n$ taken from the continuous parent distribution function F_0; that is, assume the null hypothesis $H_0: F(x) = F_0(x)$ is true. Then

$$\lim_{n \to \infty} \hat{F}_n(x) = F_0(x) \qquad (8.40)$$

Furthermore, the p quantiles of the empirical distribution function converge to the p quantiles of the parent distribution. That is,

$$\lim_{n \to \infty} Q_n(p) = Q(p) \qquad (8.41)$$

where $Q_n(p)$ and $Q(p)$ denote the p quantiles of $\hat{F}_n(x)$ and $F_0(x)$, respectively.

■

---

[2] A detailed study of the sample correlation coefficient is given in Chap. 10.

Proof. We shall not give a detailed proof here except to remark that it is a consequence of the law of large numbers applied to the empirical distribution function [see Eq. (5.39), which is a weaker version of Eq. (8.40)]. Intuitively, Eq. (8.40) states that if the random sample really comes from the distribution $F_0(x)$, then, for large values of n, the proportion of observations in the sample whose values are less than or equal to x is approximately equal to $F_0(x)$, which is a very reasonable result.

Applying this result to the special case where $F_0(x) = \Phi((x - \mu)/\sigma)$, we have

$$\lim_{n\to\infty} \hat{F}_n(x) = \Phi\left(\frac{x - \mu}{\sigma}\right) \tag{8.42}$$

Equation (8.41) implies that the p quantiles $Q_n(p)$ of the empirical distribution function should be approximately equal to the p quantiles $Q(p)$ of the normal distribution. This suggests that a plot of the quantiles of the empirical distribution function against the quantiles of the normal distribution should produce a set of points that appear to lie on a straight line. Intuitively, the jth order statistic is approximately the $100(j/n)$th percentile of the data. More precisely, we now claim that the jth order statistic is the $[100(j - 0.5)/n]$th percentile of the empirical distribution function; that is,

$$Q_n\left(\frac{j - 0.5}{n}\right) = x_{(j)} \tag{8.43}$$

Proof. To derive Eq. (8.43) we use Eq. (1.6) to verify that

$$\text{if } p = (j - 0.5)/n, \text{ then } j - 1 < np = j - 0.5 < j$$

Consequently,

$$Q_n(p) = x_{(j)}$$

Similarly, the $[100(j - 0.5)/n]$th percentiles of the normal distribution $N(\mu, \sigma^2)$ are given by

$$Q\left(\frac{j - 0.5}{n}\right) = \mu + \sigma z_j \tag{8.44}$$

where

$$\Phi(z_j) = \frac{j - 0.5}{n} \tag{8.45}$$

The n quantities $z_1, \ldots, z_n$ defined by Eq. (8.45) are called the *standardized normal scores;* they are the $(j - 0.5)/n$ quantiles of the standard normal distribution. Since $P(X_i \le \mu + \sigma z_j) = \Phi(z_j) = (j - 0.5)/n$, it follows at once that $\mu + \sigma z_j$ is the $(j - 0.5)/n$ quantile of the normal distribution $N(\mu, \sigma^2)$, as stated in Eq. (8.44).

Putting these results together yields

$$x_{(j)} \approx \mu + \sigma \times z_j \tag{8.46}$$

since, as noted earlier, the two quantiles should be nearly equal.

The content of Eq. (8.46) is this: If the data come from a normal distribution, then the points $(z_j, x_{(j)}), j = 1, \ldots, n$, will appear to lie on a straight line; this graph, as noted earlier, is called a normal probability plot or a Q-Q plot.

▨ DEFINITION 8.1

To construct a Q-Q plot of the data set $\{x_1, \ldots, x_n\}$, compute (1) the normal scores $z_j = \Phi^{-1}[(j - 0.5)/n]$ and (2) the order statistics $x_{(j)}$, and then plot (3) $(z_j, x_{(j)}), j = 1, \ldots, n$.   ∎

Constructing a Q-Q plot and looking at the graph to see if the points lie on a straight line is one of the simplest methods for testing normality. Because computing the normal scores z_j can be quite tedious, it is advisable to carry out the computations via a statistical software package. To illustrate the method, we construct the Q-Q plot for the rocket motor data (Table 8.14).

Since the data points of the Q-Q plot appear to lie on a straight line, we do not reject the hypothesis that the original data come from a normal distribution.

If you do not have access to a computer, the computations can be carried out using a scientific calculator as illustrated in Table 8.15. The first column (P) displays the order statistics of the data set of Table 8.14, and the second column (Z) displays the normal scores z_j defined by $\Phi(z_j) = (j-0.5)/24, j = 1, 2, \ldots, 24$.

In addition to plotting and displaying the normal probability plot, a statistical software package will print out the value and P-value of the Shapiro-Wilk statistic, as in the following excerpt from a SAS printout:

$$\text{W:Normal} \qquad 0.961116 \qquad \text{Pr} < \text{W} \qquad 0.4674 \qquad (8.47)$$

The value of the Shapiro-Wilk statistic and its P-value are $W = 0.961116$ and $P = 0.4674$, respectively; so the hypothesis of normality is not rejected.

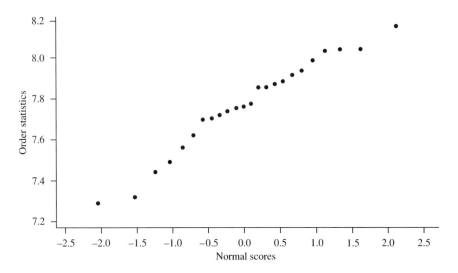

Figure 8.9 Q-Q plot of rocket motor data

TABLE 8.15 ORDER STATISTICS
AND NORMAL SCORES FOR ROCKET
MOTOR DATA (TABLE 8.14)

P	Z	P	Z
7.29040	−2.03683	7.79810	0.05225
7.31960	−1.53412	7.87640	0.15731
7.44720	−1.25816	7.87785	0.26415
7.49650	−1.05447	7.89525	0.37410
7.57190	−0.88715	7.91120	0.48878
7.63570	−0.74159	7.94310	0.61029
7.71835	−0.61029	7.96195	0.74159
7.72270	−0.48878	8.01705	0.88715
7.74010	−0.37410	8.06055	1.05447
7.75750	−0.26415	8.07070	1.25816
7.77490	−0.15731	8.07360	1.53412
7.77925	−0.05225	8.19250	2.03683

PROBLEMS

8.40 Refer to the data set of Prob. 1.6. Test the data for normality by constructing a Q-Q plot of the data. Comment on whether or not the points $(z_j, x_{(j)})$ appear to lie along a line. Begin by constructing a table similar to Table 8.15. Then plot the points $(z_j, x_{(j)})$, $j = 1, \ldots, 17$.

8.41 Refer to the lead absorption data set of Prob. 1.12. Test the first column (the exposed group) for normality by constructing a Q-Q plot of the data. Comment on whether or not the points $(z_j, x_{(j)})$ appear to lie along a line. Begin by constructing a table similar to Table 8.15. Then plot the points $(z_j, x_{(j)})$, $j = 1, \ldots, 33$.

8.42 Refer to the lead absorption data set of Prob. 1.12. Test the second column (control) for normality by constructing a Q-Q plot of the data. Comment on whether or not the points $(z_j, x_{(j)})$ appear to lie along a line. Begin by constructing a table similar to Table 8.15. Then plot the points $(z_j, x_{(j)})$, $j = 1, \ldots, 33$.

8.43 Refer to the lead absorption data set of Prob. 1.12. The paired-sample t test assumes that the data in the difference column are normally distributed. Test the third column (difference) for normality by constructing a Q-Q plot of the data. Comment on whether or not it is reasonable to assume that the data come from a normal distribution.

8.44 Refer to the radiation data of Table 1.3. The histogram (Fig. 1.8) suggests that these data are not normally distributed. Test for normality by constructing a Q-Q plot of the data. Summarize your conclusions in a brief sentence or two.

8.45 Refer to the cadmium levels data of Prob. 1.13. Test the data for normality by constructing a Q-Q plot of the data and state your conclusions.

8.46 Refer to the capacitor data set of Prob. 1.14. Test the data for normality by constructing a Q-Q plot of the data and state your conclusions.

8.5 CHAPTER SUMMARY

Hypothesis testing (or significance testing) is a statistical method, based on the careful collection of experimental data, for evaluating the plausibility of a scientific model or for comparing two conflicting models. A statistical hypothesis is a claim about the distribution of a variable X defined on a population; it is called (somewhat arbitrarily) the null hypothesis. The denial of the claim is also a statistical hypothesis; it is called the alternative hypothesis. When the distribution of X is normal, the null hypothesis is usually stated in the form of an equality between the mean μ and a particular value μ_0 called the null value. That is, we assume H_0: $\mu = \mu_0$. In this chapter we studied primarily the following three types of alternative hypotheses:

$$H_1\colon \mu > \mu_0 \quad \text{(Example 8.2)}$$
$$H_1\colon \mu < \mu_0 \quad \text{(Example 8.4)}$$
$$H_1\colon \mu \neq \mu_0 \quad \text{(Example 8.5)}$$

The alternative hypothesis H_1: $\mu > \mu_0$ or H_1: $\mu < \mu_0$ is said to be one-sided since it consists of those values of μ that deviate from μ_0 in one direction only. The alternative hypothesis H_1: $\mu \neq \mu_0$ is said to be two-sided since it consists of those values of μ that deviate from μ_0 in either of two directions, that is, either $\mu > \mu_0$ or $\mu < \mu_0$.

A test of a hypothesis is a procedure based on the value of a statistic, usually the sample mean $\overline{X}$, for accepting or rejecting H_0. A type I error occurs when one rejects H_0 when H_0 is true. A type II error occurs when one accepts H_0 when H_0 is false. The performance of the test is measured by how well it controls the type I and type II error probabilities. A good test will reject H_0 with small probability when it is true and will reject H_0 with high probability when it is false. An important tool for analyzing the performance of a test is the power function, which is the probability of rejecting H_0 as a function of the parameter μ.

To Probe Further. There is no "one size fits all" approach to statistical inference. The confidence interval approach to hypothesis testing is favored by a growing number of practitioners of statistics, including G. E. P. Box, W. G. Hunter, and J. S. Hunter (*Statistics for Experimenters*, New York, John Wiley & Sons, 1978), who write, "Significance testing in general has been a greatly overworked procedure, and in many cases where significance statements have been made it would have been better to provide an interval within which the value of the parameter would be expected to lie." The Wilcoxon rank sum test is an example of a *distribution-free* or *nonparametric* statistical method. A useful introduction to and survey of these techniques is given in R. H. Randles and D. A. Wolfe, *Introduction to the Theory of Nonparametric Statistics* (New York, John Wiley & Sons, 1979).

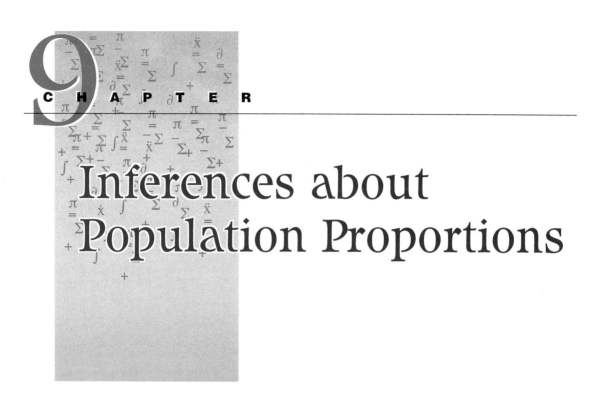

Inferences about Population Proportions

In God we trust; all others bring data.

Statistician's motto

9.1 ORIENTATION

The effectiveness of a medical procedure is usually determined by the number of patients whose symptoms are relieved by the treatment. Because treatments using powerful drugs may have severe, if not fatal, side effects, the null hypothesis is that the treatment is ineffective; that is, it is up to the drug manufacturer to demonstrate the effectiveness of its drug to the Food and Drug Administration (FDA). It is known, for example, that approximately 25 percent of babies born to HIV-infected mothers are themselves infected with the virus. There was therefore justifiable excitement when the National Institute of Allergy and Infectious Diseases announced (22 Feb 1994) that giving the drug AZT to infected mothers reduced the infection rate to 8 percent—a threefold reduction in the rate of infection! To put it another way, approximately 75 percent of babies born to HIV-infected mothers do not have the HIV virus, but for those mothers treated with AZT the percentage increased to 92 percent. The results were regarded as so significant that clinics participating in the trials were advised to offer AZT to the women and babies in the control group.

We now place these results in the context of hypothesis testing. Suppose the natural recovery rate from the disease is p_0, and let p denote the recovery rate

327

due to the treatment. Then the null hypothesis and its alternative will be of the form

$$H_0: p = p_0 \quad \text{against} \quad H_1: p > p_0$$

We test this hypothesis by using the sample total X, where X is the number of patients out of n whose condition is improved as a result of the treatment. Under reasonable assumptions the null distribution of X is binomial with parameters n, p_0, so that evaluating the results of a clinical trial is equivalent to making inferences about the parameter p of a binomial distribution. In the clinical trials of AZT, for instance, we have $\hat{p} = 0.92 > 0.75 = p_0$, so the evidence against the null hypothesis appears to be quite strong.

In this chapter we apply the general concepts of statistical inference as developed in the previous chapter to testing hypotheses concerning the parameters of one or more multinomial distributions—the binomial distribution being a special case.

Organization of Chapter

9.2   TESTS CONCERNING THE PARAMETER p OF A BINOMIAL DISTRIBUTION

Suppose we wish to make inferences concerning the (unknown) proportion of defective items produced by some manufacturing process or the proportion of patients who respond favorably to a new medical procedure. If the ith item is defective, we set the variable $X_i = 1$ and set $X_i = 0$ otherwise. Thus, the sample total $X = \sum_{1 \le i \le n} X_i$ counts the total number of defective items. The sample proportion $\hat{p} = X/n$ is an estimate of the proportion of defective items. If one can assume that conditions on the production line do not depend on the time at which the ith item is produced, then we can regard this as an example of a Bernoulli trial process (Definition 5.4); that is, it is an example of n independent repetitions of an experiment with only two outcomes, denoted by 0 and 1, respectively. Consequently, the sample total has a binomial distribution with parameters $n, p = P(X_i = 1)$. It is for this reason that a Bernoulli trials process is also called a *binomial experiment*. More formally, we can also regard $X_1, \ldots, X_n$ as a random sample taken from the Bernoulli distribution:

$$f(x; p) = p^x (1 - p)^{1-x}, \quad x = 0, 1 \quad \text{with parameter space } \Theta = \{p : 0 \le p \le 1\}$$

One-Sided Upper-Tailed Test of a Population Proportion

We suppose that our data come from a random sample $X_1, \ldots, X_n$ of size n taken from a Bernoulli distribution $f(x; p)$. We are interested in testing

$$H_0: p = p_0 \quad \text{against} \quad H_1: p > p_0 \tag{9.1}$$

Intuitively, a large value of X provides evidence against the null hypothesis that the drug is ineffective. This suggests that we choose our rejection region to be the set $\mathcal{C} = \{x : x > c\}$.

We have thus arrived at the following upper-tailed test of the hypothesis (9.1):

$$\text{accept } H_0 \text{ if } X \le c \tag{9.2}$$

$$\text{reject } H_0 \text{ if } X > c \tag{9.3}$$

The power function of this test is given by

$$\pi(p) = P(X > c \mid p) = \sum_{c+1 \le x \le n} b(x; n, p) = 1 - B(c; n, p) \tag{9.4}$$

To compute the type I error probabilities, we use the fact that the power function $\pi(p)$ defined by Eq. (9.4) is an *increasing* function of p (Proposition 4.4). Theorem 9.1 is a useful adaptation of this fact to the context of hypothesis testing.

▅▅ THEOREM 9.1

Suppose our test statistic X is binomial with parameters n, p and the power function $\pi(p)$ is given by

$$\pi(p) = P(X > c \mid p) = \sum_{c+1 \le x \le n} b(x; n, p) = 1 - B(c; n, p)$$

Then $\pi(p)$ is an increasing function of p, so $p \le p_0$ implies $\pi(p) \le \pi(p_0)$. The significance level of the test is given by

$$\alpha = \pi(p_0) = P(X > c \mid p_0) = 1 - B(c; n, p_0) \tag{9.5}$$

Thus, to compute the significance level of the test we use Eq. (9.5) and the binomial tables for $B(x; n, p_0)$. ∎

We now put these formulas to good use, but first we make a few preliminary remarks. Testing a hypothesis about a population proportion is complicated by the fact that the computation of the power function, significance levels, and type I and type II errors depends on the sample size n; in particular, when $n \le 25$ one can use the table of binomial probabilities; on the other hand, when $n > 25$ one must use the normal approximation, which yields only approximate values for the type I and type II errors. Another complication arises from the fact that X has a discrete distribution, so it will not always be possible to construct a test with a prescribed significance level α.

Upper-Tailed Test of a Population Proportion (Small Sample)

example 9.1 Assume that the natural recovery rate for a certain disease is 40 percent, so $p_0 = 0.4$. A new drug is developed, and to test the manufacturer's claim that it boosts the recovery rate, a random sample of $n = 10$ patients, called the treatment group, is given the drug and the total number of recoveries X is recorded. Our problem is to find the cutoff value c that produces a test with significance level $\alpha \leq 0.05$.

Solution. According to Eq. (9.5) the test with rejection region $\mathscr{C} = \{x : x > c\}$ has significance level 0.05 when c satisfies the equation

$$0.05 = \pi(0.40) = 1 - B(c; 10, 0.40)$$

From the binomial tables we see that

$$B(6; 10, 0.40) = 0.945 < 0.95 < 0.988 = B(7; 10, 0.40)$$

The choice of $c = 7$ yields $\alpha = 1 - 0.988 = 0.012$, and the choice of $c = 6$ yields $\alpha = 1 - 0.945 = 0.055$. If we insist on a significance level $\alpha \leq 0.05$, then we choose $c = 7$. Clearly, an exact 5 percent level test does not exist.

The type II error probability at the alternative $p = 0.7$ equals

$$\beta(0.7) = P(X \leq 7 \,|\, p = 0.7) = 1 - B(2; 10, 0.3)$$

$$= 1 - 0.383 = 0.617$$

This is quite large because the sample size is small. ∎

Upper-Tailed Test of a Population Proportion (Large Sample)

We turn now to the problem of testing H_0: $p = p_0$ against H_1: $p > p_0$ when the sample size n is large enough so that the normal approximation is valid. The method is very similar to that just used except that it is more convenient to use the sample proportion $\hat{p} = X/n$ as our test statistic. It is intuitively clear that $\hat{p} > p_0$ provides evidence in favor of H_1, and this suggests, at least when the sample size is large enough, that we use a rejection region of the form

$$\mathscr{C} = \left\{ \hat{p} : \hat{p} > p_0 + z(\alpha)\sqrt{\frac{p_0(1 - p_0)}{n}} \right\} \tag{9.6}$$

to obtain an approximate level α test.

We can, of course, always use the standardized statistic

$$Z_0 = \frac{\hat{p} - p_0}{\sqrt{\dfrac{p_0(1 - p_0)}{n}}} \tag{9.7}$$

This yields the following large-sample, approximate level α test of H_0:

$$\text{accept } H_0 \text{ if } \frac{\hat{p} - p_0}{\sqrt{\dfrac{p_0(1 - p_0)}{n}}} \leq z(\alpha) \tag{9.8}$$

$$\text{reject } H_0 \text{ if } \frac{\hat{p} - p_0}{\sqrt{\dfrac{p_0(1 - p_0)}{n}}} > z(\alpha) \tag{9.9}$$

---

example 9.2   Consider the following modification of the clinical trials experiment described in Example 9.1: The natural recovery rate from the disease is 40 percent and the treatment group now consists of $n = 200$ patients, 90 of whom recovered, so $\hat{p} = 0.45$. The question is this: Can we conclude, at the 5 percent significance level, that the drug is effective?

Solution. Substituting $p_0 = 0.40, \hat{p} = 0.45, n = 200$ into the left-hand side of Eq. (9.9) yields the value

$$\frac{\hat{p} - p_0}{\sqrt{\dfrac{p_0(1 - p_0)}{n}}} = \frac{0.05}{\sqrt{0.4 \times 0.6/200}} = 1.44 < 1.645 = z(0.05)$$

Since $Z_0 = 1.44 < 1.645$, the null hypothesis is not rejected. We arrive at the same conclusion using Eq. (9.6) to construct the rejection region:

$$\{\hat{p} : \hat{p} > c\} = \{\hat{p} : \hat{p} > 0.457\}$$

Since $\hat{p} = 0.45 < 0.457$, we do not reject H_0.   ∎

The normal approximation to the binomial may also be used to approximate the power function and the type II error probability at the alternative p. The (approximate) power function and type II error probability for a one-sided test with rejection region $\{\hat{p} : \hat{p} > c\}$ are given by

$$\pi(p) \approx 1 - \Phi\left(\frac{c - p}{\sqrt{p(1 - p)/n}}\right) \tag{9.10}$$

$$\beta(p) \approx \Phi\left(\frac{c - p}{\sqrt{p(1 - p)/n}}\right) \tag{9.11}$$

Derivation of Eq. (9.10)

$$\pi(p) = P(\hat{p} > c \mid p)$$

$$= P\left(\frac{\hat{p} - p}{\sqrt{p(1 - p)/n}} > \frac{c - p}{\sqrt{p(1 - p)/n}}\right)$$

$$\approx 1 - \Phi\left(\frac{c - p}{\sqrt{p(1 - p)/n}}\right)$$

example 9.3

Compute (1) the type II error at the alternative $p = 0.5$ and (2) the P-value of the observed test statistic $\hat{p} = 0.45$ of Example 9.2.

Solution

1. The type II error probability at the alternative $p = 0.5$ is obtained by substituting $p = 0.5, n = 200, c = 0.457$ into Eq. (9.11); thus,

$$\beta(0.5) = \Phi\left(\frac{0.4582 - 0.5}{\sqrt{0.5 \times 0.5/200}}\right) = \Phi(-1.22) = 0.1112$$

2. We recall that the P-value is the probability that the sample proportion $\hat{p}$ would be as large as the observed value $\hat{p} = 0.45$ if H_0 is true. We obtain the P-value by setting $c = \hat{p} = 0.45$, $p = 0.4$, and $n = 200$ in Eq. (9.10); thus,

$$P(\hat{p} > 0.45 \mid p_0 = 0.4) = \pi(0.45) = 1 - \Phi(1.44) = 0.0749$$

Consequently, the P-value $= 0.0749$. ∎

Table 9.1 gives the level α tests and the corresponding rejection regions for the one-sided and two-sided tests based on the standardized test statistic $Z_0 = (\hat{p} - p_0)/(\sqrt{p_0(1 - p_0)/n})$.

TABLE 9.1

Alternative hypothesis	Rejection region
$H_1: p < p_0$	$\dfrac{\hat{p} - p_0}{\sqrt{\dfrac{p_0(1 - p_0)}{n}}} < -z(\alpha)$
$H_1: p > p_0$	$\dfrac{\hat{p} - p_0}{\sqrt{\dfrac{p_0(1 - p_0)}{n}}} > z(\alpha)$
$H_1: p \neq p_0$	$\left\lvert \dfrac{\hat{p} - p_0}{\sqrt{\dfrac{p_0(1 - p_0)}{n}}} \right\rvert > z(\alpha/2)$

Controlling the Type II Error Probabilities. In Sec. 8.2.2 we showed how to choose the sample size n so that the corresponding level α test has type II error probability β at a specified alternative [Eqs.(8.21) and (8.23)]. Using similar methods here one can derive the following equations for the sample size n that has type II error probability β at the alternative p.

$$n = \left[\frac{z(\alpha)\sqrt{p_0(1 - p_0)} + z(\beta)\sqrt{p(1 - p)}}{p - p_0}\right]^2 \quad \text{(one-sided test)} \quad (9.12)$$

$$n = \left(\frac{z(\alpha/2) \sqrt{p_0(1 - p_0)} + z(\beta) \sqrt{p(1 - p)}}{p - p_0} \right)^2 \quad \text{(two-sided test)} \quad (9.13)$$

example 9.4 Suppose we have to test H_0: $p \leq 0.1$ against H_1: $p > 0.1$ at the level $\alpha = 0.01$. What is the minimum sample size needed so that the type II error probability at the alternative $p = 0.25$ is $\beta = 0.1$?

Solution. Since this is a one-sided test, we use Eq. (9.12) and the given data, $z(0.01) = 2.33, p_0 = 0.1, p = 0.25, z(0.1) = 1.28$. The result is $n = 8.355^2$, which, rounded up to the nearest integer, yields $n = 70$. ∎

9.2.1 Tests of Hypotheses Concerning Two Binomial Distributions (Large Sample)

Suppose we are interested in testing the null hypothesis that the proportion of defective items produced by two different manufacturing processes is the same. Denoting the corresponding proportions by p_1 and p_2, we are interested in testing H_0: $p_1 = p_2$ against H_1: $p_1 \neq p_2$.

When the samples are taken from two independent populations, it follows that the sample proportions $\hat{p}_1$ and $\hat{p}_2$ are independent and approximately normal; consequently, $\hat{p}_1 - \hat{p}_2$ is also approximately normal, with variance $V(\hat{p}_1 - \hat{p}_2)$, standard error $\sigma(\hat{p}_1 - \hat{p}_2)$, and estimated standard error $s(\hat{p}_1 - \hat{p}_2)$ given by

$$V(\hat{p}_1 - \hat{p}_2) = \frac{p_1(1 - p_1)}{n_1} + \frac{p_2(1 - p_2)}{n_2}$$

$$\sigma(\hat{p}_1 - \hat{p}_2) = \sqrt{\frac{p_1(1 - p_1)}{n_1} + \frac{p_2(1 - p_2)}{n_2}} \quad (9.14)$$

$$s(\hat{p}_1 - \hat{p}_2) = \sqrt{\frac{\hat{p}_1(1 - \hat{p}_1)}{n_1} + \frac{\hat{p}_2(1 - \hat{p}_2)}{n_2}}$$

To test this hypothesis, we use the confidence interval approach; that is, we construct the $100(1 - \alpha)$ percent two-sided confidence interval $[L, U]$ for $p_1 - p_2$, where L and U are given by (see Sec. 7.5.1)

$$L = \hat{p}_1 - \hat{p}_2 - z(\alpha/2)s(\hat{p}_1 - \hat{p}_2)$$
$$U = \hat{p}_1 - \hat{p}_2 + z(\alpha/2)s(\hat{p}_1 - \hat{p}_2)$$

We reject the null hypothesis if 0 is not in $[L, U]$ and we accept it otherwise.

example 9.5 This is a continuation of Example 7.13, in which the following data were collected: Of 400 items produced by machine 1, 16 were defective; of 600 produced by machine 2, 60 were defective. Test the the null hypothesis that there is no difference in the proportion of defectives produced by the two machines; use $\alpha = 0.05$.

Solution. From the data, we compute

$$\hat{p}_1 - \hat{p}_2 = -0.06$$

$$s(\hat{p}_1 - \hat{p}_2) = \sqrt{\frac{0.04 \times 0.96}{400} + \frac{0.1 \times 0.9}{600}} = 0.0157$$

$$L = -0.06 - 1.96 \times 0.0157 = -0.0908$$

$$U = -0.06 + 1.96 \times 0.0157 = -0.0292$$

Since 0 is not in the confidence interval $[-0.0908, -0.0292]$, we reject the null hypothesis. ∎

The two-sided and one-sided tests can also be expressed in terms of the standardized test statistic $Z_1 = (\hat{p}_1 - \hat{p}_2)/s(\hat{p}_1 - \hat{p}_2)$. The results are given in Table 9.2.

TABLE 9.2

Alternative hypothesis	Rejection region
$H_1: p_1 < p_2$	$\dfrac{\hat{p}_1 - \hat{p}_2}{s(\hat{p}_1 - \hat{p}_2)} < -z(\alpha)$
$H_1: p_1 > p_2$	$\dfrac{\hat{p}_1 - \hat{p}_2}{s(\hat{p}_1 - \hat{p}_2)} > z(\alpha)$
$H_1: p_1 \neq p_2$	$\left\lvert \dfrac{\hat{p}_1 - \hat{p}_2}{s(\hat{p}_1 - \hat{p}_2)} \right\rvert > z(\alpha/2)$

PROBLEMS

9.1 Let X be the sample total from a random sample of size $n = 10$ from a Bernoulli distribution with parameter p. To test H_0: $p = 0.25$ against H_1: $p > 0.25$, we choose the rejection region $\mathscr{C} = \{x : x > c\}$.
   (a) What is the significance level of this test for $c = 4$? Determine the type II error if $p = 0.30, p = 0.50$.
   (b) What is the significance level of this test for $c = 5$? Determine the type II error if $p = 0.30, p = 0.50$.

9.2 Let X be the sample total from a random sample of size $n = 20$ from a Bernoulli distribution with parameter p. To test H_0: $p = 0.40$ against H_1: $p > 0.40$, we choose the rejection region $\mathscr{C} = \{x : x > c\}$.
   (a) Find the value of c such that the type I error probability $\alpha \leq 0.05$. What is the type II error if, in fact, $p = 0.7$?
   (b) Find the value of c such that the type I error probability $\alpha \leq 0.1$. What is the type II error if, in fact, $p = 0.7$?

9.3 Let X be the sample total from a random sample of size $n = 15$ taken from a Bernoulli distribution with parameter p. To test H_0: $p = 0.90$ against H_1: $p < 0.90$, we choose the rejection region $\mathscr{C} = \{x : x < c\}$, where the value of c is to be determined. How should we choose c so that the type I error probability $\alpha \leq 0.05$? Determine the type II error if, in fact, $p = 0.60$. [*Hint:* To compute $B(x; n, p)$, $0.5 < p < 1$, we use Eq. (3.40): $B(x; n, p) = 1 - B(n - 1 - x; n, 1 - p)$.]

9.4 An urn contains two red balls, two white balls, and a fifth ball that is either red or white. Let p denote the probability of drawing a red ball. The null hypothesis is that the fifth ball is red.
(a) Describe the parameter space Θ.
(b) To test this hypothesis we draw 10 balls at random, one at a time (with replacement), and let X equal the number of red balls drawn. Using a rejection region of the form $\mathscr{C} = \{X < c\}$, compute the power function $\pi(p)$.
(c) Use the result of part (*b*) to compute the type I and type II errors when $c = 3$.

9.5 Suppose the natural recovery rate from a disease is $p_0 = 0.4$. A new drug is developed that, it is claimed, significantly improves the recovery rate. Consequently, the null hypothesis and its alternative are given by

$$H_0: p = 0.4 \quad \text{against} \quad H_1: p > 0.4$$

Consider the following test: A random sample of 15 patients are given the drug, and the number of recoveries is denoted by X. The null hypothesis will be rejected if $X > 8$; i.e.,

$$\text{accept } H_0 \text{ if } X \leq 8$$
$$\text{reject } H_0 \text{ if } X > 8$$

(a) Derive the formula for the power function $\pi(p)$ of this test.
(b) Compute the significance level of the test.
(c) Compute the type II error β if, in fact, the drug boosts the recovery rate to 0.6.

9.6 A garden supply company claims that its flower bulbs have an 80 percent germination rate. To challenge this claim, a consumer protection agency decides to test the following hypothesis:

$$H_0: p = 0.80 \quad \text{against} \quad H_1: p < 0.80$$

The following data were collected: $n = 100$ flower bulbs were planted, and 77 of them germinated. Compute the P-value of your test statistic and state whether or not it is significant at the 5 percent level.

9.7 Before proposition A was put to the voters for approval, a random sample of 1300 voters produced the following data: 625 in favor of proposition A and 675 opposed. Test the null hypothesis that a majority of the voters will approve proposition A. Use $\alpha = 0.05$. What is the alternative hypothesis? What is the P-value of your test statistic?

9.8 Suppose that we have to test $H_0: p = 0.2$ against $H_1: p > 0.2$ at the level $\alpha = 0.05$. What is the minimum sample size needed so that the type II error at the alternative $p = 0.3$ is 0.1?

9.9 The natural recovery rate from a disease is 75 percent. A new drug is developed that, according to its manufacturer, increases the recovery rate to 90 percent. Suppose we test this hypothesis at the level $\alpha = 0.01$. What is the minimum sample size needed to detect this improvement with a probability of at least 0.90?

9.10 In a comparison study of two machines for manufacturing memory chips, the following data were collected: Of 500 chips produced by machine 1, 20 were defective; of 600 chips produced by machine 2, 40 were defective. Can you conclude from these results that there is no significant difference (at the level $\alpha = 0.05$) between the proportions of defective items produced by the respective machines? What is the P-value of your test statistic?

9.11 A random sample of 250 items from lot A contains 10 defective items, and a random sample of 300 items from lot B is found to contain 18 defective items. What conclusions can you draw concerning the quality of the two lots?

9.12 The number of defective items out of a total production of 1125 is 28. The manufacturing process is then modified, and the number of defective items out of a total production of 1250 is 22. Can you reasonably conclude that the modifications reduced the proportion of defective items?

9.3   CHI-SQUARE TEST

The chi-square test is used to test hypotheses about the parameters of one or more multinomial distributions. Recall that a multinomial experiment is the mathematical model of n repetitions of an experiment with k mutually exclusive and exhaustive outcomes, denoted $C_1, \ldots, C_k$, with probabilities $p_i = P(C_i)$, where $\sum_i p_i = 1$. The outcomes C_i are called *categories* or *cells*; their probabilities $p_i = P(C_i)$ are called *cell probabilities*. In the simplest case the cell probabilities are completely specified, so the null hypothesis is given by

$$H_0: P(C_i) = p_i, \quad i = 1, \ldots, k \tag{9.15}$$

We test the null hypothesis (9.15) by carrying out n independent repetitions of the experiment and then recording the observed frequencies $Y_1, \ldots, Y_k$ of each category. In Sec. 5.3 we showed that the random variables $Y_1, \ldots, Y_k$ have the multinomial distribution defined by

$$P(Y_1 = y_1, \ldots, Y_k = y_k) = \binom{n}{y_1, \ldots, y_k} p_1^{y_1} \cdots p_k^{y_k}$$

$$= \frac{n!}{y_1! \cdots y_k!} p_1^{y_1} \cdots p_k^{y_k}$$

Consequently, the cell frequency Y_i has a binomial distribution with parameters n, p_i; therefore, the *expected frequency* of the ith category is given by $E(Y_i) = np_i$.

To test the null hypothesis that the observed data come from a multinomial distribution with given cell probabilities $P(C_i) = p_i, i = 1, \ldots, k$, we use the following important theorem (we omit the proof) due to the statistician K. Pearson (1857–1936).

▓ THEOREM 9.2

Let $Y_1, \dots, Y_k$ have a multinomial distribution with parameters $n, p_1, \dots, p_k$. Then the distribution of the random variable χ^2, defined by

$$\chi^2 = \sum_i \frac{(Y_i - np_i)^2}{np_i} \tag{9.16}$$

has an approximate chi-square distribution with $k - 1$ degrees of freedom. The accuracy of the approximation improves as n increases. ∎

Looking at Eq. (9.16), we see that a large difference between the observed and expected frequencies makes χ^2 large. Thus, a large χ^2 value indicates a poor fit between the model and the data. This leads directly to the χ^2 *goodness-of-fit* test.

Pearson's χ^2 Goodness-of-Fit Test (Large Sample)

1. Null hypothesis: H_0: $P(C_i) = p_i$, $i = 1, \dots, k$. Sample size: n.
2. Record the observed frequencies Y_i and compute the expected cell frequencies np_i. Then compute the χ^2 test statistic:

$$\chi^2 = \sum_{1 \le i \le k} \frac{(Y_i - np_i)^2}{np_i}$$

3. Reject H_0 if $\chi^2 > \chi^2_{k-1}(\alpha)$, where k is the number of cells and α is the significance level of the test.

example 9.6

One of the most famous multinomial experiments in the history of science was performed by G. Mendel, who classified $n = 556$ peas according to two traits: shape (round or wrinkled) and color (yellow or green). Each seed was put into one of $k = 4$ categories: $C_1 = ry$, $C_2 = rg$, $C_3 = wy$, $C_4 = wg$, where $r, w, g,$ and y denote *round, wrinkled, green,* and *yellow,* respectively. The results of Mendel's breeding experiment are displayed in Table 9.3.

TABLE 9.3 MENDEL'S DATA

Seed type	Observed frequency
ry	315
wy	101
rg	108
wg	32
	$n = 556$

Mendel's theory predicted that the frequency counts of the seed types produced from his experiment should occur in the ratio $9 : 3 : 3 : 1$. Thus, the ratio of the number of *ry* peas to *rg* peas should be 9 : 3, and similarly for the other ratios. We can rewrite these proportions in terms of probabilities as follows:

$$p_1 = P(ry) = \frac{9}{16}, \qquad p_2 = P(rg) = \frac{3}{16}$$

$$p_3 = P(wy) = \frac{3}{16}, \qquad p_4 = P(wg) = \frac{1}{16}$$

The null hypothesis corresponding to Mendel's theory is therefore

$$H_0: p_1 = \frac{9}{16}, p_2 = \frac{3}{16}, p_3 = \frac{3}{16}, p_4 = \frac{1}{16}$$

Do Mendel's theoretical predictions provide a good fit to the data?

Solution. We test Mendel's predictions at the level $\alpha = 0.01$ by using the χ^2 test. Looking at Table 9.3, we see that the observed frequencies are $Y_1 = 315$, $Y_2 = 101$, $Y_3 = 108$, $Y_4 = 32$, and the sample size equals sum of the cell frequencies: $n = 315 + 101 + 108 + 32 = 556$. Table 9.4 displays all the data needed to compute the χ^2 test statistic. The second, third, and fourth columns list the observed frequencies, cell probabilities, and expected cell frequencies, respectively.

TABLE 9.4 CELL PROBABILITIES AND EXPECTED CELL FREQUENCIES FOR MENDEL'S DATA

Seed type	Observed frequency	Cell probability	Expected frequency
ry	315	9/16	$556 \times 9/16 = 312.75$
wy	101	3/16	$556 \times 3/16 = 104.25$
rg	108	3/16	$556 \times 3/16 = 104.25$
wg	32	1/16	$556 \times 1/16 = 34.75$

Computation of the χ^2 Statistic

$$\chi^2 = \frac{(315 - 312.75)^2}{312.75} + \frac{(101 - 104.25)^2}{104.25} + \frac{(108 - 104.25)^2}{104.25} + \frac{(32 - 34.75)^2}{34.75}$$

$$= 0.47$$

Since $k - 1 = 3$, it follows that $\chi^2 \approx \chi_3^2$. From the chi-square table (Table A.5), we see that $\chi_3^2(0.01) = 11.345 > \chi^2 = 0.47$; consequently, we do not reject the null hypothesis. ■

Some Remarks on the Chi-Square Approximation

The accuracy of the chi-square approximation depends on the values of the cell probabilities. The following conservative criteria are recommended.

1. To use this approximation, it is recommended that the expected cell frequency $np_i \geq 5$ for each i. This condition holds for Mendel's genetics data in Table 9.4.
2. Cells with expected frequencies less than 5 should be combined to meet this rule.
3. k is the number of cells *after* cells with small expected frequencies have been combined.

9.3.1 Chi-Square Test When the Cell Probabilities Are Not Completely Specified

In the previous examples the cell probabilities were completely specified. We now consider the case when the cell probabilities depend on one or more parameters that are estimated from the data.

---

example 9.7   A computer engineer conjectures that the number of packets X per unit of time arriving at a node in a computer network has a Poisson distribution with parameter λ, so $P(X = i) = e^{-\lambda}\lambda^i/i!$. To validate this model the number of arriving packets for each of 150 time intervals is recorded; the results are displayed in columns 1 and 2 of Table 9.5. Are these data consistent with the model assumption that X has a Poisson distribution?

Solution. Let C_i denote the event that i packets arrived in a time interval, and let f_i denote the number of times the event C_i occurred. Looking at Table 9.5, we see that C_0 occurred twice, so $f_0 = 2$; C_1 occurred 10 times, so $f_1 = 10$; and so forth. The total number of packet arrivals equals $600 = \sum_i if_i$. The maximum-likelihood estimator of the arrival rate λ equals the total number of arrivals

TABLE 9.5 NUMBER OF PACKET ARRIVALS TO A NODE
IN A COMPUTER NETWORK

No. of arrivals, i	Observed no. of time intervals with i arrivals, f_i	Expected no. of intervals with i arrivals, np_i	$i \times f_i$	Cell probability, p_i
0	2	2.74735	0	0.01832
1	10	10.9894	10	0.0733
2	25	21.9788	50	0.1465
3	25	29.3050	75	0.1954
4	39	39.3050	156	0.1954
5	15	23.4440	75	0.1563
6	14	15.6293	84	0.1042
7	13	8.9311	91	0.0595
8	5	4.4655	40	0.02977
9	1	1.9847	9	0.01323
≥ 10	1	1.2150	10	0.0081
Total	150		600	

divided by the total number of intervals, so $\hat{\lambda} = 600/150 = 4$. The estimated cell probabilities are given by

$$\hat{P}(X = i) = e^{-\hat{\lambda}}\hat{\lambda}^i/i! = e^{-4}4^i/i!, \quad 0 \leq i \leq 9$$
$$\hat{P}(X \geq 10) = 0.0081$$

and are displayed in the rightmost column of Table 9.5. Observe that the last entry in this column is $\hat{P}(X \geq 10) = 0.0081$ and not $\hat{P}(X = 10) = 0.00529$; this must be so, for otherwise the categories would not be exhaustive. The expected number of arrivals for the cell is computed by multiplying the cell probability by the sample size $n = 150$; these results are displayed in column 3.

Observe that the expected cell frequencies

$$np_0 = 2.75, \qquad np_8 = 4.47, \qquad np_9 = 1.98, \qquad np_{10} = 1.215$$

are each less than 5; we therefore combine the first two cells and then the last three cells so that each expected cell frequency satisfies the condition $np_i \geq 5$.

The expected frequencies and cell probabilities corresponding to the grouped data are displayed in Table 9.6.

TABLE 9.6 TABLE 9.5 AFTER GROUPING OF THE DATA

No. of arrivals	Observed no. of time intervals with i arrivals	Expected no. of intervals with i arrivals	Cell probability
≤ 1	12	13.7367	0.0916
2	25	21.9788	0.1465
3	25	29.305	0.1954
4	39	29.305	0.1954
5	15	23.444	0.1563
6	14	15.6293	0.1042
7	13	8.93105	0.0595
≥ 8	7	7.665	0.0511

Computing the χ^2 Statistic and the Degrees of Freedom (DF). In this example the cell probabilities and the expected cell frequencies are functions of the unknown parameter λ. Nevertheless, the χ^2 test is still applicable, *provided the unknown parameter is replaced by its maximum-likelihood estimate.* More precisely, we have the following theorem:

■ **THEOREM 9.3**

When the cell probabilities depend on one or more unknown parameters, the χ^2 test is still applicable provided the unknown parameters are replaced by their maximum-likelihood estimates and provided the degrees of freedom are reduced by the number of parameters estimated, i.e.,

$$\text{DF of } \chi^2 = k - 1 - (\text{no. of parameters estimated}) \qquad (9.17)$$

■

We compute the chi-square statistic exactly as before except that we use the grouped data in Table 9.6. This yields the result $\chi^2 = 9.597$. The degrees of freedom, however, must be reduced by the number of parameters estimated from the data (Theorem 9.3). In the present instance,

$$\text{DF of } \chi^2 = k - 1 - (\text{no. of parameters estimated}) = 8 - 1 - 1 = 6$$

So $\chi^2 \approx \chi_6^2$. Because $\chi^2 = 9.597 < 12.592 = \chi_6^2(0.05)$, we conclude that the Poisson distribution provides a reasonably good fit to the data. ∎

9.4 CONTINGENCY TABLES

The concept of a *contingency table* (also called a *two-way table*) is most easily understood in the context of a specific example.

example **9.8**   The frequency data in Table 9.7 were obtained in a study of the causes of failure of vacuum tubes (VTs) used in artillery shells during World War II. Each failure was classified according to two criteria or *attributes*: (1) *position* of the VT in the shell ($A_1 = $ top, $A_2 = $ bottom) and (2) *type of failure* (B_1, B_2, B_3). The position of a VT in the shell is an example of a *categorical variable*, with two categories; that is, the outcome of the experiment is a nonnumerical value, *top* or *bottom*, as opposed to the numerical variables we have been studying heretofore. The type of failure is also a categorical variable with three categories corresponding to the three different failure types. The two classification criteria when applied simultaneously produce a partition of the sample space into the $2 \times 3 = 6$ cells denoted by $A_i \cap B_j$ ($i = 1, 2; j = 1, 2, 3$). For the 115 shells that had a type B_1 failure, 75 VTs were in the top position and 40 were in the bottom position, so the cell frequency of $A_1 \cap B_1$ is 75; similarly, the cell frequency of $A_2 \cap B_1$ is 40; and so on. Summing the cell frequencies across the rows gives the *marginal row frequencies*, and summing the cell frequencies down the columns gives the *marginal column frequencies*. For instance, the number of VTs that failed in the top position (category A_1) is 100, and there were 115 VTs classified as type B_1 failures. Table 9.7 is an example of a 2×3 *contingency table*.

TABLE 9.7 FREQUENCY DATA ON THE CAUSES OF
FAILURE OF $n = 180$ VACUUM TUBES

Position in shell	Type of failure			
	B_1	B_2	B_3	Total
$A_1 = $ top	75	10	15	100
$A_2 = $ bottom	40	30	10	80
Total	115	40	25	180

Source: M. G. Natrella, *Experimental Statistics,* National Bureau of Standards Handbook 91, 1963.

∎

The General $r \times c$ Contingency Table

Table 9.7 consists of two rows and three columns because the categorical variables A (position) and B (failure type) have two and three categories, respectively. An $r \times c$ contingency table is a straightforward generalization; it consists of r rows and c columns that correspond to the r categories of the variable A and the c categories of the variable B. We denote the r categories of A by $A_1, \ldots, A_r$ and the c categories of B by $B_1, \ldots, B_c$. We let y_{ij} denote the observed frequency of the observations falling in the cell $A_i \cap B_j$. The resulting frequency counts are displayed in the $r \times c$ *contingency table* shown in Table 9.8. The marginal row and column frequencies are denoted by *dot subscript notation*, where a dot in place of the subscript i, say, indicates summation with respect to i.

$$\text{Dot subscript notation:} \quad y_{i.} = \sum_{1 \leq j \leq c} y_{ij}$$

$$y_{.j} = \sum_{1 \leq i \leq r} y_{ij}$$

TABLE 9.8 AN $r \times c$ CONTINGENCY TABLE

	B_1	$\ldots$	B_c	Total
A_1	y_{11}	y_{1j}	y_{1c}	$y_{1.}$
$\vdots$				$y_{i.}$
A_r	y_{r1}	$\ldots$	y_{rc}	$y_{r.}$
Total	$y_{.1}$	$y_{.j}$	$y_{.c}$	n

We use the data in a contingency table to test the hypothesis that the categorical variables A and B are independent. Looking at the data in Table 9.7, for example, we can ask whether the type of failure depends on the position of the vacuum tube in the shell. The null hypothesis is that the two variables are independent, that is,

$$H_0: P(A_i \cap B_j) = P(A_i)P(B_j), \quad i = 1, \ldots, r, \quad j = 1, \ldots, c \quad (9.18)$$

Estimating the Parameters $p_{i.}$ and $p_{.j}$

The marginal probabilities $P(A_i)$ and $P(B_j)$ are obtained from the joint probabilities $p_{ij} = P(A_i \cap B_j)$ by computing the row and column sums in the usual way. Thus,

$$P(A_i) = p_{i.} = \sum_{1 \leq j \leq c} p_{ij} \quad (9.19)$$

$$P(B_j) = p_{.j} = \sum_{1 \leq i \leq r} p_{ij} \quad (9.20)$$

Using this notation, the null hypothesis of independence can be written as

$$H_0: p_{ij} = p_{i.}p_{.j}, \quad i = 1, \ldots, r, \quad j = 1, \ldots, c \tag{9.21}$$

We estimate the probabilities $p_{i.} = P(A_i)$ and $p_{.j} = P(B_j)$ by computing their observed relative frequencies. These estimates are given by

$$\hat{p}_{i.} = \frac{y_{i.}}{n} \quad \text{and} \quad \hat{p}_{.j} = \frac{y_{.j}}{n} \tag{9.22}$$

Consequently, the estimated expected frequency count $\hat{e}_{ij}$ for the (i, j)th cell, when H_0 is true, is given by

$$\hat{e}_{ij} = n\hat{p}_{i.}\hat{p}_{.j} = \frac{y_{i.}y_{.j}}{n}$$

We note that the marginal probabilities satisfy the equations

$$\sum_{1 \leq i \leq r} p_{i.} = 1 \quad \text{and} \quad \sum_{1 \leq j \leq c} p_{.j} = 1 \tag{9.23}$$

This means that we have to estimate $r - 1 + c - 1 = r + c - 2$ parameters from the data. The number of degrees of freedom of χ^2 is given by [see Theorem 9.3 and Eq. (9.17)]

$$\text{DF of } \chi^2 = rc - 1 - (r + c - 2) = (r - 1)(c - 1) \tag{9.24}$$

Consequently, χ^2 has an approximate chi-square distribution with $(r - 1)(c - 1)$ degrees of freedom.

χ^2 Test of Independence

1. Null hypothesis: $H_0: p_{ij} = p_{i.}p_{.j}$ $(i = 1, \ldots, r; j = 1, \ldots, c)$. First, compute the ith row sum $y_{i.}$ and the jth column sum $y_{.j}$. Then compute the estimated expected frequency of the i, jth cell using the formula $\hat{e}_{ij} = y_{i.}y_{.j}/n$.
2. Compute the test statistic χ^2:

$$\chi^2 = \sum_{1 \leq i \leq r} \sum_{1 \leq j \leq c} \frac{(y_{ij} - \hat{e}_{ij})^2}{\hat{e}_{ij}}$$

3. Reject H_0 if $\chi^2 > \chi^2_{(r-1)(c-1)}(\alpha)$, where α is the significance level of the test.

---

example 9.9

This is a continuation of Example 9.8. Consider the following statistical problem arising from the data in Table 9.7: Is the type of VT failure independent of the position of the tube in the shell?

Solution. The null hypothesis is that the VT position and its failure type are mutually independent, so we use the χ^2 test of independence. The observed and estimated expected frequencies derived from Table 9.7 are displayed in Table 9.9. The expected frequencies (or expected counts) are printed directly below the observed frequencies.

TABLE 9.9 OBSERVED AND EXPECTED FREQUENCIES FOR THE 2×3 CONTINGENCY TABLE 9.7 (MINITAB OUTPUT)

Expected counts are printed below observed counts

	B1	B2	B3	Total
1	75	10	15	100
	63.89	22.22	13.89	
2	40	30	10	80
	51.11	17.78	11.11	
Total	115	40	25	180

ChiSq = 1.932 + 6.722 + 0.089 +
        2.415 + 8.403 + 0.111 = 19.673

df = 2

We illustrate the method for computing the expected frequencies by showing that $\hat{e}_{11} = 63.89$. From Table 9.7 we have

$$y_{1.} = 100 \quad \text{and} \quad y_{.1} = 115$$

so

$$\frac{y_{1.}y_{.1}}{180} = 63.89$$

The MINITAB output tells us that $\chi^2 = 19.673$ and that the number of degrees of freedom (df) is 2. This follows from Eq. (9.24), which states that $(r-1)(c-1) = 1 \times 2 = 2$. Thus $\chi_2^2 = 19.673 > 9.21 = \chi_2^2(0.01)$; in other words, the type of failure and position in the tube are not independent. ∎

χ^2 *Test of Homogeneity*

In the previous example we considered a single population, the elements of which were classified according to two different criteria or attributes; we now consider the situation where the rows of the contingency table consist of independent samples taken from r different populations, with each population partitioned into c categories. We are, in effect, comparing two or more multinomial distributions since each row of the contingency table corresponds to a random sample taken from a multinomial distribution.

example 9.10   The data in Table 9.10 were obtained in a study of the relationship between smoking and emphysema. The patients were classified according to two criteria: the number of packs of cigarettes smoked per day (A) and the severity of the patient's emphysema (B). The categorical variable A has three levels: A_1, A_2, and A_3, where A_1 denotes those patients who smoked less than one pack a day, A_2 denotes those patients who smoked between one and two packs a day, and A_3 denotes those patients who smoked more than two packs a day. In this experiment 100 patients were randomly selected from category A_1, 200 patients were randomly selected from category A_2, and 100 patients were randomly selected from category A_3. The categorical variable B has four levels denoted by B_1, B_2, B_3, and B_4, where B_1 denotes the mildest form and B_4 denotes the most severe form of emphysema. The statistical problem is to determine if there is a link between the number of packs of cigarettes smoked per day and the severity of the emphysema. More precisely, are the proportions of patients in the four emphysema categories the same (homogeneous) for the three classes of smokers?

TABLE 9.10   CIGARETTE SMOKING AND EMPHYSEMA DATA

| No. of packs of cigarettes smoked | Severity of emphysema | | | | |
	B_1	B_2	B_3	B_4	Total
A_1	41	28	25	6	$n_1 = 100$
A_2	24	116	46	14	$n_2 = 200$
A_3	4	50	34	12	$n_3 = 300$
Total	69	194	105	132	$n = 400$

Source: P. G. Hoel, *Introduction to Mathematical Statistics,* 5th ed., New York, John Wiley & Sons, 1984. Used with permission.

Solution. In this example, the sample size for each level of the categorical variable A is predetermined by the researcher, which is why this is called a contingency table with one margin fixed. We denote the sample size for the ith population by n_i; thus, $n_1 = 100$, $n_2 = 200$, $n_3 = 100$, and the total sample size $n = n_1 + n_2 + n_3 = 400$. The four column totals, 69, 194, 105, and 132, are the cell frequencies of the four categories $B_1, \ldots, B_4$.

   To proceed we first formulate the null hypothesis. Denote by p_{ij} the probability that an observation from the ith population falls into category B_j. The null hypothesis of homogeneity implies that p_{ij} depends not on the population (equivalently, it does not depend on i), but only on the category B_j. Our problem, then, is to test the hypothesis

$$H_0: p_{1j} = p_{2j} = p_{3j} = p_j \quad \text{for } j = 1, 2, 3, 4$$

Here p_j denotes the probability that a smoker with emphysema falls into severity category B_j. The null hypothesis is that this probability does not depend on the number of packs of cigarettes smoked per day. A hypothesis-testing problem of this type is solved by the χ^2 *test of homogeneity*, which will now be explained.

1. First estimate the parameters p_j by computing the cell frequency of each category B_j and then dividing by the sample size:

$$\hat{p}_1 = \frac{69}{400}, \quad \hat{p}_2 = \frac{194}{400}, \quad \hat{p}_3 = \frac{105}{400}, \quad \hat{p}_4 = \frac{132}{400}$$

2. Next compute the expected cell frequencies via the formula

$$\hat{e}_{ij} = n_i \hat{p}_j = \frac{n_i y_{.j}}{n} = \frac{i\text{th row sum} \times j\text{th column sum}}{n}$$

The observed and expected cell frequencies are displayed in Table 9.11.

3. Compute the test statistic χ^2:

$$\chi^2 = \sum_{1 \le i \le r} \sum_{1 \le j \le c} \frac{(y_{ij} - \hat{e}_{ij})^2}{\hat{e}_{ij}}$$

With reference to the data in Table 9.10, we obtain

$$\chi^2 = \frac{(41 - 17.25)^2}{17.25} + \cdots + \frac{(12 - 8)^2}{8} = 64.408$$

4. Compute the number of degrees of freedom via Eq. (9.24) as before: DF $=$ $(r - 1)(c - 1) = 2 \times 3 = 6$. Since $\chi^2 = 64.408 > 12.59 = \chi^2_6(0.05)$, the null hypothesis is rejected. [*Note:* The assertion that DF $= (r - 1)(c - 1)$ will be justified next.]

TABLE 9.11 MINITAB OUTPUT: EXPECTED FREQUENCY COUNTS FOR THE CONTINGENCY TABLE 9.10.

Expected counts are printed below observed counts

	B1	B2	B3	B4	Total
1	41	28	25	6	100
	17.25	48.50	26.25	8.00	
2	24	116	46	14	200
	34.50	97.00	52.50	16.00	
3	4	50	34	12	100
	17.25	48.50	26.25	8.00	
Total	69	194	105	32	400

```
ChiSq = 32.699 +   8.665 +   0.060 +   0.500 +
         3.196 +   3.722 +   0.805 +   0.250 +
        10.178 +   0.046 +   2.288 +   2.000 = 64.408
df = 6
```

The $r \times c$ Contingency Table for the χ^2 Test of Homogeneity: General Case. Before taking up the general case we make the following summary remarks. Table 9.10 is an example of a 3×4 contingency table with one margin fixed, because, as noted earlier, specifying the sample size for each population determines the marginal row frequencies *prior* to the experiment. In particular, each row of Table 9.10 represents the results of a random sample, with predetermined sample size, taken from a population with each observation classified into one of four categories. We note that the Table 9.10 is just a special case of the $r \times c$ contingency table, Table 9.12.

TABLE 9.12

	B_1	$\ldots$	B_c	**Total**
A_1	y_{11}	y_{1j}	y_{1c}	n_1
$\vdots$				n_i
A_r	y_{r1}	$\cdots$	y_{rc}	n_r
Total	$y_{.1}$	$y_{.j}$	$y_{.c}$	$n = \sum_i n_i$

The entry y_{ij} in the ith row represents the number of observations from the ith population that fall in category j, where $\sum_j y_{ij} = n_i$ is fixed. Denoting by p_{ij} the probability that an observation from the ith population falls into category j, the null hypothesis of homogeneity takes the following form:

$$H_0: \ p_{ij} = p_j \quad i = 1, \ldots, r, \quad j = 1, \ldots, c$$

Consequently, the cell frequencies $Y_{i1}, \ldots, Y_{ic}$ have a multinomial distribution with parameters $(p_1, \ldots, p_c; n_i)$, and therefore

$$\sum_{1 \le j \le c} \frac{(Y_{ij} - n_i p_j)^2}{n_i p_j} \approx \chi^2_{c-1}$$

Thus, each row contributes $(c-1)$ degrees of freedom, and summing over the r rows contributes $r(c-1)$ degrees of freedom. This last assertion may be justified by noting that the row sums are mutually independent and approximately χ^2_{c-1} distributed; consequently, their sum has an approximate χ^2 distribution with degrees of freedom equal to their sum, which equals $(c-1) + \cdots + (c-1) = r(c-1)$ (Theorem 6.5). Finally, note that we estimate $(c-1)$ independent parameters since

$$\sum_{1 \le j \le c} \hat{p}_j = 1$$

Thus DF $= r(c-1) - (c-1) = (r-1)(c-1)$, as claimed.

PROBLEMS

9.13 We define a fair die to mean that each face is equally likely; thus, there are six categories, and the null hypothesis we wish to test is

$$H_0: \ p_i = 1/6, \quad i = 1, \ldots, 6$$

To determine if a die is fair, as opposed to the alternative that it is not, a die was thrown $n = 60$ times, with the results displayed in the following table. Complete columns 3 and 4, compute the χ^2 statistic, and test the null hypothesis at the level $\alpha = 0.05$.

Face	Observed frequency, y_i	Expected frequency, np_{i0}	$y_i - np_{i0}$
1	10		
2	9		
3	11		
4	11		
5	8		
6	11		

9.14 Let X denote the number of heads that appear when five coins are tossed. It is clear that X has a binomial distribution with $P(X = i) = b(i; 5, p)$, $i = 0, \ldots, 5$. The following data record the frequency distribution of the number of heads in $n = 40$ tosses of the five coins.

Number of heads	Observed frequency, y_i	Expected frequency, np_{i0}	$y_i - np_{i0}$
0	1		
1	17		
2	15		
3	10		
4	6		
5	1		

The null hypothesis is H_0: $p = 0.5$.

(a) Compute the expected cell frequencies, and complete columns 3 and 4.

(b) After combining cells with small expected frequencies, compute the χ^2 statistic and compute the number of degrees of freedom.

(c) Use the value of χ^2 obtained in part (b) to test H_0 at the 5 percent level of significance.

9.15 Show that the multinomial distribution reduces to the binomial distribution when $k = 2$. Here Y_1 and Y_2 denote the number of successes and failures, respectively; use the notation $p_1 = p, p_2 = 1 - p$.

9.16 The following 2×2 contingency table gives the pass/fail results for the boys and girls who took the advanced-level test in pure mathematics with statistics in Great Britain for the 1974–75 year.

	Pass	Fail	Total
Boys	891	569	1460
Girls	666	290	956
Total	1557	859	2416

Source: J. R. Green and D. Margerison, *Statistical Treatment of Experimental Data*, New York, Elsevier, 1978. Used with permission.

Use the data and the appropriate chi-square test to test the hypothesis that the passing rate does not depend on the sex of the candidate; use the level $\alpha = 0.01$.

9.17 A political consultant surveyed 275 urban, 250 suburban, and 100 rural voters and obtained the following data concerning the public policy issue that is the most important to them.

Voter	Health care	Crime	Economy	Taxes	Total
Urban	113	62	75	25	275
Suburban	88	39	75	50	250
Rural	60	10	20	10	100
Total	261	109	170	85	625

Do these data indicate any differences in the responses of the three classes of voters with respect to the importance of these four public policy issues?

9.18 To determine the possible effects of the length of a heating cycle on the brittleness of nylon bars, 400 bars were subjected to a 30-second heat treatment and 400 were subjected to a 90-second heat treatment. The nylon bars were then classified as brittle or nonbrittle.

Length of heating cycle (s)	Brittle	Nonbrittle	Total
30	77	323	400
60	177	223	400
Total	254	546	800

Source: A. J. Duncan, *Quality Control and Industrial Statistics,* 4th ed., Homewood, Ill., Richard D. Irwin, 1974.

Test the hypothesis that the brittleness of the nylon bars is independent of the heat treatment.

9.19 For each of 75 rockets fired, the lateral deflection and range in yards were measured, with the results displayed in the following 3×3 contingency table.

Range	Lateral deflection (yards)			Total
	−250 to −51	−50 to 49	50 to 199	
0–1199	5	9	7	21
1200–1799	7	3	9	19
1800–2699	8	21	6	35
Total	20	33	22	75

Source: NAVORD Report 3369 (1955); quoted in I. Guttman, S. S. Wilks, and J. S. Hunter, *Introductory Engineering Statistics,* 3d ed., New York, John Wiley & Sons, 1982.

Test at the 5 percent level of significance that lateral deflection and range are independent.

9.20 Each of 6805 pieces of molded vulcanite made from a resinous powder was classified according to two criteria: porosity and dimension. The results are displayed in the following 2×2 contingency table.

	Porous	Nonporous	Total
With defective dimensions	142	331	473
Without defective dimensions	1233	5099	6332
Total	1375	5430	6805

Source: A. Hald, *Statistical Theory with Engineering Applications*, New York, John Wiley & Sons, 1952.

Test the hypothesis that the two classification criteria are independent.

9.21 In a comparison study of two processes for manufacturing steel plates, each plate is classified into one of three categories: no defects, minor defects, major defects. The results of the study are displayed in the following 2×3 contingency table.

Process	No defects	Minor defects	Major defects
A	40	10	5
B	35	12	8

Test the hypothesis that there is no difference between the two processes; use $\alpha = 0.01$.

9.22 To test the effectiveness of drug treatment for the common cold, 164 patients with colds were selected; half were given the drug treatment (the treatment group), and the other half (the control group) were given a sugar pill. The patients' reactions to the treatment and sugar pill are recorded in the following table.

	Helped	Harmed	No effect	Total
Treatment	50	10	22	82
Control	42	12	28	82
Total	92	22	50	164

Source: P. G. Hoel, *Introduction to Mathematical Statistics*, 5th ed., New York, John Wiley & Sons, 1984. Used with permission.

Test the hypothesis that there is no difference between the treatment and control groups; use $\alpha = 0.10$.

9.5 CHAPTER SUMMARY

In this chapter we focused on the problem of testing hypotheses concerning the parameters of one or more multinomial distributions. We began with the important special case of the binomial distribution and then considered the problem of comparing two binomial distributions. We then introduced the χ^2 test and contingency tables as the basic tools for making inferences about multinomial distributions.

To Probe Further. We omitted a formula for the (approximate) power function for the tests listed in Table 9.2, since it requires more technical details than are suitable for a first course in statistics. Another omitted topic is Fisher's exact test, which concerns the critical region when one or both of the sample sizes are small (so that the central limit theorem cannot be applied). For further information on these topics, we recommend one of the standard references, such as A. Hald, *Statistical Theory with Engineering Applications* (New York, John Wiley & Sons, 1952, section 21.13).

10

Linear Regression and Correlation

Most real-life statistical problems have one or more nonstandard features. There are no routine statistical questions; only questionable statistical routines.

D. Cox, British statistician

10.1 ORIENTATION

Scientists are frequently interested in studying the functional relationship between two variables y and x. For example, in an agricultural experiment, the crop yield y may depend on the amount of fertilizer x; in a metallurgical experiment one is interested in the shear strength y of a spot weld as a function of the weld diameter x. The variable x is variously called the *explanatory variable*, the *regressor variable*, or the *predictor variable*; y is called the *dependent variable* or the *response variable*.

We assume that the relationship between the explanatory and response variables is given by a function $y = f(x)$, where f is unknown. In some cases the function is specified by a theoretical model. For example, Hooke's law asserts that the force required to stretch a spring by a distance x from its relaxed length is given by $y = \beta x$, where β is a constant that is characteristic of the spring. In this example the relationship between the variables is derived from the basic principles of mechanics. In most cases, however, the scientist is uncertain as to the exact form of the functional relationship $y = f(x)$ and proceeds to study it by performing a series of experiments and drawing the scatter plot of the observed values $(x_i, y_i), i = 1, \ldots, n$, where y_i is the value of the response corresponding to the chosen value x_i. The shape of the scatter plot provides some insight into the functional relationship between y and x. Regression analysis is a method for determining the function that best fits the observed data.

Organization of Chapter

10.2 METHOD OF LEAST SQUARES

The first step in fitting a straight line to a set of bivariate data is to construct the scatter plot; that is, plot the points $(x_i, y_i), i = 1, \ldots, n$, to see if the points appear to approximate a straight line. If they do not, the scientist must consider an alternative model for the data.

example **10.1** Consider the data in Table 10.1, which records the shear strengths (y) of spot welds of two gauges (thicknesses) of steel, 0.040″ and 0.064″, corresponding to the weld diameter (x), measured in units of thousandths of an inch. Notice that the response variable y is in the first column and the regressor variable is in the second column.

TABLE 10.1 SHEAR STRENGTH AND WELD DIAMETER FOR TWO GAUGES OF STEEL

0.040″ gauge		0.064″ gauge	
y	x	y	x
350	140	680	190
380	155	800	200
385	160	780	209
450	165	885	215
465	175	975	215
485	165	1025	215
535	195	1100	230
555	185	1030	250
590	195	1175	265
605	210	1300	250

Source: A. J. Duncan, *Quality Control and Industrial Statistics,* 5th ed., Homewood, Ill., Richard D. Irwin, 1985. Used with permission.

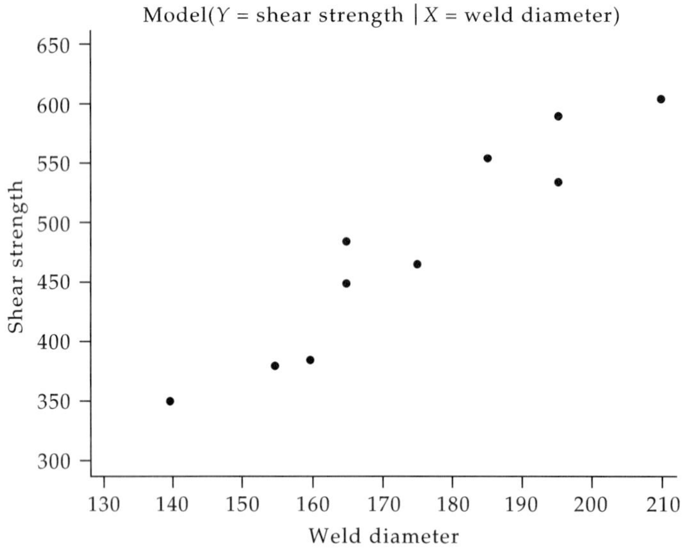

Figure 10.1 Scatter plot of the data in Table 10.1 from 0.040″ gauge steel

Notation. We use the *model statement* format

Model(Y = shear strength | X = weld diameter)

to specify the response variable Y and the explanatory variable X of the *regression model*.

Draw the scatter plot for the 0.040″ gauge steel.

Solution. The scatter plot for the 0.040″ gauge steel is shown in Fig. 10.1.

From Fig. 10.1 it appears that the shear strength y varies almost linearly with respect to the weld diameter x. We say *almost* because it is clear that it is impossible to construct a line of the form $y = b_0 + b_1 x$ that passes through all the observations (x_i, y_i), $i = 1, \ldots, n$. This leads us to the following problem: How do we choose b_0 and b_1 so that the line $y = b_0 + b_1 x$ "best fits" the observations (x_i, y_i)? In Sec. 10.2.1 we solve this problem via the *method of least squares.* ∎

---

example **10.2** After the price of petroleum tripled in the early 1970s, the fuel efficiencies of automobiles became an important concern for consumers, manufacturers, and governments. Table 10.2 records the weight in units of 1000 lb, fuel consumption in miles per gallon (mpg), engine displacement (disp), and fuel consumption in gallons per 100 miles (gpm) for 12 1992-model-year automobiles. (Displacement is a measure of the size of the automobile engine.)

There are two different ways to measure fuel efficiency: miles per gallon, which is the method used in the United States, and gallons per hundred miles, as is used in some European countries. Another important goal of regression analysis is to determine the explanatory variable that best explains the observed response. Specifically, which is the better choice for the explanatory variable, weight or engine displacement? To study this question we draw the scatter plots for the two models

$$\text{Model}(Y = \text{mpg} \mid X = \text{weight}) \quad \text{and} \quad \text{Model}(Y = \text{gpm} \mid X = \text{weight})$$

TABLE 10.2 FUEL EFFICIENCY DATA FOR 12 1992-MODEL-YEAR AUTOMOBILES

Automobile	Weight	mpg	Disp	gpm
Saturn	2.495	32	1.9	3.12
Escort	2.530	30	1.8	3.34
Elantra	2.620	29	1.6	3.44
Camry V6	3.395	25	3.0	4.00
Camry4	3.030	27	2.2	3.70
Taurus	3.345	28	3.8	3.58
Accord	3.040	29	2.2	3.44
LeBaron	3.085	27	3.0	3.70
Pontiac	3.495	28	3.8	3.58
Ford	3.950	25	4.6	4.00
Olds88	3.470	28	3.8	3.58
Buick	4.105	25	5.7	4.00

Solution. Figure 10.2 is the scatter plot for Model($Y = \text{mpg} \mid X = \text{weight}$), and Fig. 10.3 is the scatter plot for Model($Y = \text{gpm} \mid X = \text{weight}$).

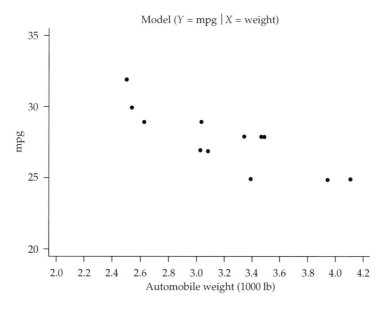

Figure 10.2 Scatter plot of the data in Table 10.2

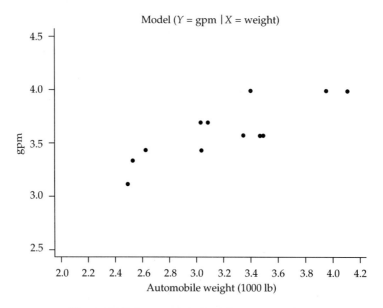

Figure 10.3 Scatter plot of the data in Table 10.2   ◼

The choice of gpm as a measure of fuel efficiency is justified by the following reasoning. The basic principles of mechanics suggest that the amount of gasoline G consumed is proportional to the work W expended in moving the vehicle (so $G = c_1 \times W$). Since Work $(W) = $ Force $(F) \times$ Distance (D), it follows that $G = c_1 \times F \times D$; thus $G/D = c_1 \times F$. In addition, the force F is proportional to the weight (wgt), so

$$\text{gpm} = 100 \times \frac{G}{D} = 100 \times c_1 \times \text{wgt} = c_2 \times \text{wgt}$$

These theoretical considerations imply that the functional relation between gpm and weight is more likely to be a linear one. We will continue our study of this data set with the goal of exploring the following question: Which model, Model $(Y = \text{mpg} \mid X = \text{weight})$ or Model $(Y = \text{gpm} \mid X = \text{weight})$, best fits the data? We will develop some criteria to help us to answer this question later in this chapter.

10.2.1   Fitting a Straight Line via Ordinary Least Squares

We now consider the problem of constructing a straight line with equation $y = b_0 + b_1 x$ that "best fits" the data $(x_1, y_1), \ldots, (x_n, y_n)$ in a sense that will be made precise.

The ith *fitted value,* (also called the ith *predicted value*) is denoted $\hat{y}_i$ and is defined as follows:

$$\hat{y}_i = b_0 + b_1 x_i, \quad i = 1, \ldots, n$$

It is the y coordinate of the point on the line $y = b_0 + b_1 x$ with x coordinate x_i.

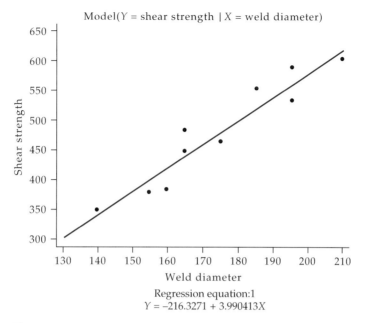

Figure 10.4 Scatter plot and regression line for the data in
Table 10.1 for 0.040″ gauge steel

The *i*th *residual* is difference between the *i*th observed value and the *i*th predicted value (see Fig. 10.4). It is denoted e_i and is given by

$$\text{ith residual:}\quad e_i = y_i - (b_0 + b_1 x_i)$$

Points of the scatter plot that lie below the line have negative residuals, and points above the line have positive residuals. The sum of the squares of the residuals, denoted SSE (*error sum of squares*), is defined as follows:

$$\text{SSE} = \sum_{1 \le i \le n} e_i^2 = \sum_{1 \le i \le n} [y_i - (b_0 + b_1 x_i)]^2 \qquad (10.1)$$

The error sum of squares is a measure of how well the line fits the data. For example, when all points of the scatter plot lie on the line $y = b_0 + b_1 x$, then each residual equals 0 and therefore SSE also equals 0.

The *method of least squares* is to find those values b_0, b_1 that minimize the sum of the squares of the residuals e_i. The error sum of squares is a function $Q(b_0, b_1)$ of the parameters b_0, b_1 and is given by the following formula:

$$\text{SSE} = \sum_{1 \le i \le n} [y_i - (b_0 + b_1 x_i)]^2 = Q(b_0, b_1) \qquad (10.2)$$

Notational Convention. From now on we adopt the convention that a sum of the form $\sum x_i$, where the limits on the indices are not specified, is a convenient shorthand for $\sum_{1 \le i \le n} x_i$.

We minimize the sum of squares $Q(b_0, b_1)$ using an algebraic method that avoids the use of calculus and offers some insight into the least squares method itself. The basic idea is to partition Q into a sum of three squares (so that each summand is nonnegative) from which the values that minimize it can be read off. The formula for the sum of squares uses the following notation, which is also useful for computing the minimizing values.

$$S_{xx} = \sum (x_i - \bar{x})^2 = \sum x_i^2 - n\bar{x}^2 \tag{10.3}$$

$$S_{yy} = \sum (y_i - \bar{y})^2 = \sum y_i^2 - n\bar{y}^2 \tag{10.4}$$

$$S_{xy} = \sum (x_i - \bar{x})(y_i - \bar{y}) = \sum x_i y_i - n\bar{x} \times \bar{y} \tag{10.5}$$

$$\text{Sample correlation coefficient:} \quad r = \frac{S_{xy}}{\sqrt{S_{xx}}\sqrt{S_{yy}}} \tag{10.6}$$

It can be shown [see the discussion following Eq. (10.10)] that

$$0 \le r^2 \le 1 \quad \text{and therefore} \quad -1 \le r \le 1$$

The Partition of Q into a Sum of Squares

The sum of squares (10.2) can be partitioned into a sum of three squares given by

$$Q(b_0, b_1) = S_{yy}(1 - r^2) + (b_1 \sqrt{S_{xx}} - r \sqrt{S_{yy}})^2 + n(\bar{y} - b_0 - b_1 \bar{x})^2 \tag{10.7}$$

The derivation is given in Sec. 10.5 at the end of this chapter.

Looking at Eq. (10.7), we see that only the second and third summands depend on the unknown parameters b_0 and b_1. We minimize $Q(b_0, b_1)$ by choosing these parameters so that these terms equal zero. Setting the second and third terms equal to zero yields the following two equations:

$$b_1 \sqrt{S_{xx}} - r \sqrt{S_{yy}} = 0$$

$$\bar{y} - b_0 - b_1 \bar{x} = 0$$

The equations for b_0 and b_1 can always be solved provided $S_{xx} \ne 0$. But this will always be true except for the uninteresting case when $x_1 = x_2 = \cdots = x_n$. Denoting the solutions by $\hat{\beta}_0$ and $\hat{\beta}_1$, we obtain the slope-intercept formula for the line that best fits the data in the sense of the least squares criterion:

$$\text{Least squares line:} \quad y = \hat{\beta}_0 + \hat{\beta}_1 x \tag{10.8}$$

where

$$\hat{\beta}_1 = \frac{r \sqrt{S_{yy}}}{\sqrt{S_{xx}}} = \frac{S_{xy}}{S_{xx}} \tag{10.9}$$

and

$$\hat{\beta}_0 = \bar{y} - \hat{\beta}_1 \bar{x}. \tag{10.10}$$

The quantities $\hat{\beta}_0$ (intercept) and $\hat{\beta}_1$ (slope) are called the *regression coefficients* or the *regression parameters*.

From Eqs. (10.2) and (10.7) it follows that

$$\text{SSE} = S_{yy}(1 - r^2) \tag{10.11}$$

Equation (10.11) follows from the fact that the second and third summands on the right-hand side of the partition of the sum of squares given in Eq. (10.7) equal zero. This formula for SSE also implies that $0 \leq r^2 \leq 1$. This is because SSE and S_{yy} are always nonnegative, since they are sums of squares. Therefore, $0 \leq 1 - r^2$, so $0 \leq r^2 \leq 1$ as asserted earlier.

The fitted values and residuals associated with the least squares line, also denoted by $\hat{y}_i$ and e_i, are defined as before:

$$\hat{y}_i = \hat{\beta}_0 + \hat{\beta}_1 x_i, \quad i = 1, \ldots, n \tag{10.12}$$

$$e_i = y_i - \hat{y}_i = y_i - \hat{\beta}_0 - \hat{\beta}_1 x_i \tag{10.13}$$

The least squares line equation (10.8) is also called the *estimated regression function* of y on x. The use of the adjective *estimated* may appear strange since the calculations do not require any statistical assumptions. We will resolve this mystery in the next section when we introduce a statistical model for the bivariate data set (x_i, y_i), $i = 1, \ldots, n$. The regression parameters are then interpreted as estimators of the true regression line.

Some Suggestions for Computing the Regression Coefficients

The regression coefficients are most conveniently computed by using a statistical software package such as MINITAB or SAS. If you do not have access to a statistical software package, the regression coefficients can be computed using a calculator as shown in the next example.

---

example 10.3   Compute the estimated regression parameters and the estimated regression function for the Model($Y = $ shear strength $\mid X = $ weld diameter) problem of Example 10.1 (0.040" gauge steel).

Solution. We first compute $\bar{x}, \bar{y}, S_{xx}, S_{xy}, S_{yy}$. A routine computation using Eqs. (10.3)–(10.5) yields

$$\bar{x} = 174.5$$
$$\bar{y} = 480$$
$$S_{xx} = 4172.5$$
$$S_{xy} = 16{,}650$$
$$S_{yy} = 73{,}450$$

It follows from Eqs. (10.9) and (10.10) that the slope $\hat{\beta}_1$ and intercept $\hat{\beta}_0$ of the regression line are given by

$$\hat{\beta}_1 = \frac{S_{xy}}{S_{xx}} = 3.990413$$

$$\hat{\beta}_0 = \bar{y} - \hat{\beta}_1 \bar{x} = -216.3271$$

Note. The final results were taken from the computer printout and may differ slightly from those you obtained because of roundoff error.

The next step in the regression analysis is to plot the estimated regression line and the scatter plot on the same set of axes, as shown in Fig. 10.4.

Table 10.3 displays an SAS printout of the observations $y_1, \ldots, y_n$ (in the column headed *Dep Var*), the fitted values $\hat{y}_1, \ldots, \hat{y}_n$ (in the column headed *Predict*), and the residuals for the data displayed in Table 10.1 for 0.040" gauge steel. We obtain the first residual by computing

$$y_1 - \hat{y}_1 = 350 - (-216.3271 + 3.990413 \times 140)$$

$$= 350 - 342.33072 = 7.6692$$

Again, because of roundoff error, the final answer obtained here differs slightly from the value in the computer printout.

TABLE 10.3 SAS PRINTOUT OF OBSERVATION NUMBER, SHEAR STRENGTH, PREDICTED SHEAR STRENGTH, AND RESIDUAL

Observation	Dep var shear	Predict value	Residual
1	350.0	342.3	7.6693
2	380.0	402.2	−22.1869
3	385.0	422.1	−37.1390
4	450.0	442.1	7.9089
5	465.0	482.0	−16.9952
6	485.0	442.1	42.9089
7	535.0	561.8	−26.8035
8	555.0	521.9	33.1007
9	590.0	561.8	28.1965
10	605.0	621.7	−16.6597

Interpreting the Regression Coefficients. The estimated slope $\hat{\beta}_1 = 3.990413$ implies that each 0.001-in. increase in the weld diameter produces an increase of 3.990413 in the shear strength. The estimate $\hat{\beta}_0 = -216.3271$ states that the estimated regression line intercepts the y axis at -216.3271. This clearly is without any physical significance since the shear strength can never be negative. This should serve as a warning against using the estimated regression line to predict the response y when x lies far outside the range of the initial data, which in this case is the interval $[140, 210]$. We can, however, use the estimated regression line to predict the value of the shear strength corresponding to a weld diameter of 180; in this case the predicted value is $-216.3271 + 3.990413 \times 180 = 501.9472$.

An Interpretation of the Sign of r. When the sample correlation is positive, then so is the estimated slope of the estimated regression line; therefore, an increase in x tends to increase the value of the response variable y. A negative value for the sample correlation, on the other hand, implies that the estimated regression line has a negative slope; in other words, an increase in x tends to decrease the value of the response variable y.

The Coefficient of Determination

The square of the sample correlation coefficient is called the *coefficient of determination* and is denoted by R^2. It plays an important role in *model checking*. This refers to a variety of analytical and graphical methods for evaluating how well the estimated regression line fits the data.

We now derive an alternative formula for the coefficient of determination that gives us a deeper understanding of its role in model checking. It is based on the sum of squares partition given in the following equation, which is absolutely fundamental for understanding the results of a regression analysis.

$$\sum (y_i - \bar{y})^2 = \sum (\hat{y}_i - \bar{y})^2 + \sum (y_i - \hat{y}_i)^2 \tag{10.14}$$

The sum of squares partition (10.14) is most easily derived and understood via a matrix approach to regression given in the next chapter.

The decomposition (10.14) is also written in the form

$$\text{SST} = \text{SSR} + \text{SSE} \tag{10.15}$$

where $\text{SST} = \sum (y_i - \bar{y})^2 = S_{yy}$ [by Eq. (10.4)]
$\text{SSR} = \sum (\hat{y}_i - \bar{y})^2$
$\text{SSE} = \sum (y_i - \bar{y})^2 = \text{SST}(1 - R^2)$ [by Eq. (10.11)]

The quantity SST is called the *total sum of squares*; it is a measure of the total variability of the original observations. The quantity SSR is called the *regression sum of squares*; it is a measure of the total variability of the fitted values. It is also a measure of the total variability explained by the regression model. The quantity SSE is the error sum of squares, defined earlier; it is a measure of the unexplained variability.

The partition (10.15) of the total sum of squares leads to an alternative formula for the coefficient of determination that gives us a deeper understanding of its role in model checking. Dividing both sides of Eq. (10.15) by SST, we see that

$$1 = \frac{\text{SSR}}{\text{SST}} + \frac{\text{SSE}}{\text{SST}}$$

From Eq. (10.15) it also follows that

$$\text{SSR} = \text{SST} - \text{SSE} = \text{SST} - \text{SST}(1 - R^2) = R^2 \text{SST}$$

Dividing both sides by SST yields the following formula for R^2:

$$R^2 = \frac{\text{SSR}}{\text{SST}} \tag{10.16}$$

Thus, the quantity $R^2 = \text{SSR}/\text{SST}$ represents the *proportion of the total variability that is explained by the model*, and SSE/SST represents the proportion of the unexplained variability. A value of R^2 close to 1 implies that most of the variability is explained by the regression model; a value of R^2 close to 0 indicates that the regression model is not appropriate.

example **10.4**   Compute the regression parameters and the coefficient of determination, and graph the estimated regression line for the regression Model($Y = $ mpg $|$ $X = $ weight) of Example 10.2.

Solution

1. These quantities were computed using a computer software package; the following results (rounded to two decimal places) were obtained:

$$\bar{x} = 3.21$$
$$\bar{y} = 27.75$$
$$S_{xx} = 2.95$$
$$S_{xy} = -10.13$$
$$S_{yy} = 50.25$$

Consequently,

$$\hat{\beta}_1 = \frac{S_{xy}}{S_{xx}} = \frac{-10.13}{2.95} = -3.43$$
$$\hat{\beta}_0 = \bar{y} - \hat{\beta}_1\bar{x} = 27.75 + 3.43 \times 3.21$$
$$= 38.77$$

The slope is negative since an increase in weight leads to a decrease in mpg.

2. From the definitions of SST, SSR, and SSE just given we obtain the following values:

$$\text{SST} = 50.25$$
$$\text{SSR} = 34.77$$
$$\text{SSE} = 15.48$$

Consequently,

$$R^2 = \frac{34.77}{50.25} = 0.69$$
$$r = -0.83$$

3. The scatter plot and the estimated regression line are shown in Fig. 10.5.   ∎

Interpreting the Regression Coefficients. The estimated slope $\hat{\beta}_1 = -3.43$ implies that each 1000-lb increase in the automobile's weight produces a *decrease* of 3.43 in the fuel efficiency as measured in miles per gallon. The estimate $\hat{\beta}_0 = 38.78$ states that the estimated regression line intercepts the y axis at 38.78. Note that the predicted fuel efficiency for an automobile weighing 12,000 lb is $38.78 - 3.43 \times 12 = -2.38$ mpg, which is absurd since mpg is always nonnegative. Once again, this should serve as a warning against using the estimated regression line to "predict" the response y when x lies far outside the range of the initial data. On the other hand, the fitted mpg value for an automobile weighing 2495 lb is given by $38.78 - 3.43 \times 2.495 = 30.22$ mpg. The residual equals $32 - 30.22 = 1.78$.

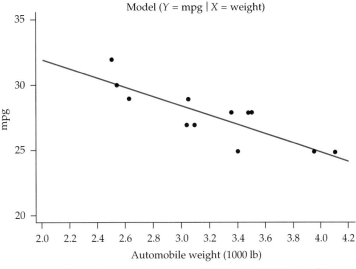

Model (Y = mpg | X = weight)

Regression equation: mpg = $38.78474 - 3.434047 \times$ weight

Figure 10.5 Scatter plot and estimated regression line for data in
Table 10.2

Some Properties of the Residuals. It can be shown that the residuals and
the fitted values satisfy the following conditions:

$$\sum e_i = \sum (y_i - \hat{y}_i) = 0 \tag{10.17}$$

$$\sum e_i x_i = \sum (y_i - \hat{y}_i)x_i = 0 \tag{10.18}$$

$$\frac{1}{n} \sum \hat{y}_i = \bar{y} \tag{10.19}$$

The derivations of these equations are left as Prob. 10.10.

PROBLEMS

10.1 (a) Draw the scatter plot for the following artificial data.

x	y
0	0
0	1
1	1
1	3
2	2
2	3

(b) Sketch a straight line that seems to satisfactorily fit these points. Denoting the equation of your line by $y = b_0 + b_1 x$, find the values of b_0, b_1.
(c) Calculate the least squares line $y = \hat{\beta}_0 + \hat{\beta}_1 x$.

(d) Is the line you obtained in part (*b*) close to the least squares line obtained in part (*c*)?

10.2 (a) Draw the scatter plot corresponding to the shear strength data for the 0.064″ gauge steel given in Table 10.1.

(b) Calculate the least squares estimates of $\hat{\beta}_0$ and $\hat{\beta}_1$ and draw the graph of the estimated regression line $y = \hat{\beta}_0 + \hat{\beta}_1 x$. Use the same axes on which you graphed the scatter plot.

(c) Compute the sample correlation, coefficient of determination, and error sum of squares.

(d) Compute the predicted values and the residuals, and display the results in a format similar to that of Table 10.3.

(e) Verify that $\sum e_i = 0$.

Save your computations.

10.3 The following data represent the number of disk I/Os and CPU times for seven programs.

Disk I/Os	CPU time
14	2
16	5
27	7
42	9
39	10
50	13
83	20

Source: R. Jain, *The Art of Computer Systems Performance Analysis*, New York, John Wiley & Sons, 1991. Used with permission.

(a) Draw the scatter plot for the regression

$$\text{Model}(Y = \text{CPU time} \mid X = \text{Disk I/Os})$$

(b) Calculate the least squares estimates $\hat{\beta}_0$ and $\hat{\beta}_1$ and draw the graph of the estimated regression line $y = \hat{\beta}_0 + \hat{\beta}_1 x$. Use the same axes on which you graphed the scatter plot.

(c) Compute the sample correlation, coefficient of determination, and error sum of squares.

(d) Compute the predicted values and the residuals, and display the results in a format similar to that of Table 10.3.

(e) Verify that $\sum e_i = 0$.

Save your computations.

10.4 The following data record the amount of water (*x*), in centimeters, and the yield of hay (*y*), in metric tons per hectare, on an experimental farm.

Water (*x*)	30	45	60	75	90	105	120
Yield (*y*)	2.11	2.27	2.5	2.88	3.21	3.48	3.37

(a) Draw the scatter plot (x_i, y_i).

(b) Calculate the least squares estimates $\hat{\beta}_0$ and $\hat{\beta}_1$ and draw the graph of the estimated regression line $y = \hat{\beta}_0 + \hat{\beta}_1 x$. Use the same axes on which you graphed the scatter plot.

(c) Compute the sample correlation, coefficient of determination, and error sum of squares.

(d) Compute the predicted value and the residual for $x = 30$, $x = 75$.

10.5 Refer to the fuel efficiency data (Table 10.2).

(a) Draw the scatter plot for the regression

$$\text{Model}(Y = \text{mpg} \mid X = \text{disp})$$

(b) Calculate the least squares estimates of $\hat{\beta}_0$ and $\hat{\beta}_1$, and draw the graph of the estimated regression line $y = \hat{\beta}_0 + \hat{\beta}_1 x$. Use the same axes on which you graphed the scatter plot.

(c) Compute the sample correlation, coefficient of determination, and error sum of squares.

(d) Compute the predicted value and residual for $x = 1.6$, $x = 4.6$.

(e) Which of the two variables, weight or engine displacement, is a better predictor of fuel efficiency (mpg)? (*Hint:* Compare the coefficients of determination.)

10.6 Refer to the fuel efficiency data (Table 10.2).

(a) Calculate the least squares estimates $\hat{\beta}_0$ and $\hat{\beta}_1$ for the regression $\text{Model}(Y = \text{gpm} \mid X = \text{weight})$. Copy the scatter plot from Fig. 10.3 and draw the graph of the estimated regression line $y = \hat{\beta}_0 + \hat{\beta}_1 x$. Use the same axes on which you graphed the scatter plot.

(b) Compute the sample correlation, coefficient of determination, and error sum of squares.

(c) Compute the residuals and display the results in a format similar to that of Table 10.3.

10.7 Refer to the fuel efficiency data (Table 10.2).

(a) Draw the scatter plot for the regression

$$\text{Model}(Y = \text{gpm} \mid X = \text{disp})$$

(b) Calculate the least squares estimates $\hat{\beta}_0$ and $\hat{\beta}_1$, and draw the graph of the estimated regression line $y = \hat{\beta}_0 + \hat{\beta}_1 x$. Use the same axes on which you graphed the scatter plot.

(c) Compute the sample correlation, coefficient of determination, and error sum of squares.

(d) Compute the predicted value and residual for $x = 1.6$, $x = 4.6$.

(e) Which of the two variables, weight or engine displacement, is a better predictor of fuel efficiency (gpm)? (*Hint:* Compare the coefficients of determination.)

10.8 Specific electrical conductivity (SEC), total dissolved solids (TDS), and silica (SiO_2) are groundwater quality parameters routinely measured in a laboratory. The SEC measurement procedure is simple, rapid, and precise. Measuring TDS is a time-consuming procedure that requires several days for evaporation and drying of a known volume of filtered water under constant and standard laboratory conditions. One study used the regression Model$(Y = \text{TDS} \mid X = \text{SEC})$ to estimate TDS from SEC measurements. The following table lists 10 measurements of these variables from city wells in Fresno, California.

SEC (ms/cm)	SiO$_2$ (mg/L)	TDS (mg/L)
422	56	275
470	48	305
841	48	535
714	55	456
749	59	484
450	61	293
214	63	173
213	65	175
229	36	147
370	54	243

Source: B. Day and H. Nightingale, *Ground Water*, vol. 22, no. 1, 1984, pp. 80–85.

(a) Draw the scatter plot for the regression Model$(Y = \text{TDS} \mid X = \text{SEC})$.
(b) Calculate the least squares estimates of $\hat{\beta}_0$ and $\hat{\beta}_1$, and draw the graph of the estimated regression line $y = \hat{\beta}_0 + \hat{\beta}_1 x$. Use the same axes on which you graphed the scatter plot.
(c) Compute the sample correlation, coefficient of determination, and error sum of squares.

10.9 Suppose $y_i = a + bx_i$, $i = 1, \ldots, n$. Show that $r = +1$ when $b > 0$ and $r = -1$ when $b < 0$.

10.10 (a) Verify Eq. (10.17), which asserts that

$$\sum (y_i - \hat{y}_i) = 0$$

(b) Verify Eq. (10.18), which asserts that

$$\sum (y_i - \hat{y}_i) x_i = 0$$

(c) Verify Eq. (10.19), which asserts that the arithmetic mean of the observed values equals the arithmetic mean of the fitted values; that is,

$$\frac{1}{n} \sum y_i = \frac{1}{n} \sum \hat{y}_i$$

10.3 THE SIMPLE LINEAR REGRESSION MODEL

The *simple linear regression model* assumes that the functional relationship between the response variable and the explanatory variable is a linear function plus an error term denoted by ϵ; that is,

$$Y = \beta_0 + \beta_1 x + \epsilon \tag{10.20}$$

The variable ϵ includes all the factors other than the explanatory variable that can also produce changes in Y.

From these considerations we are naturally led to a statistical model for the bivariate data sets of the sort we have analyzed previously and that consists of the following elements:

1. Y_i denotes the value of the response corresponding to the value x_i of the explanatory variable, where $x_i, i = 1, \ldots, n$, are the scientist's particular choices of the explanatory variable x.

2. We assume that

$$Y_i = \beta_0 + \beta_1 x_i + \epsilon_i \qquad (10.21)$$

   Our statistical model assumes that the random variables $\epsilon_1, \ldots, \epsilon_n$ are mutually independent and normally distributed with zero means and variances σ^2, so

$$E(\epsilon_i) = 0 \quad \text{and} \quad \text{Var}(\epsilon_i) = \sigma^2 \qquad (10.22)$$

   The parameters β_0 and β_1, called the *intercept* and *slope*, respectively, define the *true regression function* $y = \beta_0 + \beta_1 x$.

We shall refer to conditions 1 and 2 as the *standard assumptions*.

3. The standard assumptions imply that the random variables $Y_1, \ldots, Y_n$ are mutually independent and normally distributed with

$$E(Y_i) = \beta_0 + \beta_1 x_i \quad \text{and} \quad \text{Var}(Y_i) = \sigma^2$$

Notation. To emphasize the dependence of the expected value of Y on x, we use the notation

$$E(Y \mid x) = \beta_0 + \beta_1 x$$

Some authors use the notation

$$\mu_{Y|x} = \beta_0 + \beta_1 x$$

Figure 10.6 displays the pdf of Y for two values, x_1 and x_2.

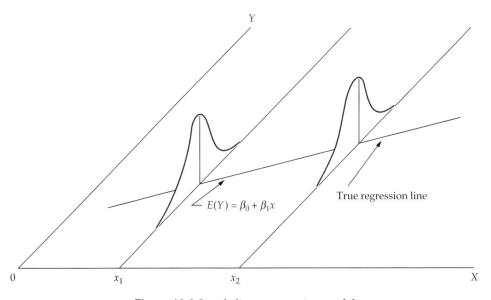

Figure 10.6 Simple linear regression model

10.3.1   The Sampling Distribution of $\hat{\beta}_1$, $\hat{\beta}_0$, SSE, and SSR

The simple linear regression model has *three* unknown parameters: β_0, β_1, and σ^2. We now estimate these parameters via the least squares method described in the previous section. Thus, we obtain the *least squares estimators* for β_0 and β_1 by substituting Y for y in Eqs. (10.9) and (10.10). Consequently, the least squares estimators of the regression parameters are given by

$$\hat{\beta}_1 = \frac{S_{xY}}{S_{xx}} = \frac{\sum (x_i - \overline{x})(Y_i - \overline{Y})}{S_{xx}} \tag{10.23}$$

$$\hat{\beta}_0 = \overline{Y} - \hat{\beta}_1 \overline{x} \tag{10.24}$$

An estimator for σ^2 will be given shortly. In the meantime, it is important to note that in the context of the simple linear regression model the least squares estimators $\hat{\beta}_0$ and $\hat{\beta}_1$ are now random variables, and that in order to make statistical inferences concerning the unknown parameters we must first derive their sampling distributions. Theorem 10.1 summarizes the essential facts; its proof, which follows directly from Eqs. (10.23) and (10.24), is sketched in Sec. 10.5 at the end of this chapter.

▬ THEOREM 10.1

1. The distribution of the least squares estimator for the slope β_1 is normal with mean, variance, and second moment [which is needed to compute $E(\text{SSR})$] given by

$$E(\hat{\beta}_1) = \beta_1 \tag{10.25}$$

$$V(\hat{\beta}_1) = \frac{\sigma^2}{S_{xx}} \tag{10.26}$$

$$E(\hat{\beta}_1^2) = \frac{\sigma^2}{S_{xx}} + \beta_1^2 \tag{10.27}$$

2. The distribution of the least squares estimator for the intercept β_0 is normal with mean and variance given by

$$E(\hat{\beta}_0) = \beta_0 \tag{10.28}$$

$$V(\hat{\beta}_0) = \sigma^2 \left[\frac{1}{n} + \frac{\overline{x}^2}{S_{xx}} \right] \tag{10.29}$$

■

It follows from Eq. (10.25) that $\hat{\beta}_1$ is an unbiased estimator of the slope. Similarly, Eq. (10.28) implies that $\hat{\beta}_0$ is an unbiased estimator of the intercept.

The parameter σ^2 is an important one because it is a measure of the variability of the response Y; a small value of σ^2 indicates that the observations (x_i, y_i) lie close to the true regression line, and a large value of σ^2 indicates that they are more widely dispersed about the true line. To estimate σ^2 we use the statistic

$$s^2 = \frac{\sum (Y_i - \hat{Y}_i)^2}{n - 2} = \frac{\text{SSE}}{n - 2} \tag{10.30}$$

The next theorem completes our description of the sampling distributions of the regression parameters; its proof is somewhat technical and lies outside the scope of this text.

■■ **THEOREM 10.2**

The statistic s^2 is an unbiased estimator of σ^2; that is,

$$E(s^2) = E\left(\frac{\text{SSE}}{n-2}\right) = \sigma^2 \tag{10.31}$$

The random variable $(n-2)s^2/\sigma^2$ has a chi-square distribution with $n-2$ degrees of freedom; that is,

$$\frac{(n-2)s^2}{\sigma^2} \sim \chi^2_{n-2}$$

In addition, s^2 is independent of both $\hat{\beta}_0$ and $\hat{\beta}_1$.   ■

A direct proof of Eq. (10.31) is given in Sec. 10.5.

We will also need the following result concerning the expected value of SSR:

$$E(\text{SSR}) = \sigma^2 + \beta_1^2 S_{xx} \tag{10.32}$$

We give the proof in Sec. 10.5.

Degrees of Freedom Associated with the Sum of Squares $\sum (Y_i - \hat{Y}_i)^2$

The divisor $n-2$ is the number of degrees of freedom associated with the sum of squares $\sum (Y_i - \hat{Y}_i)^2$. Recall [see Eqs. (10.17) and (10.18)] that the residuals satisfy the following two equations:

$$\sum e_i = \sum (Y_i - \hat{Y}_i) = 0$$

$$\sum e_i x_i = \sum (Y_i - \hat{Y}_i)x_i = 0$$

Each of these equations allows us to eliminate one summand from the sum of squares $\sum (Y_i - \hat{Y}_i)^2$; consequently, there are only $n-2$ independent quantities $Y_i - \hat{Y}_i$.

Confidence Intervals for the Regression Coefficients

We now derive confidence intervals for the parameters β_0, β_1, and σ^2 from the sampling distributions of $\hat{\beta}_0$, $\hat{\beta}_1$, and s^2. We begin with the sampling distribution of $\hat{\beta}_1$ since this is the most important parameter in the regression model. Using Theorems 10.1 and 10.2, and then standardizing the random variable $\hat{\beta}_1$ in the usual, way we obtain

$$\frac{\hat{\beta}_1 - \beta_1}{\sigma/\sqrt{S_{xx}}} = Z \sim N(0, 1)$$

$$\frac{(n-2)s^2}{\sigma^2} = V \sim \chi^2_{n-2}$$

Consequently,

$$\frac{\hat{\beta}_1 - \beta_1}{s/\sqrt{S_{xx}}} = \frac{Z}{\sqrt{V/(n-2)}} \sim t_{n-2} \tag{10.33}$$

The quantity in the denominator of Eq. (10.33) is called the *estimated standard error* of the estimate:

$$\text{Estimated standard error:} \quad s(\hat{\beta}_1) = \frac{s}{\sqrt{S_{xx}}} \tag{10.34}$$

The standard error for the estimate of β_0 is obtained by noting that

$$V(\hat{\beta}_0) = \sigma^2 \left[\frac{1}{n} + \frac{\bar{x}^2}{S_{xx}} \right]$$

Replacing σ^2 with its estimate s^2 yields the following formula for the estimated standard error for β_0:

$$s(\hat{\beta}_0) = s \sqrt{\left[\frac{1}{n} + \frac{\bar{x}^2}{S_{xx}} \right]} \tag{10.35}$$

The sampling distribution of $\hat{\beta}_0$ is derived in a similar way, leading to the result

$$\frac{\hat{\beta}_0 - \beta_0}{s(\hat{\beta}_0)} \sim t_{n-2} \tag{10.36}$$

We can now construct confidence intervals for β_1 and β_0 using methods similar to those used to derive confidence intervals in Chap. 7. Thus, a $100(1-\alpha)$ percent confidence interval for β_1 is given by

$$\hat{\beta}_1 \pm t_{n-2}(\alpha/2)s(\hat{\beta}_1) \tag{10.37}$$

Similarly, a $100(1-\alpha)$ percent confidence interval for β_0 is given by

$$\hat{\beta}_0 \pm t_{n-2}(\alpha/2)s(\hat{\beta}_0) \tag{10.38}$$

---

example **10.5**   Construct 95 percent confidence intervals for the slope and intercept parameters of the regression Model(Y = shear strength | X = weld diameter) (Example 10.1).

Solution. We computed the quantities $\bar{x}$, S_{xx}, SSE, $\hat{\beta}_0$, and $\hat{\beta}_1$ earlier (Example 10.3), and we restate them here for your convenience.

$$n = 10, \quad \bar{x} = 174.5, \quad S_{xx} = 4172.5$$

$$\text{SSE} = 7009.6165, \quad \hat{\beta}_0 = -216.3271, \quad \hat{\beta}_1 = 3.990413$$

We estimate σ^2 using Eq. (10.30); thus,

$$s^2 = \frac{\text{SSE}}{8} = \frac{7009.6165}{8} = 876.2021$$

The estimated standard error for $\hat{\beta}_1$ is therefore

$$s(\hat{\beta}_1) = \frac{s}{\sqrt{S_{xx}}} = \sqrt{\frac{876.2021}{4172.5}} = 0.4583$$

Similarly,

$$s(\hat{\beta}_0) = \sqrt{876.2021}\sqrt{\frac{1}{10} + \frac{174.5^2}{4172.5}} = 80.5109$$

To construct a 95 percent confidence interval for β_1, we choose $t_8(0.025) = 2.306$ and compute

$$3.9904 \pm 2.306 \times 0.4583 = 3.990413 \pm 1.0568$$

so the confidence interval is $[2.9336, 5.0473]$.

The 95 percent confidence interval for β_0 is given by

$$-216.3271 \pm 2.306 \times 80.5109 = -216.3271 \pm 185.6581$$

so the confidence interval is $[-401.9852, -30.6690]$. ∎

Testing Hypotheses about the Regression Parameters Using Confidence Intervals. We also use confidence intervals for the regression parameters to test hypotheses about them. Thus, we accept the null hypothesis

$$H_0: \beta_1 = 0 \quad \text{against} \quad H_1: \beta_1 \neq 0$$

if the confidence interval for β_1 contains 0; otherwise we reject it. For example, the confidence interval for the slope in Example 10.5 is $[2.9336, 5.0472]$, which does not contain 0; thus we reject $H_0: \beta_1 = 0$. We present another approach in the next section.

10.3.2   Tests of Hypotheses Concerning the Regression Parameters

The concepts and methods of hypothesis testing developed in Chap. 8 are easily adapted to make inferences about the slope β_1 (and also the intercept β_0). This is because under the standard assumptions we have shown [Eq. (10.33)] that

$$\frac{\hat{\beta}_1 - \beta_1}{s/\sqrt{S_{xx}}} \sim t_{n-2}$$

Suppose, for example, we wish to test the hypothesis $H_0: \beta_1 = 0$ against $H_1: \beta_1 \neq 0$. As noted earlier, when $\hat{\beta}_1 = 0$, then SSR $= 0$, so the regression model explains none of the variation. Thus $\beta_1 = 0$ implies that varying x has no effect on the response variable y. It follows from Eq. (10.33) that the null distribution of $\hat{\beta}_1/(s/\sqrt{S_{xx}})$ is a t distribution with $n - 2$ degrees of freedom. Consequently, a (two-sided) level α test of H_0 is

$$\text{reject } H_0 \text{ if } \frac{|\hat{\beta}_1|}{s/\sqrt{S_{xx}}} \geq t_{n-2}(\alpha/2) \tag{10.39}$$

example **10.6**   Refer back to Example 10.5. Test the null hypothesis H_0: $\beta_1 = 0$ against H_1: $\beta_1 \neq 0$ for the regression Model(Y = shear strength $\mid X$ = weld diameter) at the 5 percent significance level.

Solution. The regression parameters S_{xx}, s^2, and $\bar{x}$ were computed earlier, and the following results were obtained:

$$\hat{\beta}_0 = -216.3271, \qquad \hat{\beta}_1 = 3.990413, \qquad s^2 = 876.202$$

$$S_{xx} = 4172.5, \quad \bar{x} = 174.5, \quad n = 10$$

The value of the test statistic is given by

$$\frac{|\hat{\beta}_1|}{s/\sqrt{S_{xx}}} = \frac{3.990413}{\sqrt{876.202/4172.5}} = 8.708 \geq t_8(0.025) = 2.306$$

This is significant at the 5 percent level, so we reject the null hypothesis.   ∎

Hypothesis Testing via the Computer

The result we just obtained is contained in the edited SAS computer printout of the regression analysis displayed in Table 10.4. The column labeled *Parameter Estimate* displays the estimated regression coefficients $\hat{\beta}_0$ (INTERCEP) and $\hat{\beta}_1$ (SLOPE). The column *Standard Error* displays the estimated standard errors $s(\hat{\beta}_0)$ and $s(\hat{\beta}_1)$. The next to last column, titled *T for HO*, is an abbreviation for "test of the null hypothesis H_0: $\beta_i = 0$ ($i = 1, 2$) using the t test." It contains the value of the t statistic used to test H_0. The P-value for each test is printed in the last column, *Prob* $> |T|$.

TABLE 10.4

Parameter Estimates

Variable	Parameter Estimate	Standard Error	T for HO: Parameter=0	Prob > |T|
INTERCEP	-216.327142	80.51090203	-2.687	0.0276
SLOPE	3.990413	0.45825157	8.708	0.0001

10.3.3   Confidence Intervals and Prediction Intervals

In this section we consider two estimation problems that are easily confused because they are so closely related. The first problem is to derive point and interval estimates for the mean response $E(Y \mid x)$, which is a numerical parameter. The second problem, which appears to be similar to the first, is to *predict* the response Y corresponding to a new observation taken at the value x. Note that, in this case, Y is *not* a parameter but a random variable.

Confidence Intervals for the Mean Response $E(Y \mid x)$

In order to construct a confidence interval for $E(Y \mid x)$ we need some facts concerning the sampling distribution of the estimator $\hat{\beta}_0 + \hat{\beta}_1 x$, as summarized in the following theorem.

■■ **THEOREM 10.3**

The point estimator $\hat{\beta}_0 + \hat{\beta}_1 x$ of the mean response $\beta_0 + \beta_1 x$ is unbiased and normally distributed with mean, variance, and estimated standard error given by

$$E(\hat{\beta}_0 + \hat{\beta}_1 x) = \beta_0 + \beta_1 x$$

$$V(\hat{\beta}_0 + \hat{\beta}_1 x) = \sigma^2 \left[\frac{1}{n} + \frac{(x - \bar{x})^2}{S_{xx}} \right]$$

$$s(\hat{\beta}_0 + \hat{\beta}_1 x) = s \sqrt{\left[\frac{1}{n} + \frac{(x - \bar{x})^2}{S_{xx}} \right]} \qquad ■$$

■■ **REMARK 10.1**

Note that $s(\hat{\beta}_0 + \hat{\beta}_1 x)$ depends on x; in particular, the further x deviates from $\bar{x}$, the greater the estimated standard error. This underscores the warning about extrapolating the model beyond the range of the original data.   ■

Invoking Theorems 10.1 and 10.2 again, we can show that

$$\frac{\hat{\beta}_0 + \hat{\beta}_1 x - (\beta_0 + \beta_1 x)}{s(\hat{\beta}_0 + \hat{\beta}_1 x)} \sim t_{n-2} \qquad (10.40)$$

Proceeding in a now familiar way, we see that a $100(1 - \alpha)$ percent confidence interval for the mean response $E(Y \mid x)$ is given by

$$\hat{\beta}_0 + \hat{\beta}_1 x \pm t_{n-2}(\alpha/2) s(\hat{\beta}_0 + \hat{\beta}_1 x) \qquad (10.41)$$

---

example **10.7** Refer to the fuel efficiency data of Table 10.2. Compute a 95 percent confidence interval for the expected mpg for a car weighing 3495 lb.

Solution. We first convert the car's weight to 3.495, corresponding to the unit used in the data set. All the quantities required for the following computations can be found in Example 10.4. The point estimate (final results rounded to two decimal places) is

$$\hat{\beta}_0 + \hat{\beta}_1 x = 38.78 - 3.43 \times 3.495 = 26.78 (\text{mpg}).$$

We obtain the confidence interval by substituting the values

$$s = 1.2442, \qquad n = 12, \qquad x = 3.495, \qquad \bar{x} = 3.21$$

into the definition of $s(\hat{\beta}_0 + \hat{\beta}_1 x)$ given in the last line of Theorem 10.3. Next substitute the values

$$t_{10}(0.025) = 2.228 \quad \text{and} \quad s(\hat{\beta}_0 + \hat{\beta}_1 x) = 0.413$$

into Eq. (10.41). We obtain the confidence interval $[25.86, 27.70]$.

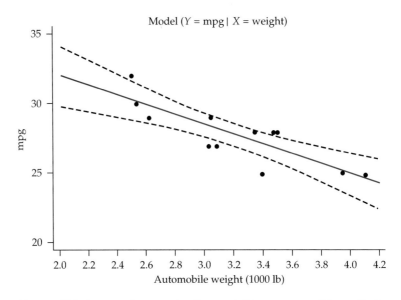

Figure 10.7 Estimated regression line and 95 percent confidence limits for mean response $E(Y \mid X = x)$ (data in Table 10.2)

The scatter plot, estimated regression line, and 95 percent confidence limits for $E(Y \mid X = x)$ for the fuel efficiency data (Table 10.2) are displayed in Fig. 10.7.   ∎

Prediction Interval

According to the hypotheses of the simple linear regression model, the future response is given by $Y = \beta_0 + \beta_1 x + \epsilon$ and the predicted future response is given by $\hat{Y} = \hat{\beta}_0 + \hat{\beta}_1 x$, where ϵ is an $N(0, \sigma^2)$ random variable that is independent of $\hat{Y}$. Consequently,

$$V(Y - \hat{Y}) = V(Y) + V(\hat{Y})$$

$$= \sigma^2 + \sigma^2 \left[\frac{1}{n} + \frac{(x - \bar{x})^2}{S_{xx}} \right]$$

$$= \sigma^2 \left[1 + \frac{1}{n} + \frac{(x - \bar{x})^2}{S_{xx}} \right]$$

Thus, the estimated standard error $s(Y - \hat{Y})$ is given by

$$s(Y - \hat{Y}) = s \sqrt{\left[1 + \frac{1}{n} + \frac{(x - \bar{x})^2}{S_{xx}} \right]} \qquad (10.42)$$

The corresponding $100(1 - \alpha)$ percent *prediction interval* is given by

$$\hat{\beta}_0 + \hat{\beta}_1 x \pm t_{n-2}(\alpha/2)s \sqrt{\left[1 + \frac{1}{n} + \frac{(x - \bar{x})^2}{S_{xx}} \right]} \qquad (10.43)$$

example 10.8   Compute a 95 percent prediction interval for the mpg for a car weighing 3495 lb. (Refer to Table 10.2 for the fuel efficiency data.)

Solution. We compute a 95 percent prediction interval for the fuel efficiency (measured in mpg) in the same way that we computed the confidence interval, except that we now use Eq. (10.43) in place of (10.41). We compute $s(Y - \hat{Y})$ by making the following substitutions into Eq. (10.42):

$$s = 1.2442, \qquad n = 12, \qquad x = 3.495, \qquad \bar{x} = 3.21, \qquad S_{xx} = 2.95$$

Therefore,

$$1.2442 \sqrt{\left[1 + \frac{1}{12} + \frac{(3.495 - 3.21)^2}{2.95}\right]} = 1.3111$$

Consequently, the prediction interval is given by

$$26.78 \pm 2.228 \times 1.311 = [23.86, 29.70]$$

The scatter plot, estimated regression line, and 95 percent prediction intervals for the fuel efficiency data of Table 10.2 are displayed in Fig. 10.8.

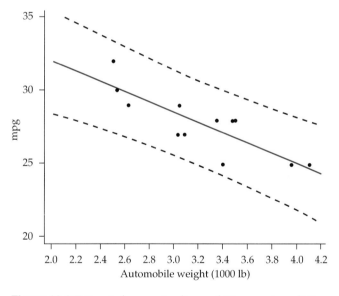

Figure 10.8 Estimated regression line and 95 percent prediction limits for future response Y (data in Table 10.2)   ∎

10.3.4   Displaying the Output of a Regression Analysis in an ANOVA Table

Regression analysis is a computationally intensive process that is best carried out via a statistical software package. The computer output, however, is displayed in the form of an *analysis of variance (ANOVA)* table, a concept involving some new ideas of independent interest. The ANOVA approach to regression analysis is based on the sum of squares decomposition SST = SSR + SSE [Eq. (10.15)] and a partition of the degrees of freedom (DF) associated with the sum of squares SST. The following elementary example illustrates the basic idea: Suppose $X_1, \ldots, X_n$ are iid $N(0, 1)$ random variables. Then it is well known (see Theorems 6.4 and 6.9) that

1. $\sum X_i^2$ has a chi-square distribution with n degrees of freedom.
2. $\sum (X_i - \overline{X})^2$ has a chi-square distribution with $n - 1$ degrees of freedom.
3. $n\overline{X}^2$ has a chi-square distribution with 1 degree of freedom.

The partition of the sum of squares

$$\sum X_i^2 = n\overline{X}^2 + \sum (X_i - \overline{x})^2$$

yields the following partition of the degrees of freedom

$$n = 1 + (n - 1)$$

Similarly, the partition of the degrees of freedom corresponding to the sum of squares partition

$$\sum (y_i - \overline{y})^2 = \sum (\hat{y}_i - \overline{y})^2 + \sum (y_i - \hat{y}_i)^2$$

is given by

$$n - 1 = 1 + (n - 2)$$

The SST term on the left hand side has $n - 1$ degrees of freedom, and the SSE term on the right has $n - 2$ degrees of freedom. This is a consequence of the fact that the random variables SSR and SSE are mutually independent. (We omit the technical details.)

A *mean square* (MS) is a sum of squares divided by its degrees of freedom. Thus, the *mean square due to regression* (MSR) and *mean square error* (MSE) are calculated by dividing SSR and SSE by their respective degrees of freedom. Their ratio, MSR/MSE, is a random variable denoted F because it has an F distribution, as will be explained shortly. These quantities play an important role in statistical inferences about the regression coefficients.

The sum of squares partition and the partition of the degrees of freedom are conveniently summarized in an ANOVA table, as shown in Table 10.5.

TABLE 10.5 ANOVA TABLE FOR THE SIMPLE LINEAR REGRESSION MODEL

Source	DF	Sum of squares	Mean square	F value	Prob > F
Model	1	SSR	MSR = SSR/1	MSR/MSE	
Error	$n - 2$	SSE	MSE = SSE/$(n - 2)$		
Total	$n - 1$	SST			

We use the F ratio MSR/MSE to test the null hypothesis H_0: $\beta_1 = 0$ against H_1: $\beta_1 \neq 0$. This follows from the fact that MSR = SSR in the simple linear regression model and from Eq. (10.32), which states that

$$E(\text{MSR}) = \sigma^2 + \beta_1^2 S_{xx}$$

Consequently, when H_1: $\beta_1 \neq 0$ is true, we have

$$E(\text{MSR}) = \sigma^2 + \beta_1^2 S_{xx} > \sigma^2 = E(\text{MSE})$$

We therefore reject H_0 at the level α when $F > F_{1,n-2}(\alpha)$. This test is in fact equivalent to the one based on the t distribution discussed previously [Eq. (10.39)]. This is a consequence of the result that $t_\nu^2 = F_{1,\nu}$ (Corollary 6.2).

---

example **10.9** Display the results of the regression analysis of Example 10.1 in the format of an ANOVA table.

Solution. The ANOVA table of the edited SAS printout is shown in Table 10.6.

TABLE 10.6

Analysis of Variance

Source	DF	Sum of Squares	Mean Square	F Value	Prob>F
Model	1	66440.38346	66440.38346	75.828	0.0001
Error	8	7009.61654	876.20207		
C Total	9	73450.00000			

Root MSE	29.60071	R-square	0.9046

Interpreting the Output. In Table 10.6 the names for SSR, SSE, and SST are listed in the first column, titled *Source*, which stands for the source of the variation. SSR is called *Model*, SSE is called *Error*, and SST is called *C Total*, which is short for "corrected total."[1] The second column, titled DF, indicates the degrees of freedom. The values of SSR, SSE, and SST appear in column 3, *Sum of Squares*. The value of the F ratio appears in the column titled F *value*, and the last column (*Prob > F*) contains the P-value, which in this case equals $P(F_{1,8} > 75.828) < 0.001$. Refer back to Table 10.4, where the value of the t statistic for testing H_0: $\beta_1 = 0$ is calculated as 8.708, and note that $75.828 = (8.708)^2$, a consequence of the fact that $F_{1,\nu} = t_\nu^2$. In other words, the F test in the ANOVA table and the t test for $\beta_1 = 0$ are equivalent.

The term *Root MSE* in the table is the square root of the mean square error; that is, $\sqrt{\text{MSE}} = \sqrt{876.20207} = 29.60071$. The coefficient of determination is printed as *R-square* in the table; in this case $R^2 = 0.9046$.

---

[1]In SAS terminology the sum $\sum (y_i - \bar{y})^2$ is called *C Total* to distinguish it from the sum $\sum y_i^2$, the uncorrected sum of squares.

example **10.10**   Display the results of the regression analysis of Model(Y = mpg $|$ X = weight) of Example 10.2 in the format of an ANOVA table.

Solution. The SAS printout of the ANOVA table is displayed in Table 10.7.

TABLE 10.7

Analysis of Variance

Source	DF	Sum of Squares	Mean Square	F Value	Prob>F
Model	1	34.76972	34.76972	22.461	0.0008
Error	10	15.48028	1.54803		
C Total	11	50.25000			

Root MSE		1.24420	R-square	0.6919	

∎

10.3.5   Curvilinear Regression

It sometimes happens that the points of the scatter plot appear to lie on a curve, ruling out the simple linear regression model. Nevertheless, it may still be possible to fit a curve to these data by transforming them so that the least squares method can be brought to bear on the transformed data. The following example illustrates the technique.

example **10.11**   The data in Table 10.8 come from a series of experiments to determine the relationship between the viscosity v (measurement units omitted) of the compound heptadecane as a function of the temperature t (measured in kelvins). The scatter plot of (t_i, v_i) shown in Fig. 10.9 indicates clearly that the data points lie on a curve, so the simple linear regression model is not appropriate.

Chemists believe that a curve defined by either $v = a\exp(b/t)$ or $v = at^b$ (where a and b are to be determined) provides a good fit to the observed data. We convert this curve-fitting problem to a linear regression problem by taking logarithms:

$$v = a\exp(b/t) \quad \text{so} \quad \ln v = \ln a + b\frac{1}{t} \tag{10.44}$$

$$v = at^b \quad \text{so} \quad \ln v = \ln a + b\ln t \tag{10.45}$$

Thus, to fit the curve defined by Eq. (10.44) we note that the transformed variables $y = \ln v$ and $x = 1/t$ satisfy the linear equation

$$y = \beta_0 + \beta_1 x, \quad (\beta_0 = \ln a, \beta_1 = b)$$

TABLE 10.8

t	v	$y = \ln v$	$1/t$	$\ln t$
303.15	3.291	1.19119	0.0032987	5.71423
313.50	2.652	0.97531	0.0031898	5.74780
323.15	2.169	0.77427	0.0030945	5.77812
333.15	1.829	0.60377	0.0030017	5.80859
343.15	1.557	0.44276	0.0029142	5.83817
353.15	1.340	0.29267	0.0028317	5.86689
363.15	1.161	0.14928	0.0027537	5.89482
373.15	1.014	0.01390	0.0026799	5.92198
393.15	0.794	−0.23067	0.0025436	5.97419
413.15	0.655	−0.42312	0.0024204	6.02381
433.15	0.546	−0.60514	0.0023087	6.07108
453.15	0.460	−0.77653	0.0022068	6.11622
473.15	0.392	−0.93649	0.0021135	6.15941
493.15	0.339	−1.08176	0.0020278	6.20081
513.15	0.296	−1.21740	0.0019487	6.24057
533.15	0.260	−1.34707	0.0018756	6.27880
553.15	0.229	−1.47403	0.0018078	6.31563
573.15	0.203	−1.59455	0.0017447	6.35115

Source: D. S. Viswanath and G. Natarjan, *Data Book on the Viscosity of Liquids*, New York, Hemisphere, 1989, p. 363. (Originally from I.F. Golubev, "Viscosity of Gases and Gas Mixtures," Moscow, Fizmat Press, 1959).

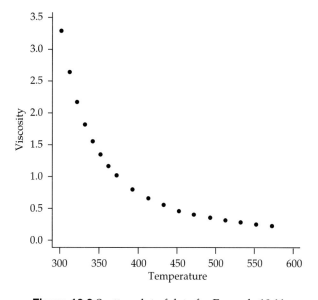

Figure 10.9 Scatter plot of data for Example 10.11

(Models of the type $v = a \exp bt$ or $v = at^b$ are called *intrinsically linear* because they can be transformed into linear models by means of a suitable change of variables.) Therefore, we use the least squares method to fit a straight line to the transformed data, shown in columns 3 and 4 of Table 10.8. That is, we solve the curvilinear regression problem by first fitting a straight line to the transformed data $y_i = \ln v_i$, $x_i = 1/t_i$, as in Sec. 10.2.1, and then estimate the parameters of the curve $v = a \exp(b/t)$ using the inverse transformation $a = \exp \beta_0$ and $b = \beta_1$; thus,

$$\hat{a} = \exp(\hat{\beta}_0) \quad \text{and} \quad \hat{b} = \hat{\beta}_1$$

Solution. The regression model statement for the transformed variables is Model($Y = \ln v \mid x = 1/t$). That is, we assume

$$Y_i = \beta_0 + \beta_1 x_i + \epsilon_i$$

where
$$Y_i = \ln v_i \quad \text{and} \quad x_i = \frac{1}{t_i}$$

In other words, we assume the simple linear regression model of Eq. (10.20) is valid for the transformed data. The scatter plot and estimated regression line for the transformed data, shown in Fig. 10.10, suggest a strong linear relation between the transformed variables. This is confirmed by the formal results of the regression analysis as summarized in Table 10.9, which shows the ANOVA table followed by the estimates for the intercept and slope:

$$\hat{\beta}_0 = -4.654467, \qquad \hat{\beta}_1 = 1754.551654$$
$$\hat{a} = \exp(-4.654467) = 0.00961273, \qquad \hat{b} = 1754.551654$$

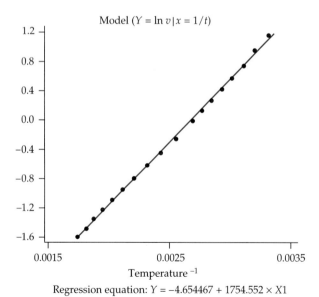

Regression equation: $Y = -4.654467 + 1754.552 \times X1$

Figure 10.10 Scatter plot and estimated regression line for transformed data of Example 10.11

TABLE 10.9 ANALYSIS OF VARIANCE AND PARAMETER ESTIMATES FOR MODEL($Y = \ln v \mid x = 1/t$)

Analysis of Variance

Source	DF	Sum of Squares	Mean Square	F Value	Prob>F
Model	1	13.36161	13.36161	21593.985	0.0001
Error	16	0.00990	0.00062		
C Total	11	13.37151			

Root MSE	0.02487	R-square	0.9993

Parameter Estimates

Variable	Parameter Estimate	Standard Error	T for H0: Parameter=0	Prob > |T|
INTERCEP	-4.654467	0.03026500	-153.790	0.0001
SLOPE	1754.551654	11.93987456	146.949	0.0001

The coefficient of determination is 0.9993. ∎

We now mention several problems arising in curvilinear regression that will not be dealt with here, but are discussed in R. H. Myers, *Classical and Modern Regression with Applications,* 2d ed., Boston, PWS-Kent, 1990:

1. There is in general no "best" solution to fitting a curve to a set of empirical data. For example, one can sometimes find a polynomial of sufficiently high degree that passes through all the points of the scatter plot.
2. It is important to note that fitting a straight line to the transformed data is not equivalent to applying the least squares method to the original data. For instance, minimizing the sum of squares

$$\sum_i [v_i - a\exp(bu_i)]^2$$

produces estimates for a and b that are not equal to the transforms of the least squares estimates. Thus, $\exp(\hat{\beta}_0) \neq \hat{a}$ and $\hat{\beta}_1 \neq \hat{b}$.
3. The form of the error term for an intrinsically linear model is also transformed. Suppose, for example, the transformed model

$$\ln y = \beta_0 + \beta_1 x + \epsilon$$

satisfies the standard conditions of the simple linear regression model. Taking exponentials of both sides yields the following model for the original data:

$$y = a\exp(bx)\exp(\epsilon), \quad a = \exp(\beta_0), \quad b = \beta_1$$

This is not the same as the model

$$y = a\exp(bx) + \epsilon$$

PROBLEMS

10.11 In Table 10.1 the weld diameter is the same for observations 4 and 6 ($x = 165$), but the shear strengths Y are different. How does the regression model (10.20) account for this?

10.12 A student gave the following expression for the fitted regression line of Fig. (10.4):

$$Y = -216.3271 + 3.990413 \times x + \epsilon$$

Explain why this expression is incorrect.

10.13 (a) Display the results of the regression analysis of Prob. 10.2 in the format of an ANOVA table (Table 10.5).
(b) What proportion of the variability is explained by the model?
(c) Compute a 95 percent confidence interval for the regression coefficients.
(d) Compute a 95 percent confidence interval for the mean shear strength when $x = 220$.

10.14 (a) Display the results of the regression analysis of Prob. 10.3 in the format of an ANOVA table.
(b) What proportion of the variability is explained by the model?
(c) Compute a 95 percent confidence interval for the regression coefficients.
(d) Compute a 95 percent confidence interval for the mean CPU time when $x = 30$.
(e) Compute a 95 percent prediction interval for the CPU time when $x = 30$.

10.15 (This is a continuation of Example 10.2.)
(a) Using the computer output displayed in Table 10.7 (and the result that $\bar{x} = 3.21$, $S_{xx} = 2.95$), construct 95 percent confidence intervals for β_0 and β_1.
(b) Compute a 95 percent confidence interval for the mpg rate for an automobile weighing 2530 lb.
(c) Compute a 95 percent prediction interval for the mpg Y of a 1992 Escort that also weighs 2530 lb. Compare the predicted value to the actual value, which is 30 mpg.

10.16 (This is a continuation of the Example 10.2.) The ANOVA table for Model($Y = $ gpm $\mid X = $ weight) follows.

```
                      Analysis of Variance

                        Sum of          Mean
Source          DF      Squares         Square      F Value       Prob>F

Model           1       0.59506        0.59506      23.916        0.0006
Error           10      0.24881        0.02488
C Total         11      0.84387

     Root MSE       0.15774      R-square       0.7052

                      Parameter Estimates

                  Parameter       Standard     T for H0:
Variable   DF     Estimate        Error      Parameter=0     Prob > |T|

INTERCEP   1      2.179754     0.29867670       7.298          0.0001
SLOPE      1      0.449247     0.09186264       4.890          0.0006
```

(a) Using the computer output in the ANOVA table (and the result that $\bar{x} = 3.21$, $S_{xx} = 2.95$), construct 95 percent confidence intervals for β_0 and β_1.

(b) Compute a 95 percent confidence interval for the mean gpm for an automobile that weighs 2530 lb.

(c) Compute a 95 percent prediction interval for the gpm of a 1992 Escort that also weighs 2530 lb. Compare the predicted value to the actual value, which is 3.34 gpm.

10.17 The following data were obtained in a comparative study of a remote procedure call (RPC) on two computer operating systems, UNIX and ARGUS.

UNIX		ARGUS	
Data bytes	Time	Data bytes	Time
64	26.4	92	32.8
64	26.4	92	34.2
64	26.4	92	32.4
64	26.2	92	34.4
234	33.8	348	41.4
590	41.6	604	51.2
846	50.0	860	76.0
1060	48.4	1074	80.8
1082	49.0	1074	79.8
1088	42.0	1088	58.6
1088	41.8	1088	57.6
1088	41.8	1088	59.8
1088	42.0	1088	57.4

Source: R. Jain, *The Art of Computer Systems Performance Analysis,* New York, John Wiley & Sons, 1991. Used with permission.

(a) Draw the scatter plot for Model(Y = time | X = data bytes) (UNIX data).

(b) Display the results of the regression analysis in the format of Table 10.5.

(c) What proportion of the variability is explained by the model?

(d) Compute 95 percent confidence intervals for the regression coefficients. Save your computations.

10.18 (Continuation of Prob. 10.17)

(a) Draw the scatter plot for Model(Y = time | X = data bytes) (ARGUS data).

(b) Display the results of the regression analysis in the format of Table 10.5.

(c) What proportion of the variability is explained by the model?

(d) Compute 95 percent confidence intervals for the regression coefficients. Save your computations.

10.19 Twelve batches of plastic are made, and from each batch one test item was molded and its Brinell hardness y was measured at time x. The results are recorded in the following table.

Time (x)	Hardness (y)	Time (x)	Hardness (y)
32	230	40	248
48	262	48	279
72	323	48	267
64	298	24	214
48	255	80	359
16	199	56	305

Source: J. Neter, W. Wasserman, and M. H. Kutner, *Applied Linear Statistical Models,* 2d ed., Homewood, Ill., Richard D. Irwin, 1985. Used with permission.

(a) Draw the scatter plot for Model(Y = hardness | X = time).

(b) Display the results of your regression analysis in the format of the ANOVA table (Table 10.5).

(c) Plot the estimated regression line and comment on how well the estimated regression line fits the data.

(d) What proportion of the variability is explained by the model?

(e) Compute 95 percent confidence intervals for the regression coefficients.

(f) Give a point estimate of the mean hardness when x = 48. Save your computations.

10.20 Refer to Example 10.11 and Table 10.8. Fit the curve of Eq. (10.45) by carrying out a linear regression analysis on the transformed data $y = \ln v$, $x = \ln t$. Graph the scatter plot and estimated regression line for the transformed data.

10.21 Kepler's third law of planetary motion states that the square of the period (P) of revolution of a planet is proportional to the cube of its mean distance (D) from the sun; that is, $P^2 = kD^3$. Taking the square root of both sides yields the formula $P = cD^{3/2}$; $c = (2\pi/\sqrt{\gamma M})$, where M is the mass of the sun and γ is the gravitational constant. We simplify Kepler's third law by taking the unit period as one earth year and the unit of distance to be the earth's mean distance to the sun (these are also called *astronomical units*). Thus, $P = D^{3/2}$. The periods of revolution and the distances for all nine planets are shown in the following table. Use the transformation $y = \ln P$, $x = \ln D$ to find the curve $P = aD^b$ that best fits the data. Are your results consistent with Kepler's predicted values: $a = 1$, $b = 1.5$?

Planet	Distance	Period
Mercury	0.387	0.241
Venus	0.723	0.615
Earth	1.0	1.0
Mars	1.524	1.881
Jupiter	5.202	11.862
Saturn	9.555	29.458
Uranus	19.218	84.01
Neptune	30.109	164.79
Pluto	39.44	248.5

Source: H. Skala, *College Mathematics Journal*, vol. 27, no. 3, May 1996, pp. 220–223. *Note:* There is a typo in Professor Skala's table: The headings for the periods and distances have been interchanged.

10.22 The data in the following table come from measuring the average speeds (s) (in meters/minute) for racing shells seating different numbers of rowers (n). From calculations based on fluid mechanics, it is known that the speed is proportional to the 1/9th power of the number of rowers; that is, $s = cn^{1/9}$. Use the transformation $y = \ln s$, $x = \ln n$ to find the curve $s = an^b$ that best fits the data. Compare your results with the theoretical prediction $b = 1/9$.

n	s	n	s
1	279	4	316
1	276	4	312
1	275	4	309
1	279	4	326
2	291	8	341
2	289	8	338
2	287	8	343
2	295	8	349

Source: H. Skala, *College Mathematics Journal*, vol. 27, no. 3, May 1996, pp. 220–223.

10.23 (a) The data in the next table are pressures, in atmospheres, of oxygen gas kept at 25°C when made to occupy various volumes, measured in liters. Draw the scatter plot of (v_i, p_i). Is the simple linear regression model appropriate?

(b) Chemists believe that the relation between the pressure and volume of a gas at constant temperature is given by $p = av^b$. Draw the scatter plot for the transformed data (x_i, y_i), where $y = \ln p$ and $x = \ln v$. Is the linear regression model appropriate for the transformed data?

(c) Fit a straight line to the transformed data by the method of least squares and determine the point estimates for the parameters a and b.

(d) Are these results consistent with Boyle's law for an ideal gas, which states that $pv = k$? That is, the product of pressure by volume is constant at constant temperature.

Volume (v)	Pressure (P)
3.25	7.34
5.00	4.77
5.71	4.18
8.27	2.88
11.50	2.07
14.95	1.59
17.49	1.36
20.35	1.17
22.40	1.06

Source: R. G. D. Steel and J. H. Torrie, *Principles and Procedures of Statistics,* 2d ed., New York, McGraw-Hill, 1980, p. 458. Used with permission.

10.24 (a) A biologist believes that the curve defined by the function

$$v = \frac{u}{a + bu}$$

provides a good fit to a data set. Show that the transformed variables $x = 1/u$, $y = 1/v$ lie on a straight line $y = \beta_0 + \beta_1 x$, and express β_0 and β_1 in terms of a and b.

(b) Sketch the curve $v = u/(1 + u)$ for $0 \leq u \leq 2$.

10.4 MODEL CHECKING

An important part of any regression analysis is determining whether or not the standard assumptions of the simple linear regression model are satisfied. We recall that the relationship between the response and explanatory variables is assumed to be of the form $Y = f(x) + \epsilon$, where $f(x) = \beta_0 + \beta_1 x$ is a (nonrandom) function of x, and ϵ represents the experimental error. It is instructive to think of $f(x)$ as a signal, ϵ as the noise, and Y as the observed, or received, signal. The *goal of regression analysis is to extract the signal f(x) from the noise. Model checking* refers to a collection of graphical displays, called *residual plots*, that help the statistician evaluate the validity of the linear regression model.

Residual Plots

It can now be shown that the residuals are normally distributed with mean, variance, and estimated standard error given by

$$E(e_i) = 0$$

$$V(e_i) = \sigma^2 \left(1 - \left[\frac{1}{n} + \frac{(x_i - \bar{x})^2}{S_{xx}} \right] \right)$$

$$s(e_i) = s \sqrt{\left(1 - \left[\frac{1}{n} + \frac{(x_i - \bar{x})^2}{S_{xx}} \right] \right)}$$

The residuals are not, however, mutually independent since $\sum_i e_i = 0$. It can be shown, however, that the vector of residuals $(e_1, \ldots, e_n)$ and the vector of fitted values $(\hat{y}_1, \ldots, \hat{y}_n)$ are mutually independent. These results imply that when the regression model is valid, the residual plot $(\hat{y}_i, e_i)$ should appear to be a set of points randomly scattered about the line $y = 0$ and such that approximately 95 percent of them lie within $\pm 2s$ of zero. Another plot, closely related to the one just described, is the *studentized residual plot* consisting of $(\hat{y}_i, e_i^*)$, where

$$\text{Studentized residual:} \quad e_i^* = \frac{e_i}{s(e_i)}$$

The studentized residuals should appear to be randomly scattered about the line $y = 0$ and approximately 95 percent of them should lie within ± 2 of zero. Because of the tedious nature of the computations, you are advised to graph these plots using a computer.

---

example **10.12**   Draw the plots of (1) the residuals and (2) the studentized residuals for the regression Model(Y = shear strength $\mid X$ = weld diameter) of Example 10.1. Are the residual plots consistent with the hypotheses of the simple linear regression model?

Solution. The residual and studentized residual plots are shown in Figs. 10.11 and 10.12. Looking at the plot of the studentized residuals (Fig. 10.12), we note that the points appear to be randomly scattered within a horizontal band of width 2 centered at about zero; the same is true for the residuals plot, Fig. 10.11, since $\pm 2s = \pm 2 \times 29.6 = \pm 59.2$. This is consistent with the scatter plot and estimated regression line shown in Fig. 10.4. We conclude that the data do not conflict with the hypotheses of the simple linear regression model. ∎

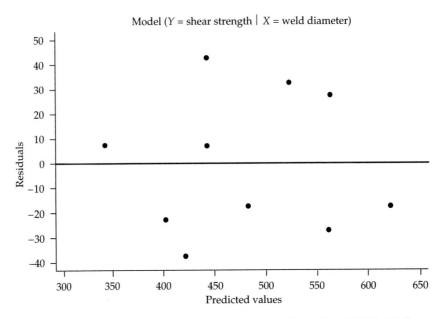

Figure 10.11 Plot of residuals against predicted values (data in Table 10.1)

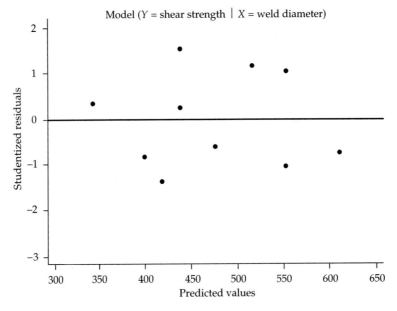

Figure 10.12 Plot of studentized residuals against predicted values (data in Table 10.1)

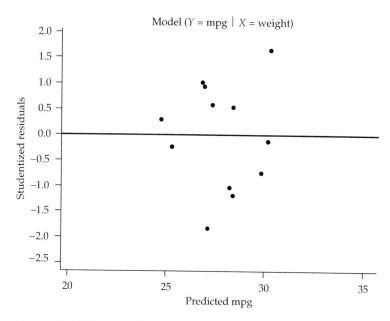

Figure 10.13 Plot of studentized residuals against predicted mpg (data in Table 10.2)

example **10.13** Draw the plot of the studentized residuals for the regression Model(Y = mpg | X = weight) of Example 10.2. Do the assumptions of the simple linear regression model appear to be valid?

Solution. The plot of the studentized residuals for Model(Y = mpg | X = weight) is shown in Fig. 10.13. All studentized residuals appear to be randomly scattered within a horizontal band of width 2 centered about 0, so the model appears to be consistent with the hypotheses of the simple linear regression model. ∎

example **10.14** Plot the studentized residuals for Model(Y = time | X = data bytes) (ARGUS data) of Prob. 10.17. What departures from the simple linear regression model are suggested by the residual plot?

Solution. The residual plot is shown in Fig. 10.14. We see that the variances of the residuals seem to increase with $\hat{y}_i$; the condition that the variances be constant appears to be violated in this case. This suggests that a linear regression model is not appropriate here. Additional evidence for this conclusion comes from a normal probability plot of the studentized residuals, as will now be explained. ∎

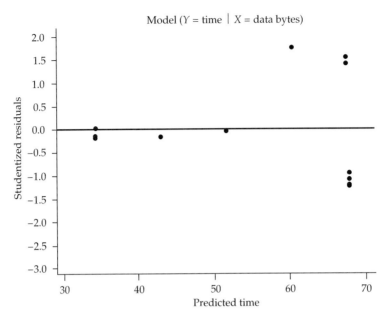

Figure 10.14 Plot of studentized residuals against predicted time (Example 10.14)

Normal Probability Plots of the Residuals

We use normal probability plots (defined in Sec. 8.4) of the studentized residuals to check the assumption that the error terms $\epsilon_1, \ldots, \epsilon_n$ are iid normal random variables.

---

example 10.15   Draw the normal probability plot of the studentized residuals for Model($Y =$ time | $X =$ data bytes) (ARGUS data) of Prob. 10.17. What departures from the simple linear regression model are suggested by the normal probability plot?

Solution. The normal probability plot is shown in Fig. 10.15. The points do not appear to lie on a straight line, which strengthens the conclusion obtained earlier that the simple linear regression model is inappropriate.   ■

We conclude with a brief summary of the criteria that statisticians recommend to assess the validity of the simple linear regression model.

The simple linear regression model is inappropriate if

1. The scatter plot does not exhibit, at least approximately, a linear relationship.
2. The residuals exhibit a pattern; that is, if the residuals do not appear to be randomly scattered about the line $y = 0$ or approximately 95 percent of them do not lie within $\pm 2s$ of zero, then the linear regression model might not be appropriate.
3. The error variances $V(\epsilon_1), \ldots, V(\epsilon_n)$ are not, at least approximately, equal. For an example of where this might not be the case, see Fig. 10.14.

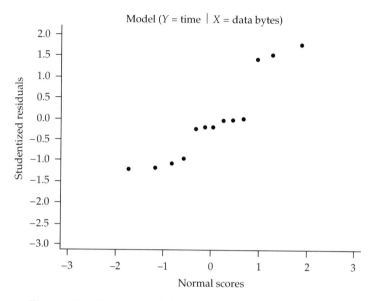

Figure 10.15 Normal probability plot of studentized residuals (Example 10.15)

4. The error terms $\epsilon_1, \ldots, \epsilon_n$ are not normally distributed; use normal probability plots.
5. The error terms are not independent. This may occur if the measurements are taken in a time sequence during which the experimental apparatus is affected by the physical conditions under which previous measurements were taken. To detect this, statisticians recommend a plot of the residuals versus the *time order* in which the observations are taken. If the independence assumption is valid, there should be no pattern, e.g., quadratic or linear, in the plot.

10.4.1 Correlation Analysis

In the simple linear regression model we assume that the response Y is a random variable that is related to the explanatory variable x by a linear relationship of the form $Y = \beta_0 + \beta_1 x + \epsilon$, where ϵ denotes an unobservable random variable representing the experimental error. As noted earlier, the strength of this linear relationship is measured by the coefficient of determination $R^2 = r^2$, where r is the sample correlation coefficient. Note that in the regression model Y is a random variable and x is not. Correlation analysis, in contrast, studies the linear relationship between two random variables X and Y, where the strength of the linear relationship is now measured by the (population) correlation coefficient ρ, defined by

$$\rho = \frac{\text{Cov}(X, Y)}{\sigma_X \sigma_Y}$$

where
$$\text{Cov}(X, Y) = E((X - \mu_X)(Y - \mu_Y))$$

If we want to emphasize the fact that the value of ρ depends on the specific random variables X and Y, we write $\rho(X, Y)$; in practice, however, this is seldom necessary.

We estimate the parameter ρ by the sample correlation coefficient r [Eq. (10.6)]. Computing r by hand is not recommended because it requires a series of tedious, error-prone computations that can be avoided through the use of either the statistical function keys on your calculator or a statistical software package in your computer laboratory.

example 10.16   Refer to Example 1.11. Compute the sample correlation coefficient between the Peabody scores X and the Raven scores Y.

Solution. We compute the sample correlation coefficient using a computer and obtain the result $0 < r = 0.38$. This confirms our preliminary conclusions (refer back to the discussion following the scatter plot in Fig. 1.7); that is, low Peabody scores appear to be associated with low Raven scores, and high Peabody scores appear to be associated with high Raven scores. The association is not a strong one, as there are a few data points that do not follow this pattern.   ■

example 10.17   The data set in Table 10.10 records the size, in billions of dollars, and the five-year rate of return of 25 General Equity mutual funds. Does the size of a mutual fund influence its rate of return? In particular, do larger funds tend to have larger rates of return?

TABLE 10.10   SIZE AND FIVE-YEAR RATE
OF RETURN FOR 25 MUTUAL FUNDS

Size	Return	Size	Return
12.325	104	2.328	79
6.534	62	2.198	92
5.922	97	2.171	62
5.606	88	2.154	60
4.356	61	2.1172	93
3.925	48	2.06	99
3.759	51	1.908	111
3.426	77	1.897	84
3.21	74	1.729	120
3.191	86	1.69	98
2.786	78	1.578	97
2.525	68	1.575	65
2.514	73		

Source: New York Times, 3 Feb 1991.

Solution. To answer this question, we plot the rate of return against the size and then compute the sample correlation coefficient. The scatter plot shown in

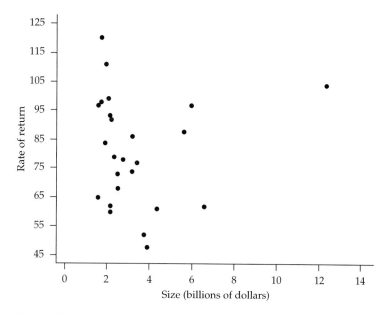

Figure 10.16 Scatter plot of rate of return of mutual fund against size (data in Table 10.10)

Fig. 10.16 does not indicate a positive, or a negative, association between the size of a mutual fund and its rate of return. The sample correlation equals 0.029, which means that the size of the fund should not be an important factor for an investor. ∎

A Statistical Model for Correlation Analysis

The statistical analysis of a bivariate data set, such as the scores of 36 girls on the Peabody and Raven mental tests (Table 1.5), assumes that the data come from a random sample drawn from the much larger population of all girls who take both exams. Specifically, we assume that the observations $(X_1, Y_1), \dots, (X_n, Y_n)$ come from a bivariate normal distribution with parameters μ_X, μ_Y, σ_X, σ_Y, ρ as defined by Eq. (5.65). The assumption that (X, Y) has a bivariate normal distribution can be justified on mathematical grounds (e.g., the so-called multidimensional central limit theorem) and empirical data since the frequency histograms of test scores often follow a bell-shaped curve. Recall that, in this case, the regression of Y on X (Definition 5.10) is given by

$$E(Y \mid X = x) = \mu_{Y|x} = \mu_Y + \rho \frac{\sigma_Y}{\sigma_X}(x - \mu_X) \qquad (10.46)$$

To estimate ρ, we use the sample correlation coefficient r, defined by

$$\text{Sample correlation:} \quad r = \frac{S_{XY}}{\sqrt{S_{XX}}\sqrt{S_{YY}}} \qquad (10.47)$$

where

$$S_{XX} = \sum (X_i - \overline{X})^2 = \sum X_i^2 - n\overline{X}^2$$

$$S_{YY} = \sum (Y_i - \overline{Y})^2 = \sum Y_i^2 - n\overline{Y}^2$$

$$S_{XY} = \sum (X_i - \overline{X})(Y_i - \overline{Y}) = \sum X_i Y_i - n\overline{XY} \qquad (10.48)$$

As noted earlier, the point estimate for the correlation coefficient between the Peabody and Raven test scores is $r = 0.38$ (Example 10.16), and the correlation coefficient between the size and return of 25 mutual funds is $r = 0.029$ (Example 10.17). Obtaining a confidence interval for ρ is complicated since the exact distribution of r is known only in the case when $\rho = 0$. For this reason, we content ourselves with discussing how to test the null hypothesis $H_0: \rho = 0$.

The Sampling Distribution of r**.** The following theorem gives the distribution of the sample correlation coefficient under the null hypothesis.

■■■ **THEOREM 10.4**

When $H_0: \rho = 0$ is true, then

$$T = \frac{r\sqrt{n-2}}{\sqrt{1-r^2}} \qquad (10.49)$$

has a t distribution with $n - 2$ degrees of freedom.   ■

---

example **10.18**   Based on the data of Example 10.17 and the T statistic of Eq. (10.49), does the size of a mutual fund determine its rate of return?

Solution. In this case there are several choices for the alternative hypothesis. For example, we might conjecture that a larger fund has a tendency to perform better, which suggests that we choose the alternative hypothesis to be $H_1: \rho > 0$; that is, we will reject the null hypothesis at the level α if $T > t_{n-2}(\alpha)$. Choosing $\alpha = 0.05$ and $n = 25$ produces the critical region $\{T > t_{23}(0.05) = 1.714\}$. The value of the T statistic in this case is given by

$$T = \frac{0.029\sqrt{23}}{\sqrt{1 - 0.029^2}} = 0.1391 < 1.714$$

Consequently, the null hypothesis cannot be rejected in this case. In other words, a larger fund size does not necessarily produce a larger rate of return.   ■

10.4.2   The Shapiro-Wilk Test for Normality

In Sec. 8.4 we introduced the Q-Q plot as a useful graphical tool for checking the normality of a given data set. Recall that the idea behind the Q-Q plot is this: If the data $\{x_1, \dots, x_n\}$ come from a normal distribution, then the points of the Q-Q plot $(z_j, x_{(j)}), j = 1, \dots, n$, will appear to lie on a straight line; here $x_{(1)}, \dots, x_{(n)}$ denote the order statistics and z_j denotes the normal scores, i.e.,

$\Phi(z_j) = (j - 0.5)/n$. We remark that the mean and variance of the order statistics are the same as the mean and variance of the original data set, since the order statistics are just a permutation of the original data set. Now, one way of measuring the straightness of the Q-Q plot is to compute its correlation coefficient, denoted in the SAS computer package by W; i.e., compute

$$W = \frac{\sum (x_{(j)} - \bar{x})(z_j - \bar{z})}{\sqrt{S_{xx}} \sqrt{S_{zz}}}$$

We reject the hypothesis of normality if $W < w$. The idea of basing a test of normality on this statistic is due to Shapiro and Wilk, and this test is available as an option in most computer software packages (e.g., PROC UNIVARIATE of SAS).

---

example **10.19**   Draw the normal probability plots and compute the Shapiro-Wilk statistic for the studentized residuals of (1) Model(Y = mpg | X = weight) and (2) Model(Y = gpm | X = weight).

Solution

1. The normal probability plot is shown in Fig. 10.17. Using SAS (PROC UNIVARIATE), we obtain the value $W = 0.9793$ with a P-value $P(W < 0.9793) = 0.9519$. The hypothesis of normality is not rejected.

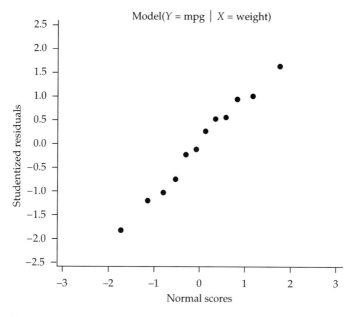

Figure 10.17 Normal probability plot of studentized residuals (data in Table 10.2)

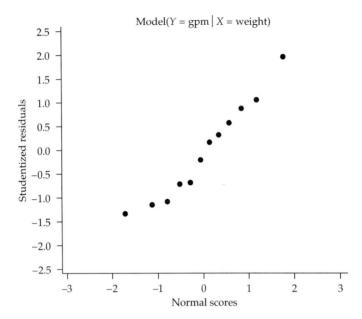

Figure 10.18 Normal probability plot of studentized residuals (data in Table 10.2)

2. The normal probability plot for the regression Model($Y = $ gpm $\mid X = $ weight) is shown in Fig. 10.18. Using SAS (PROC UNIVARIATE), we obtain the value $W = 0.9549$ for the Shapiro-Wilk statistic with a P-value $P(W < 0.9549) = 0.6579$. This is not strong evidence against the hypothesis of normality.

  We conclude that both models appear to satisfy the conditions of the simple linear regression model. So which one is better? Comparing the coefficients of determination of the two models gives a slight edge to Model($Y = $ gpm $\mid X = $ weight), since its coefficient of determination equals 0.7052 whereas the coefficient for Model($Y = $ mpg $\mid X = $ weight) equals 0.6919.   ∎

PROBLEMS

10.25 Refer back to Prob. 10.19.
  (a) Plot the studentized residuals against predicted hardness for the regression Model($Y = $ hardness $\mid X = $ time).
  (b) Draw the normal probability plots of the studentized residuals.
  (c) Are these plots consistent with the hypotheses of the simple linear regression model? Justify your conclusions.

10.26 Consider the following two bivariate data sets, each consisting of 11 (x, y) pairs. The two data sets have the same x values but different y values.

x	y	
	Data set I	Data set II
4	4.26	3.10
5	5.68	4.74
6	7.24	6.13
7	4.82	7.26
8	6.95	8.14
9	8.81	8.77
10	8.04	9.14
11	8.33	9.26
12	10.84	9.13
13	7.58	8.74
14	9.96	8.10

Source: F. J. Anscombe, "Graphs in Statistical Analysis," *American Statistician,* vol. 27, 1973, pp. 17–21. Used with permission.

(a) Draw the scatter plots of data sets I and II. Using only the information contained in these scatter plots, comment on the suitability of the simple linear regression model for each of the data sets.

(b) Compute the estimated regression line, coefficient of determination, and sample correlation coefficient for each of the data sets.

(c) Draw the residual plot for each of the data sets.

(d) Explain why only one of the linear regression models is appropriate even though the regression analyses for data sets I and II lead to the same parameter estimates, the same coefficient of determination, and the same standard errors.

10.27 Refer back to Prob. 10.18. Do you agree with the authors' conclusion that TDS can be estimated from the regression model with acceptable error? In particular, are the data consistent with the hypotheses of the simple linear regression model? Justify your conclusions by model checking; that is, write a brief report based upon an analysis of the scatter, residual, and normal probability plots for the regression Model($Y = $ TDS $| X = $ SEC).

10.28 Refer back to the UNIX data of Prob. 10.17.

(a) Draw the studentized residual and normal probability plots for Model($Y = $ time $| X = $ data bytes) (UNIX data).

(b) Do these plots and the scatter plot indicate any departures from the simple linear regression model? Summarize your conclusions in a brief paragraph or two.

10.29 Calculate the sample correlation coefficient r for the following artificial data:

x	1	2	3	4	5	6
y	2	2	3	5	4	6

10.30 Refer to the fathers' heights X and weights Y shown in the StatLab data set (Table 1.5).
(a) Draw the scatter plot and guess the value of r.
(b) Calculate r.

10.31 Refer back to of the CO_2 emissions data in Prob. 1.8. In this problem we explore the relationships between some of the variables $X_1, \ldots, X_4$.
(a) Draw the scatter plot of (X_2, X_3). In this context, would you expect a positive, zero, or negative correlation? Justify your answer.
(b) Compute the sample correlation coefficient r between X_2 and X_3, and discuss its usefulness in explaining the relationship between X_2 and X_3.
(c) Draw the scatter plot of (X_3, X_4). In this context, would you expect a positive, zero, or negative correlation? Justify your answer.
(d) Compute the sample correlation coefficient r between X_3 and X_4, and discuss its usefulness in explaining the relationship between X_3 and X_4.

10.32 The following table lists the number of vehicles per 10,000 population (VPP), the number of deaths per 10,000 vehicles (DPV), and the number of deaths per 10,000 population (DPP) for 12 countries for the year 1987. In this problem we apply the techniques of correlation analysis to explore the relationships between these variables.

Country	VPP, X_1	DPV, X_2	DPP, X_3
USA	7322	2.6	1.9
New Zealand	5950	4.0	2.4
UK	3701	2.5	0.9
Federal Republic of Germany	4855	2.7	1.3
Japan	4078	1.9	0.8
Finland	2219	3.0	1.2
Denmark	3367	3.7	1.4
Kuwait	3014	4.8	1.5
Jordan	771	17.7	1.4
Thailand	50	46.7	0.2
Niger	278	13.0	0.4
Pakistan	40	121.4	0.5

Source: K. S. Jadaan, "An Overview of Road Safety in New Zealand," *ITE Journal,* April 1993.

(a) Draw the scatter plot of (X_1, X_2). In this context, would you expect a positive, zero, or negative correlation? Justify your answer.
(b) Compute the correlation coefficient r between X_1 and X_2, and discuss its usefulness in explaining the relationship between X_1 and X_2.
(c) Draw the scatter plot of (X_1, X_3). In this context, would you expect a positive, zero, or negative correlation? Justify your answer.
(d) Compute the correlation coefficient r between X_1 and X_3, and discuss its usefulness in explaining the relationship between X_1 and X_3.

10.33 The liver disease cirrhosis is a side effect of alcohol consumption. The following table, based on statistics compiled by the Commodity Board of the Spirits Industry, records the rates of consumption of pure alcohol per capita for various countries and the cirrhosis death rates per 100,000 men in 1992 (or the most recent year of data).

Country	Alcohol consumption (quarts per capita)	Cirrhosis deaths per 100,000 men in 1992 (or most recent data)
Luxembourg	13.3	27.0
France	12.2	23.9
Austria	11.1	41.6
Germany	11.0	33.1
Portugal	11.0	40.6
Hungary	10.8	104.5
Denmark	10.6	18.8
Switzerland	10.6	28.9
Spain	10.6	28.9
Greece	9.7	13.2
Belgium	9.6	14.5
Czechoslovakia	9.4	14.5
Italy	9.1	34.8
Bulgaria	8.8	28.9
Ireland	8.8	3.2
Rumania	8.5	50.1
Netherlands	8.4	6.2
Australia	7.9	8.8
Argentina	7.8	15.0
Britain	7.7	7.2
NewZealand	7.7	5.1
Finland	7.2	13.7
USA	7.2	13.7
Japan	7.0	19.1
Canada	6.9	11.1
Poland	6.7	15.6
Sweden	5.6	10.0

Draw the scatter plot of deaths Y against alcohol consumption X. Do the data support the existence of a positive association between the two variables? Are there any data points that depart from this pattern? Discuss.

10.5   MATHEMATICAL DETAILS AND DERIVATIONS

The details of the derivations of Eq. (10.7) and the sampling distributions for $\hat{\beta}_0$, $\hat{\beta}_1$, and related random variables will now be given.

Derivation of the Sum of Squares Partition (10.7)

We begin with the following algebraic identity:

$$y_i - b_0 - b_1 x_i = (y_i - \bar{y}) - b_1(x_i - \bar{x}) + (\bar{y} - b_0 - b_1\bar{x})$$

The corresponding sum of squares $Q(b_0, b_1)$ takes the form

$$Q(b_0, b_1) = \sum [(y_i - \bar{y}) - b_1(x_i - \bar{x}) + (\bar{y} - b_0 - b_1\bar{x})]^2$$

This can be simplified by noting that

$$\sum (x_i - \bar{x}) = \sum (y_i - \bar{y}) = 0$$

and therefore

$$\sum (y_i - \bar{y})\bar{y} = 0$$

$$\sum (y_i - \bar{y})(\bar{y} - b_0 - b_1\bar{x}) = 0$$

$$\sum b_1(x_i - \bar{x})(\bar{y} - b_0 - b_1\bar{x}) = 0$$

Consequently,

$$
\begin{aligned}
Q(b_0, b_1) &= \sum (y_i - \bar{y})^2 - 2\sum b_1(x_i - \bar{x})(y_i - \bar{y}) + b_1^2 \sum (x_i - \bar{x})^2 \\
&\quad + n(\bar{y} - b_0 - b_1\bar{x})^2 \\
&= S_{yy} - 2b_1 S_{xy} + b_1^2 S_{xx} + n(\bar{y} - b_0 - b_1\bar{x})^2 \\
&= S_{yy}(1 - r^2) + (b_1\sqrt{S_{xx}} - r\sqrt{S_{yy}})^2 + n(\bar{y} - b_0 - b_1\bar{x})^2
\end{aligned}
$$

where the second summand in the last equation comes from "completing the square" of the term $-2b_1 S_{xy} + b_1^2 S_{xx}$.   ■

Proof of Theorem 10.1

The most efficient way to derive the sampling distribution for $\hat{\beta}_1$ is to rewrite it as a linear combination of mutually independent, normally distributed random variables Y_i of the sort discussed in conjunction with Theorem 6.1. The key idea is the representation

$$\hat{\beta}_1 = \sum d_i Y_i \tag{10.50}$$

where

$$d_i = \frac{x_i - \bar{x}}{S_{xx}} \tag{10.51}$$

The representation (10.50) is a consequence of the following alternative expression for S_{xy}:

$$S_{xy} = \sum (x_i - \bar{x})y_i \tag{10.52}$$

To derive Eq. (10.52), we perform the following sequence of algebraic manipulations:

$$S_{xy} = \sum (x_i - \bar{x})(y_i - \bar{y})$$

$$= \sum (x_i - \bar{x})y_i - \sum (x_i - \bar{x})\bar{y}$$

$$= \sum (x_i - \bar{x})y_i$$

since

$$\sum (x_i - \bar{x})\bar{y} = \bar{y}\sum (x_i - \bar{x}) = 0$$

Incidentally, the same argument shows that $S_{xx} = \sum x_i(x_i - \bar{x})$. Thus, $S_{xY} = \sum (x_i - \bar{x})Y_i$ and Eq. (10.23) together imply that

$$\hat{\beta}_1 = \frac{S_{xY}}{S_{xx}}$$

$$= \sum \frac{(x_i - \bar{x})}{S_{xx}} Y_i$$

This completes the derivation of Eq. (10.50).

Applying Theorem 6.1 and using the facts that $E(Y_i) = \beta_0 + \beta_1 x_i$ and $V(Y_i) = \sigma^2$ yield the result that $\hat{\beta}_1$ is normally distributed with mean

$$E(\beta_1) = \sum \left(\frac{x_i - \bar{x}}{S_{xx}} \right)(\beta_0 + \beta_1 x_i)$$

$$= \frac{\beta_0}{S_{xx}} \sum (x_i - \bar{x}) + \frac{\beta_1}{S_{xx}} \sum x_i(x_i - \bar{x})$$

$$= \frac{\beta_0}{S_{xx}} \times 0 + \frac{\beta_1}{S_{xx}} \times S_{xx} = \beta_1$$

and variance

$$V(\hat{\beta}_1) = \sum \left(\frac{x_i - \bar{x}}{S_{xx}} \right)^2 \times \sigma^2$$

$$= \frac{S_{xx}}{S_{xx}^2} \times \sigma^2 = \frac{\sigma^2}{S_{xx}}$$

The sampling distribution of $\hat{\beta}_0$ is derived in a similar way; we only sketch the outlines, leaving the details as an exercise.

A routine calculation shows that

$$E(\bar{Y}) = \beta_0 + \beta_1 \bar{x}$$

and therefore,

$$E(\hat{\beta}_0) = E(\bar{Y} - \hat{\beta}_1 \bar{x})$$

$$= E(\bar{Y}) - E(\hat{\beta}_1)\bar{x}$$

$$= \beta_0 + \beta_1 \bar{x} - \beta_1 \bar{x} = \beta_1$$

This shows that $\hat{\beta}_0$ is an unbiased estimator of β_0, as claimed. The next step is to represent $\hat{\beta}_0$ as a linear combination of $Y_1, \ldots, Y_n$:

$$\hat{\beta}_0 = \overline{Y} - \hat{\beta}_1 \overline{x} = \sum c_i Y_i \tag{10.53}$$

where

$$c_i = \frac{1}{n} - \overline{x} \left(\frac{x_i - \overline{x}}{S_{xx}} \right) \tag{10.54}$$

Since the random variables $Y_1, \ldots, n$ are independent, it follows that

$$V \left(\sum c_i Y_i \right) = \sum c_i^2 \sigma^2 = \sigma^2 \sum c_i^2$$

We leave it to you as a problem to show that

$$\sum c_i^2 = \sum \left[\frac{1}{n} - \overline{x} \left(\frac{x_i - \overline{x}}{S_{xx}} \right) \right]^2$$

$$= \frac{1}{n} + \frac{\overline{x}^2}{S_{xx}} \tag{10.55}$$

Therefore,

$$V(\hat{\beta}_0) = \sigma^2 \left[\frac{1}{n} + \frac{\overline{x}^2}{S_{xx}} \right] \tag{10.56}$$

This completes the proof of Theorem 10.1.

Derivation of Eq. (10.32)

Using the fact that the fitted values and $\overline{y}$ are given by

$$\hat{y}_i = \hat{\beta}_0 + \hat{\beta}_1 x_i \quad \text{and} \quad \overline{y} = \hat{\beta}_0 + \hat{\beta}_1 \overline{x}$$

we see that

$$E(\text{SSR}) = E \left(\sum (\hat{y}_i - \overline{y})^2 \right)$$

$$= E \left(\sum \hat{\beta}_1^2 (x_i - \overline{x})^2 \right)$$

$$= E(\hat{\beta}_1^2) S_{xx} = \left(\frac{\sigma^2}{S_{xx}} + \beta_1^2 \right) S_{xx}$$

$$= \sigma^2 + \beta_1^2 S_{xx}$$

The Sampling Distribution of $\hat{\beta}_0 + \hat{\beta}_1 x^*$

It follows from Eqs. (10.53) and (10.50) that

$$\hat{\beta}_0 + \hat{\beta}_1 x^* = \sum (c_i + d_i x^*) Y_i$$

$$= \sum \left(\frac{1}{n} + \frac{(x^* - \bar{x})(x_i - \bar{x})}{S_{xx}} \right) Y_i \qquad (10.57)$$

where x^* is a specified value of the explanatory variable. Thus, we are able to represent the random variable $\hat{\beta}_0 + \hat{\beta}_1 x^*$ as a linear combination of mutually independent, normally distributed random variables Y_i of the sort discussed in the context of Theorem 6.1.

Applying Theorem 6.1 and using the facts that $E(Y_i) = \beta_0 + \beta_1 x_i$ and $V(Y_i) = \sigma^2$ yields the result that $\hat{\beta}_0 + \hat{\beta}_1 x^*$ is normally distributed with mean, variance, and estimated standard error given by

$$E(\hat{\beta}_0 + \hat{\beta}_1 x^*) = \beta_0 + \beta_1 x^* \qquad (10.58)$$

$$V(\hat{\beta}_0 + \hat{\beta}_1 x^*) = \sigma^2 \left[\frac{1}{n} + \frac{(x^* - \bar{x})^2}{S_{xx}} \right] \qquad (10.59)$$

$$s(\hat{\beta}_0 + \hat{\beta}_1 x^*) = s \sqrt{\left[\frac{1}{n} + \frac{(x^* - \bar{x})^2}{S_{xx}} \right]} \qquad (10.60)$$

Derivation of Eq. (10.31)

Equation (10.31) asserts that $s^2 = \text{SSE}/(n - 2)$ is an unbiased estimator of σ^2. It is equivalent to

$$E(\text{SSE}) = E \left(\sum (Y_i - \hat{Y}_i)^2 \right) = (n - 2)\sigma^2 \qquad (10.61)$$

Derivation of Eq. (10.61). We begin with the representation (the proof is omitted)

$$\text{SSE} = S_{YY} - \hat{\beta}_1^2 S_{xx}$$

Taking expectations of both sides yields

$$E(\text{SSE}) = E(S_{YY}) - E(\hat{\beta}_1^2) S_{xx}$$

where

$$E(\hat{\beta}_1^2) = V(\hat{\beta}_1) + E(\hat{\beta}_1)^2 = \frac{\sigma^2}{S_{xx}} + \beta_1^2$$

Similarly,

$$E(S_{YY}) = E \left(\sum (Y_i - \bar{Y})^2 \right)$$

$$= E \left(\sum Y_i^2 \right) - n E(\bar{Y})^2$$

where

$$E(Y_i^2) = \sigma^2 + (\beta_0 + \beta_1 x_i)^2$$

$$E(\overline{Y}^2) = \frac{\sigma^2}{n} + (\beta_0 + \beta_1 \overline{x})^2$$

Therefore,

$$E(S_{YY}) = n\sigma^2 + \sum (\beta_0 + \beta_1 x_i)^2 - n\left[\frac{\sigma^2}{n} + (\beta_0 + \beta_1 \overline{x})^2\right]$$

$$= (n-1)\sigma^2 + \sum (\beta_0 + \beta_1 x_i)^2 - n(\beta_0 + \beta_1 \overline{x})^2$$

$$= (n-1)\sigma^2 + \sum [(\beta_0 + \beta_1 x_i) - (\beta_0 + \beta_1 \overline{x})]^2$$

$$= (n-1)\sigma^2 + \beta_1^2 \sum (x_i - \overline{x})^2$$

$$= (n-1)\sigma^2 + \beta_1^2 S_{xx}$$

Consequently,

$$E(\text{SSE}) = E(S_{YY}) - E(\hat{\beta}_1^2) S_{xx}$$

$$= (n-1)\sigma^2 + \beta_1^2 S_{xx} - \left(\frac{\sigma^2}{S_{xx}} + \beta_1^2\right) S_{xx}$$

$$= (n-2)\sigma^2$$

PROBLEMS

10.34 Show that $E(\overline{Y}) = \beta_0 + \beta_1 \overline{x}$.

10.35 Verify the representation for $\hat{\beta}_0$ displayed in Eq. (10.53).

10.36 Perform the necessary algebraic manipulations to derive Eq. (10.55).

10.6   CHAPTER SUMMARY

The regression model is used to study the relationship between a response variable y and an explanatory variable x. We assume that the relationship between the explanatory and response variables is given by a function $y = f(x)$, where f is unknown. A linear regression model assumes that $f(x) = \beta_0 + \beta_1 x$, and a curvilinear regression model assumes that the function $f(x)$ is nonlinear. When the function is intrinsically linear, we can reduce a curvilinear regression model to a linear one by making a suitable change of variables. The validity of the model hypotheses is examined by graphing scatter plots and residual plots. Correlation analysis studies the linear relationship between two random variables X and Y. The Shapiro-Wilk test for normality is a correlation test for assessing the linearity of the points of a Q-Q plot.

To Probe Further. The following well-written texts contain a wealth of additional information on regression analysis, including a more careful discussion of model checking and nonlinear regression.

1. R. H. Myers, *Classical and Modern Regression with Applications,* 2d ed., Boston, PWS-Kent, 1990.
2. J. Neter, M. H. Kutner, C. J. Nachtseim, and W. Wasserman, *Applied Linear Statistical Models*, 4th ed., Homewood, Ill., Richard D. Irwin, 1996.

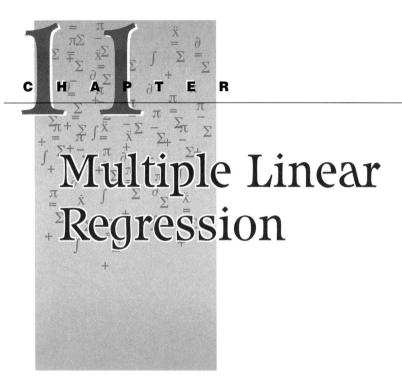

Multiple Linear Regression

Data analysis without guidance is foolish, but uncritical belief in a specific model is dangerously foolish.

J. W. Tukey and M. B. Wilk, American statisticians

11.1 ORIENTATION

Sometimes the fit of a simple linear regression model can be improved by adding one or more explanatory variables. For instance, the crop yield may depend not only on the amount of fertilizer but also on the amount of water. In this instance we have one response variable, the yield Y, and two explanatory variables, the amount of fertilizer x_1 and the amount of water x_2. Denote the relation between the expected yield and x_1, x_2 by $f(x_1, x_2)$. The object of regression analysis is to describe the unknown response function $f(x_1, x_2)$. In many cases, the response function can be approximated by a function that is linear in the variables; that is, we assume that the functional relation between the expected crop yield and the amounts of fertilizer and water is

$$E(Y \mid \text{fertilizer} = x_1; \text{water} = x_2) = \beta_0 + \beta_1 x_1 + \beta_2 x_2 = f(x_1, x_2)$$

Here the values of the variables x_1 and x_2 are known to the researcher, and the *regression parameters* β_0, β_1, and β_2 are not. *Multiple linear regression* is a statistical procedure for estimating and making inferences about the regression parameters. Since this procedure leads to tedious and complex calculations when done by hand, we assume that the student has access to a computer equipped with a standard statistical software package. Our primary goal in this chapter is to present those basic concepts of multiple linear regression that are essential for understanding the computer printouts. The most convenient and efficient way to reach this goal is to use vectors and matrices to represent the response variable, the explanatory variables, and the relations among them. For this reason we assume that the student is familiar with the following basic concepts from linear algebra: vector space, subspace, vectors, matrices, matrix multiplication, and additional concepts that will be introduced as needed. It is instructive to begin with the matrix formulation of the simple linear regression model studied in Chap. 10. This not only will ease the transition to the more complicated multiple linear regression models discussed later but will also give us additional insights into the simple linear regression model itself.

Organization of Chapter

11.2 THE MATRIX APPROACH TO SIMPLE LINEAR REGRESSION

In the simple linear regression model, discussed in Chap. 10, we assume that the relation between the value of the response variable Y_i and the value x_i of the explanatory variable is given by

$$Y_i = \beta_0 + \beta_1 x_i + \epsilon_i, \quad i = 1, \ldots, n$$

where

$$\epsilon_i \sim N(0, \sigma^2)$$

and the random variables $\epsilon_1, \ldots, \epsilon_n$ are mutually independent.

In the matrix approach to the simple linear regression model we represent the n responses $Y_1, \ldots, Y_n$, the two regression parameters β_0 and β_1, and the n random variables $\epsilon_1, \ldots, \epsilon_n$ as the column vectors[1]

$$\mathbf{Y} = \begin{pmatrix} Y_1 \\ \vdots \\ Y_n \end{pmatrix}, \qquad \mathbf{\beta} = \begin{pmatrix} \beta_0 \\ \beta_1 \end{pmatrix}, \qquad \mathbf{\epsilon} = \begin{pmatrix} \epsilon_1 \\ \vdots \\ \epsilon_n \end{pmatrix}$$

The matrix formulation of the simple linear regression model is

$$\mathbf{Y} = \mathbf{X}\mathbf{\beta} + \mathbf{\epsilon} \tag{11.1}$$

---

[1]Vectors and matrices are denoted in boldface type.

where **X** is the $(n \times 2)$ matrix

$$\mathbf{X} = \begin{pmatrix} 1 & x_1 \\ \vdots & \vdots \\ 1 & x_n \end{pmatrix}$$

The matrix **X** is called the *design matrix*. Observe that the number of columns in the design matrix equals the number of regression parameters and the number of rows equals the number of observations.

The n fitted values $\hat{Y}_i = \hat{\beta}_0 + \hat{\beta}_1 x_i$ $(i = 1, \ldots, n)$ and residuals $e_i = Y_i - \hat{Y}_i = Y_i - (\hat{\beta}_0 + \hat{\beta}_1 x_i)$ $(i = 1, \ldots, n)$ can also be represented using matrix notation as follows:

$$\hat{\mathbf{Y}} = \begin{pmatrix} \hat{Y}_1 \\ \vdots \\ \hat{Y}_n \end{pmatrix} = \begin{pmatrix} \hat{\beta}_0 + \hat{\beta}_1 x_1 \\ \vdots \\ \hat{\beta}_0 + \hat{\beta}_1 x_n \end{pmatrix} = \mathbf{X}\hat{\boldsymbol{\beta}}$$

$$\mathbf{e} = \begin{pmatrix} Y_1 - \hat{Y}_1 \\ \vdots \\ Y_n - \hat{Y}_n \end{pmatrix} = \mathbf{Y} - \mathbf{X}\hat{\boldsymbol{\beta}}$$

We observed in Chap. 10 [Eqs. (10.17) and (10.18)] that the residuals satisfy the following linear equations:

$$\sum_i e_i = 0, \qquad \sum_i x_i e_i = 0 \tag{11.2}$$

These two equations can be written as the single matrix equation

$$\begin{pmatrix} 1 & \cdots & 1 \\ x_1 & \cdots & x_n \end{pmatrix} \begin{pmatrix} e_1 \\ \vdots \\ e_n \end{pmatrix} = \begin{pmatrix} 0 \\ 0 \end{pmatrix}$$

The matrix **X'** defined by

$$\mathbf{X'} = \begin{pmatrix} 1 & \cdots & 1 \\ x_1 & \cdots & x_n \end{pmatrix}$$

is obtained from the design matrix by interchanging its rows and columns; it is called the *transpose* of **X**. Similarly, the transpose of a column vector **x**, a $(n \times 1)$ matrix, is the row vector $\mathbf{x'} = (x_1, \ldots, x_n)$, a $(1 \times n)$ matrix. In general, the transpose of a matrix with n rows and p columns is a matrix with p rows and n columns. Using this notation, we write Eq. (11.2) in the following, more useful form:

$$\mathbf{X'}(\mathbf{Y} - \mathbf{X}\hat{\boldsymbol{\beta}}) = 0 \tag{11.3}$$

Applying the distributive law for matrix multiplication to the left-hand side of Eq. (11.3) we see that $\hat{\boldsymbol{\beta}}$ satisfies the *normal equations*

$$\mathbf{X'X}\hat{\boldsymbol{\beta}} = \mathbf{X'Y} \tag{11.4}$$

Additional Properties of the Transpose and Inverse of a Matrix

We list for future reference some useful facts concerning the transpose and inverse of a matrix. Proofs are omitted.

$$(\mathbf{A}')' = \mathbf{A} \tag{11.5}$$

$$(\mathbf{AB})' = \mathbf{B}'\mathbf{A}' \tag{11.6}$$

$$(\mathbf{A}^{-1})^{-1} = \mathbf{A} \tag{11.7}$$

$$(\mathbf{AB})^{-1} = \mathbf{B}^{-1}\mathbf{A}^{-1} \tag{11.8}$$

$$(\mathbf{A}')^{-1} = (\mathbf{A}^{-1})' \tag{11.9}$$

Matrix Solution to the Normal Equations

In the simple linear regression case the determinant of $\mathbf{X}'\mathbf{X}$ is nS_{xx} (the proof is left as a problem). Because $S_{xx} \neq 0$, except for the uninteresting case when $x_1 = \cdots = x_n$, we can always assume that $\det(\mathbf{X}'\mathbf{X}) \neq 0$. Consequently, the inverse matrix $(\mathbf{X}'\mathbf{X})^{-1}$ exists and the solution to the normal equation in matrix form is given by

$$\hat{\boldsymbol{\beta}} = (\mathbf{X}'\mathbf{X})^{-1}\mathbf{X}'\mathbf{Y} \tag{11.10}$$

The matrix representation of the vector of fitted values $\mathbf{Y}$ is given by

$$\hat{\mathbf{Y}} = \mathbf{X}\hat{\boldsymbol{\beta}} = \mathbf{X}(\mathbf{X}'\mathbf{X})^{-1}\mathbf{X}'\mathbf{Y} = \mathbf{HY} \tag{11.11}$$

where

$$\mathbf{H} = \mathbf{X}(\mathbf{X}'\mathbf{X})^{-1}\mathbf{X}' \tag{11.12}$$

The matrix $\mathbf{H}$ is called the *hat matrix;* it plays an important role in describing the sampling distribution of the least squares estimators and in model checking. The residual vector $\mathbf{e}$ can also be written in terms of the hat matrix as follows:

$$\mathbf{e} = \mathbf{Y} - \hat{\mathbf{Y}} = \mathbf{Y} - \mathbf{HY} = (\mathbf{I} - \mathbf{H})Y \tag{11.13}$$

where $\mathbf{I}$ denotes the identity matrix. The hat matrix plays an important role in detecting violations of the regression model assumptions. For future reference we list some useful properties of the hat matrix. We leave the derivations as an exercise (Prob. 11.7).

$$\mathbf{H}' = \mathbf{H} \quad \text{(symmetric matrix)} \tag{11.14}$$

$$\mathbf{H}^2 = \mathbf{H} \quad \text{(idempotent matrix)} \tag{11.15}$$

$$\mathbf{H}I - \mathbf{H}) = \mathbf{0} \tag{11.16}$$

$$(\mathbf{I} - \mathbf{H})^2 = (\mathbf{I} - \mathbf{H}) \tag{11.17}$$

---

example **11.1**   Rederive the formulas (10.9) and (10.10) for the regression coefficients obtained in Chap. 10 by solving the normal equations (11.4).

Solution. It suffices to show that

$$\hat{\beta}_0 + \hat{\beta}_0\bar{x} = \overline{Y} \quad \text{and} \quad \hat{\beta}_1 = \frac{S_{xY}}{S_{xx}}$$

In this case the normal equations are two linear equations in two unknowns. To see this, compute the matrix products $\mathbf{X'X}$ and $\mathbf{X'Y}$:

$$\mathbf{X'X} = \begin{pmatrix} n & \sum x_i \\ \sum x_i & \sum x_i^2 \end{pmatrix}, \qquad \mathbf{X'Y} = \begin{pmatrix} \sum_i Y_i \\ \sum_i x_i Y_i \end{pmatrix} \tag{11.18}$$

We therefore obtain two equations for the two unknowns $\hat{\beta}_0$ and $\hat{\beta}_1$:

$$n\hat{\beta}_0 + \left(\sum_i x_i\right)\hat{\beta}_1 = \sum_i Y_i$$

$$\left(\sum_i x_i\right)\hat{\beta}_0 + \left(\sum_i x_i^2\right)\hat{\beta}_1 = \sum_i x_i Y_i$$

Using the relations $\sum_i x_i = n\bar{x}$ and $\sum_i Y_i = n\bar{Y}$, we transform these equations into the equivalent form

$$n\hat{\beta}_0 + n\bar{x}\hat{\beta}_1 = n\bar{Y}$$

$$n\bar{x}\hat{\beta}_0 + \left(\sum_i x_i^2\right)\hat{\beta}_1 = \sum_i x_i Y_i$$

Dividing both sides of the first equation by n yields the familiar relation

$$\hat{\beta}_0 + \hat{\beta}_1\bar{x} = \bar{Y}$$

Solving for $\hat{\beta}_0 = \bar{Y} - \hat{\beta}_1\bar{x}$ and substituting into the second equation yields the previously derived equation for $\hat{\beta}_1$:

$$n\bar{x}(\bar{Y} - \hat{\beta}_1\bar{x}) + \left(\sum_i x_i^2\right)\hat{\beta}_1 = \sum_i x_i Y_i$$

Therefore,

$$\hat{\beta}_1\left(\sum_i x_i^2 - n\bar{x}^2\right) = \sum_i x_i Y_i - n\bar{x}\bar{Y}$$

so

$$\hat{\beta}_1 S_{xx} = S_{xY}$$

Consequently,

$$\hat{\beta}_1 = \frac{S_{xY}}{S_{xx}}$$ ∎

---

example 11.2   Refer to Table 10.2 (fuel efficiency data for 12 automobiles). Write the response variable as a column vector, and compute the design matrix for Model($Y =$ mpg $\mid X =$ weight). Write down, but do not solve, the matrix form of the normal equations.

Solution. There are $n = 12$ observations, so the response vector has 12 rows and the design matrix has 12 rows and 2 columns:

$$
\mathbf{Y} = \begin{pmatrix} 32 \\ 30 \\ 29 \\ 25 \\ 27 \\ 28 \\ 29 \\ 27 \\ 28 \\ 25 \\ 28 \\ 25 \end{pmatrix} \qquad \mathbf{X} = \begin{pmatrix} 1 & 2.495 \\ 1 & 2.530 \\ 1 & 2.620 \\ 1 & 3.395 \\ 1 & 3.030 \\ 1 & 3.345 \\ 1 & 3.040 \\ 1 & 3.085 \\ 1 & 3.495 \\ 1 & 3.950 \\ 1 & 3.470 \\ 1 & 4.105 \end{pmatrix}
$$

The matrix product $\mathbf{X}'\mathbf{X}$ is

$$
\mathbf{X}'\mathbf{X} = \begin{pmatrix} 12 & 38.56 \\ 38.56 & 126.8546 \end{pmatrix}
$$

and the vector $\mathbf{X}'\mathbf{Y}$ is

$$
\mathbf{X}'\mathbf{Y} = \begin{pmatrix} 333 \\ 1059.915 \end{pmatrix}
$$

Computational Details. To compute $\mathbf{X}'\mathbf{X}$ we use Eq. (11.18) and the fact that $\sum_i x_i = 38.56$ and $\sum_i x_i^2 = 126.8546$. We compute the vector $\mathbf{X}'\mathbf{Y}$ in the same way, using Eq. (11.18) and the fact that $\sum_i Y_i = 333$ and $\sum_i x_i Y_i = 1059.915$. Therefore, the matrix form of the normal equations is given by

$$
\begin{pmatrix} 12 & 38.56 \\ 38.56 & 126.8546 \end{pmatrix} \times \begin{pmatrix} \hat{\beta}_0 \\ \hat{\beta}_1 \end{pmatrix} = \begin{pmatrix} 333 \\ 1059.915 \end{pmatrix} \tag{11.19}
$$

∎

In practice we never solve the normal equations by inverting the matrix $\mathbf{X}'\mathbf{X}$; instead, we use one of the standard statistical software packages.

Random Vectors and Random Matrices

A vector with components that are random variables is called a *random vector*. The response vector $\mathbf{Y} = \mathbf{X}\boldsymbol{\beta} + \boldsymbol{\epsilon}$ in the regression model is an example of a random vector. The *mean vector* $E(\mathbf{Y})$ is the vector of means, defined by

$$
E(\mathbf{Y}) = \begin{pmatrix} E(Y_1) \\ \vdots \\ E(Y_n) \end{pmatrix} \tag{11.20}
$$

The mean vector for the response variable is

$$
E(\mathbf{Y}) = \mathbf{X}\boldsymbol{\beta} \tag{11.21}
$$

The Variance-Covariance Matrix of a Random Vector

The *variance-covariance matrix* of the random vector $\mathbf{Y}' = (Y_1, \ldots, Y_n)$ is the $n \times n$ matrix defined as follows:

$$\text{Cov}(\mathbf{Y}) = \begin{pmatrix} V(Y_1) & \cdots & \text{Cov}(Y_1, Y_n) \\ \text{Cov}(Y_2, Y_1) & \cdots & \text{Cov}(Y_2, Y_n) \\ \vdots & \vdots & \vdots \\ \text{Cov}(Y_n, Y_1) & \cdots & V(Y_n) \end{pmatrix} \qquad (11.22)$$

The diagonal elements of the matrix are the variances, and the off-diagonal elements are the covariances. Recall that $V(Y_i) = \text{Cov}(Y_i, Y_i)$; consequently, the (i, j) entry is $\text{Cov}(Y_i, Y_j)$. The matrix $\text{Cov}(\mathbf{Y})$ is symmetric because $\text{Cov}(Y_i, Y_j) = \text{Cov}(Y_j, Y_i)$.

---

example 11.3   Compute the variance-covariance matrix of the response vector in the simple linear regression model.

Solution. Because Y_i and Y_j are independent random variables when $i \neq j$, it follows that $\text{Cov}(Y_i, Y_j) = 0$ $(i \neq j)$; that is, the off-diagonal elements are 0. Moreover, $V(Y_i) = \sigma^2$, so the diagonal elements are σ^2. Therefore,

$$\text{Cov}(\mathbf{Y}) = \begin{pmatrix} \sigma^2 & 0 & \cdots & 0 \\ 0 & \sigma^2 & \cdots & 0 \\ 0 & \vdots & \vdots & 0 \\ 0 & 0 & \cdots & \sigma^2 \end{pmatrix} = \sigma^2 \mathbf{I} \qquad (11.23) \quad \blacksquare$$

A matrix whose entries are random variables is called a *random matrix*. The product of the column vector $\mathbf{Y} - E(\mathbf{Y})$ and the row vector $(\mathbf{Y} - E(\mathbf{Y}))'$ is the $n \times n$ symmetric random matrix

$$[\mathbf{Y} - E(\mathbf{Y})][\mathbf{Y} - E(\mathbf{Y})]' =$$

$$\begin{pmatrix} [Y_1 - E(Y_1)][Y_1 - E(Y_1)] & \cdots & [Y_1 - E(Y_1)][Y_n - E(Y_n)] \\ [Y_2 - E(Y_2)][Y_1 - E(Y_1)] & \cdots & [Y_2 - E(Y_2)][Y_n - E(Y_n)] \\ \vdots & \vdots & \vdots \\ [Y_n - E(Y_n)][Y_1 - E(Y_1)] & \cdots & [Y_n - E(Y_n)][Y_n - E(Y_n)] \end{pmatrix}$$

Observe that the expected value of the random variable in the ith row and jth column is the covariance of the random variables Y_i and Y_j; that is, $\text{Cov}(Y_i, Y_j) = E[(Y_i - E(Y_i))(Y_j - E(Y_j))]$. It follows that the variance-covariance matrix $\text{Cov}(\mathbf{Y})$ can be expressed as the expected value of the random matrix $(\mathbf{Y} - E(\mathbf{Y}))(\mathbf{Y} - E(\mathbf{Y}))'$. That is,

$$\text{Cov}(\mathbf{Y}) = E[(\mathbf{Y} - E(Y))(\mathbf{Y} - E(Y))'] \qquad (11.24)$$

11.2.1 Sampling Distribution of the Least Squares Estimators

The sampling distributions of the least squares estimators can also be derived via matrix methods. The following theorem (we omit the proof) is useful for computing mean vectors and variance-covariance matrices for the vector of fitted values, the regression parameters, and the residual vector.

▦ THEOREM 11.1

Suppose the random vector $\mathbf{W}$ is obtained by multiplying the random vector $\mathbf{Y}$ by the matrix $\mathbf{A}$; that is, $\mathbf{W} = \mathbf{AY}$. Then

$$E(\mathbf{W}) = \mathbf{A}E(\mathbf{Y}) \tag{11.25}$$

$$\text{Cov}(\mathbf{W}) = \mathbf{A}\,\text{Cov}(\mathbf{Y})A' \tag{11.26}$$

■

Our first application of this theorem is to compute the mean vector and variance-covariance matrix of $\hat{\boldsymbol{\beta}}$.

▦ THEOREM 11.2

Under the assumptions of the linear regression model,

$$E(\hat{\boldsymbol{\beta}}) = \boldsymbol{\beta} \tag{11.27}$$

$$\text{Cov}(\hat{\boldsymbol{\beta}}) = \sigma^2(\mathbf{X'X})^{-1} \tag{11.28}$$

■

Proof. It follows from Eq. (11.25) [with $\mathbf{A} = (\mathbf{X'X})^{-1}\mathbf{X'}$ and $E(\mathbf{Y}) = \mathbf{X}\boldsymbol{\beta}$] that

$$E(\hat{\boldsymbol{\beta}}) = (\mathbf{X'X})^{-1}\mathbf{X'}E(\mathbf{Y})$$
$$= (\mathbf{X'X})^{-1}(\mathbf{X'X})\boldsymbol{\beta} = \mathbf{I}\boldsymbol{\beta} = \boldsymbol{\beta}$$

In other words, the least squares estimator $\hat{\boldsymbol{\beta}} = (\mathbf{X'X})^{-1}\mathbf{X'Y}$ is an unbiased estimator of the vector of regression parameters $\boldsymbol{\beta}$.

It follows from Eq. (11.26) [with $\mathbf{A} = (\mathbf{X'X})^{-1}\mathbf{X'}$, $\mathbf{A'} = \mathbf{X}(\mathbf{X'X})^{-1}$, and $\text{Cov}(\mathbf{Y}) = \sigma^2 I$] that

$$\text{Cov}(\hat{\boldsymbol{\beta}}) = \mathbf{A}\sigma^2 I\mathbf{A'}$$
$$= \sigma^2(\mathbf{X'X})^{-1}\mathbf{X'X}(\mathbf{X'X})^{-1} = \sigma^2(\mathbf{X'X})^{-1}$$

When we substitute the unbiased estimator $s^2 = \text{MSE} = \text{SSE}/(n-2)$ for σ^2 in Eq. (11.28), we obtain the *estimated variance-covariance matrix*:

$$s^2(\hat{\boldsymbol{\beta}}) = s^2(\mathbf{X'X})^{-1} \tag{11.29}$$

---

example **11.4**

Refer to Example 11.2. Compute the estimated standard errors $s(\hat{\beta}_0)$ and $s(\hat{\beta}_1)$ from the diagonal entries of the matrix $s^2(\mathbf{X'X})^{-1}$.

Solution. The error sum of squares was computed in Example 10.4. Using the definition $s^2 = \text{SSE}/(n-2)$, it follows that $s^2 = 1.54803$. (*Note:* The following calculations were performed on a computer, and the final values were rounded to four places. Because of roundoff errors, these answers will differ slightly from those obtained using a scientific calculator.) The inverse matrix is

$$(\mathbf{X}'\mathbf{X})^{-1} = \begin{pmatrix} 3.5854 & -1.0899 \\ -1.0899 & 0.3392 \end{pmatrix}$$

Consequently,

$$s^2(\mathbf{X}'\mathbf{X})^{-1} = \begin{pmatrix} 5.5503 & -1.6871 \\ -1.6871 & 0.5250 \end{pmatrix}$$

Thus, $s(\hat{\beta}_0) = \sqrt{5.5503} = 2.3559$ and $s(\hat{\beta}_1) = \sqrt{0.5250} = 0.7246$. (Problem 11.6 asks you to compute the estimated standard errors directly using the methods of Chap. 10.)   ∎

Matrix Expressions for Confidence Intervals and Prediction Intervals

Point estimates and confidence intervals for the mean response and prediction intervals for a future response can also be expressed using matrix notation. The mean response for a specified value x_0 of the explanatory variable is $E(Y \mid x_0) = \beta_0 + \beta_1 x_0$. The estimated mean response, denoted $\hat{Y}(x_0)$, can be written as the matrix product

$$\hat{Y}(x_0) = \hat{\beta}_0 + \hat{\beta}_1 x_0 = (1, x_0)\begin{pmatrix} \hat{\beta}_0 \\ \hat{\beta}_1 \end{pmatrix} = \mathbf{x}_0'\hat{\boldsymbol{\beta}}$$

where $\mathbf{x}_0' = (1, x_0)$. It follows from Eq. (11.10) that

$$\mathbf{x}_0'\hat{\boldsymbol{\beta}} = \mathbf{A}\mathbf{Y}$$

where

$$\mathbf{A} = \mathbf{x}_0'(\mathbf{X}'\mathbf{X})^{-1}\mathbf{X}' \qquad (11.30)$$

$$\mathbf{A}' = \mathbf{X}(\mathbf{X}'\mathbf{X})^{-1}\mathbf{x}_0' \qquad (11.31)$$

It follows from Eq. (11.26) [with $\mathbf{A}$ defined as in Eq. (11.30) and $\text{Cov}(\mathbf{Y}) = \sigma^2 I$] that

$$V(\hat{Y}(x_0)) = \sigma^2 \mathbf{x}_0'(\mathbf{X}'\mathbf{X})^{-1}\mathbf{x}_0 \qquad (11.32)$$

We obtain the estimated *standard error of prediction*, denoted $s(\hat{Y}(x_0))$, by replacing σ^2 with its estimate $s^2 = \text{MSE} = \text{SSE}/(n-2)$; thus,

$$s(\hat{Y}(x_0)) = s\sqrt{\mathbf{x}_0'(\mathbf{X}'\mathbf{X})^{-1}\mathbf{x}_0} \qquad (11.33)$$

A $100(1 - \alpha)$ percent confidence interval for the mean response at the value x_0 is given by

$$\mathbf{x}_0'\hat{\boldsymbol{\beta}} \pm t_{n-2}(\alpha/2)s\sqrt{\mathbf{x}_0'(\mathbf{X}'\mathbf{X})^{-1}\mathbf{x}_0} \qquad (11.34)$$

Similarly, the matrix expression for the estimated standard error of the difference between a future response $Y(x_0) = \beta_0 + \beta_1 x_0 + \epsilon$ and the predicted future response $\hat{Y}(x_0) = \hat{\beta}_0 + \hat{\beta}_1 x_0$ at the value x_0 of the explanatory variable, denoted $s(Y(x_0) - \hat{Y}(x_0))$, is given by

$$s(Y(x_0) - \hat{Y}(x_0)) = s\sqrt{1 + x_0'(X'X)^{-1}x_0} \qquad (11.35)$$

The corresponding prediction interval for a future response at the value x_0 is

$$x_0'\hat{\beta} \pm t_{n-2}(\alpha/2)s\sqrt{1 + x_0'(X'X)^{-1}x_0} \qquad (11.36)$$

Matrix Representation of the Sums of Squares SST, SSR, and SSE

Recall that the computer output of a regression analysis is displayed in an ANOVA table that is based on the sum of squares decomposition

$$\text{SST} = \text{SSR} + \text{SSE} \qquad (11.37)$$

We now give a derivation of Eq. (11.37) using a matrix method that extends without change to the multiple linear regression context.

We begin by deriving matrix expressions for the sums of squares SST, SSR, and SSE.

$$\text{SST} = \sum_i (Y_i - \overline{Y})^2 = \sum_i Y_i^2 - n\overline{Y}^2 = Y'Y - n\overline{Y}^2$$

$$\text{SSR} = \sum_i \hat{Y}_i^2 - n\overline{Y}^2 = \hat{Y}'\hat{Y} - n\overline{Y}^2$$

$$\text{SSE} = \sum_i e_i^2 = e'e$$

Recall that the residual vector is orthogonal to the column space of the design matrix [Eq. (11.3)]. It follows that

$$\hat{Y}'e = (X\hat{\beta})'e = \hat{\beta}'X'e = 0$$
$$e'\hat{Y} = (\hat{Y}'e)' = 0' = 0$$

Consequently,

$$Y'Y = (\hat{Y} + e)'(\hat{Y} + e)$$
$$= \hat{Y}'\hat{Y} + e'\hat{Y} + \hat{Y}'e + e'e$$
$$= \hat{Y}'\hat{Y} + e'e$$

Therefore,

$$Y'Y - n\overline{Y}^2 = (\hat{Y}'\hat{Y} - n\overline{Y}^2) + e'e$$

Thus,

$$\text{SST} = \text{SSR} + \text{SSE}$$

11.2.2   Geometric Interpretation of the Least Squares Solution

The essence of the least squares method depends on the concept of the distance between two points in n-dimensional space. We begin with the concept of the *length* of a vector, denoted by $|x|$ and defined as follows:

$$|\mathbf{x}| = \sqrt{\sum_i x_i^2}$$

The *distance* between vectors $\mathbf{x}$ and $\mathbf{y}$ is the length of their difference $\mathbf{x} - \mathbf{y} = (x_1 - y_1, \ldots, x_n - y_n)$, which is

$$|\mathbf{x} - \mathbf{y}| = \sqrt{\sum_i (x_i - y_i)^2}$$

To free ourselves from the tyranny of subscripts, it is useful to know that the square of the length of a vector can also be expressed in terms of matrix multiplications. For instance, the matrix product $\mathbf{x}'\mathbf{x}$ equals the square of the length of $\mathbf{x}$, since

$$\mathbf{x}'\mathbf{x} = \sum_i x_i^2 = |\mathbf{x}|^2$$

More generally, the matrix product of a row vector and column vector, called the *scalar product* and denoted $\mathbf{x}'\mathbf{y}$, is a real number. Recall that in two- and three-dimensional space the condition $\mathbf{x}'\mathbf{y} = 0$ implies that the angle between the two vectors is a right angle. In the more general context of linear algebra, we say that two vectors are *orthogonal* when $\mathbf{x}'\mathbf{y} = 0$. The matrix form of the normal equations [Eq. (11.3)] implies that the residual vector $\mathbf{Y} - \mathbf{X}\hat{\boldsymbol{\beta}}$ is orthogonal to each column of the design matrix. This leads to a geometric characterization of the least squares solution that provides additional insight and understanding, as will now be explained.

Recall that the least squares solution for the best-fitting line was obtained by determining the values of the parameters $(b_0, b_1) = \mathbf{b}'$ that minimize the following sum of squares:

$$Q(b_0, b_1) = \sum_{1 \le i \le n} [Y_i - (b_0 + b_1 x_i)]^2 \tag{11.38}$$

To proceed further, we use matrix notation to represent the sum of squares $Q(b_0, b_1)$ as the square of the distance between the vectors $\mathbf{Y}$ and $\mathbf{X}\mathbf{b}$. Thus,

$$\mathbf{X}\mathbf{b} = \begin{pmatrix} b_0 + b_1 x_1 \\ \vdots \\ b_0 + b_1 x_n \end{pmatrix}, \quad \text{so} \quad \mathbf{Y} - \mathbf{X}\mathbf{b} = \begin{pmatrix} Y_1 - (b_0 + b_1 x_1) \\ \vdots \\ Y_n - (b_0 + b_1 x_n) \end{pmatrix}$$

Therefore,

$$|\mathbf{Y} - \mathbf{X}\mathbf{b}|^2 = \sum_{1 \le i \le n} [Y_i - (b_0 + b_1 x_i)]^2 = Q(b_0, b_1) \tag{11.39}$$

The set of vectors $\mathbf{y} = \mathbf{X}\mathbf{b}$ is called the *column space* of the design matrix. It is the *subspace* consisting of all linear combinations of the columns of the design matrix; that is,

$$\mathbf{y} = \mathbf{X}\mathbf{b} = \begin{pmatrix} b_0 + b_1 x_1 \\ \vdots \\ b_0 + b_1 x_n \end{pmatrix} = b_0 \begin{pmatrix} 1 \\ \vdots \\ 1 \end{pmatrix} + b_1 \begin{pmatrix} x_1 \\ \vdots \\ x_n \end{pmatrix}$$

The *least squares problem* is to find the vector $\mathbf{Xb}$ in the column space of the design matrix that minimizes the distance to $\mathbf{Y}$. The least squares solution now has a simple and elegant geometric interpretation as the minimum distance from the response vector to the column space of the design matrix. Recall from Euclidean geometry that the minimum distance from a point to a given line is obtained by measuring the length of the perpendicular line segment from the point to the line. This perpendicular line segment is the geometric representation of the residual vector. In the regression context, we interpret this to mean that $\hat{\boldsymbol{\beta}}$ is the least squares solution provided the residual vector $\mathbf{e} = \mathbf{Y} - \mathbf{X}\hat{\boldsymbol{\beta}}$ is orthogonal to the column space of the design matrix. We obtain the condition, expressed in matrix form, that $\mathbf{X}'(\mathbf{Y} - \mathbf{X}\hat{\boldsymbol{\beta}}) = 0$, which is Eq. (11.3).

To repeat: The normal equations are a consequence of the fact that the vector of residuals must be orthogonal to the columns of the design matrix. No calculus is required to obtain this fundamental result. In Sec. 11.4 we show that any solution of the normal equations minimizes the sum of squares.

PROBLEMS

11.1 Refer to Example 11.2. Show by direct substitution that $\hat{\beta}_0 = 38.78474$, $\hat{\beta}_1 = -3.434047$ is a solution to the normal equations (11.19).

11.2 Refer to Example 10.3 of Chap. 10.
   (a) State the matrix formulation of Model(Y = shear strength $\mid X$ = weld diameter).
   (b) Write down the normal equations for this model in the matrix format (11.4).
   (c) Verify that $\hat{\beta}_0 = -216.325$, $\hat{\beta}_1 = 3.9904$ satisfy the normal equations.
   (d) Compute the estimated variance-covariance matrix $s^2(\mathbf{X}'\mathbf{X})^{-1}$.

11.3 Refer to Prob. 10.3 of Chap. 10.
   (a) State the matrix formulation of Model(Y = CPU time $\mid X$ = Disk I/Os).
   (b) State the normal equations for this model in the matrix format (11.4).
   (c) Verify that $\hat{\beta}_0 = -0.008282$ and $\hat{\beta}_1 = 0.243756$ satisfy the normal equations given in part (*b*).
   (d) Compute the estimated variance-covariance matrix $s^2(\mathbf{X}'\mathbf{X})^{-1}$.

11.4 Refer to Prob. 10.8 of Chap. 10.
   (a) State the matrix formulation of Model(Y = TDS $\mid X$ = SEC).
   (b) State the normal equations for this model in the matrix format (11.4).
   (c) Verify that $\hat{\beta}_0 = 29.268569$ and $\hat{\beta}_1 = 0.597884$ satisfy the normal equations given in part (*b*).
   (d) Compute the estimated variance-covariance matrix $s^2(\mathbf{X}'\mathbf{X})^{-1}$.

11.5 Show that in the case of simple linear regression,
$$\det(\mathbf{X}'\mathbf{X}) = nS_{xx}$$

11.6 Refer to Example 11.4. Compute the estimated standard errors using the equations (derived in Chap. 10)

$$s(\hat{\beta}_0) = s\sqrt{\frac{1}{n} + \frac{\bar{x}^2}{S_{xx}}}$$

$$s(\hat{\beta}_1) = \frac{s}{\sqrt{S_{xx}}}$$

Your results should agree with those obtained in the solution to Example 11.4.

11.7 Derive the properties of the hat matrix **H** listed in Eqs. (11.14)–(11.17).

11.8 Refer to Prob. 10.19 of Chap. 10.
  (a) State the matrix formulation of Model(Y = hardness | X = time).
  (b) State the normal equations for this model in the matrix format (11.4).
  (c) Verify that $\hat{\beta}_0 = 153.916667$ and $\hat{\beta}_1 = 2.416667$ satisfy the normal equations given in part (b).
  (d) Compute the estimated variance-covariance matrix $s^2(\mathbf{X}'\mathbf{X})^{-1}$.

11.3   THE MATRIX APPROACH TO MULTIPLE LINEAR REGRESSION

The multiple linear regression model with two explanatory variables assumes that the statistical relationship between the response variable and the explanatory variables is of the form

$$Y_i = \beta_0 + \beta_1 x_{i1} + \beta_2 x_{i2} + \epsilon_i$$
$$\epsilon_i \sim N(0, \sigma^2) \tag{11.40}$$

where ϵ_i denotes an iid sequence of independent normal random variables with zero means and common variance σ^2. The matrix representation for this regression model is exactly the same as in the simple linear regression case except that a column representing the new variable is adjoined to the design matrix; that is,

$$\mathbf{Y} = \mathbf{X}\boldsymbol{\beta} + \boldsymbol{\epsilon} \tag{11.41}$$

where

$$\mathbf{Y} = \begin{pmatrix} Y_1 \\ \vdots \\ Y_n \end{pmatrix}, \qquad \mathbf{X} = \begin{pmatrix} 1 & x_{11} & x_{12} \\ \vdots & \vdots & \vdots \\ 1 & x_{n1} & x_{n2} \end{pmatrix}$$

$$\boldsymbol{\beta} = \begin{pmatrix} \beta_0 \\ \beta_1 \\ \beta_2 \end{pmatrix}, \qquad \boldsymbol{\epsilon} = \begin{pmatrix} \epsilon_1 \\ \vdots \\ \epsilon_n \end{pmatrix}$$

Notation. We use the model statement format

$$\text{Model}(Y = y \mid X_1 = x_1, X_2 = x_2)$$

to specify the response variable Y and the explanatory variables X_1 and X_2.

example **11.5**   Refer to Table 10.2 of Example 10.2 (fuel efficiency data for 12 automobiles). It was shown (see Example 10.4) that for Model($Y = $ mpg $| X_1 = $ weight) the coefficient of determination $R^2 = 0.69$. To fit a better model to the data, the statistician decided to fit the multiple linear regression Model($Y = $ mpg $| X_1 = $ weight, $X_2 = $ displacement). Write the response variable as a column vector, and compute the design matrix for this model.

Solution

$$
Y = \begin{pmatrix} 32 \\ 30 \\ 29 \\ 25 \\ 27 \\ 28 \\ 29 \\ 27 \\ 28 \\ 25 \\ 28 \\ 25 \end{pmatrix} \quad
X = \begin{pmatrix}
1 & 2.495 & 1.9 \\
1 & 2.530 & 1.8 \\
1 & 2.620 & 1.6 \\
1 & 3.395 & 3.0 \\
1 & 3.030 & 2.2 \\
1 & 3.345 & 3.8 \\
1 & 3.040 & 2.2 \\
1 & 3.085 & 3.0 \\
1 & 3.495 & 3.8 \\
1 & 3.950 & 4.6 \\
1 & 3.470 & 3.8 \\
1 & 4.105 & 5.7
\end{pmatrix}
$$ ■

Polynomial Regression Models

Polynomial regression deals with the problem of fitting a polynomial to a data set. Suppose the statistical relation between the response Y and the explanatory variable x is of the form

$$Y = \beta_0 + \beta_1 x + \beta_2 x^2 + \epsilon \qquad (11.42)$$

where $\epsilon \sim N(0, \sigma^2)$

It follows that the mean response

$$E(Y \mid x) = \beta_0 + \beta_1 x + \beta_2 x^2$$

is a quadratic polynomial of the explanatory variable x. Although the mean response is a nonlinear function, we can reduce it to a multiple linear regression problem by defining the explanatory variables $x_1 = x$ and $x_2 = x^2$. The polynomial regression model Eq. (11.42) is transformed to the multiple linear regression model

$$Y = \beta_0 + \beta_1 x_1 + \beta_2 x_2 + \epsilon \qquad (11.43)$$

where $\epsilon \sim N(0, \sigma^2)$

example **11.6**   The data listed in Table 11.1 came from an experiment to determine the fuel efficiency of a light truck equipped with an experimental overdrive gear. The response variable Y is miles per gallon (mpg), and the explanatory variable x is the constant speed, in miles per hour (mph), on the test track. Graph the scatter plot of mpg (Y) against speed (x) to see if the simple linear regression model is appropriate. Calculate the design matrix for the quadratic polynomial Model(Y = mpg $|$ x_1 = speed, x_2 = speed^2).

TABLE 11.1

Miles per gallon (Y)	Speed (x)	Miles per gallon (Y)	Speed (x)
22	35	41	50
20	35	39	50
28	40	34	55
31	40	37	55
37	45	27	60
38	45	30	60

Source: J. Neter, M. H. Kutner, C. J. Nachtseim, and W. Wasserman, *Applied Linear Statistical Models,* 4th ed., Chicago, Richard D. Irwin, 1996. Used with permission.

Solution. The scatter plot is shown in Fig. 11.1. Clearly, the simple linear regression model is inappropriate.

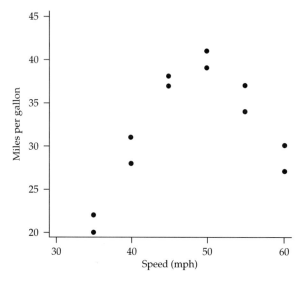

Figure 11.1 Scatter plot for the data in Table 11.1

The design matrix is

$$\mathbf{X} = \begin{pmatrix} 1 & 35 & 1225 \\ 1 & 35 & 1225 \\ 1 & 40 & 1600 \\ 1 & 40 & 1600 \\ 1 & 45 & 2025 \\ 1 & 45 & 2025 \\ 1 & 50 & 2500 \\ 1 & 50 & 2500 \\ 1 & 55 & 3025 \\ 1 & 55 & 3025 \\ 1 & 60 & 3600 \\ 1 & 60 & 3600 \end{pmatrix}$$

■

11.3.1   Normal Equations, Fitted Values, and ANOVA Table for the Multiple Linear Regression Model

Consider the general multiple linear regression model

$$\mathbf{Y} = \mathbf{X}\boldsymbol{\beta} + \boldsymbol{\epsilon}$$

where

$$\mathbf{Y} = \begin{pmatrix} Y_1 \\ \vdots \\ Y_n \end{pmatrix}, \qquad \mathbf{X} = \begin{pmatrix} 1 & x_{11} & \cdots & x_{1k} \\ \vdots & \vdots & \vdots & \vdots \\ 1 & x_{n1} & \cdots & x_{nk} \end{pmatrix}$$

$$\boldsymbol{\beta} = \begin{pmatrix} \beta_0 \\ \beta_1 \\ \vdots \\ \beta_k \end{pmatrix}, \qquad \boldsymbol{\epsilon} = \begin{pmatrix} \epsilon_1 \\ \vdots \\ \epsilon_n \end{pmatrix}$$

and ϵ_i denotes, as usual, an iid sequence of independent normal random variables with zero means and common variance σ^2. Thus, the ith response, Y_i, is

$$Y_i = \beta_0 + \beta_1 x_{i1} + \cdots + \beta_k x_{ik} + \epsilon_i \qquad (11.44)$$

It follows that the mean response as a function of the explanatory variables is

$$E(Y \mid x_1, \ldots, x_k) = \beta_0 + \beta_1 x_1 + \cdots + \beta_k x_k \qquad (11.45)$$

Interpretation of the Regression Coefficients

The regression coefficient β_i measures the change in the expected response of Y when the explanatory variable x_i is increased by one unit and all the other explanatory variables are held constant.

Notation. We use the model statement format Model($Y = y \mid X_1 = x_1, \ldots,$ $X_k = x_k$) to specify the response variable Y and the k explanatory variables

$X_1, \ldots, X_k$. The multiple linear regression model has $k + 1$ regression parameters $\beta_0, \beta_1, \ldots, \beta_k$, and the design matrix has n rows and $p = k + 1$ columns, corresponding to the number of regression parameters.

The Least Squares Solution

The least squares solution and its geometric interpretation are the same as for the simple linear regression case; that is, the residual vector $\mathbf{e} = \mathbf{Y} - \mathbf{X}\hat{\boldsymbol{\beta}}$ is orthogonal to the columns of the design matrix. Consequently, Eq. (11.3),

$$\mathbf{X}'(\mathbf{Y} - \mathbf{X}\hat{\boldsymbol{\beta}}) = 0$$

remains valid, as does the matrix form of the normal equations derived earlier [Eq. (11.4)],

$$\mathbf{X}'\mathbf{X}\hat{\boldsymbol{\beta}} = \mathbf{X}'\mathbf{Y}$$

When the inverse matrix $(\mathbf{X}'\mathbf{X})^{-1}$ exists, the solutions to the normal equations, the vector of fitted values, and the residual vector in matrix form are given by

$$\hat{\boldsymbol{\beta}} = (\mathbf{X}'\mathbf{X})^{-1}\mathbf{X}'\mathbf{Y}$$

$$\hat{\mathbf{Y}} = \mathbf{X}\hat{\boldsymbol{\beta}} = \mathbf{X}(\mathbf{X}'\mathbf{X})^{-1}\mathbf{X}'\mathbf{Y} = \mathbf{H}\mathbf{Y}$$

$$\mathbf{e} = \mathbf{Y} - \hat{\mathbf{Y}} = \mathbf{Y} - \mathbf{H}\mathbf{Y} = (\mathbf{I} - \mathbf{H})Y$$

where

$$\mathbf{H} = \mathbf{X}(\mathbf{X}'\mathbf{X})^{-1}\mathbf{X}'$$

So much for the theory. In practice, the sums of squares SST, SSR, and SSE, the estimated regression parameters $\hat{\beta}_i$, and their estimated standard errors $s(\hat{\beta}_i)$ are listed in the computer printouts. For this reason we omit the tedious computational details needed for calculating these terms by hand.

ANOVA Table for the Multiple Linear Regression Model

The computer printout of a regression analysis is displayed in the format of the ANOVA table in Table 11.2. The entries in the third column are computed from the sum of squares decomposition SST = SSR + SSE derived earlier [see the discussion following Eq. (11.37)], and the entries in the other columns are computed using theorems concerning the distributions of SSR and SSE (we omit the derivations).

TABLE 11.2 ANOVA TABLE FOR THE MULTIPLE LINEAR REGRESSION MODEL

Source	DF	Sum of squares	Mean square	F value	prob > F
Model	$p - 1$	SSR	MSR = SSR/$(p - 1)$	MSR/MSE	
Error	$n - p$	SSE	MSE = SSE/$(n - p)$		
Total	$n - 1$	SST			

Degrees of Freedom Associated with SSE

It can be shown that the number of degrees of freedom associated with the error sum of squares is $(n - p)$ and the *mean square error* MSE $= \text{SSE}/(n - p)$ (the error sum of squares divided by its degrees of freedom) is an unbiased estimator of σ^2; that is,

$$E(\text{MSE}) = E\left(\frac{\text{SSE}}{n - p}\right) = \sigma^2$$

The terms $n - p$, SSE, and MSE are listed in the second, third, and fourth columns of Table 11.2.

Estimated Standard Errors of the Estimated Regression Parameters

Equations (11.28) and (11.29) for the variance-covariance and estimated variance-covariance matrices remain valid for the multiple linear regression model; that is,

$$\text{Cov}(\hat{\boldsymbol{\beta}}) = \sigma^2(\mathbf{X}'\mathbf{X})^{-1}$$
$$s^2(\hat{\boldsymbol{\beta}}) = s^2(\mathbf{X}'\mathbf{X})^{-1} \quad (s^2 = \text{MSE})$$

We use the ANOVA table to test the null hypothesis

$$H_0\colon \beta_0 = \beta_1 = \cdots = \beta_k = 0$$

against

$$H_1\colon \text{not all } \beta_i = 0 \quad (i = 0, 1, \ldots, k)$$

When H_0 is true, there is no regression relation between the response variable Y and the explanatory variables $X_1, \ldots, X_k$; consequently, the regression model is not a useful one. In addition, assuming H_0 is true, it can be shown that the distribution of SSR/σ^2 is chi-square with $(p - 1)$ degrees of freedom, so the number of degrees of freedom associated with SSR is $(p - 1)$. This number is listed in the second column (DF) of Table 11.2. The *mean square due to regression*, denoted MSR, is $\text{SSR}/(p - 1)$. The SSR and MSR values are listed in the third and fourth columns, respectively. The SST term, as usual, has $n - 1$ degrees of freedom. The partition of the number of degrees of freedom

$$n - 1 = (p - 1) + (n - p) \tag{11.46}$$

(displayed in the second column of Table 11.2) corresponds to the decomposition SST $=$ SSR $+$ SSE.

Suppose we wish to test the null hypothesis

$$H_0\colon \beta_0 = \beta_1 = \cdots = \beta_k = 0$$

It can be shown that, when H_0 is true, the ratio MSR/MSE has an F distribution with $(p - 1)$ degrees of freedom in the numerator and $(n - p)$ degrees in the denominator; that is,

$$F = \frac{\text{MSR}}{\text{MSE}} \sim F_{p-1, n-p}$$

The value of the F ratio appears in the fifth column. The P-value of the F test is

$$P\text{-value} = P(F_{p-1,n-p} > F)$$

and appears in the last column of Table 11.2.

Coefficient of Multiple Determination

The *coefficient of multiple determination*, denoted R^2, is defined as follows:

$$R^2 = \frac{\text{SSR}}{\text{SST}} \tag{11.47}$$

We obtain an interpretation of R^2 by noting that

$$1 = \frac{\text{SSR}}{\text{SST}} + \frac{\text{SSE}}{\text{SST}}$$

[This equation is obtained by dividing both sides of Eq. (11.37) by SST.] It follows that the term $R^2 = \text{SSR}/\text{SST}$ represents (just as in the simple linear regression model) the proportion of the total variability that is explained by the model, and SSE/SST represents the proportion of the unexplained variability. A value of R^2 close to 1 implies that most of the variability is explained by the regression model; a value of R^2 close to zero indicates that the regression model is not appropriate. It can be shown (the details are left as a problem) that

$$F = \frac{\text{MSR}}{\text{MSE}} = \left(\frac{n-p}{p-1}\right)\frac{R^2}{1-R^2} \tag{11.48}$$

Consequently, the F ratio is large when R^2 is close to 1.

---

example 11.7   Refer to Model($Y = \text{mpg} \mid X_1 = \text{weight}, X_2 = \text{displacement}$) of Example 11.5. Fit a multiple linear regression model to the data using a computer software package. State the estimated regression function. Compute the coefficient of multiple determination and test the null hypothesis

$$H_0: \beta_0 = \beta_1 = \beta_2 = 0$$

against

$$H_1: \text{not all } \beta_i = 0 \quad (i = 0, 1, 2)$$

Solution. Table 11.3 is the SAS computer printout. The estimated regression parameters are referred to by name; thus, $\hat{\beta}_0$ is called INTERCEP (which is the word *intercept* reduced to eight characters), $\hat{\beta}_1$ is WGT (weight), and $\hat{\beta}_2$ is DISP (displacement). The estimated regression parameters, estimated regression function [denoted $\hat{E}(Y \mid x_1, x_2)$], and coefficient of multiple of determination ("R-square" in the computer printout) are

$$\hat{\beta}_0 = 46.351255, \qquad \hat{\beta}_1 = -7.477042, \qquad \hat{\beta}_2 = 1.740633$$

$$\hat{E}(Y \mid x_1, x_2) = 46.351255 - 7.477042 \times x_1 + 1.740633 \times x_2$$

$$R^2 = 0.7878$$

The estimated standard errors for the estimated regression parameters $\hat{\beta}_i$ ($i = 0, 1, 2$) appear in the third column. The entries in the last two columns are the results of testing hypotheses about the individual regression parameters and are discussed in the next section.

There are $n = 12$ observations and $p = 3$ regression parameters. Consequently, the F ratio has $p - 1 = 2$ degrees of freedom in the numerator and $n - p = 9$ degrees of freedom in the denominator. The F ratio is $16.706 > F_{2,9}(0.01) = 8.02$. The P-value is 0.0009, so we reject the null hypothesis.

TABLE 11.3

Dependent Variable: MPG

Analysis of Variance

Source	DF	Sum of Squares	Mean Square	F Value	Prob>F
Model	2	39.58691	19.79346	16.706	0.0009
Error	9	10.66309	1.18479		
C Total	11	50.25000			

	Root MSE	1.08848	R-square	0.7878

Parameter Estimates

Variable	DF	Parameter Estimate	Standard Error	T for H0: Parameter=0	Prob > |T|
INTERCEP	1	46.351255	4.28124784	10.827	0.0001
WGT	1	-7.477042	2.10287560	-3.556	0.0062
DISP	1	1.740633	0.86323760	2.016	0.0746

■

11.3.2   Testing Hypotheses about the Regression Model

When the null hypothesis of no regression relation is rejected, we would like to determine which regression parameters differ from zero. Suppose, for example, we want to test

$$H_0: \beta_i = 0 \quad \text{against} \quad H_1: \beta_i \neq 0$$

It can be shown that

$$\frac{\hat{\beta}_i - \beta_i}{s(\hat{\beta}_i)} \sim t_{n-p} \tag{11.49}$$

Confidence intervals for the regression parameters can now be derived in the usual way. For example, a $100(1 - \alpha)$ percent confidence interval for $\hat{\beta}_i$ is

$$\hat{\beta}_i \pm s(\hat{\beta}_i)t_{n-p}(\alpha/2)$$

When the null hypothesis H_0: $\beta_i = 0$ is true, it follows from Eq. (11.49) that $\hat{\beta}_i / s(\hat{\beta}_i)$ has a t distribution with $(n - p)$ degrees of freedom. We therefore reject H_0 at the level α if

$$\left| \frac{\hat{\beta}_i}{s(\hat{\beta}_i)} \right| > t_{n-p}(\alpha/2)$$

---

example 11.8

Refer to Example 11.7. Compute 95 percent confidence intervals for the regression parameters β_1 and β_2. Test at the 5 percent level the null hypotheses H_0: $\beta_i = 0$ $(i = 1, 2)$.

Solution. From Table 11.3 we see that $s(\hat{\beta}_1) = 2.10287560$ and $s(\hat{\beta}_2) = 0.86323760$. Moreover, $n - p = 12 - 3 = 9$, so $t_{n-p}(0.025) = t_9(0.025) = 2.262$. It follows that the 95 percent confidence interval for β_1 (answers are rounded to four decimal places) is

$$-7.477042 \pm 2.262 \times 2.10287560 = [-12.2337, -2.7203]$$

Similarly, the 95 percent confidence interval for β_2 is

$$1.740633 \pm 2.262 \times 0.86323760 = [-0.2120, 3.6933]$$

Notice that the confidence interval for β_2 contains 0; this suggests that we cannot reject (at the 5 percent level) the null hypothesis that $\beta_2 = 0$. This is confirmed by computing the ratio

$$\left| \frac{\hat{\beta}_2}{s(\hat{\beta}_2)} \right| = \frac{1.740633}{0.86323760} = 2.0164 < 2.262 = t_9(0.025)$$

The ratio $\hat{\beta}_2 / s(\hat{\beta}_2) = 2.016$ is listed in column 5 ("T for HO:") of Table 11.3, and the P-value $P(|t_9| > 2.016) = 0.0746$ is listed in column 6 ("Prob > |T|"). ∎

Building the Regression Model

Adding a variable to a regression model can lead to results that appear strange and that cast doubt on the validity of the model itself. To illustrate, let us pursue further the multiple linear regression Model($Y = $ mpg $| X_1 = $ weight, $X_2 = $ displacement) of Example 11.7. The estimated regression coefficient $\hat{\beta}_1 = -7.477042$ implies that an increase of 1000 lb in the car's weight (with the other variable held constant) results in a mean decrease of 7.477042 miles per gallon. Similarly, a one-liter increase in the displacement of the engine produces a mean increase of 1.74063 miles per gallon. The latter result is unusual because a larger engine should decrease fuel efficiency. The problem is that the car's weight and engine size are highly correlated [it can be shown that the sample correlation $\rho(X_1, X_2) = 0.95348$]; in particular, heavier cars tend to have larger engines, so it is not really possible to increase one variable and hold the other constant. This is an example of *multicollinearity*, which occurs when an explanatory variable is highly correlated with one or more other variables already in the model, or when the estimated regression coefficient, as in Example 11.7, has the wrong sign. Detecting and grappling with multicollinearity is a serious problem in multiple regression; it is treated at great length in the references listed at the end of this chapter.

The *partial F test* is a general statistical procedure for evaluating the contribution of additional explanatory variables to the regression model. To fix our ideas, let us continue our study of some of the problems that arise when we add the explanatory variable X_2 (displacement) to the simple linear regression Model$(Y = \text{mpg} \mid X_1 = \text{weight})$; we call this the *reduced model*. We call Model$(Y = \text{mpg} \mid X_1 = \text{weight}, X_2 = \text{displacement})$ the *full model*. The sums of squares associated with the full model are denoted SSR(f) and SSE(f); similarly, SSR(r) and SSE(r) denote the sums of squares associated with the reduced model. The total sum of squares is the same for all models:

$$SST = SSR(f) + SSE(f) = SSR(r) + SSE(r)$$

Therefore,

$$SSR(f) - SSR(r) = SSE(r) - SSE(f)$$

The difference $SSR(f) - SSR(r)$ is called the *extra sum of squares*. It measures the increase in the regression sum of squares when the explanatory variable X_2 is added to the model. This difference is always nonnegative, since adding more variables to the model always results in an increase in the total variability explained by the model. It follows that adding an explanatory variable reduces the unexplained variability. Thus, the extra sum of squares also measures the reduction in the error sum of squares that results when one adds an explanatory variable to the model.

---

example 11.9

Compute the extra sum of squares when the variable X_2 (displacement) is added to the simple linear regression Model$(Y = \text{mpg} \mid X = \text{weight})$ of Example 11.2.

Solution. The analysis of variance for the reduced and full models are shown in Tables 10.7 and 11.3, respectively. It follows that

$$SSR(f) - SSR(f) = 39.58691 - 34.76792 = 4.81719$$

Observe that

$$SSE(r) - SSE(f) = 15.48028 - 10.66309 = 4.81719$$

as predicted by the theory.   ∎

Consider the problem of determining when the extra sum of squares obtained by adding an explanatory variable to the model is large enough to justify including it in the model; that is, we want to test

$$H_0: \beta_2 = 0 \quad \text{against} \quad H_1: \beta_2 \neq 0$$

When H_0 is accepted, this means that we have not found it useful to add X_2 to the model; that is, the full model performs no better (and perhaps actually worse) than the reduced model. Denote the number of degrees of freedom of

SSE(f) and SSE(r) by DF(f) and DF(r), respectively. Under the null hypothesis, it can be shown that the ratio

$$F = \frac{[SSE(r) - SSE(f)]/[DF(r) - DF(f)]}{SSE(f)/DF(f)} \qquad (11.50)$$

has an F distribution with $DF(r) - DF(f)$ degrees of freedom in the numerator and $DF(f)$ degrees of freedom in the denominator. Consequently, the decision rule is

$$\text{accept } H_0: \text{ if } F \leq F_{DF(r)-DF(f),DF(f)}(\alpha) \qquad (11.51)$$

$$\text{reject } H_0: \text{ if } F > F_{DF(r)-DF(f),DF(f)}(\alpha) \qquad (11.52)$$

---

example **11.10**   Refer to Example 11.7. Use the partial F test to determine whether or not the variable $X_2 = $ displacement should be dropped from the model.

Solution. In the solution to Example 11.9 we showed that the extra sum of squares $SSE(r) - SSE(f)$ is 4.81719. It follows that

$$SSE(r) - SSE(f) = 4.81719$$

$$DF(r) - DF(f) = 10 - 9 = 1$$

Thus,

$$F = \frac{[SSE(r) - SSE(f)]/[DF(r) - DF(f)]}{SSE(f)/DF(f)}$$

$$= \frac{4.81719/1}{10.66309/9} = 4.065867$$

Because $F = 4.065867 < 5.12 = F_{1,9}(0.05)$, we do not reject $H_0: \beta_2 = 0$. It follows from Corollary 6.2 (which states that $t_n^2 = F_{1,n}$) that, in this case, the partial F test is equivalent to the t test. Indeed, $t_9^2 = (2.016)^2 = 4.0643$, which equals the F ratio (except for roundoff error). ∎

Adjusted R^2 Criterion

Although adding a variable to the model always reduces the unexplained variability, it does not necessarily reduce the mean square error because the number of degrees of freedom of the error sum of squares for the full model is less than the corresponding number of degrees of freedom for the reduced model. More precisely,

$$DF(r) - DF(f) = \text{number of additional variables in the full model}$$

Consequently $MSE(f) = SSE(f)/DF(f)$ could be larger than $MSE(r) = SSE(r)/DF(r)$. This leads to the *adjusted* R^2, denoted R_a^2 and defined as follows:

$$R_a^2 = 1 - \left(\frac{n-1}{n-p}\right)R^2 = 1 - \frac{MSE}{SST/(n-1)} \qquad (11.53)$$

Choosing a model with the smallest adjusted R^2 is equivalent to choosing the model with the smallest mean square error, since $\text{SST}/(n-1)$ is the same for all models. If the reduced model has a smaller adjusted R^2, then it will have a smaller estimated standard error of prediction. Consequently, the reduced model has shorter confidence and prediction intervals. Consult the references at the end of the chapter for a detailed discussion of various methods for selecting the best model.

11.3.3 Model Checking

The assumptions of the multiple linear regression model [refer back to Eq. (11.44)] should always be checked by analyzing the residuals. When the model is valid, the statistical properties of the residuals should resemble those of the error terms ϵ_i $(i = 1, \ldots, n)$, where $E(\epsilon_i) = 0$ and $V(\epsilon_i) = \sigma^2$. Two particularly useful methods for detecting violations of model assumptions are the residual plots and normal probability plots described in Chapter 10.

Residual Plots

It can be shown that the mean and variance of the ith residual are given by

$$E(e_i) = 0 \quad \text{and} \quad V(e_i) = \sigma^2(1 - h_{ii}) \tag{11.54}$$

where h_{ii} is the ith diagonal element of the hat matrix [Eq. (11.12)]. We obtain the *estimated standard error for the ith residual*, denoted $s(e_i)$, by replacing σ^2 with its estimate s^2. Thus,

$$s(e_i) = s \sqrt{1 - h_{ii}} \tag{11.55}$$

It follows that the residuals do not have constant variance. For this reason it is more useful to plot the *studentized residuals*, denoted e_i^* and defined as follows:

$$e_i^* = \frac{e_i}{s(e_i)} \tag{11.56}$$

When the assumptions of the multiple linear regression model hold, the joint distribution of the studentized residuals is approximately equal to a random sample taken from a normal population with zero mean and variance 1. Moreover, it can be shown that the fitted values are independent of the studentized residuals. Therefore, the points $(\hat{Y}_i, e_i^*)$ $(i = 1, \ldots, n)$ of the residual plot should appear to be randomly scattered about the line $y = 0$, with most of the points within ± 2 of the line $y = 0$. The plot of the studentized residuals shown in Fig. 11.2 appear to be randomly scattered within a horizontal band of width 2 centered about the line $y = 0$.

Normal Probability Plots

Normal probability plots of the studentized residuals are useful for checking the assumption that the error terms ϵ_i $(i = 1, \ldots, n)$ are iid normal random variables with zero means. To construct a *normal probability plot of the studentized*

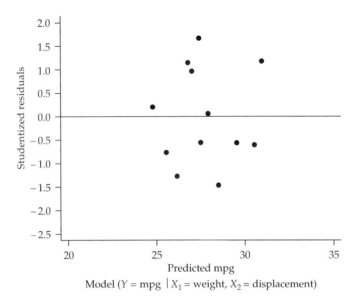

Model (Y = mpg $\mid X_1$ = weight, X_2 = displacement)

Figure 11.2 Plot of studentized residuals against predicted mpg (data in Table 10.2)

residuals, we follow the procedure outlined in Definition 8.1:

1. Compute the normal scores $z_i = \Phi^{-1}((i - 0.5)/n)$ $(i = 1, \ldots, n)$.
2. Compute the order statistics $e_{(i)}^*$ $(i = 1, \ldots, n)$.
3. Plot the points $(z_i, e_{(i)}^*)$ $(i = 1, \ldots, n)$.

The model assumption that ϵ_i $(i = 1, \ldots, n)$ are iid normal random variables with zero means is doubtful when the points of the normal probability plot do not appear to lie nearly on a straight line.

The normal probability plot of the studentized residuals shown in Fig. 11.3 reveals no serious violation of the normality assumption.

It would take too much time and space to fully exploit all the information we can obtain from a careful analysis of the residuals. Consult the references at the end of the chapter for a more comprehensive treatment of this subject.

11.3.4 Confidence Intervals and Prediction Intervals in Multiple Linear Regression

Matrix expressions for confidence intervals on the mean response and prediction intervals on a future response for the simple linear regression model were derived earlier in this chapter [refer to Eqs. (11.34) and (11.36) in Sec. 11.2.1]. The formulas for the confidence intervals and prediction intervals in a multiple linear regression (assuming the error terms ϵ_i are iid normal random variables with zero means) are similar except that the number of degrees of freedom for

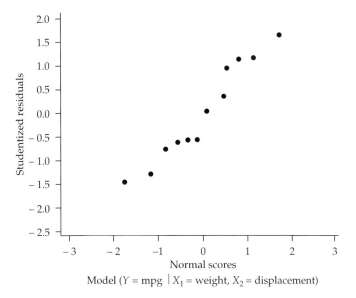

Model (Y = mpg $\mid X_1$ = weight, X_2 = displacement)

Figure 11.3 Normal probability plot of studentized residuals (data in Table 10.2)

the t distribution is $(n - p)$. We omit the derivations. The vector
$$\mathbf{x}_0' = (1, x_{01}, \ldots, x_{0k})$$
denotes the values of the explanatory variables; that is,
$$X_1 = x_{01}, \ldots, X_k = x_{0k}$$
A $100(1 - \alpha)$ percent confidence interval for the mean response is given by
$$\mathbf{x}_0'\hat{\boldsymbol{\beta}} \pm t_{n-p}(\alpha/2)s\sqrt{\mathbf{x}_0'(\mathbf{X}'\mathbf{X})^{-1}\mathbf{x}_0} \tag{11.57}$$

The term
$$s\sqrt{\mathbf{x}_0'(\mathbf{X}'\mathbf{X})^{-1}\mathbf{x}_0}$$
is called the *standard error of prediction* and is usually listed in the computer printouts.

Similarly, a $100(1 - \alpha)$ percent prediction interval for a future response
$$Y(\mathbf{x}_0) = \beta_0 + \beta_1 x_{01} + \cdots + \beta_k x_{0k} + \epsilon_0$$
is given by
$$\mathbf{x}_0'\hat{\boldsymbol{\beta}} \pm t_{n-p}(\alpha/2)s\sqrt{1 + \mathbf{x}_0'(\mathbf{X}'\mathbf{X})^{-1}\mathbf{x}_0}, \quad (p = k + 1) \tag{11.58}$$

---

example 11.11   Refer to the multiple linear regression model of Example 11.7. Compute 95 percent confidence intervals and prediction intervals for each observation using a computer software package.

Solution. These values are shown in Table 11.4 and were obtained using SAS. The second column lists the observed value Y_i. The third column ("Predict Value") lists the fitted value

$$\hat{Y}_i = \mathbf{x}'\hat{\boldsymbol{\beta}} = \hat{\beta}_0 + \hat{\beta}_1 x_1 + \cdots + \hat{\beta}_k x_k$$

The fourth column ("Std Err Predict") lists the standard error of prediction. Columns 5 and 6 list the lower and upper limits for the 95 percent confidence intervals, and columns 7 and 8 list the lower and upper limits for the prediction intervals. Consider observation 1 (refer to the design matrix shown in the solution to Example 11.5). The values of the explanatory variables are $X_1 = 2.495$ and $X_2 = 1.9$, so

$$\mathbf{x}_0' = (1, 2.495, 1.9)$$

The observed value Y_1 is 32.00, and the fitted value $\hat{Y}_1$ is 31.0032. The standard error of prediction is 0.677. The 95 percent confidence interval is [29.4720, 32.5345], and the 95 percent prediction interval is [28.1036, 33.9029].

TABLE 11.4

Obs	Dep Var MPG	Predict Value	Std Err Predict	Lower95% Mean	Upper95% Mean	Lower95% Predict	Upper95% Predict
1	32.0000	31.0032	0.677	29.4720	32.5345	28.1036	33.9029
2	30.0000	30.5675	0.584	29.2467	31.8883	27.7733	33.3617
3	29.0000	29.5464	0.504	28.4052	30.6876	26.8325	32.2603
4	25.0000	26.1886	0.573	24.8927	27.4845	23.4061	28.9711
5	27.0000	27.5252	0.540	26.3033	28.7471	24.7764	30.2740
6	28.0000	27.9550	0.460	26.9136	28.9963	25.2815	30.6284
7	29.0000	27.4504	0.555	26.1956	28.7053	24.6868	30.2141
8	27.0000	28.5065	0.360	27.6912	29.3217	25.9127	31.1002
9	28.0000	26.8334	0.362	26.0139	27.6529	24.2383	29.4285
10	25.0000	24.8239	0.596	23.4752	26.1725	22.0164	27.6313
11	28.0000	27.0203	0.362	26.2020	27.8387	24.4256	29.6151
12	25.0000	25.5796	0.783	23.8074	27.3518	22.5458	28.6134

PROBLEMS

11.9 Refer to the groundwater quality data set of Prob. 10.8.

(a) Fit the multiple linear regression Model(Y = TDS | X_1 = SEC, X_2 = SiO$_2$) and state the estimated regression function.

(b) Estimate the change in the total dissolved solids (TDS) when the specific electrical conductivity (SEC) is increased by 100 ms/cm and the amount of silica (SiO$_2$) is increased by 5.

(c) What is the fitted value and residual corresponding to $X_1 = 841$ and $X_2 = 48$?

(d) Compute the ANOVA table, R^2, and the adjusted R^2.

(e) Use the partial F test to decide if X_2 should be dropped from the regression model.

11.10 Refer to Example 11.6 and the associated data set, Table 11.1.

(a) Fit the quadratic polynomial Model(Y = mpg | X_1 = speed, X_2 = speed^2) and state the estimated regression function.

(b) Draw the scatter plot and the graph of the fitted polynomial on the same coordinate axes.

(c) Compute the fitted value, 95 percent confidence interval, and 95 percent prediction interval for x_0' = (1, 35, 1225).

11.11 The following data come from an experiment to determine the relation between Y, the shear strength (psi) of a rubber compound, and X, the cure temperature (°F).

(a) Draw the scatter plot to see if the simple linear regression model is appropriate.

(b) Fit the quadratic polynomial Model(Y = shear strength | X_1 = x, X_2 = x^2) to these data.

(c) Draw the graph of the fitted polynomial on the same set of axes you used for the scatter plot.

(d) Test the hypothesis $\hat{\beta}_2$ = 0.

(e) Check the assumptions of the model by (1) graphing the studentized residual plot and (2) graphing the normal probability plot of the studentized residuals.

y	x	y	x
770	280	735	298
800	284	640	305
840	292	590	308
810	295	560	315

Source: A. F. Dutka and F. J. Ewens, *A Method for Improving the Accuracy of Polynomial Regression Analysis,* Journal of Quality Technology, 1971, pp. 149–155. Used with permission.

11.12 The data in the following table were obtained from a forestry study of certain characteristics of a stand of pine trees, including A, the age of a particular pine stand; HD, the average height of dominant trees, in feet; N, the number of pine trees per acre at age A; and MDBH, the average diameter measured at breast height (4.5 feet above the ground). A theoretical analysis suggested that the following model is appropriate:

$$\text{Model}(Y = \text{MDBH} \mid X_1 = \text{HD}, X_2 = A \times N, X_3 = \text{HD}/N)$$

(a) Fit this model to the data. Compute R^2 and R_a^2 [the adjusted R^2; see Eq. (11.53)].

(b) Test the hypothesis

$$H_0: \beta_3 = 0 \quad \text{against} \quad H_1: \beta_3 \neq 0$$

State your conclusions.

(c) Fit the reduced Model(Y = MDBH | X_1 = HD, X_2 = $A \times N$). Compute R^2 and R_a^2 for the reduced model.

(d) For each model prepare a table of fitted values, standard error of prediction, and so on, similar in format to Table 11.4. Compare the full and reduced models by comparing the standard errors of prediction. Which model performs better?

A	HD	N	MDBH	A	HD	N	MDBH
19	51.5	500	7.0	13	37.3	800	5.4
14	41.3	900	5.0	21	54.2	650	6.4
11	36.7	650	6.2	11	32.5	530	5.4
13	32.2	480	5.2	19	56.3	680	6.7
13	39.0	520	6.2	17	52.8	620	6.7
12	29.8	610	5.2	15	47.0	900	5.9
18	51.2	700	6.2	16	53.0	620	6.9
14	46.8	760	6.4	16	50.3	730	6.9
20	61.8	930	6.4	14	50.5	680	6.9
17	55.8	690	6.4	22	57.7	480	7.9

Source: R. H. Myers, *Classical and Modern Regression with Applications,* 2d ed., Boston, PWS-Kent, 1990. Used with permission.

11.13 The data in the following table come from an experiment to study the effects of three environmental variables on exhaust emissions of light-duty diesel trucks. Fit Model($Y = NO_x \mid X_1 = $ humidity, $X_2 = $ temperature, $X_3 = $ barometric pressure). Test the hypothesis

$$H_0: \beta_i = 0 \ (i = 1, 2, 3) \quad \text{against} \quad H_1: \text{at least one } \beta_i \neq 0$$

NO_x (ppm)	Humidity (%)	Temperature (°F)	Barometric pressure (in. Hg)
0.70	96.5	78.1	29.08
0.79	108.72	87.93	29.98
0.95	61.37	68.27	29.34
0.85	91.26	70.63	29.03
0.79	96.83	71.02	29.05
0.77	95.94	76.11	29.04
0.76	83.61	78.29	28.87
0.79	75.97	69.35	29.07
0.77	108.66	75.44	29.00
0.82	78.59	85.67	29.02
1.01	33.85	77.28	29.43
0.94	49.20	77.33	29.43
0.86	75.75	86.39	29.06
0.79	128.81	86.83	28.96
0.81	82.36	87.12	29.12
0.87	122.60	86.20	29.15
0.86	124.69	87.17	29.09
0.82	120.04	87.54	29.09
0.91	139.47	87.67	28.99
0.89	105.44	86.12	29.21

Source: R. L. Mason, R. F. Gunst, and J. L. Hess, *Statistical Design and Analysis of Experiments,* New York, John Wiley & Sons, 1989. Used with permission.

11.14 Show that the variance-covariance matrix of the residual vector is

$$\text{Cov}(\mathbf{e}) = \sigma^2(\mathbf{I} - \mathbf{H})$$

where $\mathbf{H}$ is the hat matrix. [*Hint:* Use the representation $\mathbf{e} = (\mathbf{I} - \mathbf{H})Y$, Eq. (11.26), and Eq. (11.16).]

11.4 MATHEMATICAL DETAILS AND DERIVATIONS

We now show that any solution of the normal equations minimizes the sum of squares. More precisely, we show that for any $\mathbf{b} \neq \hat{\boldsymbol{\beta}}$ the following inequality holds:

$$|\mathbf{Y} - \mathbf{Xb}|^2 \geq |\mathbf{Y} - \mathbf{X}\hat{\boldsymbol{\beta}}|^2 \qquad (11.59)$$

Proof. Let $\mathbf{b} = \hat{\boldsymbol{\beta}} + \mathbf{d}, \mathbf{d} \neq \mathbf{0}$. The inequality (11.59) is a consequence of the following equation:

$$|\mathbf{Y} - \mathbf{Xb}|^2 = |\mathbf{Y} - \mathbf{X}\hat{\boldsymbol{\beta}}|^2 + |\mathbf{Xd}|^2 \qquad (11.60)$$

We obtain Eq. (11.60) by substituting $\hat{\boldsymbol{\beta}} + \mathbf{d} = \mathbf{b}$ into the left-hand side of Eq. (11.60), expanding the matrix products, and noting that the cross-product terms vanish; that is,

$$(\mathbf{Xd})'(\mathbf{Y} - \mathbf{X}\hat{\boldsymbol{\beta}}) = 0, \qquad (\mathbf{Y} - \mathbf{X}\hat{\boldsymbol{\beta}})'\mathbf{Xd} = 0$$

In detail,

$$
\begin{aligned}
|\mathbf{Y} - \mathbf{Xb}|^2 = |\mathbf{Y} - \mathbf{X}(\hat{\boldsymbol{\beta}} + \mathbf{d})|^2 &= [\mathbf{Y} - \mathbf{X}(\hat{\boldsymbol{\beta}} + \mathbf{d})]'[\mathbf{Y} - \mathbf{X}(\hat{\boldsymbol{\beta}} + \mathbf{d})] \\
&= (\mathbf{Y} - \mathbf{X}\hat{\boldsymbol{\beta}})'(\mathbf{Y} - \mathbf{X}\hat{\boldsymbol{\beta}}) - (\mathbf{Xd})'(\mathbf{Y} - \mathbf{X}\hat{\boldsymbol{\beta}}) - (\mathbf{Y} - \mathbf{X}\hat{\boldsymbol{\beta}})'\mathbf{Xd} + (\mathbf{Xd})'(\mathbf{Xd}) \\
&= (\mathbf{Y} - \mathbf{X}\hat{\boldsymbol{\beta}})'(\mathbf{Y} - \mathbf{X}\hat{\boldsymbol{\beta}}) + (\mathbf{Xd})'(\mathbf{Xd}) \\
&= |\mathbf{Y} - \mathbf{X}\hat{\boldsymbol{\beta}}|^2 + |\mathbf{Xd}|^2
\end{aligned}
$$

We complete the proof by showing that the cross-product terms vanish. This follows from Eq. (11.3) for the residual vector, which implies that

$$(\mathbf{Xd})'(\mathbf{Y} - \mathbf{X}\hat{\boldsymbol{\beta}}) = \mathbf{d}'\mathbf{X}'(\mathbf{Y} - \mathbf{X}\hat{\boldsymbol{\beta}}) = 0$$
$$(\mathbf{Y} - \mathbf{X}\hat{\boldsymbol{\beta}})'\mathbf{Xd} = [(\mathbf{Xd})'(\mathbf{Y} - \mathbf{X}\hat{\boldsymbol{\beta}})]' = 0$$

11.5 CHAPTER SUMMARY

A multiple linear regression model is used to study the relationship between a response variable y and two or more explanatory variables $x_1, \dots, x_k$. We assume that the relationship between the explanatory and response variables is given by a function $y = f(x_1, \dots, x_k) + \epsilon$, where f is unknown. A multiple linear regression model assumes that $f(x_1, \dots, x_k) = \beta_0 + \beta_1 x + \cdots + \beta_k x_k$. It is difficult, although not impossible, to explain the concepts and methodology of multiple linear regression without using matrix algebra. In addition to freeing ourselves from the tyranny of subscripts, the geometric interpretation of the response variable and the design matrix (as a vector and subspace of a vector space) gives us a deeper understanding of the least squares method itself, even in the case of simple linear regression.

To Probe Further. The following texts contain a wealth of additional information on multiple linear regression, including multicollinearity, model selection, model checking, and regression diagnostics.

1. R. H. Myers, *Classical and Modern Regression with Applications,* 2d ed., Boston, PWS-Kent, 1990.
2. J. Neter, M. H. Kutner, C. J. Nachtseim, and W. Wasserman, *Applied Linear Statistical Models,* 4th ed., Chicago, Richard D. Irwin, 1996.

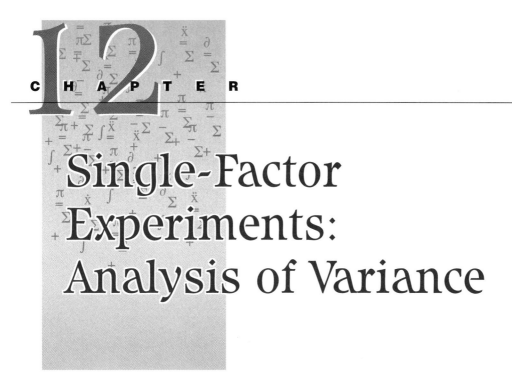

12

Single-Factor Experiments: Analysis of Variance

To consult a statistician after an experiment is finished is often merely to ask him to conduct a post mortem examination. He can perhaps say what the experiment died of.

Sir Ronald A. Fisher (1890–1962), British statistician

12.1 ORIENTATION

In this chapter we introduce the *analysis of variance* (ANOVA) method, a powerful technique of great generality for estimating and comparing the means of two or more populations, and particularly for testing the null hypothesis that the means are the same. This is an extension of the problem of testing hypotheses about the means of two populations discussed in Sec. 8.3. When the null hypothesis of no difference among the means is rejected, the next step in the analysis is to determine which means differ significantly from one another. This naturally leads us to study the problem of multiple comparisons of the population means.

Organization of Chapter

12.2   THE SINGLE-FACTOR ANOVA MODEL

We develop the basic concepts and methods of the single-factor ANOVA model in the context of an experiment to compare different methods of teaching arithmetic.

example **12.1**   Table 12.1 lists the numerical grades on a standard arithmetic test given to 45 students divided randomly into five equal-sized groups. Groups 1 and 2 were taught by the current method. Groups 3, 4, and 5 were taught together for a number of days; on each day group 3 students were praised publicly for their previous work, group 4 students were criticized publicly, and group 5 students, while hearing the praise and criticism of groups 3 and 4, were ignored.

TABLE 12.1   ARITHMETIC TEST GRADES FOR FIVE GROUPS
OF STUDENTS

Method	Grades									Mean
1 (control)	17	14	24	20	24	23	16	15	24	19.67
2 (control)	21	23	13	19	13	19	20	21	16	18.33
3 (praised)	28	30	29	24	27	30	28	28	23	27.44
4 (criticized)	19	28	26	26	19	24	24	23	22	23.44
5 (ignored)	21	14	13	19	15	15	10	18	20	16.11

Source: G. B. Wetherill, *Elementary Statistical Methods*, London, Methuen, 1967, p. 263.

The response variable in this experiment is the student's grade, and the explanatory variable is the teaching group, which is an example of a nonnumerical variable called a *factor*. A particular value of the factor is called a *factor level*; it is also called a *treatment*. The different groups correspond to five factor levels. For each treatment there are nine students; we say that there are nine *replicates* for each treatment. Since the treatments are distinguished from one another by the different levels of a single factor, the experiment is called a *single-factor experiment*; the subsequent analysis is called *one-way ANOVA*.

We define the model using the statement format

$$\text{Model}(Y = \text{arithmetic grade} \,|\, X = \text{group}) \qquad\blacksquare$$

An experiment is called a *completely randomized design* when, as in Example 12.1, the different treatments are randomly assigned to the experimental units. The purpose of randomization is to guard against bias. For instance, if one put the best students into the third group, then their better-than-average performance could not be attributed primarily to the teaching method. This is an example of an *experimental design*.

The goal of ANOVA is to study the mean of the response variable as a function of the factor levels. In this sense ANOVA is similar to regression analysis, except that the explanatory variable may be nonnumeric. In the context of Example 12.1, there are two questions of primary interest: (1) What effect does the teaching method have on the students' performance? (This is an estimation problem.)

(2) Are there significant differences among the groups? (This is, of course, a hypothesis-testing problem in which we are interested in comparing the means of five populations based on independent samples drawn from each of them.)

The null hypothesis here is that there is no difference in the mean grades produced by these teaching methods. Thus, the null and alternative hypotheses are given by

$$H_0: \mu_1 = \mu_2 = \mu_3 = \mu_4 = \mu_5$$

H_1: At least two of the means are different

A Visual Comparison of the Teaching Methods: Side-by-Side Box Plots

Before turning to more sophisticated methods of analysis, we first look at the data using some of the exploratory data analysis techniques discussed in Chap. 1. The side-by-side box plot shown in Fig. 12.1 is a very informative graphic display that reveals several important features of the data:

1. The width of each box gives a rough indication of the extent to which the variability of the grades depends on the teaching method.
2. The "+" signs (which represent the sample medians) indicate the extent to which the row medians depend on the teaching method. The box plots indicate some variability in the medians, particularly for method 3, which appears to be significantly better than the other four. (The * in the box plot for method 3 indicates an outlier, corresponding to the observation 23.)
3. The box plots suggest that it is very unlikely that these samples were drawn from the same population; in other words, the null hypothesis that these teaching methods are equally effective appears to be highly unlikely. We return to this point shortly.

Notation and Model Assumptions

Before proceeding further we introduce some notation and assumptions.

1. **Treatments and replications.** It is assumed that there are I treatments and that the number of replications, denoted J, is the same for each treatment. The total number of observations, then, is $n = IJ$. The jth observation at the ith factor level is denoted by y_{ij}. The data are displayed in the format of Table 12.2.

TABLE 12.2   DATA FORMAT FOR A ONE-WAY ANOVA

Treatment	Observation			
1	y_{11}	y_{12}	$\cdots$	y_{1J}
2	y_{21}	y_{22}	$\cdots$	y_{2J}
$\vdots$	$\vdots$	$\vdots$	$\vdots$	$\vdots$
I	y_{I1}	y_{I2}	$\cdots$	y_{IJ}

Method

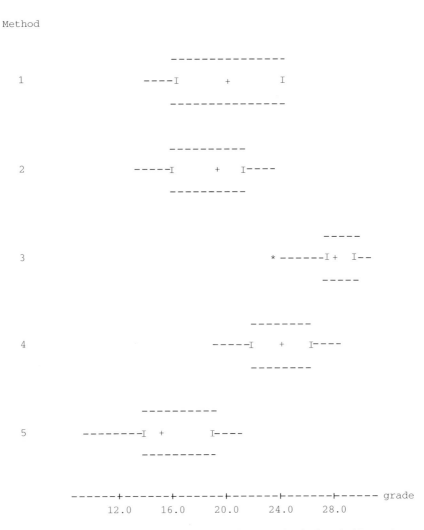

Figure 12.1 Side-by-side box plots of teaching methods data (Table 12.1)

For an example, refer back to Table 12.1. The five teaching methods correspond to factor levels 1, 2, 3, 4, and 5, respectively. In particular, $y_{1,1} = 17$, $y_{1,2} = 14$, $y_{5,1} = 21$, etc. Since the sample sizes corresponding to each treatment are equal, the experimental design is said to be *balanced*. The case of unequal sample sizes will be dealt with later.

2. **Random allocations.** We assume that the experimental units have been randomly allocated to the treatments.

3. **The statistical model for the response variable.** We assume that the response of the jth experimental unit to the ith factor level is a random variable Y_{ij} defined by

$$Y_{ij} = \mu_i + \epsilon_{ij} \tag{12.1}$$

where the random variables ϵ_{ij} are iid $N(0, \sigma^2)$, so $E(Y_{ij}) = \mu_i$ and $V(Y_{ij}) = \sigma^2$. In particular, we assume that the variance of the response variable is the same for all factor levels. The parameter μ_i is called the *ith treatment mean*.

4. **The overall mean and treatment effects.** It is convenient to specify the statistical model in terms of new parameters μ, called the *overall mean*, and $\alpha_i = \mu_i - \mu$, which is the difference between the *i*th treatment mean and the overall mean; it is called the *i*th *treatment effect*. In detail:

$$\text{Overall mean:} \quad \mu = \frac{1}{I} \sum_{i=1}^{I} \mu_i$$

$$\text{ith treatment effect:} \quad \alpha_i = \mu_i - \mu$$

Consequently,

$$\sum_{i=1}^{I} \alpha_i = 0$$

In terms of the new parameters, the response variables are

$$Y_{ij} = \mu + \alpha_i + \epsilon_{ij}$$

Thus

$$E(Y_{ij}) = \mu_i = \mu + \alpha_i$$

5. **Dot subscript notation.** Dot notation is useful for expressing summation with respect to one index with the other fixed. Replacing the subscript *j* with a dot means summation with respect to the subscript *j*. Replacing each subscript *i* and *j* with a dot means summation with respect to both subscripts *i* and *j*. Thus,

$$y_{i.} = \sum_{1 \le j \le J} y_{ij} \qquad \bar{y}_{i.} = \frac{y_{i.}}{J}$$

$$y_{..} = \sum_{i} \sum_{j} y_{ij} \qquad \bar{y}_{..} = \frac{y_{..}}{n}$$

The quantity $\bar{y}_{..}$ is called the *grand mean*, and the quantity $\bar{y}_{i.}$ is called the *ith row mean*.

---

example **12.2**   Refer back to Example 12.1, comparing the five methods for teaching arithmetic. Compute the grand mean and the five treatment means. Express the results using dot subscript notation.

Solution. With reference to Table 12.1, it can be shown that the grand mean $\bar{y}_{..} = 21$, which means that average grade of all students tested was 21. Similarly, it can be shown that the average grades for teaching methods 1–5 are

$$\bar{y}_{1.} = 19.67, \quad \bar{y}_{2.} = 18.33, \quad \bar{y}_{3.} = 27.44, \quad \bar{y}_{4.} = 23.44, \quad \bar{y}_{5.} = 16.11 \quad \blacksquare$$

The results indicate that the average grade for method 3 is greater than the average grades for the other four methods. Notice, however, that $y_{1,9} = 24 >$

$23 = y_{3,9}$; not every student in group 3 graded higher than all students in group 1. The fact that the box plots of groups 3 and 5 do not overlap indicates that there are major differences between these two teaching methods. On the other hand, there is considerable overlap between the box plots for groups 1 and 2, indicating that the differences between their means is probably not statistically significant. We will discuss these preliminary inferences shortly from a more sophisticated perspective.

12.2.1   Estimating the ANOVA Model Parameters

In this section we use the method of least squares to compute the values of the parameters $\mu, \alpha_1, \ldots, \alpha_I$ of the ANOVA model that best fits the data; the method is similar to the one used in regression analysis. We then develop a criterion for measuring how well the model fits the data; it is similar to the coefficient of determination used in regression analysis, and is in fact obtained from a similar partition of the total variability into two parts: the variability explained by the model and the unexplained variability.

We recall that the least squares method is to find those values $\mu, \alpha_1, \ldots, \alpha_I$ that minimize the error sum of squares

$$Q(\mu, \alpha_1, \ldots, \alpha_I) = \sum_i \sum_j e_{ij}^2 = \sum_i \sum_j (y_{ij} - \mu - \alpha_i)^2$$

We minimize the function Q using calculus; that is, we first compute the partial derivatives of Q with respect to each of the parameters $\mu, \alpha_1, \ldots, \alpha_I$ and set them equal to zero. This yields the following $I + 1$ linear equations for the parameters:

$$\frac{\partial Q}{\partial \mu} = -2 \sum_i \sum_j (y_{ij} - \mu - \alpha_i) = 0$$

$$\frac{\partial Q}{\partial \alpha_i} = -2 \sum_j (y_{ij} - \mu - \alpha_i) = 0, \quad i = 1, \ldots, I$$

subject to the constraint

$$\sum_i \alpha_i = 0$$

Solving these linear equations yields the least squares estimates for the model parameters denoted by $\hat{\mu}$ and $\hat{\alpha}_i$:

$$\hat{\mu} = \overline{Y}_{..} \tag{12.2}$$
$$\hat{\alpha}_i = \overline{Y}_{i.} - \overline{Y}_{..} \tag{12.3}$$

Consequently,

$$\hat{\mu}_i = \hat{\mu} + \hat{\alpha}_i = \overline{Y}_{i.} \tag{12.4}$$

These results are intuitively reasonable. The overall mean is estimated by the grand mean, the ith treatment effect is estimated by the difference between the ith row mean and the grand mean, and the ith treatment mean is estimated by the ith row mean.

Notice that the assumption of normality was not used to derive the least squares estimates; the normality assumption, however, will be used to derive confidence intervals for the model parameters.

---

example **12.3**   Refer back to the teaching methods data (Table 12.1) of Example 12.1. Compute the least squares estimators for the grand mean, treatment means, and treatment effects.

Solution. Using Eqs. (12.2)–(12.4), we obtain the following results:

$$\text{Grand mean:} \quad \hat{\mu} = \bar{y}_{..} = 21$$

$$\text{Treatment means:} \quad \hat{\mu}_1 = \bar{y}_{1.} = 19.67$$
$$\hat{\mu}_2 = \bar{y}_{2.} = 18.33$$
$$\hat{\mu}_3 = \bar{y}_{3.} = 27.44$$
$$\hat{\mu}_4 = \bar{y}_{4.} = 23.44$$
$$\hat{\mu}_5 = \bar{y}_{5.} = 16.11$$

$$\text{Treatment effects:} \quad \hat{\alpha}_1 = \bar{y}_{1.} - \bar{y}_{..} = -1.33$$
$$\hat{\alpha}_2 = \bar{y}_{2.} - \bar{y}_{..} = -2.67$$
$$\hat{\alpha}_3 = \bar{y}_{3.} - \bar{y}_{..} = 6.44$$
$$\hat{\alpha}_4 = \bar{y}_{4.} - \bar{y}_{..} = 2.44$$
$$\hat{\alpha}_5 = \bar{y}_{5.} - \bar{y}_{..} = -4.89$$

These results indicate that the average score of the students taught using method 3 (praised) was 6.44 points above the grand mean (21), and the average score of the students taught using method 5 (ignored) was 4.89 points below. ■

Fitted Values and Residuals. As in regression analysis, we define the fitted (predicted) values and the residuals as follows:

$$\text{Fitted value:} \quad \hat{y}_{ij} = \bar{y}_{i.}$$
$$\text{Residual:} \quad e_{ij} = y_{ij} - \hat{y}_{ij} = y_{ij} - \bar{y}_{i.}$$

Partitioning the Total Variability of the Data

To study how well the model produced by the least squares estimates fits the data, we use a partition of the total sum of squares $\text{SST} = \sum\sum(y_{ij} - \bar{y}_{..})^2$ analogous to that used in regression analysis. Adding and subtracting $\bar{y}_{i.}$, we obtain the algebraic identity

$$y_{ij} - \bar{y}_{..} = (\bar{y}_{i.} - \bar{y}_{..}) + (y_{ij} - \bar{y}_{i.}) \tag{12.5}$$

That is,

$$\text{Observation} - \frac{\text{Grand}}{\text{mean}} = \frac{\text{Difference due}}{\text{to treatment}} + \text{Residual}$$

This corresponds to the model

$$Y_{ij} - \mu = \alpha_i + e_{ij}$$

Proceeding in a manner similar to that used in regression analysis [refer to Eq. (10.15)], we partition the total sum of squares SST into the sum

$$\text{SST} = \text{SSTr} + \text{SSE} \tag{12.6}$$

where

$$\text{SST} = \sum_{i=1}^{I} \sum_{j=1}^{J} (y_{ij} - \bar{y}_{..})^2$$

$$= \sum_{i=1}^{I} \sum_{j=1}^{J} y_{ij}^2 - \frac{y_{..}^2}{n}$$

$$\text{SSTr} = \sum_{i=1}^{I} \sum_{j=1}^{J} (\bar{y}_{i.} - \bar{y}_{..})^2$$

$$= J \sum (\bar{y}_{i.} - \bar{y}_{..})^2$$

$$= \frac{1}{J} \sum_{i=1}^{I} y_{i.}^2 - \frac{y_{..}^2}{n}$$

$$\text{SSE} = \sum_{i=1}^{I} \sum_{j=1}^{J} (y_{ij} - \bar{y}_{i.})^2$$

$$= \text{SST} - \text{SSTr} = \sum_{i=1}^{I} \sum_{j=1}^{J} e_{ij}^2$$

SSTr is called the *treatment sum of squares*; it represents the variability explained by the ANOVA model. SSE, the *error sum of squares*, represents the unexplained variability. The *proportion of the total variability explained by the model* is denoted R^2, where

$$R^2 = \frac{\text{SSTr}}{\text{SST}} \tag{12.7}$$

Derivation of Eq. (12.6). We obtain Eq. (12.6) by squaring both sides of Eq. (12.5), summing over the indices i, j, and using the fact that the cross-product term vanishes, that is,

$$\sum_{i} \sum_{j} (\bar{y}_{i.} - \bar{y}_{..})(y_{ij} - \bar{y}_{i.}) = 0$$

---

example 12.4 Refer back to the teaching methods data of Example 12.1. Compute the sums of squares SST, SSTr, SSE, and R^2.

Solution. Based on the side-by-side box plots (Fig. 12.1), it is clear that there are two sources of variability: the variability due to the different teaching methods and the variability arising from the differences between the individual students. The sums of squares and the proportion of variability explained by the model are given by

$$SST = 1196.00$$
$$SSTr = 722.67$$
$$SSE = 473.33$$
$$R^2 = \frac{722.67}{1196} = 0.60$$

The model explains 60 percent of the total variability of the data, although you must keep in mind that the results of a statistical analysis of this type of experiment cannot be reduced to a single number.   ■

Computing the Degrees of Freedom

Associated with each sum of squares appearing in Eq. (12.6) is its number of degrees of freedom (DF), a count of the independent summands appearing in its definition. As a general rule,

$$\begin{pmatrix} \text{DF of a} \\ \text{SS} \end{pmatrix} = \begin{pmatrix} \text{Number of summands} \\ \text{in SS} \end{pmatrix} - \begin{pmatrix} \text{Number of linear equations} \\ \text{satisfied by the summands} \end{pmatrix}$$

In particular,

1. The number of degrees of freedom associated with SST is $n - 1$ since $SST = \sum_i \sum_j (Y_{ij} - \overline{Y}_{..})^2$ contains $n = IJ$ summands and the single constraint $\sum_i \sum_j (Y_{ij} - \overline{Y}_{..}) = 0$.
2. The number of degrees of freedom associated with SSTr is $I - 1$ since $SSTr = J\sum_i (\overline{Y}_{i.} - \overline{Y}_{..})^2$ contains I summands satisfying the one constraint $\sum_i (\overline{Y}_{i.} - \overline{Y}_{..}) = 0$.
3. The number of degrees of freedom associated with SSE is $n - I$ since $SSE = \sum_i \sum_j (Y_{ij} - \overline{Y}_{i.})^2$ contains $n = IJ$ summands satisfying the I constraints $\sum_j (Y_{ij} - \overline{Y}_{i.}) = 0$.
4. Notice that the partition (12.6) of the sum of squares SST yields a corresponding partition of the degrees of freedom:

$$n - 1 = I - 1 + (n - I)$$

We will use this result shortly.

12.2.2   Testing Hypotheses about the Parameters

In addition to estimating the treatment means, the researcher will be interested in determining the differences among them. For instance, do the data on the five methods for teaching arithmetic indicate significant differences among them? Let us reformulate this as a hypothesis-testing problem. Recall that the ANOVA model assumes that the response Y_{ij} is of the form

$$Y_{ij} = \mu + \alpha_i + \epsilon_{ij}$$

where the random variables ϵ_{ij} are iid $N(0, \sigma^2)$; thus $E(Y_{ij}) = \mu + \alpha_i$ and $V(Y_{ij}) = \sigma^2$. This suggests that we formulate the null hypothesis as follows:

$$H_0: \alpha_1 = \cdots = \alpha_5 = 0$$

$$H_1: \alpha_i \neq 0 \text{ for at least one } i = 1, \ldots, 5$$

The single-factor ANOVA test of H_0 is based on the partition of the sum of squares SST given in Eq. (12.6) and some results concerning the sampling distribution of SSE, SSTr, and SST, as will now be explained.

Sampling Distribution of SSE, SSTr, and SST

The statistic on which we base our test of H_0 is defined by the *F ratio*:

$$F = \frac{\text{SSTr}/(I-1)}{\text{SSE}/(n-I)} = \frac{\text{MSTr}}{\text{MSE}}$$

where

1. MSE is the *mean square error,* computed as the error sum of squares divided by its degrees of freedom:

$$\text{MSE} = \frac{SSE}{n-I}$$

It can be shown that MSE is always an unbiased estimator of σ^2 whether H_0 is true or not. Thus

$$E(\text{MSE}) = E\left(\frac{SSE}{n-I}\right) = \sigma^2 \qquad (12.8)$$

2. MSTr is the *mean square treatment,* computed as the treatment sum of squares divided by its degrees of freedom:

$$\text{MSTr} = \frac{SSTr}{I-1}$$

It can also be shown that

$$E(\text{MSTr}) = \sigma^2 + \frac{J \sum \alpha_i^2}{I-1} \qquad (12.9)$$

Consequently, when H_0 is true, so that $\alpha_1 = \cdots = \alpha_I = 0$, then $E(\text{MSTr}) = \sigma^2 = E(\text{MSE})$; however, when H_0 is false, which means that some of the α_i are different from zero, then Eq. (12.9) implies that $E(\text{MSTr}) > E(\text{MSE})$. This suggests that we use the ratio $F = \text{MSTr}/\text{MSE}$ to test H_0; in particular, large values of F make H_0 less plausible. (Theorem 12.1, to follow, is a more precise version of this statement.)

3. Advanced statistical theory tells us that, when H_0 is true, the treatment sum of squares and error sum of squares are mutually independent χ^2 distributed random variables with $I-1$ and $n-I$ degrees of freedom, respectively;

that is,

$$\frac{\text{SSTr}}{\sigma^2} \sim \chi^2_{I-1}$$

$$\frac{\text{SSE}}{\sigma^2} \sim \chi^2_{n-I}$$

Thus, the distribution of the F ratio MSTr/MSE is $F_{I-1,n-I}$.

Our test of the equality of the treatment means is based on the following theorem.

■ **THEOREM 12.1**

The null distribution of the ratio MSTr/MSE has an F distribution with $I - 1$ degrees of freedom in the numerator and $n - I$ degrees of freedom in the denominator; that is,

$$F \text{ ratio:} \quad F = \frac{\text{MSTr}}{\text{MSE}} \sim F_{I-1,n-I}$$

where F_{ν_1,ν_2} denotes the F distribution with ν_1 and ν_2 degrees of freedom, respectively. The critical points of the F distribution are given in Table A.6 of the appendix.   ■

The F Test for the Equality of Treatment Means

We now show how to test

$$H_0: \mu_1 = \mu_2 = \cdots = \mu_I$$

against

$$H_1: \text{At least two of the means are different}$$

using the F ratio.

The Decision Rule. When H_0 is true, $F = \text{MSTr/MSE}$ is distributed as $F_{I-1,n-I}$, so large values of F support the alternative H_1. Consequently, our decision rule is

$$\text{accept } H_0 \text{ if } F \leq F_{I-1,n-I}(\alpha)$$
$$\text{reject } H_0 \text{ if } F > F_{I-1,n-I}(\alpha)$$

The P-value of the test equals

$$P\text{-value} = P(F_{I-1,n-I} > F)$$

and usually appears in the computer printout of the results.

---

example **12.5**   Use the F ratio to test the null hypothesis that there is no difference among the five teaching methods in Example 12.1.

Solution. We previously computed SSTr = 722.67 and SSE = 473.33. We compute the F ratio by noting that $I = 5, J = 9$, so the degrees of freedom are given by $I - 1 = 4$ and $n - I = 45 - 5 = 40$. Therefore,

$$\text{MSTr} = \frac{\text{SSTr}}{I - 1} = \frac{\text{SSTr}}{4} = 180.67$$

$$\text{MSE} = \frac{\text{SSE}}{n - I} = \frac{\text{SSE}}{40} = 11.83$$

$$F = \frac{\text{MSTr}}{\text{MSE}} = 15.27$$

Since $15.27 > F_{4,40}(0.01) = 3.83$, we see that the P-value is less than 0.01; this is pretty strong evidence against the null hypothesis. ∎

These calculations are conveniently summarized in the single-factor ANOVA table (also called the one-way ANOVA table) shown in Table 12.3.

TABLE 12.3 SINGLE-FACTOR ANOVA TABLE

Source	DF	SS	MS	F
Model	$I - 1$	SSTr	$\text{MSTr} = \dfrac{\text{SSTr}}{I - 1}$	$\dfrac{\text{MSTr}}{\text{MSE}}$
Error	$n - I$	SSE	$\text{MSE} = \dfrac{\text{SSE}}{n - I}$	
Total	$n - 1$	SST		

example **12.6**   Construct the the one-way ANOVA table for the teaching methods data (Table 12.1) of Example 12.1.

Solution. The ANOVA table is shown in Table 12.4.

TABLE 12.4 SINGLE-FACTOR ANOVA TABLE
FOR EXAMPLE 12.1

Source	DF	SS	MS	F
Model	4	722.67	180.67	15.27
Error	40	473.33	11.83	
Total	44	1196.00		

∎

12.2.3 Model Checking via Residual Plots

The assumptions of equal variances and normality should always be checked with various residual plots similar to those used in Chap. 10. Figure 12.2 [the graph of $(\hat{y}_{ij}, e_{ij})$] is a plot of the residuals against the predicted values. Looking at this graph, we do not see any evidence that the variances of the residuals depend on the predicted values.

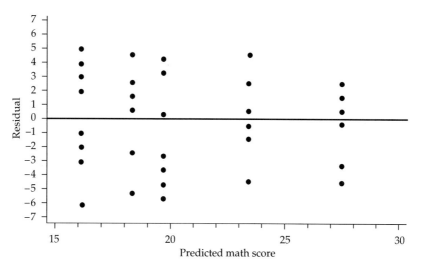

Figure 12.2 Plot of residuals against predicted math scores

The normal probability plot for the residuals is shown in Fig. 12.3. It suggests that the residuals are not normally distributed. Moreover, the P-value of the Shapiro-Wilk statistic equals 0.0107, which provides additional evidence against the normality hypothesis.

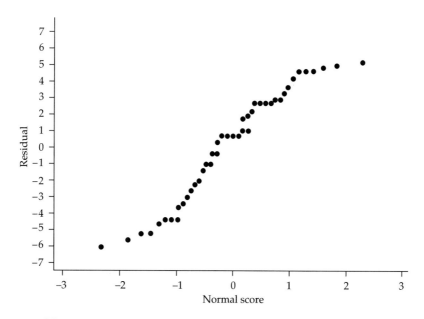

Figure 12.3 Normal probability plot of residuals for Example 12.1

Fortunately, many of the statistical tests for comparing means, such as the *t* test and ANOVA test, are relatively insensitive to moderate violations of the normality assumption (see G. E. P. Box, W. G. Hunter, and J. S. Hunter, *Statistics for Experimenters*, New York, John Wiley & Sons, 1978, p. 91).

12.2.4 Unequal Sample Sizes

It is not always possible to arrange an experiment so that there is the same number of observations for each treatment.

---

example **12.7**   Consider the data in Table 12.5, which records the silver content (percent Ag) of 27 coins discovered in Cyprus. The coins were grouped into four classes, corresponding to four different coinages during the reign of King Manuel I Comnenus (1143–1180). Archaeologists were interested in whether there were significant differences in the silver content of coins minted early and late in King Manuel's reign.

TABLE 12.5

Coinage			
1	2	3	4
5.9	6.9	4.9	5.3
6.8	9.0	5.5	5.6
6.4	6.6	4.6	5.5
7.0	8.1	4.5	5.1
6.6	9.3		6.2
7.7	9.2		5.8
7.2	8.0		5.8
6.9			
6.2			

Source: M. F. Hendy and J. A. Charles, "The Production Techniques, Silver Content and Circulation History of the Twelfth-Century Byzantine Trachy," *Archaeometry,* vol. 12, 1970, pp. 13–21. ∎

In the preceding example there are $I = 4$ treatments, corresponding to the four different coinages. Note that the sample sizes are *not* equal; this is an example of an *unbalanced design*. The ANOVA procedure in this case is a straightforward modification of the equal-sample-size case.

The F test for the Equality of Treatment Means (Unbalanced Design)

Let J_i $(i = 1, \ldots, I)$ denote the sample size corresponding to the *i*th treatment; the total number of observations is then $n = J_1 + \cdots + J_I$. Looking at Table 12.5, we see that $J_1 = 9, J_2 = 7, J_3 = 4, J_4 = 7$, so the total number of observations $n = 27$. The sum of squares partition

$$SST = SSTr + SSE$$

still holds, where

$$\text{SST} = \sum_{i=1}^{I} \sum_{j=1}^{J_i} (y_{ij} - \bar{y}_{..})^2$$

$$\text{SSTr} = \sum_{i=1}^{I} \sum_{j=1}^{J_i} (\bar{y}_{i.} - \bar{y}_{..})^2$$

$$= \sum_{i=1}^{I} J_i (\bar{y}_{i.} - \bar{y}_{..})^2$$

$$\text{SSE} = \sum_{i=1}^{I} \sum_{j=1}^{J_i} (y_{ij} - \bar{y}_{i.})^2$$

The mean square treatment, mean square error, and F ratio are defined by

$$\text{MSTr} = \frac{\text{SSTr}}{I-1}$$

$$\text{MSE} = \frac{\text{SSE}}{n-I}$$

$$F = \frac{\text{MSTr}}{\text{MSE}}$$

The quantities $I - 1$ and $n - I$ are the number of degrees of freedom associated with the sum of squares SSTr and SSE, respectively. (A more detailed explanation of this point will be given shortly.) There is also an analogue of Theorem 12.1:

▬ THEOREM 12.2

The null distribution of the F ratio (unbalanced design) is $F_{I-1,n-I}$; that is,

$$F = \frac{\text{MSTr}}{\text{MSE}} \sim F_{I-1,n-I}$$

where $n = \sum_i J_i$. ■

When H_0 is true, $F = \text{MSTr}/\text{MSE}$ is distributed as $F_{I-1,n-I}$. A large F value provides evidence against the null hypothesis; consequently, our decision rule is

$$\text{accept } H_0 \text{ if } F \leq F_{I-1,n-I}(\alpha)$$
$$\text{reject } H_0 \text{ if } F > F_{I-1,n-I}(\alpha)$$

The P-value of the test is

$$\text{P-value} = P(F_{I-1,n-I} > F)$$

---

example **12.8**   Refer back to the coinage data in Table 12.5. (1) Compute the grand mean, the treatment means, and the components SSTr and SSE of the sum of squares partition of SST. (2) Test the null hypothesis (no difference in the percentage of silver in the coins during the reign of King Manuel I) by computing the F ratio.

CHAPTER **12**   Single-Factor Experiments: Analysis of Variance

Solution

1. The quantities $\bar{y}_{..}, \bar{y}_{i.}$, SSTr, SSE, and SSE are

$$\bar{y}_{..} = 6.56, \qquad \bar{y}_{1.} = 6.74, \qquad \bar{y}_{2.} = 8.24, \qquad \bar{y}_{3.} = 4.88, \qquad \bar{y}_{4.} = 5.61$$
$$\text{SSTr} = 37.75 \qquad \text{SSE} = 11.02 \qquad \text{SST} = 48.77$$

2. We compute the F ratio after computing the mean square treatment and the mean square error:

$$\text{MSTr} = \frac{\text{SSTr}}{I - 1} = 12.58$$

$$\text{MSE} = \frac{\text{SSE}}{n - I} = 0.48$$

$$F = \frac{\text{MSTr}}{\text{MSE}} = 26.27$$

In this case $F \sim F_{3,23}$; since $F = 26.27 > F_{3,23}(0.01) = 4.76$, the P-value of the test is < 0.01; the null hypothesis is not accepted. ∎

The Single-Factor ANOVA Table (Unbalanced Design)

To construct the ANOVA table it is necessary to calculate the degrees of freedom associated with the sums of squares SSE and SSTr.

1. The number of degrees of freedom associated with SSTr is $I - 1$ since $\text{SSTr} = \sum J_i (\bar{Y}_{i.} - \bar{Y}_{..})^2$ contains I summands satisfying the one constraint $\sum J_i (\bar{Y}_{i.} - \bar{Y}_{..}) = 0$.
2. The number of degrees of freedom associated with SSE is $n - I$ since $\text{SSE} = \sum \sum_{j=1}^{J_i} (Y_{ij} - \bar{Y}_{i.})^2$ contains n summands satisfying the I constraints $\sum_{j=1}^{J_i} (Y_{ij} - \bar{Y}_{i.}) = 0$.
3. The number of degrees of freedom associated with SST is $n - 1$ since $\text{SST} = \sum \sum (Y_{ij} - \bar{Y}_{..})^2$ contains n summands and the single constraint $\sum_{i=1}^{I} \sum_{j=1}^{J_i} (Y_{ij} - \bar{Y}_{..}) = 0$.
4. Notice once again that the partition SST = SSTr + SSE yields a corresponding partition of the degrees of freedom:

$$n - 1 = (I - 1) + (n - I)$$

The single-factor ANOVA table for unequal sample sizes is displayed in Table 12.6.

TABLE 12.6   SINGLE-FACTOR ANOVA
TABLE (UNEQUAL SAMPLE SIZES)

Source	DF	SS	MS	F
Model	$I - 1$	SSTr	$\text{MSTr} = \dfrac{\text{SSTr}}{I - 1}$	$\dfrac{\text{MSTr}}{\text{MSE}}$
Error	$n - I$	SSE	$\text{MSE} = \dfrac{\text{SSE}}{n - I}$	
Total	$n - 1$	SST		

The single-factor ANOVA table for the data set in Table 12.5 is displayed in Table 12.7.

TABLE 12.7 SINGLE-FACTOR ANOVA TABLE
FOR THE DATA IN TABLE 12.5

Source	DF	SS	MS	F
Model	3	37.75	12.58	26.27
Error	23	11.02	0.48	
Total	26	48.76		

PROBLEMS

12.1 The side-by-side box plot displayed in Fig. 12.1 reveals an outlier in the group of students taught by method 3 (praised). (Outliers are discussed in Sec. 1.5.2.) Verify that 23 is the outlier by showing that it lies outside the interval

$$(Q_1 - 1.5 \times IQR, Q_3 + 1.5 \times IQR)$$

where IQR is the interquartile range and Q_1, Q_3 are the lower and upper quartiles of the arithmetic scores of the students taught by method 3.

12.2 The following table lists the tensile strengths of a rubber compound produced by three different methods, denoted A, B, and C.

Treatment	Tensile strength			
A	4038	4364	4410	4242
B	4347	3859	3432	3720
C	3652	3829	4071	4232

(a) Compute the three treatment means and the grand mean $(\bar{y}_1, \bar{y}_2, \bar{y}_3, \bar{y}_{..})$.
(b) Compute SSTr, SSE, and SST, and verify that SST = SSTr + SSE.
(c) Construct the ANOVA table and test the null hypothesis, at the 5 percent level, that there are no significant differences among the three methods of producing the rubber compound.

12.3 (a) Table 12.1 indicates that the first two students taught by method 3 obtained different scores (28 and 30) on the arithmetic test. What feature of the statistical model (12.1) accounts for this?
(b) As shown in Table 12.1, three students taught by method 1 scored a 24 on the arithmetic test and one student in group 3 scored a 23, even though the group 3 mean (27.44) is larger than the group 1 mean (19.67). What feature of the statistical model (12.1) accounts for this?

12.4 The effect of four different mixing techniques on the compressive strength of concrete is investigated and the following data are collected:

Mixing technique	Compressive strength			
1	2868	2913	2844	2887
2	3250	3262	3110	2996
3	2838	3049	3030	2832
4	2847	2721	2530	2598

Does the mixing technique affect the compressive strength? State the null hypothesis and the alternative hypothesis. Construct the ANOVA table. Test the null hypothesis at the level $\alpha = 0.05$. Summarize your conclusions. Save your computations.

12.5 The data in the following table are the results of a test in which the burst strengths of cardboard boxes made with four different types of wood pulp were studied; three measurements were taken for each wood pulp mixture.

Wood pulp	Burst strength (psi)		
1	237	219	209
2	205	202	201
3	144	220	167
4	256	246	245

(a) Compute the four treatment means and the grand mean $(\bar{y}_{1.}, \ldots, \bar{y}_{4.}, \bar{y}_{..})$.
(b) Compute SSTr, SSE, and SST.
(c) Construct the ANOVA table, and test at the level $\alpha = 0.05$ whether the four types of boxes have the same mean burst strength.

12.6 The following data give the burn times, in seconds, of four fabrics tested for flammability.

Fabric	Burn time (s)					
1	18	17	18	17	14	18
2	12	11	11	11	12	11
3	15	9	13	7	12	9
4	14	12	8	13	15	9

Construct the ANOVA table and test, at the level $\alpha = 0.05$, the null hypothesis that there is no difference in the flammability of the four fabrics. What conclusions can you draw from this experiment? Save your computations.

12.7 The following data give four measurements of the ionization potential, measured in electron volts (eV), of a gaseous ion for three different methods: photoionization (PI), extrapolated voltage difference (EVD), and semilog plot (SL).

Method	Ionization potential (eV)			
PI	10.68	9.22	8.47	8.22
EVD	10.53	9.92	11.10	10.22
SL	15.03	14.72	15.39	15.04

Test the null hypothesis that there is no significant difference among the three measurement methods by constructing the ANOVA table and performing the F test. Use $\alpha = 0.05$.

12.8 The following data give the results of an experiment to study the effects of three different word-processing programs on the length of time (measured in minutes) to type a text file. A group of 24 secretaries were randomly divided into three groups of 8 secretaries each, and each group was randomly assigned to one of the three word-processing programs.

Program	Typing time (min)							
1	15	13	11	19	14	19	16	19
2	13	12	9	18	13	17	15	20
3	18	17	13	18	15	22	16	22

Test the hypothesis that the word-processing program has no effect on the length of time it takes to type a text file; use $\alpha = 0.05$. Summarize your results in an ANOVA table.

12.9 In a weight loss experiment men were given one of three different diets: 10 men followed diet A, 7 men followed diet B, and 8 men followed diet C. Their weight losses after one month are displayed in the following table. Notice that the observations at each factor level now appear as columns instead of rows. The question to be decided is whether or not the mean weight loss is the same for each of these diets.

Weight loss (lb)		
Diet A	Diet B	Diet C
15.3	3.4	11.9
2.1	10.9	13.1
8.8	2.8	11.6
5.1	7.8	6.8
8.3	0.9	6.8
10.0	5.2	8.8
8.3	2.5	12.5
9.4		17.5
12.5		
11.1		

(a) Identify the factor, the number of levels, and the response variable.

(b) State the null hypothesis.

(c) Construct the ANOVA table and test at the level $\alpha = 0.05$ whether the three diets produce the same weight loss.

Save your computations.

12.10 The Analytical Methods Committee of the Royal Society of Chemistry reported the following results on the levels of tin recovered after samples (taken from the same product) were boiled with hydrochloric acid for 30, 45, 60, and 75 minutes, respectively.

Refluxing time (min)	Tin found (mg/kg)					
30	55	57	59	56	56	59
45	57	59	58	57	57	56
60	58	57	56	57	58	59
75	57	55	58	59	59	59

Source: "The Determination of Tin in Organic Matter," *Analyst,* vol. 108, 1983, pp. 109–115, Table VIII. Used with permission.

Test the hypothesis that the amount of tin recovered is the same for all four boiling times. Use $\alpha = 0.05$. Summarize your results in an ANOVA table.

12.11 (a) Show that $E(\bar{y}_{i.}) = \mu + \alpha_i$.

(b) Show that $E(\bar{y}_{i.} - \bar{y}_{..}) = \alpha_i$; i.e., $\bar{y}_{i.} - \bar{y}_{..}$ is an unbiased estimator of the ith treatment mean.

12.3 CONFIDENCE INTERVALS FOR THE TREATMENT MEANS; CONTRASTS

In Sec. 12.2.1 we derived point estimates for the treatment means $\mu, \mu_1, \ldots, \mu_I$. We now turn our attention to deriving confidence intervals for these parameters and their differences $\mu_i - \mu_k$. For instance, when the null hypothesis of no difference among the treatment means is rejected, the experimenter would still like to know which means, or linear combinations of means, are significantly different from one another. In particular, the experimenter might be interested in constructing a confidence interval for the parameter $\theta = \mu_i - \mu_k$ with the goal of testing the hypothesis

$$H_0:\ \mu_i - \mu_k = 0 \quad \text{against} \quad H_1:\ \mu_i - \mu_k \neq 0$$

Of course, this is equivalent to testing

$$H_0:\ \mu_i = \mu_k \quad \text{against} \quad H_1:\ \mu_i \neq \mu_k$$

The experimenter will conclude that the treatment means are significantly different if and only if the confidence interval does not contain 0. Before proceeding further, however, we assume that there is the same number of replicates, denoted J, for each treatment; this will simplify the calculations. The modifications to be made when the design is unbalanced will be mentioned briefly at the end of this section.

The pairwise comparison $\mu_i - \mu_k$ of two treatment means is an example of a *contrast*, which refers to any linear combination of two or more treatment means of the form

$$\theta = \sum_i c_i \mu_i, \quad \text{where} \quad \sum_i c_i = 0 \quad \text{(balanced design)} \qquad (12.10)$$

---

example **12.9**   Refer again to the teaching methods study (Example 12.1). In this experiment groups 1 and 2 (the control groups) were taught using the current method; the other three groups were taught with three different methods. This suggests that we study the differences between the test scores for the current teaching method and those for the other three. Express this comparison as a contrast.

Solution. The null hypothesis in this case is that there is no difference between the current method of teaching arithmetic and the alternative methods. Consequently, we test the null hypothesis

$$H_0: \frac{\mu_1 + \mu_2}{2} = \frac{\mu_1 + \mu_2 + \mu_3}{3}$$

The contrast in this case is

$$\frac{1}{2}\mu_1 + \frac{1}{2}\mu_2 - \frac{1}{3}\mu_3 - \frac{1}{3}\mu_4 - \frac{1}{3}\mu_5 = 0$$

Thus,

$$c_1 = c_2 = \frac{1}{2} \quad \text{and} \quad c_3 = c_4 = c_5 = -\frac{1}{3} \qquad \blacksquare$$

Confidence Intervals for Treatment Means and Their Differences

To test a null hypothesis of the form

$$H_0: \sum_i c_i \mu_i = 0, \quad \text{where} \quad \sum_i c_i = 0 \qquad (12.11)$$

we construct a confidence interval for the parameter $\theta = \sum_i c_i \mu_i$ and reject H_0 if the confidence interval does not contain 0.

To illustrate the method we consider the contrast $\theta = \mu_i - \mu_k$. We use the statistics $\overline{Y}_i$ and $\overline{Y}_{i.} - \overline{Y}_{k.}$ to construct $100(1 - \alpha)$ percent confidence intervals for the parameters μ_i and their differences $\mu_i - \mu_k$. Using Eq. (12.8) we obtain the following estimate for σ:

$$\hat{\sigma} = \sqrt{\text{MSE}} = \sqrt{\frac{\text{SSE}}{n - I}} \qquad (12.12)$$

We note that

$$V(\overline{Y}_{i.}) = \sigma^2/J \quad \text{and} \quad V(\overline{Y}_{i.} - \overline{Y}_{k.}) = 2\sigma^2/J$$

Therefore, the corresponding estimated standard errors are given by

$$s(\overline{Y}_{i.}) = \sqrt{\frac{MSE}{J}}$$

$$s(\overline{Y}_{i.} - \overline{Y}_{k.}) = \sqrt{\frac{2MSE}{J}}$$

Combining these results in the usual way yields the following formulas for the $100(1 - \alpha)$ percent confidence intervals for μ_i and $\mu_i - \mu_k$:

$100(1 - \alpha)$ percent confidence limits for μ_i:

$$\overline{Y}_{i.} \pm t_{n-I}(\alpha/2)\sqrt{\frac{MSE}{J}} \tag{12.13}$$

$100(1 - \alpha)$ percent confidence limits for $\mu_i - \mu_k$:

$$\overline{Y}_{i.} - \overline{Y}_{k.} \pm t_{n-I}(\alpha/2)\sqrt{\frac{2MSE}{J}} \tag{12.14}$$

If the design is unbalanced, the corresponding formulas are given by

$$\overline{Y}_{i.} \pm t_{n-I}(\alpha/2)\sqrt{\frac{MSE}{J_i}}$$

$$\overline{Y}_{i.} - \overline{Y}_{k.} \pm t_{n-I}(\alpha/2)\sqrt{MSE\left(\frac{1}{J_i} + \frac{1}{J_k}\right)}$$

The *least significant difference*, abbreviated LSD, is defined by the following equation:

$$LSD = t_{n-I}(\alpha/2)\sqrt{2MSE/J} \tag{12.15}$$

example **12.10**   Refer back to the study of the five methods of teaching arithmetic (Example 12.1). Using the F test, we concluded that these teaching methods are not equally effective. More precisely, we did not accept, at the 5 percent level, the null hypothesis that the mean score on the arithmetic test is the same for each of the five teaching methods. This leads to the next step in the analysis: Determine which of these teaching methods differ significantly from one another and by how much. In particular, determine a 95 percent confidence interval for the difference between teaching methods 3 and 4.

Solution. We compute the confidence interval using Eq. (12.14). Refer to Table 12.4 for the arithmetic grades for groups of children taught by the five methods. In this case $n - I = 40, J = 9$, and $MSE = 11.83$; consequently, choosing $\alpha = 0.05$ yields the values

$$t_{n-I}(\alpha/2)\sqrt{2MSE/J} = 2.021 \times 1.62 = 3.28$$

Using this criterion, we see that the difference between the mean responses for methods 3 and 4 is significant at the 5 percent level because

$$\bar{y}_3 - \bar{y}_4 = 27.44 - 23.44 = 4 > 3.28$$ ∎

Testing the Significance of a General Contrast

In the preceding discussion we considered contrasts of the form $\theta = \mu_i - \mu_k$. We now consider testing the significance of the contrast $\theta = \sum_i^I c_i \mu_i$, where $\sum_i^I c_i = 0$. To construct a confidence interval for the contrast we use the statistic

$$\theta = \sum_i^I c_i \bar{Y}_{i.} \tag{12.16}$$

which is an unbiased estimate of θ. Any linear combination of the sample treatment means of the form (12.16) is also called a *contrast*. The null and alternative hypotheses corresponding to a contrast are

$$H_0: \sum_i^I c_i \mu_i = 0 \quad \text{against} \quad H_1: \sum_i^I c_i \mu_i \neq 0$$

where

$$\sum_i^I c_i = 0$$

To test this hypothesis we use the result that the random variable $\hat{\theta}$ defined by

$$\hat{\theta} = \sum_i^I c_i \bar{Y}_i$$

is normally distributed with mean and variance given by

$$E(\hat{\theta}) = \theta \tag{12.17}$$

$$\sigma^2(\hat{\theta}) = \sigma^2 \sum_i^I \frac{c_i^2}{J} \tag{12.18}$$

An unbiased estimator of σ^2 is MSE; consequently,

$$s^2(\hat{\theta}) = \text{MSE} \sum_i \frac{c_i^2}{J}$$

is an unbiased estimator of $\sigma^2(\hat{\theta})$. It follows that

$$\frac{\hat{\theta} - \theta}{s(\hat{\theta})} \quad \text{is } t_{n-I} \text{ distributed}$$

and the $100(1 - \alpha)$ percent confidence limits for θ are

$$\hat{\theta} \pm t_{n-I}(\alpha/2)s(\hat{\theta}) \tag{12.19}$$

---

example 12.11   Refer back to the teaching methods study (Example 12.1). Construct a 95 percent confidence interval for the contrast

$$\theta = \frac{1}{2}\mu_1 + \frac{1}{2}\mu_2 - \frac{1}{3}\mu_3 - \frac{1}{3}\mu_4 - \frac{1}{3}\mu_5$$

This corresponds to comparing the averages of groups 1 and 2, who were taught by the current method, to the averages of groups 3, 4, and 5, who were taught by different methods.

Solution. A straightforward calculation using Eq. (12.19) yields

$$\hat{\theta} = -3.33, \qquad s(\hat{\theta}) = 1.0467, \qquad t_{40}(0.025) = 2.021$$

The 95 percent confidence interval is therefore

$$-3.33 \pm 2.021 \times 1.0467 = -3.33 \pm 2.12$$

Since the interval does not include 0, we conclude that there is a significant difference between teaching methods 1 and 2 and teaching methods 3, 4, and 5.
■

12.3.1   Multiple Comparisons of Treatment Means

In the previous section we constructed $100(1 - \alpha)$ percent confidence intervals of the form

$$\overline{Y}_{i.} - \overline{Y}_{k.} - t_{n-I}(\alpha/2)\sqrt{\frac{2\text{MSE}}{J}} \le \mu_i - \mu_k \le \overline{Y}_{i.} - \overline{Y}_{k.} + t_{n-I}(\alpha/2)\sqrt{\frac{2\text{MSE}}{J}}$$

for each of the $C_{I,2}$ paired differences $\mu_i - \mu_k$ that can be formed from the I treatment means. We then used these confidence intervals to test whether the differences between ith and kth treatment means are statistically significant. For example, the difference between ith and kth treatment means are considered significant at level α if

$$|\overline{y}_{i.} - \overline{y}_{k.}| \ge t_{n-I}(\alpha/2)\sqrt{2\text{MSE}/J} = \text{LSD} \qquad (12.20)$$

There are, however, serious objections to using the least significant difference as a criterion for testing whether or not the difference between two treatment means is statistically significant. To see why, consider the case of five treatments with $C_{5,2} = 10$ paired comparisons. Suppose, in fact, that the null hypothesis is true, so that all treatment means are equal. Then, as Snedecor and Cochran point out (see reference at the end of this chapter), the probability of at least one of the paired differences exceeding the LSD bound 12.20 (with $\alpha = 0.05$) is approximately 0.30. Thus, when this experiment is repeated many times, the largest observed paired difference will exceed the LSD bound nearly 30 percent of the time! Consequently, the significance level is really 0.30 and not, as the researcher might think, 0.05. In general, the probability under the null hypothesis of finding a pair of treatment means whose difference is significant increases with I; in other words, using this method to detect differences in the means increases the type I error probabilities.

Tukey's Simultaneous $100(1 - \alpha)$ *Percent Confidence Intervals for the Difference between Two Means (Balanced Design)*

We now present a method for comparing the treatment means that avoids the aforementioned difficulties; it is called *Tukey's multiple comparison procedure*. We give the formula for the $100(1 - \alpha)$ percent confidence interval first, followed by the level α test for the difference between two means.

$$\overline{Y}_{i.} - \overline{Y}_{k.} - q(\alpha; I, n - I)\sqrt{\frac{\text{MSE}}{J}} \leq \mu_i - \mu_k \leq \overline{Y}_{i.} - \overline{Y}_{k.} + q(\alpha; I, n - I)\sqrt{\frac{\text{MSE}}{J}}$$

(12.21)

Here $q(\alpha; I, \nu)$ denotes the critical value of the *studentized range* distribution. Table A.7 in the appendix lists some of these critical values for $\alpha = 0.05$ and selected values of I and ν.

The corresponding test for rejecting the null hypothesis that $\mu_i = \mu_k, i \neq k$, is given by

$$|\overline{y}_{i.} - \overline{y}_{k.}| \geq q(\alpha; I, n - I)\sqrt{\text{MSE}/J} = \text{HSD} \qquad (12.22)$$

The quantity HSD appearing on the right-hand side of Eq. (12.22) is called the *honest significant difference*.

---

example **12.12**   Refer back to the teaching methods study in Example 12.1. Construct a 95 percent confidence interval for the difference between teaching methods 3 and 4; use the HSD criterion.

Solution. Refer again to Table 12.1 on the arithmetic grades for groups of children taught by the five methods. In this case $n - I = 40, J = 9, \text{MSE} = 11.83$; consequently, choosing $\alpha = 0.05$ yields the values

$$\text{HSD} = q(0.05; 5, 40)\sqrt{\text{MSE}/J} = 4.04 \times \sqrt{1.3148} = 4.63$$

Notice that $\text{HSD} = 4.63 > 3.23 = \text{LSD}$; in other words, the width of the simultaneous confidence interval is larger than the one obtained via the t test. Using the HSD criterion, we see that the difference in mean response between methods 3 and 4 is not significant at the 5 percent level because

$$\overline{y}_3 - \overline{y}_4 = 27.44 - 23.44 = 4 < 4.63 \qquad \blacksquare$$

Observe that the difference between two treatment means is significant only if the corresponding confidence interval does not contain zero. A useful way to visualize the relationships between the various means is to arrange the means in increasing order and underline those means that differ by less than HSD. The following table illustrates how this is done for the comparison of the teaching methods.

$\overline{y}_{5.}$	$\overline{y}_{2.}$	$\overline{y}_{1.}$	$\overline{y}_{4.}$	$\overline{y}_{3.}$
16.11	18.33	19.67	23.44	27.44

Notice that, according to Tukey's HSD criterion, there are fewer significant differences. For instance, the difference $\mu_3 - \mu_4$ is not significant according to Tukey's criterion, but is significant based on the LSD criterion.

Interpretation of Tukey's Multiple Comparisons

1. The confidence intervals produced by Tukey's method [Eq. (12.21)] include *all* differences $\mu_i - \mu_k$ with probability at least $1 - \alpha$.
2. Similarly, if we test for differences between some or all pairs of treatment means using the criterion in Eq. (12.22), the probability of a type I error is less than α. Since the error rate α refers to all pairs of treatment means, it is also called the *experimentwise error rate*.

Tukey's Method for Unbalanced Designs

When the sample sizes for the treatments are not equal, but are not too much different from one another, then Tukey's method, with the following modifications, can still be used. The $100(1 - \alpha)$ percent confidence intervals produced by this method, however, are only approximately valid.

$$\overline{Y}_{i.} - \overline{Y}_{k.} \pm q(\alpha; I, n - I) \sqrt{\frac{\text{MSE}}{2}\left(\frac{1}{J_i} + \frac{1}{J_k}\right)} \tag{12.23}$$

PROBLEMS

12.12 The Analytical Methods Committee of the Royal Society of Chemistry reported the following results on the levels of tin recovered from samples of blended fish-meat product after boiling with hydrochloric acid for 30, 45, 60, and 70 minutes, respectively.

Refluxing time (min)	Tin found (mg/kg)							
30	57	57	55	56	56	55	56	55
45	59	56	55	55	57	51	50	50
60	58	56	52	56	57	56	54	50
70	51	60	48	32	46	58	56	51

Source: "The Determination of Tin in Organic Matter," *Analyst,* vol. 108, 1983, pp. 109–115, table VII. Used with permission.

During the experiment the authors noted "that maintaining boiling for 70 minutes without sputtering and enforced cooling was difficult. The effect is shown in a lower mean recovery and much greater variability of results." Write a brief report justifying these conclusions.

12.13 (Continuation of Prob. 12.4)

(a) Construct 95 percent confidence intervals for the treatment means.

(b) Use an appropriate t statistic to construct 95 percent confidence intervals for all differences $\mu_i - \mu_k$.

(c) Use Tukey's method to identify those differences in the treatment means that are significant. Use $\alpha = 0.05$.

12.14 (Continuation of Prob. 12.6)

(a) Construct 95 percent confidence intervals for the treatment means.

(b) Use Tukey's method to identify those differences in the treatment means that are significant. Use $\alpha = 0.05$.

12.15 (Continuation of Prob. 12.8)

(a) Construct 95 percent confidence intervals for the treatment means.

(b) Use Tukey's method to identify those differences in the treatment means that are significant. Use $\alpha = 0.05$.

12.16 A high-voltage electric cable consists of 12 wires. The following data are the tensile strengths of the 12 wires in each of nine cables.

(a) Use a computer to draw side-by-side box plots for the nine cables. Do the box plots indicate that the strength of the wires varies among the cables?

(b) Cables 1 to 4 were made from one lot of raw material, and cables 5 to 9 came from another lot. Do the box plots indicate that the tensile strength depends on the lot?

(c) Perform a one-way ANOVA to test the null hypothesis that the tensile strength does not vary among the cables.

(d) Use Tukey's method to determine those differences in the treatment means that are significant. Are these results consistent with the information that cables 1 to 4 came from one lot of raw material and cables 5 to 9 came from another lot?

(e) Express the comparison between the two lots as a contrast, and test whether the tensile strength depends on the lot.

Cable								
1	2	3	4	5	6	7	8	9
345	329	340	328	347	341	339	339	342
327	327	330	344	341	340	340	340	346
335	332	325	342	345	335	342	347	347
338	348	328	350	340	336	341	345	348
330	337	338	335	350	339	336	350	355
334	328	332	332	346	340	342	348	351
335	328	335	328	345	342	347	341	333
340	330	340	340	342	345	345	342	347
337	345	336	335	340	341	341	337	350
342	334	339	337	339	338	340	346	347
333	328	335	337	330	346	336	340	348
335	330	329	340	338	347	342	345	341

Source: A. Hald, *Statistical Theory with Engineering Applications*, New York, John Wiley & Sons, 1952. Used with permission.

12.17 Verify that, under the assumptions of the single-factor ANOVA model, the random variable

$$\theta = \sum_i c_i \overline{Y}_i$$

is normally distributed with mean and variance given by Eqs. (12.17) and (12.18).

12.18 (Continuation of Prob. 12.8)

(a) Construct 95 percent confidence intervals for each of the treatment means.

(b) Use Tukey's method to identify those differences in the treatment means that are significant. Use $\alpha = 0.05$.

12.19 (Continuation of Prob. 12.9)

(a) Construct 95 percent confidence intervals for each of the treatment means.

(b) Use Tukey's method to determine those differences in the treatment means that are significant. Use $\alpha = 0.05$.

12.4   RANDOM EFFECTS MODEL

In the fixed effects model the researcher chooses the factor levels in order to compare the means of the response variable as a function of these specific factor levels. In many situations, however, the factor levels are not chosen in this way, but are instead randomly selected from a much larger population. The next example illustrates how such problems arise.

---

example 12.13   Total dissolved solids (TDS), measured in milligrams per liter, is a groundwater quality parameter that is routinely measured in the laboratory. There is a justifiable concern that the measured TDS value depends on the laboratory performing the test. Excessive variability in the results reported by different laboratories would support the need for regulating the conditions under which these tests are performed. In addition to determining whether or not the labs differ in their measurements of TDS levels, researchers are interested in studying how much of the variability in these measurements is due to the variability among labs.

To study the variability in the measurements, the following experiment is performed: $I = 4$ laboratories are selected at random from the large population of federal, state, and commercial water quality laboratories. One sample of well water is divided into four parts—one for each lab—and each lab makes $J = 5$ measurements of the TDS level in the sample. The data, which are fictitious, are displayed in Table 12.8. Does the measured value of total dissolved solids in the water depend on the laboratory performing the test?

TABLE 12.8 MEASURED VALUES OF TOTAL DISSOLVED SOLIDS (TDS) RECORDED BY FOUR LABORATORIES

Laboratory	TDS level (mg/L)				
1	112	190	145	65	21
2	246	134	121	136	79
3	131	122	146	193	129
4	137	231	277	211	157

Solution. In the random effects model we assume that the jth measurement of the TDS level made by the ith laboratory is a random variable Y_{ij}, written as

$$Y_{ij} = \mu + A_i + \epsilon_{ij}$$

where the random variables A_i $(i = 1, \dots, I)$ and ϵ_{ij} are mutually independent, normally distributed random variables with zero means and variances σ_A^2, σ^2, respectively; in symbols we write

$$A_i \sim N(0, \sigma_A^2) \quad \text{and} \quad \epsilon_{ij} \sim N(0, \sigma^2)$$

Thus, $\qquad\qquad V(Y_{ij}) = \sigma_A^2 + \sigma^2$

The quantities σ_A^2 and σ^2 are called the *components of variance*.

Note that $\sigma_A^2 = 0$ implies that $A_i = 0$ $(i = 1, \dots, I)$; thus, the null hypothesis (of no treatment effects due to the different factor levels) and the alternative are written as

$$H_0: \sigma_A^2 = 0 \quad \text{against} \quad H_1: \sigma_A^2 > 0$$

Although this model is different from the fixed effects model previously discussed, the construction of the ANOVA table is carried out in exactly the same way, so we content ourselves with a sketch. The quantities SST, SSTr, MSTr, SSE, MSE, and $F = \text{MSTr}/\text{MSE}$ are defined exactly as before, and the sum of squares decomposition

$$\text{SST} = \text{SSTr} + \text{SSE}$$

remains valid. It can be shown (we omit the tedious details) that

$$E(\text{MSE}) = E\left(\frac{\text{SSE}}{IJ - I}\right) = \sigma^2 \tag{12.24}$$

$$E(\text{MSTr}) = \sigma^2 + J\sigma_A^2 \tag{12.25}$$

$$F = \frac{\text{MSTr}}{\text{MSE}} \sim F_{I-1, IJ-I} \tag{12.26}$$

From Eqs. (12.24) and (12.25) we see that if H_0 is true, then $E(\text{MSTr}) = E(\text{MSE})$; otherwise, if H_0 is false, $E(\text{MSTr}) > E(\text{MSE})$. Thus, we do not accept H_0 at the level α if $F > F_{I-1, IJ-I}(\alpha)$.

An unbiased estimate of σ_A^2 is given by

$$\hat{\sigma}_A^2 = \frac{\text{MSTr} - \text{MSE}}{J} \tag{12.27}$$

It is not too difficult to show (the details are left as a problem) that the variance and estimated variance of the grand mean are given by

$$V(\overline{Y}_{..}) = \frac{J\sigma_A^2 + \sigma^2}{n} \tag{12.28}$$

$$\hat{V}(\overline{Y}_{..}) = \frac{MSTr}{n} \tag{12.29}$$

Equation (12.29) asserts that the estimated variance of the grand mean $\overline{Y}_{..}$ is the *mean square treatment divided by the total number of observations.*

The ANOVA table derived from the TDS data in Table 12.8 is displayed in Table 12.9, from which the following conclusions can be drawn:

1. $F = 2.58 < 3.24 = F_{3,16}(0.05)$; consequently, the null hypothesis is not rejected.
2. The estimates of the variance components are

$$\hat{V}(Y_{..}) = \frac{MSTr}{n} = \frac{7878.85}{20} = 393.94$$

$$\hat{\sigma}_A^2 = \frac{MSTr - MSE}{J} = \frac{7878.85 - 3057.125}{5} = 964.35$$

$$\hat{\sigma}^2 = 3057.125$$

TABLE 12.9   ANALYSIS OF TDS DATA (TABLE 12.8)

Analysis of Variance Procedure

Dependent Variable: TDS

Source	DF	Sum of Squares	Mean Square	F Value	Pr > F
Model	3	23636.550	7878.850	2.58	0.0899
Error	16	48914.000	3057.125		
Corrected Total	19	72550.550			

REMARK 12.1

It sometimes happens that the ANOVA estimate $\hat{\sigma}_A^2$ [Eq. (12.27)] is negative! When this occurs, there is no hard and fast rule about what to do next. One possibility, but not the only one, is to conclude that the true value of σ_A^2 is 0. ■

PROBLEMS

12.20 The following data give the tensile strengths (psi) of cotton yarn produced by four different looms selected at random from a large group of such looms at a textile mill.

Loom	Tensile strength (psi)					
1	18	18	20	16	14	17
2	16	14	15	20	16	14
3	11	8	8	5	13	8
4	24	23	20	19	16	19

(a) Is this a fixed effects model or a random effects model? Justify your answer.

(b) Construct the ANOVA table to test at the level $\alpha = 0.05$ whether the tensile strength is the same for each loom.

(c) Estimate the components of the variance.

12.21 The following data give the amounts, in parts per million (ppm), of a toxic substance as measured by four randomly selected laboratories. Each laboratory made six measurements of the toxic substance. The question to be studied was how much of the variability is due to the different laboratory techniques and how much is due to random error.

Laboratory	Toxic level (ppm)					
1	53.2	54.5	52.8	49.3	50.4	53.8
2	51	40.5	50.8	51.5	52.4	49.9
3	47.4	46.2	46	45.3	48.2	47.1
4	51	51.5	48.8	49.2	48.3	49.8

Source: J. S. Milton and J. C. Arnold, *Introduction to Probability and Statistics,* 2d ed., New York, McGraw-Hill, 1990. Used with permission.

(a) Explain why the random effects model is appropriate.

(b) Construct the appropriate ANOVA table, and test at the level $\alpha = 0.05$ the null hypothesis that there is no significant difference in the toxic levels as measured by the different laboratories.

(c) Estimate the variance components.

12.22 A manufacturing engineer wants to study the amount of time required to assemble a piece of equipment; he is particularly interested in how the assembly time varies among workers. To study this question he selects four workers at random and records the assembly times for each of the workers.

Assembly time (min)			
Worker 1	Worker 2	Worker 3	Worker 4
20	16	18	21
22	23	20	17
21	20	23	18
22	18	21	17
22	19	24	18
24	19	19	23

(a) Construct the ANOVA table and test the hypothesis that there is no significant variability in assembly time among the workers; use $\alpha = 0.05$.

(b) Estimate the components of variance.

12.23 Derive Eq. (12.28).

12.24 (a) Show that $E(\overline{Y}_{i.}) = \mu + \alpha_i$.

(b) Show that $E(\overline{Y}_{i.} - \overline{Y}_{..}) = \alpha_i$, i.e., $\overline{Y}_{i.} - \overline{Y}_{..}$ is an unbiased estimator of the ith treatment mean.

12.5   CHAPTER SUMMARY

A single-factor experiment is an experimental design to study $\text{Model}(Y = \text{response variable} \mid X = \text{factor level})$, where X denotes a (nonnumeric) variable called a factor. The different values of the factor are called factor levels or treatments. Some examples of factor levels are teaching methods, types of fabric, and diet methods. The goal of ANOVA is to study the mean of the response variable as a function of the factor levels. Side-by-side box plots help us understand how the mean of the response variable depends on the factor level. ANOVA is a statistical method for estimating and comparing the means of two or more populations and, in particular, for testing the null hypothesis that the means are the same. We test the model assumptions by graphing residual plots. When the null hypothesis of no difference among the means is rejected, the next step in the analysis is to determine which means differ significantly from one another. One method to accomplish this is Tukey's method of multiple comparisons.

To Probe Further. The following references contain a wealth of useful information on the practical applications of many statistical techniques, including ANOVA.

1. G. W. Snedecor and W. G. Cochran, *Statistical Methods,* 7th ed.. Ames, Iowa State University Press, 1980.

2. G. E. P. Box, W. G. Hunter, and J. S. Hunter, *Statistics for Experimenters*, New York, John Wiley & Sons, 1978.

13

Design and Analysis of Multifactor Experiments

Probably the most important aim in experimental design is to obtain a given
degree of accuracy at the least cost.

W. G. Cochran, English statistician

13.1   ORIENTATION

We now turn our attention to the study of *two-way factorial* experiments, that is,
experiments where the mean response is a function of two factors: factor A with
a levels, denoted $i = 1, \ldots, a$, and factor B with *b* levels, denoted $j = 1, \ldots, b$.
We define the (i, j)th *cell* to be the set of observations taken at level *i* of factor
A and level *j* of factor B. We assume that *n* experiments (called *replicates*)
have been performed at each possible combination of the factor levels. For
instance, a computer engineer might be interested in the execution times (the
response variable) of five different workloads (factor A, $a = 5$) on three different
configurations of a processor (factor B, $b = 3$). The goal here is to study how
the processor performance is affected by the different processor configurations.

In a two-way factorial experiment the factor A and factor B effects are of
equal interest to the experimenter. There is, however, an important subclass of
experiments in which factor A is of primary interest and factor B represents an
extraneous variable whose effects are of little or no interest to the experimenter.
This is the *randomized complete block design* (RCBD), one of the most frequently
used experimental designs and also one of the simplest to analyze. We then

469

consider the two-factor experiment with $n > 1$ observations at each combination of factor levels; the case where the number of observations is not the same in all cells is more complicated and will not be dealt with here. Finally, we consider 2^k factorial experiments. These are experiments with k factors, each of which has two levels, denoted high and low. For instance, in an agricultural experiment the researcher might be interested in how the crop yield is affected by the amounts of herbicide, pesticide, and fertilizer applied to a field. Since there are many possible factor combinations, the researcher might begin by studying the crop yields with each factor at high and low levels. If the crop yields do not change much when the factor levels are varied from low to high, it is unlikely that these factors are important, and they can be dropped from the research program.

Organization of Chapter

13.2   RANDOMIZED COMPLETE BLOCK DESIGNS

The precision of an experiment comparing two or more population means can be adversely affected by an extraneous variable in which the scientist has no interest. As an example, consider the results of a study of lead absorption in children of employees who worked in a factory where lead is used to make batteries (described in Prob. 1.12). In this study the authors matched 33 such children (the exposed group) from different families to 33 children (the control group) of the same age and neighborhood whose parents were employed in industries not using lead. The advantages of pairing the children in this way are clear: It eliminates the variability in the data due to the influence of the extraneous variables of age and neighborhood; therefore, the differences in lead concentration between the two groups will not be masked by these variables. For instance, a child living close to a major highway may have a high lead concentration in her blood even though her father does not work in a battery factory. The partition of the experimental units into homogeneous groups so that the experimental units within the groups have certain common characteristics (such as age, sex, and weight) is an example of a method of experimental design called *blocking;* the groups are called *blocks* and the criteria used to define the blocks are called *blocking variables.* The purpose of blocking, then, is to reduce the experimental error by eliminating the variability due to an extraneous variable.

example 13.1 As an illustration of the advantages of blocking, consider the data set in Table 13.1, which lists the mortality rates (measured in percent) for four different genetic strains (factor B) of fruit flies subjected to three different dosages of an insecticide (factor A). The experimental unit is the group of flies to which the insecticide is applied; the response variable is the mortality rate. The blocking variable is the genetic strain, and there are three factor levels corresponding to the three dosage levels. Observe that while the experimenter randomly assigns the fly to one of three dosages of the insecticide, she cannot similarly assign the genetic strain. Are the mortality rates for the fruit flies influenced by the dosage levels of the insecticide?

TABLE 13.1 MORTALITY RATES (IN PERCENT) OF FOUR GENETIC STRAINS OF FRUIT FLIES RESULTING FROM THREE DOSAGE LEVELS OF INSECTICIDE

	Block (B)				
Treatment (A)	1	2	3	4	Treatment mean
1	66	55	43	32	49.00
2	71	57	44	37	52.25
3	79	63	51	44	59.25
Block mean	72.00	58.33	46.00	37.67	

Source: John L. Gill, *Design and Analysis of Experiments in the Animal and Medical Sciences,* vol. 2, Ames, Iowa State University Press, 1978, p. 25. Used with permission.

Solution. We can model this experiment in two different ways. For instance, we can ignore the genetic strain and consider the dosage level as the explanatory variable and the mortality rate as the response variable. The model format statement in this case is Model(Y = mortality rate | A = dosage level). This is, of course, a one-way ANOVA. (Problem 13.1 asks you to analyze this model.) There is also the possibility that the mortality rate is affected by the genetic strain. The model statement in this case is Model(Y = mortality rate | A = dosage level, B = genetic strain). It is this model, called a randomized complete block design (RCBD), to which we now turn our attention. ∎

A Statistical Model for the Randomized Complete Block Design. We assume that there are a levels of factor A and b blocks. Within each block we randomly assign the experimental units to each treatment, so there are a treatments within each block. This experimental design is called a *randomized complete block design* because each treatment is represented in each block. The data in Table 13.1 are an example of a randomized complete block design consisting of three factor levels and four blocks, so $a = 3$ and $b = 4$. We assume that the response of the jth experimental unit in the jth block at the ith factor level is a random variable Y_{ij} such that

$$Y_{ij} = \mu_{ij} + \epsilon_{ij}, \quad i = 1, \ldots, a, \quad j = 1, \ldots, b$$

where the random variables ϵ_{ij} are iid $N(0, \sigma^2)$. Consequently,

$$E(Y_{ij}) = \mu_{ij} \quad \text{and} \quad V(Y_{ij}) = \sigma^2, \quad i = 1, \ldots, a, \quad j = 1, \ldots, b$$

The random variables ϵ_{ij} represent all factors other than A or B that can also produce changes in the response Y_{ij}.

Data Format. In the general case of a RCBD experiment with a levels for factor A and b levels for factor B, we display the data in the format of an $a \times b$ matrix as in Table 13.2.

TABLE 13.2 DATA FORMAT FOR A RANDOMIZED COMPLETE BLOCK DESIGN

	Block (B)			
Treatment (A)	1	2	$\cdots$	b
1	y_{11}	y_{12}	$\cdots$	y_{1b}
2	y_{21}	y_{22}	$\cdots$	y_{2b}
$\vdots$	$\vdots$	$\vdots$	$\vdots$	$\vdots$
a	y_{a1}	y_{a2}	$\cdots$	y_{ab}

The Treatment Effects and the Block Effects. We use the symbols μ, $\overline{\mu}_{i.}$, $\overline{\mu}_{.j}$ to denote the overall mean, the mean response when the treatment is at level i, and the mean response when the experimental unit is in block j, respectively. In particular,

$$\text{Overall mean:} \quad \mu = \frac{1}{ab} \sum_{i=1}^{a} \sum_{j=1}^{b} \mu_{ij}$$

$$i\text{th treatment mean:} \quad \overline{\mu}_{i.} = \frac{1}{b} \sum_{j=1}^{b} \mu_{ij} (i = 1, \ldots, a)$$

$$j\text{th block mean:} \quad \overline{\mu}_{.j} = \frac{1}{a} \sum_{i=1}^{a} \mu_{ij} (j = 1, \ldots, b)$$

Similarly, we denote the treatment effects and block effects by the symbols α_i, β_j, where

$$\alpha_i = \overline{\mu}_{i.} - \mu \quad \text{and} \quad \beta_j = \overline{\mu}_{.j} - \mu$$

More precisely, α_i is the effect of the ith treatment and β_j is the effect of the jth block. It is easy to verify that these effects satisfy the following conditions:

$$\sum_{i=1}^{a} \alpha_i = 0, \qquad \sum_{j=1}^{b} \beta_j = 0$$

A Parameterization of the Randomized Complete Block Design. For the RCBD we assume that the treatment effects and the block effects are *additive*.

This means that the mean response at the (i, j)th factor combination level is additive with respect to the two factors, that is,

$$\mu_{ij} = \mu + \alpha_i + \beta_j \tag{13.1}$$

where

$$\sum_{i=1}^{a} \alpha_i = 0 \quad \text{and} \quad \sum_{j=1}^{b} \beta_j = 0 \tag{13.2}$$

[*Note:* The plausibility of the additive effects model (13.1) can be checked by plotting a suitable graph. We return to this point shortly.]

Homogeneity of the Variances. We assume that the variance of the response variable is the same for all factor levels. Serious violations of normality and homogeneity of variances can be checked by plotting the residuals. We return to this point later (see Sec. 13.2.2).

Least Squares Estimates of the Model Parameters. Using the least squares method as in Sec. 12.2.1 (we omit the details), we obtain the following least squares estimate for the overall mean, treatment effects, and block effects:

$$\hat{\mu} = \overline{Y}_{..} \quad \text{(grand mean)} \tag{13.3}$$

$$\hat{\alpha}_i = \overline{Y}_{i.} - \overline{Y}_{..}, \qquad \hat{\beta}_j = \overline{Y}_{.j} - \overline{Y}_{..} \tag{13.4}$$

In words, the overall mean is estimated by the grand mean, the factor A effect at level i is estimated by the ith row mean minus the grand mean, and the effect of the jth block is estimated by the jth column mean minus the grand mean.

The fitted values and the residuals are found by

Fitted value: $\hat{y}_{ij} = \hat{\mu} + \hat{\alpha}_i + \hat{\beta}_j$

Residual: $e_{ij} = y_{ij} - \hat{y}_{ij}$

$$= y_{ij} - \hat{\mu} - (\overline{y}_{i.} - \overline{y}_{..}) - (\overline{y}_{.j} - \overline{y}_{..})$$

$$= y_{ij} - \overline{y}_{i.} - \overline{y}_{.j} + \overline{y}_{..}$$

It can be shown (the details are left as a problem) that

$$\sum_{i=1}^{a} e_{ij} = 0, \qquad \sum_{j=1}^{b} e_{ij} = 0 \tag{13.5}$$

The corresponding estimates for the ith treatment mean and jth block mean are given by

$$i\text{th treatment mean} = \overline{Y}_{i.}$$

$$j\text{th block mean} = \overline{Y}_{.j}$$

---

example **13.2**   Refer to the fruit fly mortality rate data in Table 13.1. Compute the least squares estimates of the overall mean, treatment effects, and block effects.

Solution. We calculate the least squares estimates using Eq. (13.4) and the treatment means and block means displayed in the last column and bottom row of Table 13.1. ∎

Parameter	Least squares estimate
$\hat{\mu}$	$\bar{y}_{..} = 53.50$
$\hat{\alpha}_1$	$\bar{y}_{1.} - \bar{y}_{..} = 49 - 53.50 = -4.50$
$\hat{\alpha}_2$	$\bar{y}_{2.} - \bar{y}_{..} = 52.25 - 53.50 = -1.25$
$\hat{\alpha}_3$	$\bar{y}_{3.} - \bar{y}_{..} = 59.25 - 53.50 = 5.75$
$\hat{\beta}_1$	$\bar{y}_{.1} - \bar{y}_{..} = 72 - 53.50 = 18.50$
$\hat{\beta}_2$	$\bar{y}_{.2} - \bar{y}_{..} = 58.33 - 53.50 = 4.83$
$\hat{\beta}_3$	$\bar{y}_{.3} - \bar{y}_{..} = 46 - 53.50 = -7.50$
$\hat{\beta}_4$	$\bar{y}_{.4} - \bar{y}_{..} = 37.67 - 53.50 = -15.83$

How Well Does the Model Fit the Data? We judge the performance of the model by first partitioning the total variability SST into the sum of SSA (the variability explained by the treatments), SSB (the variability explained by the blocks), and SSE (the unexplained variability). The coefficient of determination R^2 is the proportion of the total variability explained by the model. Thus

$$\text{SST} = \text{SSA} + \text{SSB} + \text{SSE}, \qquad \text{DF} = ab - 1 \tag{13.6}$$

$$R^2 = \frac{\text{SSA} + \text{SSB}}{\text{SST}} \tag{13.7}$$

where the sums of squares and their degrees of freedom are defined by

$$\text{SSA} = \sum_{i=1}^{a}\sum_{j=1}^{b}(\bar{y}_{i.} - \bar{y}_{..})^2, \qquad \text{DF} = a - 1$$

$$= b\sum_{i=1}^{a}(\bar{y}_{i.} - \bar{y}_{..})^2$$

$$\text{SSB} = \sum_{i=1}^{a}\sum_{j=1}^{b}(\bar{y}_{.j} - \bar{y}_{..})^2, \qquad \text{DF} = b - 1$$

$$= a\sum_{j=1}^{b}(\bar{y}_{.j} - \bar{y}_{..})^2$$

$$\text{SSE} = \sum_{i}\sum_{j}(y_{ij} - \hat{y}_{ij})^2, \qquad \text{DF} = (a - 1)(b - 1)$$

Equation (13.6) for SST is a consequence of the algebraic identity

$$y_{ij} - \bar{y}_{..} = (\bar{y}_{i.} - \bar{y}_{..}) + (\bar{y}_{.j} - \bar{y}_{..}) + (y_{ij} - \bar{y}_{i.} - \bar{y}_{.j} + \bar{y}_{..}) \tag{13.8}$$

which corresponds to the model assumption

$$Y_{ij} - \mu = \alpha_i + \beta_j + \epsilon_{ij}$$

We obtain the partition (13.6) by squaring both sides of the algebraic identity (13.8), adding all the terms, and noting that the cross-product terms equal zero. In detail,

$$\sum_i \sum_j (\bar{y}_{i.} - \bar{y}_{..})(\bar{y}_{.j} - \bar{y}_{..}) = 0$$

$$\sum_i \sum_j (\bar{y}_{i.} - \bar{y}_{..})(y_{ij} - \bar{y}_{i.} - \bar{y}_{.j} + \bar{y}_{..}) = 0$$

$$\sum_i \sum_j (\bar{y}_{.j} - \bar{y}_{..})(y_{ij} - \bar{y}_{i.} - \bar{y}_{.j} + \bar{y}_{..}) = 0$$

We omit the algebraic details.

---

example 13.3   Refer to the fruit fly mortality rate data in Table 13.1. Compute the sum of squares and the proportion of the total variability explained by the RCBD model.

Solution

$$\text{SSA} = 219.5$$
$$\text{SSB} = 2017.67$$
$$\text{SSE} = 11.83$$
$$\text{SST} = 2249.00$$

The proportion of the total variability explained by the RCBD model is given by

$$R^2 = \frac{\text{SSA} + \text{SSB}}{\text{SST}} = \frac{237.17}{2249.00} = 0.99$$

Thus, 99 percent of the total variability is accounted for by the model.   ■

Computing the Degrees of Freedom. The next step in the analysis is to compute the degrees of freedom (DF) associated with the sums of the squares SSA, SSB, SSE, and SST.

1. There are $a - 1$ degrees of freedom associated with SSA since SSA $= b \sum_i (\bar{Y}_{i.} - \bar{Y}_{..})^2$ contains a summands satisfying the one constraint $\sum_i (\bar{Y}_{i.} - \bar{Y}_{..}) = 0$.
2. There are $b - 1$ degrees of freedom associated with SSB since SSB $= a \sum_j (\bar{Y}_{.j} - \bar{Y}_{..})^2$ contains b summands satisfying the one constraint $\sum_j (\bar{Y}_{.j} - \bar{Y}_{..}) = 0$.
3. There are $(a - 1)(b - 1)$ degrees of freedom associated with SSE. To see this, write the error sum of the squares in terms of the residuals as follows:

$$\text{SSE} = \sum_i \sum_j e_{ij}^2$$

Then use the fact that the residuals satisfy the conditions (13.5). Therefore, we need to specify, say, only the first $(b - 1)$ entries for each row,

$$e_{i1}, \ldots, e_{i(b-1)}$$

and the first $(a - 1)$ entries for each column:

$$e_{1j}, \ldots, e_{(a-1)j}$$

Consequently, knowing these $(a-1)(b-1)$ residuals suffices to determine all others.

4. The number of degrees of freedom associated with SST is $ab-1$ since $\text{SST} = \sum_i \sum_j (Y_{ij} - \overline{Y}_{..})^2$ contains ab summands and the single constraint $\sum_i \sum_j (Y_{ij} - \overline{Y}_{..}) = 0$.

5. Notice that the partition (13.6) of the sum of squares SST yields a corresponding partition of the degrees of freedom:

$$ab - 1 = (a-1) + (b-1) + (a-1)(b-1)$$

Consequences of the Additive Effects Hypothesis. The additive effects hypothesis $(\mu_{ij} = \mu + \alpha_i + \beta_j)$ implies

$$\mu_{i'j} - \mu_{ij} = \alpha_{i'} - \alpha_i$$

which is independent of the jth block. This means that the change in the mean response due to a change in the level of factor A is independent of the block. When this condition holds, we say that there is *no interaction* between the treatments and the blocks. The additivity of the treatment and block effects implies that the plots obtained by connecting the points (i, μ_{ij}) $(i = 1, \ldots, a)$ by straight line segments are parallel as in Fig. 13.1.

Because the values of μ_{ij} are unknown, statisticians recommend plotting the points (i, y_{ij}) $(i = 1, \ldots, a)$ and connecting them by straight line segments as in Fig. 13.2, which uses the data on fruit fly mortality rates (Table 13.1). For instance, we see that the differences in mortality rates between the second and first treatments for blocks 1, 2, 3, and 4 are given by

$$y_{21} - y_{11} = 5, \qquad y_{22} - y_{12} = 2, \qquad y_{23} - y_{13} = 1, \qquad y_{24} - y_{14} = 5$$

The fact that these graphs are not exactly parallel is due to the fact that the response is a random variable and not necessarily because the hypothesis of

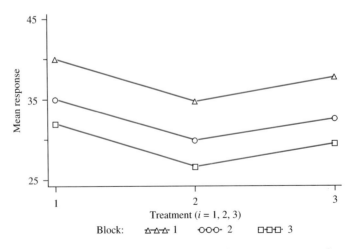

Figure 13.1 Mean response for each block versus treatment $l = 1, 2, 3$: no interaction between treatments and blocks

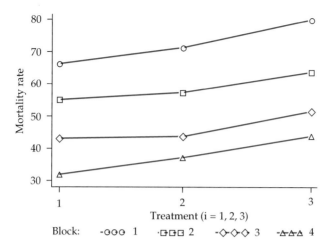

Figure 13.2 Mortality rate for each block versus treatment $l = 1, 2, 3$ (data in Table 13.1)

additivity is false. In any event, Fig. 13.2 does not indicate a gross violation of the hypothesis that the treatment and block effects are additive.

Testing Hypotheses about the Model

Let us return to the problem of testing the null hypothesis that there are no differences in the fruit fly mortality rates due to the different dosages of the insecticide (Table 13.1). That is, our main interest is in testing the equality of the treatment means; consequently, the null hypothesis and its alternative have the following form:

$$H_0: \alpha_1 = \cdots = \alpha_a = 0$$
$$H_1: \alpha_i \neq 0 \quad \text{for at least one } i = 1, \ldots, a$$

The basic idea is to compare the variability explained by factor A to the unexplained variability; that is, we compute the F ratio

$$F = \frac{\text{SSA}/(a-1)}{\text{SSE}/(a-1)(b-1)} \tag{13.9}$$

A large value of the F ratio suggests that the differences among the treatment effects are not due to chance (thus providing strong evidence against H_0); a small F value provides weak evidence against H_0. The quantities $(a-1)$ and $(a-1)(b-1)$ are the degrees of freedom associated with the sums of squares SSA and SSE. Theorem 13.1 (to follow) gives a formula for the distribution of F; the F test for the equality of the treatment means appears immediately afterward.

Sampling Distribution of SST, SSA, SSB, and SSE. Assuming that the error terms in the model are iid $N(0, \sigma^2)$, it is possible to derive the following results on the sampling distribution of SST, SSA, F, SSB, and SSE. The statistic on which we base our test of H_0 is the F ratio, defined in Eq. (13.9). To see why this is the "right" statistic, we list without proof some facts concerning the distributions of SSE, SSA, and SST.

1. The *mean square error* (MSE) is the error sum of squares divided by its number of degrees of freedom, which is $(a - 1)(b - 1)$. Consequently,

$$MSE = \frac{SSE}{(a - 1)(b - 1)}$$

It can also be shown that MSE is an unbiased estimator of σ^2; thus,

$$E(MSE) = E\left(\frac{SSE}{(a - 1)(b - 1)}\right) = \sigma^2 \tag{13.10}$$

2. The *mean square treatment* (MSA) is the sum of squares SSA divided by its number of degrees of freedom, which is $(a - 1)$; thus,

$$MSA = \frac{SSA}{a - 1}$$

It can also be shown that

$$E(MSA) = \sigma^2 + b\frac{\sum_i \alpha_i^2}{a - 1} \tag{13.11}$$

Similarly, we define $MSB = SSB/(b - 1)$ and note that

$$E(MSB) = \sigma^2 + a\frac{\sum_j \beta_j^2}{b - 1} \tag{13.12}$$

3. We next observe that when H_0 is true, that is, $\alpha_1 = \cdots = \alpha_a = 0$, then $E(MSA) = \sigma^2 = E(MSE)$. When H_0 is not true, which means that some of the α_is are different from zero, then Eq. (12.27) implies that $E(MSA) > E(MSE)$. This suggests that the ratio $F = MSA/MSE$ be used to test H_0; in particular, large values of F would make H_0 less plausible. Theorem 13.1 is a more precise version of this statement.

▨ THEOREM 13.1

The null distribution of the F ratio MSA/MSE has an F distribution with $(a - 1)$ degrees of freedom in the numerator and $(a - 1)(b - 1)$ degrees of freedom in the denominator; that is,

$$F = \frac{MSA}{MSE} \sim F_{a-1,(a-1)(b-1)} \qquad\qquad ■$$

The F Test for the Equality of Treatment Means. We now show how to use the F ratio to test

$$H_0: \mu_1 = \mu_2 = \cdots = \mu_a$$

H_1: At least two of the means are different

When H_0 is true, Theorem 13.1 tells us that $F = \text{MSA/MSE}$ is distributed as $F_{a-1,(a-1)(b-1)}$, and that large values of F support the alternative hypothesis H_1. Consequently, our decision rule is

$$\text{accept } H_0 \text{ if } F \leq F_{a-1,(a-1)(b-1)}(\alpha)$$
$$\text{reject } H_0 \text{ if } F > F_{a-1,(a-1)(b-1)}(\alpha)$$

The *P*-value of the test is

$$\text{P-value} = P(F_{a-1,(a-1)(b-1)} > F)$$

and usually appears in the computer printout of the results. Your final results should be summarized in the format of the ANOVA table shown in Table 13.3.

TABLE 13.3 ANOVA TABLE FOR A RANDOMIZED COMPLETE BLOCK DESIGN

Source	DF	SS	MS	F
A	$a-1$	SSA	$\text{MSA} = \dfrac{\text{SSA}}{a-1}$	$\dfrac{\text{MSA}}{\text{MSE}}$
B	$b-1$	SSB	$\text{MSB} = \dfrac{\text{SSB}}{b-1}$	$\dfrac{\text{MSB}}{\text{MSE}}$
Error	$(a-1)(b-1)$	SSE	$\text{MSE} = \dfrac{\text{SSE}}{(a-1)(b-1)}$	
Total	$ab-1$	SST		

example **13.4**   Refer to the the fruit fly mortality rate data in Table 13.1. Compute the ANOVA table and use the *F* ratio to test whether the mean mortality rate depends on the dosage level of the insecticide. State the null hypothesis, alternative hypothesis, decision rule, and conclusion.

Solution. The sums of squares SSA, SSB, SSE, and SST were computed in Example 13.3. The corresponding mean squares are given by

$$\text{MSA} = \frac{\text{SSA}}{a-1} = \frac{219.5}{2} = 109.75$$

$$\text{MSB} = \frac{\text{SSB}}{b-1} = \frac{2017.67}{3} = 672.56$$

$$\text{MSE} = \frac{\text{SSE}}{(a-1)(b-1)} = \frac{11.83}{6} = 1.97$$

$$F = \frac{\text{MSA}}{\text{MSE}} = \frac{126.33}{2.67} = 55.65$$

The ANOVA table for the fruit fly mortality data (Table 13.1) is displayed in Table 13.4.

The null hypothesis is that the mean mortality rates do not depend on the dosage levels of the insecticide. There are three dosage levels, so

TABLE 13.4 ANOVA TABLE FOR THE FRUIT FLY MORTALITY RATE DATA (TABLE 13.1) ANALYZED AS A RANDOMIZED COMPLETE BLOCK DESIGN

Dependent Variable: Mortality Rate

Source	DF	Sum of Squares	Mean Square	F Value	Pr > F
A	2	219.500000	109.750000	55.65	0.0001
Block (B)	3	2017.666667	672.555556	341.01	0.0001
Error	6	11.833333	1.972222		
Corrected Total	11	2249.000000			

$$H_0: \mu_1 = \mu_2 = \mu_3$$
$$H_1: \mu_i \neq \mu_j \text{ for at least one pair } i, j$$

Our decision rule is based on the P-value of the F ratio. Since $F = 55.65 > F_{2,6}(0.01) = 10.92$, we reject the null hypothesis at the 1 percent level. In fact, the computer printout shows that the P-value is less than 0.0001. Consequently, the three dosage levels differ significantly in their effects. ∎

13.2.1   Confidence Intervals and Multiple Comparison Procedures

When one rejects the null hypothesis of no differences among the treatment means, the next step in the analysis of the experiment is to obtain confidence intervals for the individual treatment means and for their differences. We first derive confidence intervals for the treatment means.

The standard error of a treatment mean for a randomized complete block design is

$$s(\overline{Y}_{i.}) = \sqrt{\frac{MSE}{b}}$$

Therefore, a $100(1 - \alpha)$ percent confidence interval for the ith treatment mean is given by

$$\overline{Y}_{i.} \pm t_\nu(\alpha/2)\sqrt{\frac{MSE}{b}} \tag{13.13}$$
$$\nu = (a-1)(b-1)$$

Tukey's Simultaneous $100(1 - \alpha)$ Percent Confidence Interval for the Differences Between Two Means (Randomized Complete Block Design). We now turn our attention to determining which pairs of treatment means are significantly different. We do this in the usual way by noting that a $100(1 - \alpha)$ percent confidence interval for the difference between the ith and kth treatment means is given by

$$\overline{Y}_{i.} - \overline{Y}_{k.} \pm q(\alpha; a, \nu)\sqrt{\frac{2MSE}{b}} \tag{13.14}$$
$$\nu = (a-1)(b-1)$$

We therefore reject the null hypothesis that $\mu_i = \mu_k, i \neq k$, if

$$|\bar{y}_{i.} - \bar{y}_{k.}| > q(\alpha; a, \nu)\sqrt{2MSE/b} \qquad (13.15)$$

---

example **13.5**   Refer to the fruit fly mortality rate data in Table 13.1 and the ANOVA table in Table 13.4. Use Tukey's method to find the significant differences between the treatment means. Use $\alpha = 0.05$.

Solution

$$\nu = 6, \quad a = 3, \quad b = 4, \quad MSE = 1.972$$

so

$$q(0.05; a, \nu)\sqrt{2MSE/b} = 4.31$$

The treatment means, in increasing order, are 49, 52.25, and 59.25. The difference between the first and second treatments is

$$|\bar{y}_{1.} - \bar{y}_{2.}| = 3.25 < 4.31$$

which is not significant. On the other hand, the difference between the third and second treatments is

$$|\bar{y}_{3.} - \bar{y}_{2.}| = 7.00 > 4.31$$

which is significant.   ∎

13.2.2   Model Checking via Residual Plots

Residual and normal probability plots are highly useful for checking model assumptions. For instance, a plot of the residuals against the fitted values is very useful for detecting a gross violation of the hypothesis of equal variances. Looking at Fig. 13.3, we see that the assumption of equal variances for Example 13.1 does not appear to be violated. The normal probability plot for the

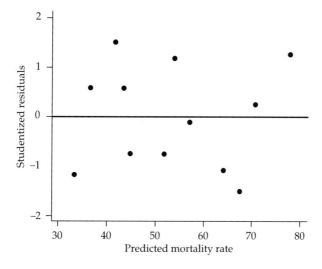

Figure 13.3 Residual plot for data in Table 13.1

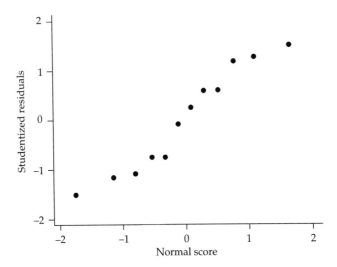

Figure 13.4 Q-Q plot of residuals for data in Table 13.1

studentized residuals is displayed in Fig. 13.4. The plot does not reveal any gross violation of normality, which is confirmed by the fact that the P-value of the Shapiro-Wilk test for normality is 0.41.

PROBLEMS

13.1 With reference to the fruit fly mortality data (Table 13.1), perform a one-way ANOVA for Model(Y = mortality rate $|A$ = dosage level). That is, drop the blocking variable (genetic strain) from the model and analyze the data using a one-way ANOVA model to determine whether or not the mortality rates of the fruit flies are affected by the dosage levels of the insecticide. Display your results in the format of a single-factor ANOVA table (Table 12.3). Compare your conclusions concerning the existence of a treatment effect with the results obtained in the text using the RCBD model. Discuss in particular the effects of blocking.

13.2 The following table lists counts of the numbers of insects (*Leptinotarsa decimlineata*) recorded after four treatments were applied to six areas of a field.

	Area					
Treatment	**1**	**2**	**3**	**4**	**5**	**6**
1	492	410	475	895	401	330
2	111	67	233	218	28	18
3	58	267	283	278	392	141
4	4	1	53	14	138	11

Source: G. Beall, *Biometrika*, vol. 32, 1942, pp. 243–262; quoted in J. W. Tukey, *Exploratory Data Analysis*, Reading, Mass., Addison-Wesley, 1977.

(a) Estimate the overall mean, the treatment means, and the area means.

(b) Compare the four treatments and test for differences among them by computing the ANOVA table. Use $\alpha = 0.05$.

(c) Check the model assumptions by plotting the residuals.

13.3 The following table lists the final weights (in grams) of 18 hamsters fed three different diets varying in fat content. The hamsters were blocked by weight in groups of three; that is, the hamsters in each block had the same weight at the start of the experiment.

(a) Estimate the overall mean, the treatment means, and the block means.

(b) Compare the three diets and test for differences among them by computing the ANOVA table. Use $\alpha = 0.05$.

(c) Check the model assumptions by plotting the residuals.

Treatment	Block 1	2	3	4	5	6
1	96	96	94	99	99	102
2	103	101	103	105	101	107
3	103	104	106	108	109	110

Source: John L. Gill, Design and Analysis of Experiments in the Animal and Medical Sciences, vol. 2, Ames, Iowa State University Press, 1978. Used with permission.

13.4 Show that the residuals satisfy both conditions in Eq. (13.5).

13.5 The following data come from a comparison study of the execution times (measured in milliseconds) of five different workloads on three different configurations of a processor. The goal was to study the effect of processor configuration on computer performance.

(a) Identify the experimental units, the response variable, and the two factors.

(b) Calculate the ANOVA table and test whether there are differences among the three configurations. Use $\alpha = 0.05$.

(c) Which treatment means are significantly different from one another? Use Tukey's multiple comparison method with $\alpha = 0.05$.

(d) Check the model assumptions by graphing suitable plots of the residuals.

Workload	Configuration Two cache	One cache	No cache
ASM	54.0	55.0	106.0
TECO	60.0	60.0	123.0
SIEVE	43.0	43.0	120.0
DRHYSTONE	49.0	52.0	111.0
SORT	49.0	50.0	108.0

Source: R. Jain, The Art of Computer Systems Performance Analysis, New York, John Wiley & Sons, 1991. Used with permission.

13.6 Two types of fertilizer are being compared against a control (no fertilizer). The response variable is the yield (tons) of sugar beets. The results are displayed in the following table.

(a) Calculate the ANOVA table and test whether there are differences among the treatments. Use $\alpha = 0.05$.

(b) Check the model assumptions.

	Fertilizer applied		
Area	**None**	**PO$_4$**	**PO$_4$ and NO$_3$**
1	12.45	6.71	6.48
2	2.25	5.44	7.11
3	4.38	4.92	5.88
4	4.35	5.23	7.54
5	3.42	6.74	6.61
6	3.27	4.74	8.86

Source: G. W. Snedecor, *Statistical Methods,* Ames, University of Iowa Press, 1946. Used with permission.

13.7 The following data come from an experiment to determine the effectiveness of blast furnace slags (A) as agricultural liming materials on three types of soil (B): sandy loam (1), sandy clay loam (2), and loamy sand (3). The treatments were all applied at 4000 lb per acre, and the response variable was the corn yield in bushels per acre.

(a) Calculate the ANOVA table and use your results to test for differences among the treatment means. Use $\alpha = 0.05$.

(b) Plot the residuals against the fitted values. Do you see any nonrandom patterns?

(c) (Computer required) Calculate the normal probability plot of the residuals and test the residuals for normality using the Shapiro-Wilk test.

(d) Plot (i, y_{ij}) $(i = 1, \ldots, 7)$ for each soil type; follow the pattern used to obtain Fig. 13.2. Does the graph support or refute the assumption that there are no treatment-block interactions?

	Soil (B)		
Furnace slag (A)	**1**	**2**	**3**
None	11.1	32.6	63.3
Coarse slag	15.3	40.8	65.0
Medium slag	22.7	52.1	58.8
Agricultural slag	23.8	52.8	61.4
Agricultural limestone	25.6	63.1	41.1
Agricultural slag + minor elements	31.2	59.5	78.1
Agricultural limestone + minor elements	25.8	55.3	60.2

Source: D. E. Johnson and F. A. Graybill, *Journal of the American Statistical Association,* vol. 67, 1972, pp. 862–868. Used with permission.

13.8 The following data come from an experiment comparing four methods for manufacturing penicillin. Note that the blocks are displayed as rows and the treatments are displayed as columns. The raw material consisted of $b = 5$ blends consisting of enough material so that all $a = 4$ methods could be tested on each blend. It was known that the variability among blends was not negligible. The response variable is the yield (the units were not given).

(a) Calculate the ANOVA table and determine whether there are differences among the four methods for manufacturing penicillin.

(b) Check the model assumptions by graphing suitable residual plots.

	Method (A)			
Blend (B)	**1**	**2**	**3**	**4**
1	89	88	97	94
2	84	77	92	79
3	81	87	87	85
4	87	92	89	84
5	79	81	80	88

Source: G. E. P. Box, W. G. Hunter, and J. S. Hunter, *Statistics for Experimenters,* New York, John Wiley & Sons, 1978, p. 209. Used with permission.

13.3 TWO-FACTOR EXPERIMENTS WITH $n > 1$ OBSERVATIONS PER CELL

We now consider experiments where the response variable depends on two factors of equal interest. The statistical analysis is more complex because the scientist must consider several competing models to determine which one best fits the data. We consider in detail the *main effects* model (no interactions) and the model with interactions. A thorough analysis requires a combination of analytical and graphical methods that are best understood in the context of a concrete example.

---

example **13.6**   Table 13.5 lists the results of a study to determine the effects of a fungicide and two insecticides on the egg production of pheasants over a five-week period.

This is a two-factor experiment with two levels of factor A (no fungicide, captan) and three levels of factor B (no pesticide, dieldrin, and Diazinon). There are eight observations at each combination of factor levels; consequently, in this experiment we have

$$a = 2, \quad b = 3, \quad n = 8$$

This is an example of a *complete a × b factorial experiment* because the experiment was performed at all possible combinations of the factor levels.   ■

TABLE 13.5 THE EFFECTS OF A FUNGICIDE AND TWO INSECTICIDES ON THE EGG PRODUCTION OF PHEASANTS OVER A FIVE-WEEK PERIOD

	Pesticide (B)					
Fungicide (A)	**None**		**Dieldrin**		**Diazinon**	
None	15	29	22	32	30	25
	18	30	24	7	28	21
	25	21	2	15	17	19
	26	24	9	18	16	30
Captan	37	12	1	13	9	7
	38	5	1	10	0	1
	24	21	6	18	0	4
	10	6	20	25	4	9

Source: John L. Gill, *Design and Analysis of Experiments in the Animal and Medical Sciences,* vol. 1, Ames, Iowa State University Press, 1978. Used with permission.

Statistical Analysis of the Two-Factor Experiment

The Statistical Model for the Response Variable. We assume that $n > 1$ experiments are performed at each combination of the factor levels. We denote the value of the kth response at the (i, j)th combination of the factor levels by Y_{ijk}, where

$$Y_{ijk} = \mu_{ij} + \epsilon_{ijk}, \quad i = 1, \ldots, a, \quad j = 1, \ldots, b, \quad k = 1, \ldots, n$$

We assume the random variables ϵ_{ijk} are iid $N(0, \sigma^2)$; thus, their means and variances are given by

$$E(Y_{ij}) = \mu_{ij} \quad \text{and} \quad V(Y_{ij}) = \sigma^2$$

The ab parameters μ_{ij} are called the *cell means*. We use the symbols $\mu, \overline{\mu}_{i.}, \overline{\mu}_{.j}$ to denote the overall mean, the mean response when factor A is at level i (row mean), and the mean response when factor B is at level j (column mean), respectively. In particular,

$$\text{Overall mean response:} \quad \mu = \frac{1}{ab} \sum_{i=1}^{a} \sum_{j=1}^{b} \mu_{ij}$$

$$\text{Mean response at } i\text{th level of A:} \quad \overline{\mu}_{i.} = \frac{1}{b} \sum_{j=1}^{b} \mu_{ij}, \quad i = 1, \ldots, a$$

$$\text{Mean response at } j\text{th level of B:} \quad \overline{\mu}_{.j} = \frac{1}{a} \sum_{i=1}^{a} \mu_{ij} \quad j = 1, \ldots, b$$

The quantities α_i and β_j, defined by the equations

$$\alpha_i = \overline{\mu}_{i.} - \mu, \quad \beta_j = \overline{\mu}_{.j} - \mu$$

are called *treatment effects*. In particular, α_i is the effect of the ith level of factor A, and β_j is the effect of the jth level of factor B. More generally, an *effect* is a change in the mean response caused by a change in the factor-level combination. It is easy to verify that the treatment effects satisfy the following conditions:

$$\sum_{i=1}^{a} \alpha_i = 0, \qquad \sum_{j=1}^{b} \beta_j = 0 \qquad\qquad (13.16)$$

Data Format. We denote the kth observation in the (i, j)th cell by

$$y_{ijk}, \quad i = 1, \ldots, a, \quad j = 1, \ldots, b, \quad k = 1, \ldots, n$$

We group these observations by cell and arrange the ab cells in an $a \times b$ matrix consisting of a rows and b columns as shown in Table 13.6.

TABLE 13.6 DATA FORMAT FOR A TWO-FACTOR DESIGN

Factor A	Factor B		
	1	...	b
1	$y_{111}, \ldots, y_{11n}$	...	$y_{1b1}, \ldots, y_{1bn}$
$\vdots$	$\vdots$	$\vdots$	$\vdots$
a	$y_{a11}, \ldots, y_{a1n}$	...	$y_{ab1}, \ldots, y_{abn}$

The Dot Subscript Notation. We denote the sample cell means, factor A (row) means, factor B (column) means, and the grand mean using dot subscript notation:

$$\bar{y}_{ij.} = \frac{1}{n} \sum_{k=1}^{n} y_{ijk}$$

$$\bar{y}_{i..} = \frac{1}{bn} \sum_{j=1}^{b} \sum_{k=1}^{n} y_{ijk}$$

$$\bar{y}_{.j.} = \frac{1}{an} \sum_{i=1}^{a} \sum_{k=1}^{n} y_{ijk}$$

$$\bar{y}_{...} = \frac{1}{abn} \sum_{i=1}^{a} \sum_{j=1}^{b} \sum_{k=1}^{n} y_{ijk} \quad \text{(grand mean)}$$

example **13.7** Refer to the data set in Table 13.5. Compute the cell means, grand mean, and factor A and factor B means. Use the dot subscript notation to express your results.

Solution. The cell means, grand mean, factor A means, and factor B means are given by

$$\bar{y}_{11.} = 23.5, \qquad \bar{y}_{12.} = 16.125, \qquad \bar{y}_{13.} = 23.25$$

$$\bar{y}_{21.} = 19.125, \qquad \bar{y}_{22.} = 11.75, \qquad \bar{y}_{23.} = 4.25$$
$$\bar{y}_{...} = 16.33, \qquad \bar{y}_{1..} = 20.96, \qquad \bar{y}_{2..} = 11.71$$
$$\bar{y}_{.1.} = 21.32, \qquad \bar{y}_{.2.} = 13.94, \qquad \bar{y}_{.3.} = 13.75 \qquad\blacksquare$$

The Main Effects and the Interaction Effects. There are two models to consider depending on whether or not *interactions* occur between the two factors.

No interactions: $\mu_{ij} = \mu + \alpha_i + \beta_j, \quad i = 1,\ldots,a, \quad j = 1,\ldots,b$

Interactions: $\mu_{ij} \neq \mu + \alpha_i + \beta_j \quad$ for at least one pair (i,j)

The *main effects model* (also called the no-interaction model) assumes that

$$\mu_{ij} = \mu + \alpha_i + \beta_j, \quad i = 1,\ldots,a, \quad j = 1,\ldots,b$$

The model statement format for the two-factor experiment with no interactions is

$$\text{Model}(Y = y \mid A = i, B = j)$$

As noted earlier, the main effects model implies that

$$\mu_{i'j} - \mu_{ij} = \alpha_{i'} - \alpha_i$$

That is, the effect of a change in a level of factor A is the same for all levels of factor B. In particular, if a change (from i to i', say) in the temperature of a reaction increases the yield at one level of the pressure, then it will produce the same increase in the yield at all pressures. Of course, if there is an interaction effect, this will not be the case. This consequence of the main effects model can be visualized by drawing the following graphs:

$$(i, \mu_{ij}), \quad i = 1,\ldots,a$$

where there are b graphs, one for each level of factor B. Figure 13.5 displays these graphs for the case $a = 2$ and $b = 3$. Looking at this figure, we see that

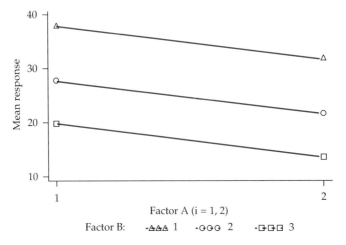

Figure 13.5 Mean response for each Factor B level versus Factor A level ($i = 1, 2$); no interaction between Factors A and B

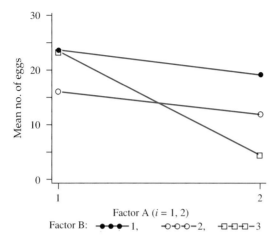

Figure 13.6 Mean response for each level of Factor B versus level of Factor A (data in Table 13.5)

the three graphs are obtained from one another by a vertical translation; in particular, the three line segments are parallel.

Of course, the cell means are not known, so we have to content ourselves with replacing them with their estimates $\bar{y}_{ij\cdot}$; that is, we graph the points

$$(i, \bar{y}_{ij\cdot}), \quad i = 1, \dots, a$$

Figure 13.6 displays the graph obtained by plotting the cell means of the data in Table 13.5. Note that the magnitude of the decrease in egg production as we change the level of factor A from no fungicide ($i = 1$) to captan ($i = 2$) is much greater when the pesticide is Diazinon (factor B at level $j = 3$) than it is for the case when no pesticide is used (factor B at level $j = 1$). An *interaction* effect occurs when the factor A effect is not the same for each level of factor B. Similarly, we say that an interaction effect exists when the magnitude or direction (up or down) of a response to a change in a level of the B factor is not the same for each level of the A factor. The graphs strongly suggest that there is an interaction effect, which the two-way ANOVA table will confirm.

The (i, j)th *interaction effect* is denoted by

$$(\alpha\beta)_{ij} = \mu_{ij} - (\mu + \alpha_i + \beta_j)$$

This quantity is a measure of the discrepancy between the no-interaction model and the one with interactions. It is easy to verify that

$$\sum_{i=1}^{a}(\alpha\beta)_{ij} = 0, \qquad \sum_{j=1}^{b}(\alpha\beta)_{ij} = 0. \tag{13.17}$$

The Two-Factor Model with Interactions. To allow for the possibility of interactions, we assume that the mean response at the (i, j)th factor level combination is given by

$$\mu_{ij} = \mu + \alpha_i + \beta_j + (\alpha\beta)_{ij}$$

The statistical model of the kth response at the (i, j)th factor level is given by

$$Y_{ijk} = \mu + \alpha_i + \beta_j + (\alpha\beta)_{ij} + \epsilon_{ijk}$$

where the random variables ϵ_{ijk} are iid $N(0, \sigma^2)$. The corresponding model statement format for the two-factor experiment with interactions is

$$\text{Model}(Y = y | A = i, B = j, A * B = (i, j))$$

Least Squares Estimates of the Model Parameters for the Two-Factor Model with Interactions. It can be shown that the least squares estimates of the overall mean, the factors A and B, and the interaction effects are

$$\hat{\mu} = \overline{Y}_{...} \tag{13.18}$$

$$\hat{\alpha}_i = \overline{Y}_{i..} - \overline{Y}_{...} \tag{13.19}$$

$$\hat{\beta}_j = \overline{Y}_{.j.} - \overline{Y}_{...} \tag{13.20}$$

$$(\hat{\alpha\beta})_{ij} = \overline{Y}_{ij.} - \overline{Y}_{i..} - \overline{Y}_{.j.} + \overline{Y}_{...} \tag{13.21}$$

In other words, the overall mean is estimated by the grand mean, the factor A effect at the level i is estimated by the ith row mean minus the grand mean, the factor B effect at the level j is estimated by the jth column mean minus the grand mean, and the (i, j)th interaction is estimated by the (i, j)th estimated cell mean minus the (i, j)th additive effect. Similarly, the fitted values and the residuals are

Fitted value: $\hat{y}_{ijk} = \hat{\mu} + \hat{\alpha}_i + \hat{\beta}_j + (\hat{\alpha\beta})_{ij} = \overline{Y}_{ij.}$

Residual: $e_{ijk} = Y_{ijk} - \overline{Y}_{ij.}$

Thus, the fitted value is the (i, j)th estimated cell mean, and the (i, j, k)th residual is the (i, j, k)th observation minus the (i, j)th estimated cell mean.
It can be shown (the details are left as a problem) that

$$\sum_{k=1}^{n} e_{ijk} = 0 \tag{13.22}$$

How Well Does the Model Fit the Data?

We analyze the model's goodness of fit by first partitioning the total variability SST into the sum of SSA (the variability explained by factor A), SSB (the variability explained by factor B), SSAB (the variability explained by the interaction of factors A and B), and SSE (the unexplained variability). The coefficient of determination R^2 is the proportion of the total variability explained by the model:

$$R^2 = \frac{\text{SSA} + \text{SSB} + \text{SSAB}}{\text{SST}}$$

The definitions of these sums of squares and their corresponding degrees of freedom will now be given. We then arrange these quantities in the format of a two-way ANOVA table. This will allow us to test hypotheses about the model parameters.

Partitioning the Total Variability of the Data. We begin with the algebraic identity

$$y_{ijk} - \bar{y}_{...} = (\bar{y}_{i..} - \bar{y}_{...}) + (\bar{y}_{.j.} - \bar{y}_{...}) + (\bar{y}_{ij.} - \bar{y}_{i..} - \bar{y}_{.j.} + \bar{y}_{...}) + (y_{ijk} - \bar{y}_{ij.})$$
$$= \hat{\alpha}_i + \hat{\beta}_j + (\widehat{\alpha\beta})_{ij} + e_{ijk}$$

which corresponds to the model

$$Y_{ijk} - \mu = \alpha_i + \beta_j + (\alpha\beta)_{ij} + \epsilon_{ijk}$$

Squaring both sides and summing over the indices i, j, k yields the following partition of the total sum of squares (SST) and their associated degrees of freedom (DF):

$$\text{SST} = \text{SSA} + \text{SSB} + \text{SSAB} + \text{SSE} \tag{13.23}$$

where

$$\text{SSA} = \sum_{i=1}^{a} \sum_{j=1}^{b} \sum_{k=1}^{n} (\bar{y}_{i..} - \bar{y}_{...})^2, \qquad \text{DF} = a - 1$$

$$= nb \sum_{i=1}^{a} (\bar{y}_{i..} - \bar{y}_{...})^2$$

$$\text{SSB} = \sum_{j=1}^{b} \sum_{i=1}^{a} \sum_{k=1}^{n} (\bar{y}_{.j.} - \bar{y}_{...})^2, \qquad \text{DF} = b - 1$$

$$= na \sum_{j=1}^{b} (\bar{y}_{.j.} - \bar{y}_{...})^2$$

$$\text{SSAB} = \sum_{i=1}^{a} \sum_{j=1}^{b} \sum_{k=1}^{n} (\bar{y}_{ij.} - \bar{y}_{i..} - \bar{y}_{.j.} + \bar{y}_{...})^2, \qquad \text{DF} = (a-1)(b-1)$$

$$= n \sum_{i=1}^{a} \sum_{j=1}^{b} (\bar{y}_{ij.} - \bar{y}_{i..} - \bar{y}_{.j.} + \bar{y}_{...})^2$$

$$\text{SSE} = \sum_{i=1}^{a} \sum_{j=1}^{b} \sum_{k=1}^{n} (y_{ijk} - \bar{y}_{ij.})^2, \qquad \text{DF} = ab(n-1)$$

$$= \text{SST} - \text{SSA} - \text{SSB} - \text{SSAB}$$

$$\text{SST} = \sum_{i=1}^{a} \sum_{j=1}^{b} \sum_{k=1}^{n} (y_{ijk} - \bar{y}_{...})^2 \qquad \text{DF} = nab - 1$$

It can be shown by means of a tedious algebraic calculation that the sums of the cross-product terms equal zero; we omit the details. We do not give computational formulas for the sums of squares because we assume the reader has access to one of the standard statistical software packages.

Sampling Distribution of SST, SSA, SSB, SSAB, SSE. The *mean square error* is the random variable MSE defined by the equation

$$\text{MSE} = \frac{\text{SSE}}{ab(n-1)}$$

[*Note:* When there is only one observation per cell ($n = 1$), we cannot estimate the experimental error, since the denominator is 0 and so the quotient is undefined. This follows from the fact that the ab parameters of the model with interactions equal the number of observations.] The random variable MSE is an unbiased estimator of σ^2; thus,

$$E(\text{MSE}) = E\left(\frac{\text{SSE}}{ab(n-1)}\right) = \sigma^2 \tag{13.24}$$

The *mean squares* MSA, MSB, and MSAB are the sums of squares divided by their associated degrees of freedom. Their expected values are given by

$$\text{MSA} = \frac{\text{SSA}}{a-1}, \qquad E(\text{MSA}) = \sigma^2 + \frac{nb\sum_i \alpha_i^2}{a-1}$$

$$\text{MSB} = \frac{\text{SSB}}{b-1}, \qquad E(\text{MSB}) = \sigma^2 + \frac{na\sum_j \beta_j^2}{b-1}$$

$$\text{MSAB} = \frac{\text{SSAB}}{(a-1)(b-1)}, \qquad E(\text{MSAB}) = \sigma^2 + \frac{n\sum_i\sum_j(\alpha\beta)_{ij}^2}{(a-1)(b-1)}$$

Partitioning the Degrees of Freedom. The total sum of squares SST has $nab - 1$ degrees of freedom. The degrees of freedom associated with the sums of squares SSA, SSB, SSAB, and SSE are $(a-1)$, $(b-1)$, $(a-1)(b-1)$, and $ab(n-1)$, respectively. We therefore have the following partition of the degrees of freedom:

$$nab - 1 = (a-1) + (b-1) + (a-1)(b-1) + ab(n-1)$$

These calculations are conveniently summarized in the two-way ANOVA table displayed in Table 13.7.

TABLE 13.7 TWO-WAY ANOVA TABLE

Source	DF	SS	MS	F
A	$a-1$	SSA	$\text{MSA} = \dfrac{\text{SSA}}{a-1}$	$\dfrac{\text{MSA}}{\text{MSE}}$
B	$b-1$	SSB	$\text{MSB} = \dfrac{\text{SSB}}{b-1}$	$\dfrac{\text{MSB}}{\text{MSE}}$
AB	$(a-1)(b-1)$	SSAB	$\text{MSAB} = \dfrac{\text{SSAB}}{(a-1)(b-1)}$	$\dfrac{\text{MSAB}}{\text{MSE}}$
Error	$ab(n-1)$	SSE	$\text{MSE} = \dfrac{\text{SSE}}{ab(n-1)}$	
Total	$abn-1$	SST		

Testing Hypotheses about the Parameters

In a multifactor experiment statisticians recommend that one first check for the presence of an interaction effect; otherwise one might erroneously conclude that there are no main effects (see Example 13.9). Our main interest, therefore, is in testing the following hypotheses:

1. There are no interactions between the A and B factors:
$$H_{(\alpha\beta)}: (\alpha\beta)_{ij} = 0, \quad i = 1, \ldots, a, \quad j = 1, \ldots, b$$
2. There are no factor A effects:
$$H_\alpha: \alpha_1 = \cdots = \alpha_a = 0$$
3. There are no factor B effects:
$$H_\beta: \beta_1 = \cdots = \beta_b = 0$$

Testing the Null Hypotheses $H_{\alpha\beta}, H_\alpha, H_\beta$. To test for the existence of interactions and the main effects, we proceed in the usual way by computing the F ratios defined by
$$\frac{\text{MSAB}}{\text{MSE}}, \quad \frac{\text{MSA}}{\text{MSE}}, \quad \frac{\text{MSB}}{\text{MSE}}$$

1. *A test for the existence of an interaction.* If there is no interaction effect, then the null hypothesis
$$H_{(\alpha\beta)}: (\alpha\beta)_{ij} = 0, \quad i = 1, \ldots, a, \quad j = 1, \ldots, b$$
is true and therefore $E(\text{MSAB}) = \sigma^2 = E(\text{MSE})$. On the other hand, if it is false, then $E(\text{MSAB}) > \sigma^2 = E(\text{MSE})$. It can be shown (under the null hypothesis) that the ratio MSAB/MSE has the F distribution with numerator and denominator degrees of freedom given by $\nu_1 = (a-1)(b-1)$, $\nu_2 = ab(n-1)$. Thus,
$$F = \frac{\text{MSAB}}{\text{MSE}} \sim F_{(a-1)(b-1), ab(n-1)}$$
Therefore, our decision rule is
$$\text{accept } H_{(\alpha\beta)} \text{ if } F \leq F_{(a-1)(b-1), ab(n-1)}(\alpha)$$
$$\text{reject } H_{(\alpha\beta)} \text{ if } F > F_{(a-1)(b-1), ab(n-1)}(\alpha)$$
2. *Tests for the existence of main effects.* To detect the existence of a factor A effect we consider the F ratio MSA/MSE, which is distributed as
$$F = \frac{\text{MSA}}{\text{MSE}} \sim F_{(a-1), ab(n-1)}$$
The corresponding decision rule is
$$\text{accept } H_\alpha \text{ if } F \leq F_{(a-1), ab(n-1)}(\alpha)$$
$$\text{reject } H_\alpha \text{ if } F > F_{(a-1), ab(n-1)}(\alpha)$$
To detect the existence of a factor B effect we proceed just as in the previous case, only we use the F ratio MSB/MSE, which is distributed as
$$F = \frac{\text{MSB}}{\text{MSE}} \sim F_{(b-1), ab(n-1)}$$
The corresponding decision rule is
$$\text{accept } H_\beta \text{ if } F \leq F_{(b-1), ab(n-1)}(\alpha)$$
$$\text{reject } H_\beta \text{ if } F > F_{(b-1), ab(n-1)}(\alpha)$$

example 13.8   Refer to the pheasant egg production data in Table 13.5. Compute the two-way ANOVA table and the proportion of the total variability explained by the model with interaction. Use the F ratio to test whether there is an interaction between the fungicide (A) and pesticide (B). State the null hypothesis, the alternatives, the decision rule, and your conclusion. What further conclusions can you draw from the ANOVA table?

Solution. The two-way ANOVA table for the data set in Table 3.5 is displayed in Table 13.8.

TABLE 13.8   TWO-WAY ANOVA TABLE FOR THE DATA SET IN TABLE 13.5

Source	DF	SS	MS	F Value	Pr > F
FUNGICIDE (A)	1	1026.7500000	1026.7500000	14.53	0.0004
PESTICIDE (B)	2	595.2916667	297.6458333	4.21	0.0215
AB	2	570.3750000	285.1875000	4.04	0.0249
Error	42	2968.2500000	70.6726190		
Corrected Total	47	5160.6666667			

The proportion of the total variability explained by the model is given by the coefficient of determination:

$$R^2 = \frac{\text{SSA} + \text{SSB} + \text{SSAB}}{\text{SST}} = \frac{2192.42}{5160.67} = 0.42$$

Consequently, 42 percent of the total variability is explained by the model. The null hypothesis is that there is no interaction effect between the levels of the fungicide and pesticide. The alternative is that there is an interaction effect. Our decision rule is to reject the no-interaction hypothesis when the P-value of the F ratio MSAB/MSE is less than 0.05.

Looking at Table 13.8 we see that the F ratio MSAB/MSE has an $F_{2,42}$ distribution, which is not in Table A.6 in the appendix. From Table A.6 it follows that $3.23 > F_{2,42}(0.05) > 3.15$ because

$$F_{2,40}(0.05) = 3.23 > F_{2,42}(0.05) > F_{2,60}(0.05) = 3.15$$

Since $F = 4.04 > 3.23$, we conclude that there is an interaction effect. The computer printout tells us that the F ratio has a P-value equal to 0.0249; that is, $P(F_{2,42} > 4.04) = 0.0249$. The P-values for the tests of fungicide effects, pesticide effects, and interaction effects are listed in the rightmost column. The conclusion is that each of these effects is significant.   ∎

Equal Error Variances

In our model we make the assumption that the variances of the error terms are equal:

$$V(\epsilon_{ijk}) = \sigma^2, \quad i = 1,\dots,a, \quad j = 1,\dots,b, \quad k = 1,\dots,n$$

Serious violations of this assumption can be detected by plotting the residuals against the fitted values. Figure 13.7 displays the graph of the residual plot for the data in Table 13.5. Recall that the fitted values are the estimated cell means $\bar{y}_{ij.}$, which we computed in Example 13.7. Listed in increasing order of magnitude, the estimated cell means are

$$4.25, 11.75, 16.125, 19.125, 23.25, 23.50$$

The funnel shape of the residual plot suggests that the variances are not constant. At this point one might well ask how our conclusions are affected by the fact that the variances appear to be unequal. When this appears to be the case, statisticians recommend a transformation of the data of the form $u = f(y)$ to stabilize the variance; see G. W. Snedecor and W. G. Cochran (*Statistical Methods*, 7th ed., Ames, Iowa State University Press, 1980) for a discussion and specific suggestions. When, as in this case, the observed values in each cell result from counting, a square root transformation is often appropriate. In detail, it is recommended that each observation y_{ijk} be replaced by $u_{ijk} = \sqrt{1 + y_{ijk}}$, with the two-way ANOVA table then calculated using the transformed data (see Prob. 13.9).

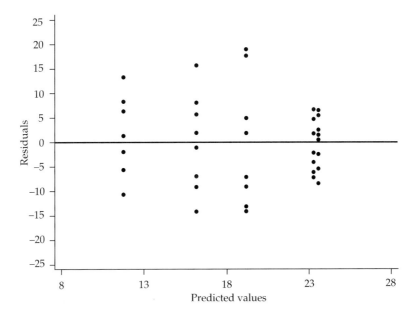

Figure 13.7 Residual plot: Model$(Y \mid A, B, A * B)$ (data in Table 13.5)

example **13.9**   The data set displayed in Table 13.9 (modified from a real data set for the purposes of simplifying the calculations) indicates how an interaction effect can mask the effect of a factor. The entries in each cell are the estimated cell means of a 3×3 complete factorial experiment. The data have been modified so that $\bar{Y}_{i..} = \bar{Y}_{...} = 55$, which implies that SSA = 0. The naive analyst might conclude that factor A has no effect on the response. That factor A does have an effect is clear once we look at the data more closely. For instance, you will note that as the level of factor A varies from 1 to 3 in column 2, the response variable increases. We say that factor A has a positive effect on the response at the second level of factor B. On the other hand, as the level of factor A varies from 1 to 3 in column 3, the response variable decreases. Consequently, factor A has a negative effect on the response at the third level of factor B. The existence of an interaction is made clear by Fig. 13.8, which is a plot of the estimated cell means $\bar{y}_{ij.}$ against the three levels of factor A.

TABLE 13.9 HYPOTHETICAL DATA SET IN WHICH THE EXISTENCE OF A FACTOR A EFFECT IS MASKED BY THE INTERACTION EFFECT

Factor A	Estimated cell mean Factor B			Row mean
	1	2	3	
1	60	40	65	55
2	50	59	56	55
3	60	63	42	55
Column mean	56.67	54.0	54.33	$\bar{y}_{...} = 55$

13.3.1 Confidence Intervals and Multiple Comparisons

When the null hypothesis of no differences among the factor A means (or among the factor B means) is rejected, the researcher then computes confidence

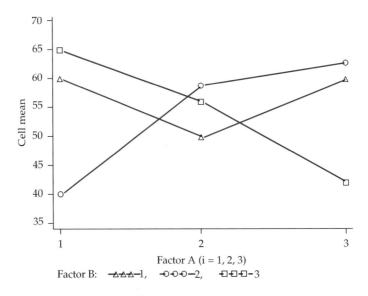

Figure 13.8 Estimated Factor B level cell means versus Factor A level (data in Table 13.9)

intervals for each of the factor means and determines which pairs of means are significantly different. Unbiased estimators of the factor A means $\overline{\mu}_{i.}$ $(i = 1, \ldots, a)$ and the factor B means $\overline{\mu}_{.j}$ $(j = 1, \ldots, b)$ are given by $\overline{Y}_{i..}$ and $\overline{Y}_{.j.}$. The $\overline{Y}_{i..}$ term is the sample mean of bn independent summands with the same variance σ^2, so $\mathrm{Var}(\overline{Y}_{i..}) = \sigma^2/bn$. Similarly, $\mathrm{Var}(\overline{Y}_{.j.}) = \sigma^2/an$. We obtain unbiased estimators for these variances by using MSE to estimate σ^2. The estimated standard errors of the estimators are

$$s(\overline{Y}_{i..}) = \sqrt{\frac{\mathrm{MSE}}{bn}}, \qquad s(\overline{Y}_{.j.}) = \sqrt{\frac{\mathrm{MSE}}{an}}$$

The confidence limits for the factor A and factor B means are given by

$$\overline{Y}_{i..} \pm t_{ab(n-1)}(\alpha/2)\sqrt{\frac{\mathrm{MSE}}{bn}} \tag{13.25}$$

$$\overline{Y}_{.j.} \pm t_{ab(n-1)}(\alpha/2)\sqrt{\frac{\mathrm{MSE}}{an}} \tag{13.26}$$

where the number of degrees of freedom in the t distribution comes from the $ab(n-1)$ degrees of freedom associated with the MSE term.

Tukey's Simultaneous Multiple Comparisons. The simultaneous confidence intervals for the differences among all pairs of factor A and factor B means are given by

$$(\overline{y}_{i..} - \overline{y}_{i'..}) \pm q(\alpha; a, ab(n-1))\sqrt{\frac{\mathrm{MSE}}{bn}}$$

$$(\overline{y}_{.j.} - \overline{y}_{.j'.}) \pm q(\alpha; b, ab(n-1))\sqrt{\frac{\mathrm{MSE}}{an}}$$

The quantities

$$\mathrm{HSD}_A = q(\alpha; a, ab(n-1))\sqrt{\frac{\mathrm{MSE}}{bn}}$$

$$\mathrm{HSD}_B = q(\alpha; b, ab(n-1))\sqrt{\frac{\mathrm{MSE}}{an}}$$

are Tukey's *honest significant difference* for factors A and B. As noted in the previous chapter, Tukey's HSD provides simultaneous $100(1 - \alpha)$ percent confidence intervals for all pairs of differences of treatment means.

---

example **13.10**   Compute Tukey's HSD for the two-factor experiment on the egg production of pheasants (Table 13.5).

Solution. In this case, $a = 2$, $b = 3$, $n = 8$, $ab(n-1) = 42$, and

$$q(0.05; 2, 42) = 2.854 \quad \text{and} \quad q(0.05; 3, 42) = 3.436$$

Consequently,

$$HSD_A = q(0.05; 2, 42)\sqrt{\frac{70.673}{24}} = 4.897$$

$$HSD_B = q(0.05; 3, 42)\sqrt{\frac{70.673}{16}} = 7.221$$

Because

$$|\bar{y}_{.2.} - \bar{y}_{.3.}| = 5.811 < 7.221$$

we conclude that the difference between the dieldrin and Diazinon treatments is not significant. ■

PROBLEMS

13.9 Refer to the data of Table 13.5. The residuals plot (Fig. 13.7) indicates that the hypothesis of equal variances is invalid. To stabilize the variance we use the transformation

$$u = \sqrt{1 + y}$$

The transformed data set is given in the following table.
(a) Compute the two-way ANOVA table for the transformed data, and test for the existence of interactions and main effects.
(b) Plot the residuals against the fitted values and compare this plot with the one obtained using the original data set (Fig. 13.7). Do you detect any improvement?

	$u = \sqrt{1 + y}$					
	Pesticide (B)					
Fungicide (A)	None		Dieldrin		Diazinon	
None	4.00	5.48	4.80	5.74	5.57	5.10
	4.36	5.57	5.00	2.83	5.39	4.69
	5.10	4.69	1.73	4.00	4.24	4.47
	5.20	5.00	3.16	4.36	4.12	5.57
Captan	6.16	3.61	1.41	3.74	3.16	2.83
	6.24	2.45	1.41	3.32	1.00	1.41
	5.00	4.69	2.65	4.36	1.00	2.24
	3.32	2.65	4.58	5.10	2.24	3.16

13.10 Refer to Prob. 13.9.
(a) Compute Tukey's HSD_A and HSD_B. Use $\alpha = 0.05$.
(b) Using Tukey's HSD, determine all pairs of means that are significantly different.

13.11 Show that $\bar{Y}_{i..}$ and $\bar{Y}_{.j.}$ are unbiased estimators of the factor A means $\bar{\mu}_{i.}$ ($i = 1, \ldots, a$) and the factor B means $\bar{\mu}_{.j}$ ($j = 1, \ldots, b$), respectively.

13.12 (a) Show that $\overline{Y}_{ij.}$ is an unbiased estimator of the (i,j)th cell mean μ_{ij} $(i = 1,\ldots,a; j = 1,\ldots,b)$.
(b) Compute $\mathrm{Var}(\overline{Y}_{ij.})$.

13.13 Verify that the interaction effects $(\alpha\beta)_{ij}$ satisfy Eq. (13.17).

13.14 Verify Eq. (13.22).

13.15 The data in the following table are the results of an experiment to investigate the sources of variability in tests of the strength of Portland cement. A sample of cement was divided into small samples for testing. The cement was "gauged" (mixed with water) by three different men, called gaugers, and then it was cast into cubes. Three other men, called "breakers," later tested the cubes for compressive strength. The measurements are in pounds per square inch. Each gauger gauged 12 cubes, which were then divided into three sets of 4, and each breaker tested one set of 4 cubes from each gauger. All the tests were carried out on the same machine. The purpose of the experiment was to identify the source of and measure the variability among gaugers and breakers.
(a) Compute the estimated cell means and display your results in the format of Table 13.9.
(b) Plot the estimated cell means of the breakers (factor B) against the three levels of gaugers (factor A) (see Fig. 13.6 for an example). Does the graph suggest the existence of an interaction?
(c) Test for the existence of main effects and interactions by constructing the ANOVA table.
(d) Check the model assumptions by graphing suitable residual plots.

| | Breaker (B) | | | | | |
Gauger (A)	1		2		3	
1	5280	5520	4340	4400	4160	5180
	4760	5800	5020	6200	5320	4600
2	4420	5280	5340	4880	4180	4800
	5580	4900	4960	6200	4600	4480
3	5360	6160	5720	4760	4460	4930
	5680	5500	5620	5560	4680	5600

Source: O. L. Davies and P. L. Goldsmith, eds., *Statistical Methods in Research and Production,* 4th ed., Edinburgh, Oliver and Boyd, 1972, p. 154.

13.16 The data in the following table record the service time (in minutes) required to service disk drives from three different manufacturers by three different technicians. Each technician was randomly assigned to repair five disk drives of each brand.
(a) Compute the estimated cell means and display your results in the format of Table 13.9.
(b) Plot the estimated cell means $(i,\overline{y}_{ij.})$ $(i = 1,\ldots,3)$ for each brand in the format of Fig. 13.8. Does the graph support or refute the assumption

that there are no interactions between the technicians and the disk drive brands?

(c) Test for the existence of interactions and main effects by computing the two-way ANOVA table.

(d) Compute a 95 percent confidence interval for the time to service each brand of drive.

(e) Compute a 95 percent confidence interval for the time for each technician to service a drive.

Technician (A)	Brand of drive (B)		
	1	2	3
1	62	57	59
	48	45	53
	63	39	67
	57	54	66
	69	44	47
2	51	61	55
	57	58	58
	45	70	50
	50	66	69
	39	51	49
3	59	58	47
	65	63	56
	55	70	51
	52	53	44
	70	60	50

Source: J. Neter, W. Wasserman, and M. H. Kutner, *Applied Linear Statistical Models,* Homewood, Ill., Richard D. Irwin, 1990, p. 723. Used with permission.

13.17 The data in the following table come from an experiment to determine the amount of warping (the response variable) in a metal plate as a function of the following two factors: (A) temperature (measured in degrees Celsius) and (B) copper content (measured as a percentage).

(a) Compute the estimated cell means and display your results in the format of Table 13.9.

(b) Plot the estimated cell means $(i, \bar{y}_{ij.})(i = 1, \ldots, 4)$ for each concentration of copper in the format of Fig. 13.8. Does the graph support or refute the assumption that there are no interactions between the temperature and copper content of the metal plate?

(c) Test for the existence of interactions and main effects by computing the two-way ANOVA table.

Temperature (°C) (A)	*Copper content (%) (B)*							
	40		60		80		100	
50	17	20	16	21	24	22	28	27
75	12	9	18	13	17	12	27	31
100	16	12	18	21	25	23	30	23
125	21	17	23	21	23	22	29	31

Source: N. L. Johnson and F. C. Leone, *Statistics and Experimental Design,* vol. II, New York, John Wiley & Sons, 1964, p. 98.

13.18 Refer to Prob. (13.17).
   (a) Compute Tukey's HSD_A and HSD_B. Use $\alpha = 0.05$.
   (b) Using Tukey's HSD, determine all pairs of means that are significantly different.

13.4 2^k FACTORIAL DESIGNS (OPTIONAL)

A 2^k factorial experiment is an experimental design used to study the effects of k factors with two levels for each factor. We call the two levels of the factor "high" and "low." Such designs are frequently used when the number of factors and number of levels of an experiment are so large that running a complete factorial design would be prohibitively expensive in terms of time, labor, or material. A major goal of this type of experiment is to determine those factors that have a significant effect on the response variable and screen out those that do not. The total number of factor level combinations equals 2^k, and with n replicates at each factor level combination there are $2^k n$ experimental runs. However, it is not only the number of experimental runs that increases rapidly with k; the complexity of the analysis does as well. Theoretically, there are $2^k - 1$ interactions of potential importance, although in practice the higher-order interactions are usually not significant. For this reason we focus our attention primarily on the simplest cases: $k = 2, 3$.

The 2^2 Experiment

We denote the two factors by A and B. Statisticians represent the high levels of factors A and B by the symbols a and b and the low levels by 1. More precisely, we represent the factor level combinations by the formal product

$$a^{l_1} b^{l_2}, \quad l_i = 0, 1$$

where $l_i = 0$ denotes the low level of the factor and $l_i = 1$ denotes the high level. By convention we set $a^0 = b^0 = 1$. For instance, the symbol $ab = a^1 b^1$ represents the factor level combination with both factors at the high level ($l_1 = l_2 = 1$); the symbol $a = a^1 b^0$ represents the combination with factor A at the high level and factor B at the low level ($l_1 = 1, l_2 = 0$). The symbol $1 = a^0 b^0$ denotes the factor level combination with both factors at the low level. Summing up, the four factor level combinations are denoted by: $1, a, b, ab$.

Data Format. We represent the kth replicate at the i, j factor level combination by y_{ijk} ($i = 1, 2; j = 1, 2; k = 1, \ldots, n$). With this notation the factor level combination with both factors at the low level corresponds to ($i = 1, j = 1$). Similarly, ($i = 2, j = 1$) corresponds to the factor level combination with A at the high level and B at the low level.

In the context of 2^2 factorial experiments, statisticians arrange the data in columns by first writing the factor level combinations in the standard order: $1, a, b, ab$. The n replicates at each factor level combination are then listed in the next column. A third column, containing the sample totals for each factor level combination, is sometimes added to the display, as in Table 13.10. We use the special notation $(1), (a), (b), (ab)$ to denote the sample totals for each factor level combination. In terms of dot subscript notation, the sample totals and sample means for factor level combination are given by

$$y_{11.} = (1) \qquad \bar{y}_{11.} = \frac{(1)}{n}$$

$$y_{21.} = (a) \qquad \bar{y}_{21.} = \frac{(a)}{n}$$

$$y_{12.} = (b) \qquad \bar{y}_{12.} = \frac{(b)}{n}$$

$$y_{22.} = (ab) \qquad \bar{y}_{22.} = \frac{(ab)}{n}$$

example 13.11   Table 13.10 records the result of an experiment to measure the performance of a computer workstation, measured in millions of instructions per second (MIPS), as a function of memory size (factor A) and cache size (factor B). Memory size has two levels, 4 Mbytes (low) and 16 Mbytes (high), and cache size has two levels, 1 Kbyte (low) and 2 Kbytes (high). This is a 2^2 experiment with three replicates ($n = 3$) at each factor level combination, resulting in a total sample size of $2^2 \times 3 = 12$. We shall analyze these data in more detail shortly (see Examples 13.13 and 13.14).

TABLE 13.10 PERFORMANCE OF A COMPUTER WORKSTATION WITH TWO MEMORY SIZES AND TWO CACHE SIZES

Factor level combination	Replicates	Sample total
1	15, 18, 12	$(1) = 45$
a	25, 28, 19	$(a) = 72$
b	45, 48, 51	$(b) = 144$
ab	75, 75, 81	$(ab) = 231$

Source: R. Jain, *The Art of Computer Systems Performance Analysis*, New York, John Wiley & Sons, 1991, p. 294. Used with permission.   ∎

The Statistical Model for the Response Variable. We assume that n experiments are performed at each combination of the factor levels. We represent the value of the kth response at the (i, j)th combination of factor levels by Y_{ijk}, where

$$Y_{ijk} = \mu_{ij} + \epsilon_{ijk}, \quad i = 1, 2, \quad j = 1, 2, \quad k = 1, \ldots, n$$

where
$$\mu_{ij} = \mu + \alpha_i + \beta_j + (\alpha\beta)_{ij} \tag{13.27}$$

and the random variables ϵ_{ijk} are iid $N(0, \sigma^2)$.

Main Effects and Interaction Effects. The *effect of factor A at the low level of B* is the difference in the mean response when factor A is changed from the low to the high level with factor B fixed at its low level. It is denoted by $\mu_{21} - \mu_{11}$. Similarly, we call $\mu_{22} - \mu_{12}$ the *effect of A at the high level of B*. We define the *main effect* of factor A to be the average of the effects of A at the low and high levels of B; thus,

$$\text{Main effect of factor A} = \tfrac{1}{2}(\mu_{22} - \mu_{12} + \mu_{21} - \mu_{11})$$

Interchanging the roles of 1 and 2, we obtain the analogous formulas for the factor B effects. Thus, $\mu_{12} - \mu_{11}$ is the *effect of B at the low level of A* and $\mu_{22} - \mu_{21}$ is the *effect of B at the high level of A*. The *main effect* of factor B equals the average of the effects of B at the low and high levels of A; thus,

$$\text{Main effect of factor B} = \tfrac{1}{2}(\mu_{22} - \mu_{21} + \mu_{12} - \mu_{11})$$

Recall that no interaction effect exists when the effect of A is independent of the level of B; equivalently, there is no interaction when the effect of A at the high level of B equals the effect of A at the low level of B. Consequently, there is no AB interaction when $\mu_{21} - \mu_{11} = \mu_{22} - \mu_{12}$. Therefore, we define the AB interaction effect to be the average of the difference between the effect of A at the high and low levels of B; thus,

$$\text{AB interaction effect} = \tfrac{1}{2}(\mu_{22} - \mu_{12} - \mu_{21} + \mu_{11})$$

Notice that each main effect and interaction is defined by a linear combination of the cell means whose coefficients sum to zero; that is, the main effects and interactions are contrasts.

As just noted, the main effects and interactions are linear combinations of the cell means. Consequently, their estimates will also be linear combinations of the estimates of the cell means. These estimators (using the special notation) are given by

$$\hat{\mu}_{11} = \frac{(1)}{n} \qquad \hat{\mu}_{12} = \frac{(b)}{n}$$

$$\hat{\mu}_{21} = \frac{(a)}{n} \qquad \hat{\mu}_{22} = \frac{(ab)}{n}$$

This leads to the following three formulas for computing the estimated main effects for A and B, and their AB interaction effects:

$$\text{Estimator of main effect of A} = \frac{1}{2n}[(ab) - (b) + (a) - (1)]$$

$$\text{Estimator of main effect of B} = \frac{1}{2n}[(ab) + (b) - (a) - (1)]$$

$$\text{Estimator of AB interaction} = \frac{1}{2n}[(ab) - (b) - (a) + (1)]$$

It is an interesting and useful fact that each quantity in square brackets of the expressions for the main and interaction effects is a contrast in the sample totals. In particular, we represent the main effects of A and B and the AB interaction effect in terms of the contrasts L_A, L_B, and L_{AB}, defined as follows:

$$L_A = (ab) - (b) + (a) - (1) \tag{13.28}$$
$$L_B = (ab) + (b) - (a) - (1) \tag{13.29}$$
$$L_{AB} = (ab) - (b) - (a) + (1) \tag{13.30}$$

Computing the Contrast Coefficients. Table 13.11 displays the sign patterns for each contrast as a column. An algorithm for computing the sign pattern for each contrast will now be given. Begin by writing down the factor level combinations in the standard order: $(1), (a), (b), (ab)$. These are listed in the first column. We ignore for the moment the column I. In each of the A and B columns enter -1 when the factor occurs at its low level and enter $+1$ when it occurs at its high level in the corresponding contrast. For instance, the pattern $-1, +1, -1, +1$ in column A corresponds to the *low, high, low, high* coefficients of the factor level combinations in the contrast L_A. Each coefficient for the AB contrast (column AB) is the product of the corresponding coefficients in columns A and B. The column I, which consists only of $+1$s, is not a contrast since its coefficients do not sum to zero. It plays the role of the identity with respect to the operation of column multiplication; thus, $IA = A, IAB = AB$, etc. Note also that the product of each column with itself equals the identity I. Thus, $A^2 = B^2 = (AB)^2 = I$.

TABLE 13.11 CONTRAST COEFFICIENTS
FOR A 2^2 FACTORIAL EXPERIMENT

Factor level combination	I	A	B	AB
1	+1	−1	−1	+1
a	+1	+1	−1	−1
b	+1	−1	+1	−1
ab	+1	+1	+1	+1
Contrast	$\bar{y}$	L_A	L_B	L_{AB}

example **13.12** Refer to Table 13.10 (computer performance data). Compute the estimates of the main effects and interactions.

Solution. We first compute the contrasts, using Eqs. (13.28)–(13.30) and the sample totals listed in Table 13.10.

$$L_A = 231 - 144 + 72 - 45 = 114$$
$$L_B = 231 + 144 - 72 - 45 = 258$$
$$L_{AB} = 231 - 144 - 72 + 45 = 60$$

Next we compute the effects.

$$\text{Main effect of A} = \frac{L_A}{2n} = \frac{114}{6} = 19$$

$$\text{Main effect of B} = \frac{L_A}{2n} = \frac{258}{6} = 43$$

$$\text{Interaction effect AB} = \frac{L_{AB}}{2n} = \frac{60}{6} = 10$$ ∎

Partitioning the Total Variability of the Data. We analyze the model in the usual way by partitioning the total variability of the data into a portion explained by the model and the unexplained variability. These computations are much simpler than in the general case because each sum of squares SSA, SSB, and SSAB equals the corresponding contrast sum of squares divided by $4n$; that is,

$$SST = SSA + SSB + SSAB + SSE \qquad (13.31)$$

where

$$SSA = \frac{L_A^2}{4n} \qquad (13.32)$$

$$SSB = \frac{L_B^2}{4n} \qquad (13.33)$$

$$SSAB = \frac{L_{AB}^2}{4n} \qquad (13.34)$$

[Problem 13.28 asks you to derive Eqs. (13.32), (13.33), and (13.34).] Summing up, we compute SSE by first computing SST and then use the formulas:

$$SST = \sum_{i=1}^{2}\sum_{j=1}^{2}\sum_{k=1}^{n} y_{ijk}^2 - \frac{y_{...}^2}{4n}$$

$$SSE = SST - SSA - SSB - SSAB$$

---

example **13.13**   Compute the partition of the total variability of the data in Table 13.10 (computer performance data).

Solution. Using the values for the contrasts computed earlier,

$$L_A = 114, \qquad L_B = 258, \qquad L_{AB} = 60$$

a straightforward application of Eqs. (13.32), (13.33), and (13.34) yields the following values for the sums of squares:

$$SSA = \frac{(114)^2}{12} = 1083$$

$$SSB = \frac{(258)^2}{12} = 5547$$

$$SSAB = \frac{(60)^2}{12} = 300$$

$$SST = 7032$$

$$SSE = 7032 - 1083 - 5547 - 300 = 102 \qquad \blacksquare$$

Partitioning the Degrees of Freedom. A 2^2 factorial experiment with n replicates for each factor level combination is a special case of the two-way ANOVA with $a = 2$, $b = 2$, and $4n$ observations. Consequently, the sums of squares SSA, SSB, and SSAB each have 1 degree of freedom, and SSE has $4(n-1)$ degrees of freedom. We therefore have the following partition of the degrees of freedom:

$$4n - 1 = 1 + 1 + 1 + 4(n-1)$$

We have thus reduced the task of computing the two-way ANOVA table for the 2^2 experiment (and this is also true for the 2^k experiment) to computing the contrasts in the sample totals for each factor level combination.

---

example 13.14   Compute the ANOVA table for the computer performance data (Table 13.10) and test for the existence of main effects and interactions. Use $\alpha = 0.05$.

Solution. The analysis of variance is displayed in Table 13.12. The entries in the sums of squares column (SS) were computed in Example 13.13.

Since $F_{1,8}(0.01) = 11.26$, we conclude that the main effects and the AB interaction effects are all significant at the 0.01 level and, a fortiori, at the 5 percent level.

TABLE 13.12   TWO-WAY ANOVA TABLE FOR 2^2 FACTORIAL
EXPERIMENT OF EXAMPLE 13.11

Source	DF	SS	MS	F
A	1	1083	1083.00	84.94
B	1	5547	5547.00	435.06
AB	1	300	300.00	23.53
Error	8	102	12.75	
Total	11	7032		

$\blacksquare$

Measuring the Importance of an Effect

The preceding analysis is somewhat misleading because the effects are not of equal importance even though they are all judged to be significant. We measure the *importance of an effect* by computing the proportion of the total variability of the data that is explained by that effect. Thus,

$$\text{Proportion of total variability explained by A} = \frac{\text{SSA}}{\text{SST}}$$

$$\text{Proportion of total variability explained by B} = \frac{\text{SSB}}{\text{SST}}$$

$$\text{Proportion of total variability explained by AB} = \frac{\text{SSAB}}{\text{SST}}$$

---

example **13.15**   Refer to Table 13.10 (computer performance data). Compute the proportion of variability that is explained by each main effect and the interaction effect.

Solution

$$\text{Proportion of total variability explained by A} = \frac{1083}{7032} = 0.15$$

$$\text{Proportion of total variability explained by B} = \frac{5547}{7032} = 0.79$$

$$\text{Proportion of total variability explained by AB} = \frac{300}{7032} = 0.04$$

Clearly, the most important factor affecting the computer performance is factor B (cache size), since it accounts for 79 percent of the total variability.   ∎

---

example **13.16**   The manufacture of an integrated circuit is a complex process depending on many control parameters. The first step is to grow an epitaxial layer with a specified thickness on a silicon wafer. The data in Table 13.13 come from a carefully planned experimental design in which the researchers were interested in controlling the mean and reducing the variance of the thickness of the epitaxial layer. The epitaxial thickness is specified to be between 14 and 15 μm thick with a target value of 14.5 μm. Each experimental run produces 14 wafers. The epitaxial thickness is measured at five different positions on each wafer, yielding a total of $70 = 14 \times 5$ measurements of the epitaxial thickness. The Y1 column records the mean epitaxial thickness, and the Y2 column records $\log s^2$, the logarithm of the sample variance, a measure of the variation in the manufacturing process.[1] Reducing the variation Y2 while keeping Y1 on target would result in a smaller proportion of nonconforming output and thus a higher yield. The eight control parameters labeled X1–X8 are 0 or 1 according to whether the factor is at the low or high level. The two response variables are Y1 and Y2. An exploratory analysis of the data indicated that the effects of the

---

[1]We use the symbols Y1, X1, etc., instead of Y_1, X_1 to make our notation consistent with the computer printouts.

factors X1 (rotation method) and X8 (nozzle position) on Y2 were significant, but their effects on Y1 were not. Verify these conclusions by computing the two-way ANOVA table for each of the following two models:

$$\text{Model}(Y2\,|\,X1, X8, X1 * X8) \quad \text{and} \quad \text{Model}(Y1\,|\,X1, X8, X1 * X8).$$

Key to the Variables

X1 = Susceptor rotation method
X2 = Wafer code
X3 = Deposition temperature
X4 = Deposition time
X5 = Arsenic gas flow rate
X6 = Hydrochloric acid etch temperature
X7 = Hydrochloric acid flow rate
X8 = Nozzle position
Y1 = Epitaxial thickness
Y2 = $\log s^2$

TABLE 13.13 PARAMETER SETTINGS FOR THE CONTROL VARIABLES X1–X8 (0 = LOW, 1 = HIGH) FOR EACH OF 16 EXPERIMENTAL RUNS

EXPT	X1	X2	X3	X4	X5	X6	X7	X8	Y1	Y2
1	0	0	0	0	0	0	0	0	14.821	-0.4425
2	0	0	0	0	1	1	1	1	14.888	-1.1989
3	0	0	1	1	0	0	1	1	14.037	-1.4307
4	0	0	1	1	1	1	0	0	13.880	-0.6505
5	0	1	0	1	0	1	0	1	14.165	-1.4230
6	0	1	0	1	1	0	1	0	13.860	-0.4969
7	0	1	1	0	0	1	1	0	14.757	-0.3267
8	0	1	1	0	1	0	0	1	14.921	-0.6270
9	1	0	0	1	0	1	1	0	13.972	-0.3467
10	1	0	0	1	1	0	0	1	14.032	-0.8563
11	1	0	1	0	0	1	0	1	14.843	-0.4369
12	1	0	1	0	1	0	1	0	14.415	-0.3131
13	1	1	0	0	0	0	1	1	14.878	-0.6154
14	1	1	0	0	1	1	0	0	14.932	-0.2292
15	1	1	1	1	0	0	0	0	13.907	-0.1190
16	1	1	1	1	1	1	1	1	13.914	-0.8625

Source: R. N. Kackar and A. Shoemaker, "Robust Design: A Cost Effective Method for Improving Manufacturing Processes," *AT&T Technical Journal*, vol. 65, no. 2, March/April 1986, pp. 39–50.

Solution

1. **Analysis of variance for** Model(Y2 | X1, X8, X1 * X8). The first step is to drop all the variables from the model except X1, X8, Y1, and Y2 and sort the data in the standard order with A = X1 and B = X8. This is done in Table 13.14. Table 13.15 gives the data for the model rearranged in the standard order (remember that X1 and X8 play the roles of factors A and B, respectively). Table 13.16 is the analysis of variance table for this model. Since this table

TABLE 13.14 THE DATA FOR MODEL(Y2 | X1, X8, X1 * X8) SORTED IN STANDARD ORDER: (X1 = 0, X8 = 0), (X1 = 1, X8 = 0), (X1 = 0, X8 = 1), (X1 = 1, X8 = 1)

EXPT	X1	X8	Y1	Y2
1	0	0	14.821	-0.4425
4	0	0	13.880	-0.6505
6	0	0	13.860	-0.4969
7	0	0	14.757	-0.3267
9	1	0	13.972	-0.3467
12	1	0	14.415	-0.3131
14	1	0	14.932	-0.2292
15	1	0	13.907	-0.1190
2	0	1	14.888	-1.1989
3	0	1	14.037	-1.4307
5	0	1	14.165	-1.4230
8	0	1	14.921	-0.6270
10	1	1	14.032	-0.8563
11	1	1	14.843	-0.4369
13	1	1	14.878	-0.6154
16	1	1	13.914	-0.8625

has a different format from a standard two-way ANOVA table (for example, Table 13.12), a brief explanation is called for. The sums of squares SSX1, SSX8, and SSX1X8 appear in the lower portion of the table. The variability explained by the model, denoted SSm, is defined to be

$$SSm = SSX1 + SSX8 + SSX1X8$$
$$= 0.49600328 + 1.28034883 + 0.06248750$$
$$= 1.83883960$$

This sum appears in the "Sum of Squares" column. It has 3 degrees of freedom because it is a sum of three squares, each of which has 1 degree of freedom. The error sum of squares, SSE, represents the unexplained variability; it is given by SSE = SST − SSm = 0.63983382. The *overall F ratio* equals $F = (SSm/3)/(SSE/12) = 11.50$. The overall F ratio tests the null hypothesis of no X1 effects, no X8 effects, and no X1 * X8 interactions.

TABLE 13.15 THE DATA FOR MODEL(Y2 | X1, X8, X1 * X8) REARRANGED IN STANDARD ORDER

Treatment combination	Replicates	Total
1	−0.4425, −0.6505, −0.4969, −0.3267	(1) = −1.9166
a	−0.3467, −0.3131, −0.2292, −0.1190	(a) = −1.0080
b	−1.1989, −1.4307, −1.4230, −0.6270	(b) = −4.6796
ab	−0.8563, −0.4369, −0.6154, −0.8625	(ab) = −2.7710

TABLE 13.16 TWO-WAY ANOVA TABLE FOR MODEL(Y2 | X1, X8, X1 * X8)

Dependent Variable: Y2   Log of s-square

Source	DF	Sum of Squares	Mean Square	F Value	Pr > F
Model	3	1.83883960	0.61294653	11.50	0.0008
Error	12	0.63983382	0.05331948		
Corrected Total	15	2.47867342			

R-Square	C.V.	Root MSE	Y2 Mean
0.741864	-35.60921	0.230910	-0.648456

Source	DF	Anova SS	Mean Square	F Value	Pr > F
X1	1	0.49600328	0.49600328	9.30	0.0101
X8	1	1.28034883	1.28034883	24.01	0.0004
X1*X8	1	0.06248750	0.06248750	1.17	0.3003

The *P*-value for the *F* ratio is 0.0008, so we reject the null hypothesis. In other words, the method of susceptor rotation and the nozzle position have significant effects on the process variation.

2. **Analysis of variance for** Model(Y1 | X1, X8, X1 * X8). We analyze this model in the same way by first rearranging the data in the standard order, as shown in Table 13.17, and then performing the analysis of variance, which is displayed in Table 13.18.

Looking at Table 13.16, we see that for the Model(Y2 | X1, X8, X1 * X8) the main effects of X1 and X8 are significant, but the interaction effect is not. The proportion of variability accounted for is $R^2 = 0.741864$; that is, this model accounts for 74 percent of the total variability. The results for Model(Y1 | X1, X8, X1 * X8) are quite different. Looking at Table 13.18, we see that none of the main effects are significant (the *P*-value of the overall *F* ratio is 0.9407). The proportion of the total variability explained by this model is $R^2 = 0.031372$, which is negligible. This means that setting the X1

TABLE 13.17 THE DATA FOR MODEL(Y1 | X1, X8, X1 * X8) REARRANGED IN STANDARD ORDER

Treatment combination	Replicates	Total
1	14.821, 13.880, 13.860, 14.757	(1) = 57.318
a	13.972, 14.415, 14.932, 13.907	(*a*) = 57.226
b	14.888, 14.037, 14.165, 14.921	(*b*) = 58.011
ab	14.032, 14.843, 14.878, 13.914	(*ab*) = 57.667

TABLE 13.18 TWO-WAY ANOVA TABLE FOR MODEL(Y1 | X1, X8, X1 * X8)

Dependent Variable: Y1   Epitaxial thickness

Source	DF	Sum of Squares	Mean Square	F Value	Pr > F
Model	3	0.09622225	0.03207408	0.13	0.9407
Error	12	2.97090150	0.24757512		
Corrected Total	15	3.06712375			

R-Square	C.V.	Root MSE	Y1 Mean
0.031372	3.458013	0.497569	14.38888

Source	DF	Anova SS	Mean Square	F Value	Pr > F
X1	1	0.01188100	0.01188100	0.05	0.8303
X8	1	0.08037225	0.08037225	0.32	0.5793
X1*X8	1	0.00396900	0.00396900	0.02	0.9013

and X8 control parameters to minimize the process variance will have a negligible effect on the epitaxial thickness. ∎

The 2^3 Factorial Experiment

In a 2^3 experiment the researcher is interested in studying the response variable as a function of three factors with two levels for each factor.

Notation. We denote the three factors by A, B, and C and represent the 2^3 factor level combinations by the formal product

$$a^{l_1} b^{l_2} c^{l_3}$$

where $l_i = 0$ denotes the low level of the factor and $l_i = 1$ denotes the high level. Thus, $1 = a^0 b^0 c^0$ represents the treatment combination with all factors at their low levels; the symbol (1) represents the sample total at this factor level combination. Similarly, $bc = a^0 b^1 c^1$ represents the factor level combination *(low, high, high)*; the corresponding sample total is denoted (bc). With n replicates per factor level combination, there are a total of $2^3 n$ observations. There are three main effects (A, B, C), three *two*-factor interactions (AB, AC, BC), and one *three*-factor interaction (ABC).

Data Format. We arrange the data from a 2^3 factorial experiment in columns by first writing the factor level combinations in the standard order:

$$1, a, b, ab, c, ac, bc, abc$$

The sample totals for each factor level combination are denoted by

$$(1), (a), (b), (ab), (c), (ac), (bc), (abc)$$

The n replicates at each factor level combination are listed in the second column. Sometimes a third column containing the sample totals for each factor level combination is added to the display. Table 13.19 is an example of how the data from such an experiment are displayed.

TABLE 13.19 DATA FORMAT FOR A 2^3 EXPERIMENT WITH n REPLICATES FOR EACH TREATMENT COMBINATION

Factor level combination	Replicates	Sample total
1	$y_{1111}, \ldots, y_{111n}$	(1)
a	$y_{2111}, \ldots, y_{211n}$	(a)
b	$y_{1211}, \ldots, y_{121n}$	(b)
ab	$y_{2211}, \ldots, y_{221n}$	(ab)
c	$y_{1121}, \ldots, y_{112n}$	(c)
ac	$y_{2121}, \ldots, y_{212n}$	(ac)
bc	$y_{1221}, \ldots, y_{122n}$	(bc)
abc	$y_{2221}, \ldots, y_{222n}$	(abc)

Main Effects and Interactions. We estimate the main effects and interactions by constructing the corresponding contrasts, following the same procedure used to construct Table 13.11. First write down sample totals for the factor level combinations in the standard order: $(1), (a), (b), (ab), (c), (ac), (bc), (abc)$. These are listed in the first column of Table 13.20. Columns A, B, and C list the patterns of signs corresponding to the coefficients in the contrast used to estimate the main effects. We give the method for deriving these patterns shortly. Each coefficient for the AB contrast (column AB) is the product of the corresponding coefficients in columns A and B, and each coefficient for the ABC contrast (column ABC) is the product of the corresponding coefficients in columns A, B, and C. The entries for the columns AC and BC are obtained by multiplying the corresponding columns.

TABLE 13.20 TABLE OF CONTRAST COEFFICIENTS FOR A 2^3 FACTORIAL EXPERIMENT

Treatment total		Main effects and interactions						
	I	A	B	C	AB	AC	BC	ABC
(1)	+1	−1	−1	−1	+1	+1	+1	−1
(a)	+1	+1	−1	−1	−1	−1	+1	+1
(b)	+1	−1	+1	−1	−1	+1	−1	+1
(ab)	+1	+1	+1	−1	+1	−1	−1	−1
(c)	+1	−1	−1	+1	+1	−1	−1	+1
(ac)	+1	+1	−1	+1	−1	+1	−1	−1
(bc)	+1	−1	+1	+1	−1	−1	+1	−1
(abc)	+1	+1	+1	+1	+1	+1	+1	+1
Contrast	$\bar{y}$	L_A	L_B	L_C	L_{AB}	L_{AC}	L_{BC}	L_{ABC}

Derivation of the estimates for the main effects via contrasts. Looking at column A of Table 13.20, we see that the A contrast L_A is

$$L_A = -(1) + (a) - (b) + (ab) - (c) + (ac) - (bc) + (abc) \qquad (13.35)$$

The estimates for an effect and its corresponding sum of squares are given by

$$\text{Estimate of an effect} = \frac{L_{\text{effect}}}{4n} \qquad (13.36)$$

$$\text{SS(effect)} = \frac{L_{\text{effect}}^2}{8n} \qquad (13.37)$$

Derivation of Eq. (13.36). It suffices to consider the main effect A, since the derivations for the other effects are similar. In particular, we want to show that

$$\text{Estimate of the main effect of A} = \frac{L_A}{4n}$$

We first compute the effect of A at each of the four possible factor level combinations of B and C. Recall that this equals the change in the response when only factor A is changed from the low to the high level, with all the other factors held constant. These computations are listed in Table 13.21.

TABLE 13.21 COMPUTING THE EFFECTS OF A AT EACH OF THE FOUR FACTOR LEVEL COMBINATIONS OF B AND C IN A 2^3 FACTORIAL EXPERIMENT

B	C	Effect of A
Low	Low	$\dfrac{(a) - (1)}{n}$
High	Low	$\dfrac{(ab) - (b)}{n}$
Low	High	$\dfrac{(ac) - (c)}{n}$
High	High	$\dfrac{(abc) - (bc)}{n}$

The estimated effect of A is defined to be the average of these four effects, which, as you can easily verify, equals $L_A/4n$. Using the same reasoning, you can show that the estimated B effect equals $L_B/4n$, etc.

Partitioning the Total Variability of the Data. To each main effect and interaction there corresponds a sum of squares that measures the contribution of that effect to the total variability of the data. In a three-way ANOVA, the partition of SST takes the following form:

$$\text{SST} = \text{SSA} + \text{SSB} + \text{SSC} + \text{SSAB} + \text{SSAC} + \text{SSBC} + \text{SSABC} + \text{SSE} \qquad (13.38)$$

We use Eq. (13.37) to compute the sums of squares for the main effects and interactions. We compute the total sum of squares in the usual way:

$$SST = \sum_i \sum_j \sum_k \sum_{1 \le l \le n} (y_{ijkl} - \bar{y}_{....})^2$$

$$= \sum_i \sum_j \sum_k \sum_{1 \le l \le n} y_{ijkl}^2 - N\bar{y}_{....}^2$$

where $N = n2^k$ is the total number of observations. The formulas for computing the sums of squares for the general three-way ANOVA are not given because these calculations are best done via one of the standard statistical software packages such as MINITAB or SAS.

Partitioning the Degrees of Freedom. Corresponding to the partition of the total variability given in Eq. (13.38), we have the following partition of the degrees of freedom. The total sum of squares SST has $2^3 n - 1$ degrees of freedom, and there are $2^3 - 1 = 7$ single-degree-of-freedom contrasts; therefore,

$$2^3 n - 1 = (2^3 - 1) + 2^3(n - 1)$$

Consequently, the error sum of squares SSE has $2^3(n-1)$ degrees of freedom. The preceding formula for the partition of the degrees of freedom is a special case of the following, more general result, which holds for a complete 2^k experiment with n replicates for each treatment combination. The total sum of squares SST has $2^k n - 1$ degrees of freedom, and there are $2^k - 1$ single-degree-of-freedom contrasts; therefore, SSE has $2^k(n - 1)$ degrees of freedom, and the partition of the degrees of freedom is given by

$$2^k n - 1 = (2^k - 1) + 2^k(n - 1)$$

The corresponding F ratios have an F distribution with parameters $\nu_1 = 1$ and $\nu_2 = 2^k(n - 1)$.

example 13.17 The data in Table 13.22 come from a 2^3 factorial experiment with three replicates for each treatment combination. Estimate the main effects and all multifactor interactions. Then analyze the data by constructing the ANOVA table.

TABLE 13.22 DATA FROM A COMPLETE 2^3 FACTORIAL EXPERIMENT WITH THREE REPLICATES FOR EACH TREATMENT COMBINATION

Treatment combination	Replicates	Total
1	14, 24, 11	49
a	18, 26, 20	64
b	32, 20, 21	73
ab	30, 21, 36	87
c	21, 33, 27	81
ac	20, 24, 24	68
bc	32, 30, 39	101
abc	38, 33, 26	97

Solution. We first compute the contrasts, the main effects, and all multifactor interactions using Eq. (13.36). We then compute the corresponding sums of squares using Eq. (13.37). The results are summarized in Table 13.23.

TABLE 13.23 CONTRASTS AND ESTIMATES OF THE MAIN EFFECTS AND INTERACTIONS FOR THE DATA IN TABLE 13.22

Contrast	Estimate of the effect or interaction
$L_A = 12$	$\dfrac{L_A}{12} = 1.00$
$L_B = 96$	$\dfrac{L_B}{12} = 8.00$
$L_{AB} = 8$	$\dfrac{L_{AB}}{12} = 0.67$
$L_C = 74$	$\dfrac{L_C}{12} = 6.17$
$L_{AC} = -46$	$\dfrac{L_{AC}}{12} = -3.83$
$L_{BC} = 2$	$\dfrac{L_{BC}}{12} = 0.17$
$L_{ABC} = 10$	$\dfrac{L_{ABC}}{12} = 0.83$

Looking at the contrasts only, it is easy to see that the most significant main effects are due to factors B and C, with A being relatively insignificant. The higher-order interactions, with the possible exception of the AC interaction, do not appear to be significant. These preliminary findings are confirmed by examining the ANOVA table in Table 13.24. Note that the entries in column 3 (SS) are computed via the formula SS(effect) = $L^2_{\text{effect}}/8n$. Thus, SSA = $12^2/24 = 6.00$.

TABLE 13.24 ANOVA TABLE FOR A 2^3 FACTORIAL EXPERIMENT (DATA SET FROM TABLE 13.22) (EDITED VERSION OF SAS PRINTOUT)

Source	DF	SS	Mean Square	F Value	Pr > F
A	1	6.0000000	6.0000000	0.18	0.6761
B	1	384.0000000	384.0000000	11.59	0.0036
A*B	1	2.6666667	2.6666667	0.08	0.7803
C	1	228.1666667	228.1666667	6.89	0.0184
A*C	1	88.1666667	88.1666667	2.66	0.1223
B*C	1	0.1666667	0.1666667	0.01	0.9443
A*B*C	1	4.1666667	4.1666667	0.13	0.7275
Error	16	530.0000000	33.1250000		
Total	23	1243.3333333			

Looking at the last column of the ANOVA table, we see that only the P-values for the B and C main effects are significant at the 5 percent level. Another way of evaluating the various effects and interactions is to observe that factor B accounts for 31 percent [$(384/1243) \times 100$ percent] of the total variability, whereas factor A accounts for only 0.48 percent [$(6/1243) \times 100$ percent] of the total variability. This suggests that the experimenter should concentrate her future research on factors B and C because the other factors and their interactions appear to be negligible.   ∎

Fractional Replication of a 2^k Factorial Experiment

The results of the preceding analysis of the data in Table 13.22 imply that the factor A and the higher-order interactions are negligible; only factors B and C were judged to be significant. In other words, most of the observed variability is due to a small number of factors and low-order interactions. This suggests that we performed more experiments than were necessary to reach these conclusions. However, another consideration of no small importance is that a complete 2^k factorial experiment with n replicates for each treatment combination requires $2^k n$ experimental runs. This number increases rapidly even for small values of k and n. For instance, a complete factorial experiment with four factors and three replicates for each treatment combination requires $2^4 \times 3 = 48$ experimental runs; a complete factorial experiment with five factors and two replicates for each treatment combination requires $2^5 \times 2 = 64$ experimental runs. If the researcher believes (as in the previous example) that only the main effects are important and that the higher-order interactions are negligible, then useful information concerning these effects can still be obtained by performing only a fraction of the experiments needed for the complete factorial. This is the driving force behind the idea of a *fractional replication* of a 2^k factorial experiment. This topic, which is of considerable importance in modern manufacturing, is treated at length in the references cited in the chapter summary (Sec. 13.5).

We will not attempt a comprehensive introduction here since it is very easy to get lost in the intricate technical details that obscure the basic ideas. For this reason we study only the simplest example, the *half fraction of the 2^k factorial*. In this design the experimenter makes half the number of experimental runs of the complete factorial, and this is why we denote the half fraction design by the symbol 2^{k-1} (which equals one-half of 2^k). For instance, a half fraction of a complete 2^3 factorial experiment consists of $2^{3-1} = 4$ treatment combinations.

Although many possible factors can affect the final results, most scientists have a pretty good idea (based on theory and experience) which factors and higher-order interactions are more important than others. It is on this basis that the senior scientist and her staff determine which treatment combinations to include in the experimental runs.

Choosing the Treatment Combinations for a 2^{3-1} Design. A half fraction of a 2^3 factorial experiment is run with $2^{3-1} = 4$ treatment combinations. We now present an algebraic method that statisticians have devised for selecting the treatment combinations. Pick a contrast, for example,

$$L_{ABC} = -(1) + (a) + (b) - (ab) + (c) - (ac) - (bc) + (abc)$$

and choose the treatment combinations with a + sign for the experimental runs. This means that we perform the experiments on the four treatment combinations (a), (b), (c), (abc).

Table 13.25 is obtained by rearranging the rows of Table 13.20 so that the treatment combinations with a + sign in the contrast L_{ABC} occupy the first four rows.

TABLE 13.25 TABLE OF CONTRAST COEFFICIENTS
FOR A 2^3 FACTORIAL EXPERIMENT

Treatment total	Main effects and interactions							
	I	**A**	**B**	**C**	**AB**	**AC**	**BC**	**ABC**
(a)	+1	+1	−1	−1	−1	−1	+1	+1
(b)	+1	−1	+1	−1	−1	+1	−1	+1
(c)	+1	−1	−1	+1	+1	−1	−1	+1
(abc)	+1	+1	+1	+1	+1	+1	+1	+1
(ab)	+1	+1	+1	−1	+1	−1	−1	−1
(ac)	+1	+1	−1	+1	−1	+1	−1	−1
(bc)	+1	−1	+1	+1	−1	−1	+1	−1
(1)	+1	−1	−1	−1	+1	+1	+1	−1
Contrast	$\bar{y}$	L_A	L_B	L_C	L_{AB}	L_{AC}	L_{BC}	L_{ABC}

Computing the Contrasts for a 2^{3-1} Design with Defining Relation I = ABC. Looking at the first four rows of Table 13.25, we see that the contrast coefficients for column I match those of column ABC, so we write I = ABC and call it the *defining relation* of our design. ABC is called the *generator* of this design.

Let us agree to call a contrast coming from a fractional replicate a *fractional contrast* and denote it by the symbols l_A, l_B, ... Looking at Table 13.25, we see that the fractional contrasts are given by

$$l_A = (a) - (b) - (c) + (abc) = l_{BC}$$
$$l_B = -(a) + (b) - (c) + (abc) = l_{AC}$$
$$l_C = -(a) - (b) + (c) + (abc) = l_{AB}$$

When two fractional contrasts are identical, we say that they are *aliases* of one another. Thus A is aliased with BC since $l_A = l_{BC}$; we write A $\equiv$ BC. Define the *sum of the contrasts* A and BC to be the sum of the corresponding contrast coefficients in columns A and BC. This leads to another interpretation of the phenomenon of aliasing. Looking at Table 13.25, we see that

$$l_A = \tfrac{1}{2}(L_A + L_{BC})$$
$$l_B = \tfrac{1}{2}(L_B + L_{AC})$$
$$l_C = \tfrac{1}{2}(L_C + L_{AB})$$

Therefore, the corresponding estimate $l_A/2n$ of the main effect of A is really an estimate of the main effect of A *plus* the main effect of BC.

---

example 13.18   Refer to Table 13.22. Analyze these data using a 2^{3-1} design with defining relation I = ABC.

Solution. A routine computation (left to you) yields the estimates in Table 13.26 for the main effects based on the fractional contrasts (column 1) and the contrasts based on the complete factorial (column 2).

TABLE 13.26   ESTIMATES OF THE MAIN EFFECTS BASED ON FRACTIONAL CONTRASTS AND COMPLETE CONTRASTS (DATA FROM TABLE 13.22)

Fractional factorial	Complete factorial
$\dfrac{l_A}{2n} = 1.17$	$\dfrac{L_A}{4n} = 1$
$\dfrac{l_B}{2n} = 4.17$	$\dfrac{L_B}{4n} = 8$
$\dfrac{l_C}{2n} = 6.83$	$\dfrac{L_C}{4n} = 6.17$

The results indicate clearly that factor A is much less important than factors B and C. They are consistent with the results obtained by running the complete factorial, but were obtained by running four experiments instead of eight. This illustrates the cost savings that are possible using a suitable fractional factorial. A formal analysis of variance (not given here) for Model$(Y \mid A, B, C)$ uses the formulas SSA $= l_A^2/2^2n$ and so on to compute the sums of squares for the main effects. Each of these has one degree of freedom, SST has $2^2n - 1$ degrees of freedom and SSE = SST $-$ SSA $-$ SSB $-$ SSC has $2^2n - 4$ degrees of freedom. ∎

Confounding in a 2^k Factorial Experiment

Confounding occurs when it is impossible to run all 2^k factor level combinations under the same experimental conditions. Consider a 2^2 agricultural experiment to study the effects of nitrogen (factor A) and phosphorus (factor B) on crop yield. Suppose we have two separate plots of land, each of which is large enough to run only two factor level combinations. Each plot of land is a *block*, and the experimenter must decide which two treatments to assign to each block. This is an example of a factorial experiment in *incomplete blocks*, since not all factor level combinations can be run in each block. This complicates the analysis somewhat since the experimenter must include an additional parameter in the model to account for a possible block effect. Table 13.27 shows one possible way of allocating the factor level combinations to the two blocks.

TABLE 13.27 FORMAT FOR A 2^2
FACTORIAL EXPERIMENT PARTITIONED
INTO TWO INCOMPLETE BLOCKS

Block 1	Block 2
1	a
ab	b

To estimate the block effect we compute the difference between the mean response of block 1 and that of block 2; that is, we compute the following contrast:

$$\frac{(1)+(ab)}{2n} - \frac{(a)+(b)}{2n} = \frac{1}{2n}[(1)+(ab)-(a)-(b)]$$

$$= \frac{L_{AB}}{2n}$$

You will recognize that the contrast used to estimate the block effect is the same as the contrast used to estimate the AB interaction! In other words, it is not possible to separate the AB interaction effect from the block effect. We say that the AB interaction is *completely confounded with blocks.* The quantities $(1)+(ab)$ and $(a)+(b)$ are called *block totals.*

Similarly, confounding occurs in a 2^3 factorial experiment when it is impossible to run the experiment for all eight factor level combinations under the same experimental conditions. For instance, suppose each experiment requires two hours, which means that your research staff can run only four experiments per shift (assuming your staff works an eight-hour shift). This is another example of a factorial experiment in incomplete blocks. In this case each shift plays the role of a block.

The method for assigning treatments to two blocks is similar to that used for selecting the treatments in a half replicate 2^{k-1} design. In detail, choose a contrast, and assign the treatment combinations with a + sign to one block and the remaining treatments to the other block. Table 13.28 shows the block assignments corresponding to the contrast L_{ABC}.

TABLE 13.28 FORMAT FOR A 2^3
FACTORIAL EXPERIMENT DIVIDED
INTO TWO INCOMPLETE BLOCKS

Block 1	Block 2
1	a
ab	b
ac	c
bc	abc

The contrast for estimating the block effect is equal to the difference between the block totals:

$$(a) + (b) + (c) + (abc) - [(1) + (ab) + (ac) + (bc)] = L_{ABC}$$

Thus, the block effects are confounded with the ABC interaction, and it is not possible to separate the interaction from the block effect. Call the block containing the treatment combination 1 the *principal block*. It is worth noting that the treatments in the other block can be obtained by choosing any treatment combination not in the principal block and multiplying it by the treatment combinations in the principal block. For instance, the treatment combinations in block 2 of Table 13.28 are (a, b, c, abc); they can also be obtained by choosing a, which is not in the principal block, and multiplying it according to the rule that $a^2 = b^2 = c^2 = 1$; thus,

$$a \times 1 \equiv a$$
$$a \times ab \equiv a^2 b \equiv b$$
$$a \times ac \equiv a^2 c \equiv c$$
$$a \times bc \equiv abc$$

Constructing the ANOVA Table for a 2^3 Factorial Experiment Divided into Two Blocks. The formal analysis of variance is the same as for the nonconfounded case. That is, compute the sums of squares exactly as before and calculate the sum of squares due to the blocks (denoted SSBl) by adding the sums of squares of all effects confounded with the blocks, which in this case yields SSBl = SSABC.

---

example 13.19   The data in Table 13.29 came from a 2^3 factorial experiment divided into two blocks. Compute the ANOVA table and determine which effects are significant at the 5 percent level. In addition, compute the proportion of variability accounted for by each main effect.

TABLE 13.29   DATA FROM A 2^3 FACTORIAL EXPERIMENT DIVIDED INTO TWO BLOCKS WITH THREE REPLICATES PER TREATMENT COMBINATION

Block 1	Block 2
1: 7, 17, 8	a: 29, 20, 30
ab: 69, 66, 66	b: 43, 53, 51
ac: 28, 26, 32	c: 11, 8, 9
bc: 51, 54, 53	abc: 72, 77, 80

Solution. We used SAS (PROC GLM) to compute the ANOVA table in Table 13.30 in the usual way, and we then edited the output to take into account the fact that the ABC interaction is confounded with the blocks.

Only the main effects are significant at the 5 percent level; two-factor interactions are not. Factor B is clearly the most important factor since it

TABLE 13.30 ANOVA TABLE FOR A 2^3 FACTORIAL EXPERIMENT
DIVIDED INTO TWO BLOCKS (DATA SET FROM TABLE 13.29)

Source	DF	Type I SS	Mean Square	F Value	Pr > F
A	1	2204.16667	2204.16667	144.54	0.0001
B	1	10837.50000	10837.50000	710.66	0.0001
A*B	1	16.66667	16.66667	1.09	0.3114
C	1	73.50000	73.50000	4.82	0.0432
A*C	1	32.66667	32.66667	2.14	0.1627
B*C	1	54.00000	54.00000	3.54	0.0782
Block	1	1.50000	1.50000	0.10	0.7579
Error	16	244.00000	15.25000		
Total	23	13464.00000			

accounts for 80 percent [(10,837/13,464) × 100 percent] of the total variability. Factor C, even though it is judged statistically significant, accounts for only 0.55 percent of the total variability. ∎

PROBLEMS

13.19 Refer to Model(Y2 | X1, X8, X1 * X8) of Example 13.16.

  (a) Estimate the main effects and interactions. (*Hint:* First compute the contrasts L_{X1}, L_{X8}, L_{X1X8}.)

  (b) Compute SSX1, SSX8, and SSX1X8. Your answers should agree with those in Table 13.16. [*Hint:* Use Eqs. (13.32), (13.33), and (13.34).]

  (c) How would you set the parameter value X8 (0 or 1) to minimize the process variation? Justify your answer. [*Hint:* Compute the average value of Y2 at the levels X8 = 0 and X8 = 1.]

13.20 Refer to Model(Y1 | X1, X8, X1 * X8) of Example 13.16.

  (a) Estimate the main effects and interactions. (*Hint:* First compute the contrasts L_{X1}, L_{X8}, L_{X1X8}.)

  (b) Compute SSX1, SSX8 and SSX1X8. Your answers should agree with those in Table 13.18. [*Hint:* Use Eqs. (13.32), (13.33), and (13.34).]

  (c) Which variable has the largest effect on Y1 (the epitaxial thickness)?

13.21 Refer to Example 13.16. An exploratory analysis of the data indicated that the effect of factor X4 (deposition time) on Y1 is significant, but its effect on Y2 is not. This suggests that we study both Model(Y1 | X4, X8, X4 * X8) and Model(Y2 | X4, X8, X4 * X8).

  (a) Rearrange the data for Model(Y1 | X4, X8, X4 * X8) in the standard order; use the format of Table 13.17.

  (b) Perform the analysis of variance and test for the existence of main effects and interactions. Use $\alpha = 0.05$.

  (c) Consider the Model(Y2 | X4, X8, X4 * X8). Rearrange the data for this model in the standard order; use the format of Table 13.15.

  (d) Perform the analysis of variance and test for the existence of main effects and interactions. Use $\alpha = 0.05$.

(e) Combining the results from parts (b) and (d), is it reasonable to conclude that the effect of deposition time on epitaxial thickness is significant and that its effect on Y2 (process variance) is not? Justify your answer.

13.22 The following data came from a complete 2^3 factorial experiment with three replicates for each treatment combination.

Treatment combination	Replicates
1	48, 37, 58
a	94, 90, 107
b	94, 91, 92
ab	160, 156, 154
c	70, 75, 68
ac	117, 114, 110
bc	123, 105, 126
abc	187, 170, 182

(a) Prepare a table similar in format to Table 13.23 listing all the contrasts and estimates of the main effects and interactions.

(b) Using the results obtained in part (a), compute the sums of squares for each main and interaction effect.

(c) Given the information that SST = 39,811.96 and SSE = 863.33, compute the proportion of the total variability explained by each main effect and interaction. Rank the effects in decreasing order of importance as measured by SS(effect)/SST.

(d) Using the information obtained in parts (b) and (c), compute the ANOVA table in a format similar to Table 13.24. Are your final results consistent with the preliminary conclusions you obtained in part (c)? Comment.

13.23 Analyze the data of the preceding problem as a half replicate with defining relation I = ABC. That is, compute the fractional replicates l_A, l_B, l_C and the corresponding sum of squares; then compute SST and SSE for this half replicate. Determine the proportion of variability that is explained by each main effect, and comment on whether or not your results are consistent with those obtained from the complete factorial.

13.24 The following data are from an experiment to study the effects of temperature (factor A), processing time (factor B), and the rate of temperature rise (factor C) on the amount of dye (the response variable) left in the residue bath of a dyeing process. The experiment was run at two levels for each factor, with two replicates at each factor level combination.

(a) Prepare a table similar in format to Table 13.23 listing all the contrasts and estimates of the main effects and interactions.

(b) Using the results obtained in part (a), compute the sums of squares for each main and interaction effect.

(c) Given the information that SST = 313.88 and SSE = 56.14, compute the proportion of the total variability explained by each main effect

Treatment combination	Replicates
1	19.9, 18.6
a	25.0, 22.8
b	17.4, 16.8
ab	19.5, 18.3
c	14.5, 16.1
ac	27.7, 18.0
bc	16.3, 14.6
abc	28.3, 26.2

Source: J. S. Milton and J. C. Arnold, *Introduction to Probability and Statistics,* 2d ed., New York, McGraw-Hill, 1990. Used with permission.

and interaction. Rank the effects in decreasing order of importance as measured by SS(effect)/SST.

(d) Using the information obtained in parts (*b*) and (*c*), compute the ANOVA table in a format similar to Table 13.24. Are your final results consistent with the preliminary conclusions you obtained in part (*c*)? Comment.

13.25 Analyze the data of the preceding problem as a half replicate with defining relation I = ABC. That is, compute the fractional replicates l_A, l_B, l_C and the corresponding sum of squares; then compute SST and SSE for this half replicate. Determine the proportion of variability that is explained by each main effect, and comment on whether or not your results are consistent with those obtained from the complete factorial.

13.26 The data in the following table come from a complete 2^3 factorial experiment to study the effect of cutting speed (A), tool geometry (B), and cutting angle (C) on the life of a machine tool. Two levels of each factor were chosen, and there were three replicates for each factor level combination.

Treatment combination	Replicates
1	22, 31, 25
a	32, 43, 29
b	35, 34, 50
ab	55, 47, 46
c	44, 45, 38
ac	40, 37, 36
bc	60, 50, 54
abc	39, 41, 47

Source: D. C. Montgomery, *Design and Analysis of Experiments,* 2d ed., New York, John Wiley & Sons, 1984, p. 292. Used with permission.

(a) Given that SSE = 482.67 and SST = 2095.33, analyze the data by computing the ANOVA table and determine which main effects and interactions are significant at the level $\alpha = 0.05$.

(b) Using the results in part (*a*), compute the proportion of the total variability explained by each main effect and interaction. Rank the effects in decreasing order of importance as measured by SS(effect)/SST.

13.27 Consider a 2^4 agricultural experiment with four factors A, B, C, and D. The experiment is to be carried out on two plots of land, each of which is large enough to run only eight factor level combinations. The scientists choose to confound the ABCD interaction with the block effect. Prepare a table, similar to Table 13.28, listing the factor level combinations for each block. (*Hint:* First prepare the table of contrast coefficients similar to Table 13.20.)

13.28 In this problem we outline a derivation of Eq. (13.32) for the sum of squares SSA. The proof is easily modified to derive Eqs. (13.33) and (13.34) for SSB and SSAB.

(a) Show that
$$\hat{\alpha}_1 + \hat{\alpha}_2 = (\bar{y}_{1..} - \bar{y}_{...}) + (\bar{y}_{2..} - \bar{y}_{...}) = 0$$

(b) Use the result in part (*a*) to show that $SSA = 4n\hat{\alpha}_2^2$.

(c) Show that $\hat{\alpha}_2 = L_A/4n$ and therefore $SSA = L_A^2/4n$, as claimed. [*Hint:* These formulas are applications of the equations for the sums of squares displayed just after Eq. (13.23). The sums simplify when one uses the fact that $1 \le i \le 2; 1 \le j \le 2.$]

13.5 CHAPTER SUMMARY

The validity of a scientific investigation depends in a critical way on carefully designed and analyzed experiments. This means identifying and reducing sources of extraneous variation so that observed differences will be mostly due to the factor level combinations in which the researcher is interested. Excessive variation in the production process is a major cause of poor-quality output. Identifying and eliminating the source of excessive variation is an essential first step toward reducing the costs of scrap work, repair, and warranties. In this chapter we presented some of the basic statistical concepts underlying the design and analysis of experiments depending on two or more factors. The basic philosophy is to construct a statistical model for the data and to check its validity by performing an analysis of variance. A comprehensive treatment of this important topic is simply not possible within the confines of an introductory text, so we have contented ourselves with the simplest examples: randomized complete block designs, two-way ANOVA, and 2^k factorial experiments.

To Probe Further. Students interested in further applications and generalizations should consult the following references:

1. G. E. P. Box, W. G. Hunter, and J. S. Hunter, *Statistics for Experimenters*, New York, John Wiley & Sons, 1978.
2. R. L. Mason, R. F. Gunst, and J. L. Hess, *Statistical Design and Analysis of Experiments*, New York, John Wiley & Sons, 1989.
3. D. C. Montgomery, *Design and Analysis of Experiments*, 2d ed., New York, John Wiley & Sons, 1984.

14

Statistical Quality Control

The aim in production should be not just to get statistical control, but to shrink variation. Costs go down as variation is reduced. It is not enough to meet specifications.

W. E. Deming, *Out of the Crisis* (Cambridge, Mass., MIT Press, 1982, p. 334)

14.1 ORIENTATION

Fluctuations in the quality of a manufactured product, such as integrated circuits, steel bars, and cotton fabric, are a result of the variations in the raw materials, machines, and workers used to manufacture them. We measure quality by recording the value of a numerical variable X defined on the population of manufactured objects. Examples of such variables include the thickness of an epitaxial layer deposited on a silicon wafer, the compressive strength of concrete, and the tensile strength of a steel cable. Sometimes the variable is categorical, such as when we classify each item inspected as conforming or nonconforming. In all cases the product is required to meet a standard usually stated in the form of lower and upper control limits on the variable X.

The primary goal of a quality control program is to save money by reducing or eliminating the production of defective products, which, if uncorrected, can only result in declining sales, loss of market share, layoffs, and bankruptcy. The following example is cited in Kazmierski, *Statistical Problem Solving in Quality Engineering*, (New York, McGraw-Hill, 1995):

A small company had $10 million in sales over a given period. During that period, warranty costs were $1,550,000. Scrap and rework were $150,000. The inspection budget was $250,000, and the prevention budget was $50,000. The total cost of poor quality was $2 million (20% of sales). Reducing the cost of poor quality by 50% would increase profit by $1 million.

Among the most important techniques for detecting variation in the quality of a manufactured product is the *control chart*. We study several types of control charts that differ depending on whether the variable being measured is numerical or categorical. The $\bar{x}$ chart monitors the fluctuations about the target value μ, and the R chart monitors the fluctuations in the process variation. Two other charts are the p chart, which monitors the proportion of inspected items rejected as nonconforming, and the c chart, which monitors the number of defects in a nonconforming item.

Organization of Chapter

Section 14.2 $\bar{x}$ and R Control Charts
Section 14.3 p Charts and c Charts
Section 14.4   Chapter Summary

14.2 $\bar{x}$ AND R CONTROL CHARTS

The concept of the control chart rests on the assumption that the manufacturing process is in *statistical control*. Informally, this means that the variability in product quality comes from a statistical model of random variation. The formal mathematical model of a process in statistical control is as follows.

■ **DEFINITION 14.1**

The manufacturing process is said to be *in statistical control* when the measurements $X_1, X_2, \ldots, X_n$ taken on a sequence of successively drawn objects behave as if they come from a random sample of size n taken from a common distribution F. We then say that the pattern of variation is *stable*.   ■

Quality control engineers emphasize the importance of first partitioning the variation in the manufacturing process into two parts, as indicated Eq. (14.1).

$$\begin{pmatrix} \text{Total process} \\ \text{variation} \end{pmatrix} = \begin{pmatrix} \text{Common-cause} \\ \text{variation} \end{pmatrix} + \begin{pmatrix} \text{Special-cause} \\ \text{variation} \end{pmatrix} \qquad (14.1)$$

Deming's red bead experiment (described in Example 1.3) provides an entertaining example of a *common cause of variation,* which refers to the variations due to random fluctuations within the system itself, as opposed to a *special cause of variation,* which refers to variations caused by external, possibly nonrandom events such as computer software bugs, defective raw materials, misaligned gauges, and workers unfamiliar with the equipment. From the engineering point of view the variation due to special causes can be reduced or eliminated, for example, by using only fully trained workers, properly maintained equipment, and high-quality raw materials. Historically, it was Shewhart, Deming, and their disciples who were among the first to point out that improving the quality of a manufactured product depends first on identifying the causes of the observed variation. "Until special causes have been identified and eliminated," wrote Deming, "one dare not predict what the process will produce the next hour." The common-cause variation is what remains after the variation due to special causes has been eliminated.

Charts for a Process in Statistical Control with Known Mean and Variance

Assume that the manufacturing process is in statistical control and that the k samples of size n come from a normal distribution with known mean μ and known variance σ^2. Consequently, the sample mean $\overline{X}$ has a normal distribution with mean μ and standard deviation $\sigma(\overline{X}) = \sigma/\sqrt{n}$. It follows that

$$P(\mu - 3\sigma/\sqrt{n} \le \overline{X} \le \mu + 3\sigma/\sqrt{n}) = 0.9974$$

It is therefore very unlikely for a sample average to fall outside the *lower control limit* (LCL) and *upper control limit* (UCL), defined by

$$\text{LCL}_{\overline{X}} = \mu - \frac{3\sigma}{\sqrt{n}} \quad \text{and} \quad \text{UCL}_{\overline{X}} = \mu + \frac{3\sigma}{\sqrt{n}} \tag{14.2}$$

It is worth pointing out that the sample average will be approximately normal even if the parent distribution from which the sample is drawn is not normal.

Three-Sigma Control Chart for Averages ($\bar{x}$ chart). Take k successive samples of size n and compute their sample means $\bar{x}_1, \ldots, \bar{x}_k$. The $\bar{x}$ chart is obtained by first plotting the horizontal *center line* (CL) at the target value μ, and the horizontal lower and upper control lines at the values $\text{LCL}_{\overline{X}}$ and $\text{UCL}_{\overline{X}}$, defined in Eq. (14.2). We then plot the points $(i, \bar{x}_i)$, $i = 1, \ldots, k$. Points outside the control limits suggest that this is due to a special-cause variation (e.g., malfunctioning machinery, a change in personnel) and not to the system itself. We illustrate the method in Example 14.1.

Control Chart for the Sample Ranges (R chart). One of the main causes of poor quality is excessive variation in the manufacturing process. The R control chart is a graphical method for monitoring the process variation. We plot an R chart as follows. Take k successive samples of size n, and for each sample compute the range defined by

$$R = \max(X_1, \ldots, X_n) - \min(X_1, \ldots, X_n)$$

The ith sample range is denoted by R_i. It can be shown (we omit the proof) that the mean and standard deviation of the sample range R of a sequence of n independent and identically distributed normal random variables is proportional to σ; thus,

$$E(R) = d_2(n)\sigma \qquad (14.3)$$

$$\sigma(R) = d_3(n)\sigma \qquad (14.4)$$

where the constants of proportionality depend only on the sample size n. These constants are listed in Table A.8 in the appendix, where they are denoted d_2, d_3 and their explicit dependence on n is understood. Consequently, when σ is known, the center line (CL) of the R chart is defined to be the horizontal line at the value $\bar{R}$, defined by

$$\bar{R} = d_2\sigma, \quad \sigma \text{ known} \qquad (14.5)$$

The horizontal lower and upper control lines at the values LCL_R and UCL_R are defined by

$$\text{LCL}_R = D_1\sigma \quad \text{and} \quad \text{UCL}_R = D_2\sigma, \quad \sigma \text{ known} \qquad (14.6)$$

where the values of D_1 and D_2 are taken from Table A.10 in the appendix. Note that $D_1 = 0$ for $n \le 6$. The final step is to plot the points $(i, R_i), i = 1, \ldots, k$. We illustrate the method next.

---

example 14.1   At regular time intervals a sample of five parts is drawn from a manufacturing process and the width (measured in millimeters) of a critical gap in each part is recorded. The recorded values are listed in Table 14.1. The ith row lists the

TABLE 14.1   GAP WIDTH DATA, SAMPLE MEANS, AND SAMPLE RANGES FOR EXAMPLE 14.1

Sample	Gap widths					$\bar{x}$	R
1	212	206	184	198	185	197.0	28
2	196	195	199	209	207	201.2	14
3	190	203	201	186	210	198.0	24
4	204	194	208	203	207	203.2	14
5	217	196	206	204	207	206.0	21
6	194	211	219	199	204	205.4	25
7	197	200	192	200	197	197.2	8
8	206	208	184	188	195	196.2	24
9	200	196	209	207	195	201.4	14
10	207	207	202	209	194	203.8	15
11	209	181	194	206	204	198.8	28
12	200	200	192	199	192	196.6	8
13	174	192	212	211	184	194.6	38
14	180	190	194	203	216	196.6	36
15	201	212	184	196	210	200.6	28
16	212	177	195	218	203	201.0	41
17	202	196	200	201	203	200.4	7
18	219	218	206	202	204	209.8	17
19	202	208	196	200	208	202.8	12
20	207	200	175	186	196	192.8	32
21	211	207	205	214	210	209.4	9

sample number, the five measurements of the gap widths, the sample mean, and the sample range. Here, $n = 5$ and $k = 21$. When the process is in control, the distribution of the gap width is normal with mean $\mu = 200$ and $\sigma = 10$. Compute the lower and upper control limits for both the $\bar{x}$ and R charts, the sample means, and the sample ranges. Then plot the three-sigma $\bar{x}$ chart and the R chart.

Solution. We compute the lower and upper control limits by substituting $\mu = 200, \sigma = 10, n = 5$ into Eq. (14.2):

$$\text{LCL}_{\bar{X}} = 200 - \frac{3 \times 10}{\sqrt{5}} = 186.6 \quad \text{and} \quad \text{UCL}_{\bar{X}} = 200 + \frac{3 \times 10}{\sqrt{5}} = 213.4$$

The last two columns of Table 14.1 record the ith sample mean $\bar{x}_i$ and the ith sample range R_i. Let us now compute the sample average and range for the first sample of five measurements (the final results are rounded to one decimal place). We then compute $\bar{R}$ and the lower and upper control limits for the R chart.

$$\bar{x}_1 = \frac{212 + 206 + 184 + 198 + 185}{5} = 197.0$$

$$R_1 = 212 - 184 = 28$$

$$\bar{R} = d_2\sigma = 2.326 \times 10 = 23.3$$

$$D_1 = 0 \quad \text{and} \quad D_2 = 4.92$$

$$\text{LCL}_R = 0 \quad \text{and} \quad \text{UCL}_R = D_2\sigma = 4.92 \times 10 = 49.2$$

Figure 14.1 displays the $\bar{x}$ and R charts. The center line in an $\bar{x}$ chart is always denoted by $\bar{\bar{x}}$ even when μ is known. Looking at these charts, we see that all the plotted points lie within the control limits, so we conclude that the process is in control. However, a process that is in statistical control is not necessarily satisfactory from the standpoint of quality control. Suppose, for example, that the product specifications are that the gap width be 200 ± 10. Call 190 and 210 *tolerance limits*. A total of 27 parts are outside the tolerance limits in Table 14.1. That is, 25.7 percent of the parts are nonconforming, even though none of the sample averages fell outside the control limits! In this case the control chart is telling management that the excessive variation is in the system and will not be improved by firing workers, the "get tough" policy. ∎

Charts for a Process in Statistical Control with Unknown Mean and Variance

We now show how to construct control charts for the process mean and process variation when the parameters μ and σ^2 are unknown. In this case we replace the previous control limits with their estimates determined from the recorded data. In detail, we proceed as follows.

Assume that the manufacturing process is in statistical control and that the k samples of size n come from a normal distribution with unknown mean

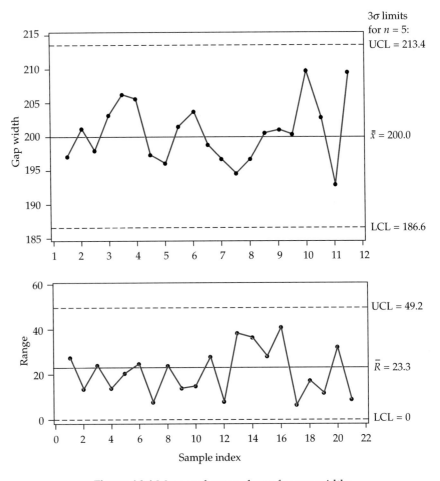

Figure 14.1 Mean and range charts for gap widths

μ and unknown variance σ^2. We estimate μ by computing the grand mean defined by

$$\bar{\bar{x}} = \frac{\sum_{1 \le j \le k} \bar{x}_j}{k} \tag{14.7}$$

Take k successive samples of size n and denote the ith sample range by R_i.

The estimated center line for the R chart, which is denoted by $\bar{R}$, is the average of the k sample ranges and is given by

$$\bar{R} = \frac{\sum_{1 \le i \le k} R_i}{k} \tag{14.8}$$

The average $\bar{R}$ of the sample ranges together with Eqs. (14.3) and (14.4) can be used to derive the following unbiased estimates of σ and the standard deviation of the sample range, denoted $\sigma(R)$.

$$\hat{\sigma} = \frac{\overline{R}}{d_2} \quad \text{and} \quad \hat{\sigma}(R) = \frac{d_3\overline{R}}{d_2} \tag{14.9}$$

It follows that $3\overline{R}/d_2\sqrt{n}$ is an unbiased estimator of $3\sigma/\sqrt{n}$. Consequently, the estimated lower and upper control limits are given by

$$\text{LCL}_{\overline{X}} = \bar{\bar{x}} - A_2\overline{R} \quad \text{and} \quad \text{UCL}_{\overline{X}} = \bar{\bar{x}} + A_2\overline{R}, \quad A_2 = \frac{3}{d_2\sqrt{n}} \tag{14.10}$$

The values A_2 are given in Table A.9.

Similarly, the lower and upper control limits for the R chart are given by

$$\text{LCL}_R = \overline{R} - 3\hat{\sigma}(R) = \overline{R}\left(1 - \frac{3d_3}{d_2}\right) = D_3\overline{R} \tag{14.11}$$

$$\text{UCL}_R = \overline{R} + 3\hat{\sigma}(R) = \overline{R}\left(1 + \frac{3d_3}{d_2}\right) = D_4\overline{R} \tag{14.12}$$

where the constants D_3 and D_4 are given in Table A.9. For $n \leq 6$ we set $D_3 = 0$ because in these cases $(1 - 3d_3/d_2) < 0$.

Three-Sigma Control Chart for Averages ($\overline{X}$ Chart). Take k successive samples of size n and compute their sample means $\bar{x}_1, \ldots, \bar{x}_k$. The $\bar{x}$ chart is obtained by first plotting the horizontal center line (CL) at the value $\bar{\bar{x}}$, and the horizontal lower and upper control lines at the values $\text{LCL}_{\overline{X}}$ and $\text{UCL}_{\overline{X}}$ defined in Eq. (14.10). We then plot the points $(i, \bar{x}_i), i = 1, \ldots, k$. We illustrate the method in Example 14.2.

Control Chart for the Ranges (R Chart). Take k successive samples of size n and compute the sample range for each sample. The R chart is obtained by plotting the horizontal center line (CL) at the value $\overline{R}$, and the horizontal lower and upper control lines at the values LCL_R and UCL_R defined by Eqs. (14.11) and (14.12). We illustrate the method in Example 14.2.

---

example 14.2   Every hour a sample of five fittings to be used in an aircraft hydraulic system is drawn from the production line and the pitch diameter of the threads is measured. The pitch diameter is specified as 0.4037 ± 0.0013. The data are coded by subtracting 0.4000 from each observation and then expressed in units of 0.0001 inch. Thus, the observation 0.4036 is recorded as 36. The measurements are displayed in Table 14.2. The ith row lists the sample number, the five measurements of the pitch diameter, the sample mean, and the sample range. Here, $n = 5$ and $k = 20$. We now drop the assumption that the parameters μ and σ^2 are known. Compute the lower and upper control limits for both the $\bar{x}$ and R charts. Then plot the three-sigma $\bar{x}$ chart and the R chart.

TABLE 14.2 SAMPLE NUMBER, PITCH DIAMETER OF THREADS ON AIRCRAFT FITTINGS (FIVE PER HOUR), SAMPLE MEAN, AND SAMPLE RANGE

Sample	Pitch diameters					Average	Range
1	36	35	34	33	32	34.0	4
2	31	31	34	32	30	31.6	4
3	30	30	32	30	32	30.8	2
4	32	33	33	32	35	33.0	3
5	32	34	37	37	35	35.0	5
6	32	32	31	33	33	32.2	2
7	33	33	36	32	31	33.0	5
8	23	33	36	35	36	32.6	13
9	43	36	35	24	31	33.8	19
10	36	35	36	41	41	37.8	6
11	34	38	35	34	38	35.8	4
12	36	38	39	39	40	38.4	4
13	36	40	35	26	33	34.0	14
14	36	35	37	34	33	35.0	4
15	30	37	33	34	35	33.8	7
16	28	31	33	33	33	31.6	5
17	33	30	34	33	35	33.0	5
18	27	28	29	27	30	28.2	3
19	35	36	29	27	32	31.8	9
20	33	35	35	39	36	35.6	6

Source: E. L. Grant and R. S. Leavenworth, *Statistical Quality Control,* 7th ed., New York, McGraw-Hill, 1996. Used with permission.

Solution. We compute the lower and upper control limits for the $\bar{x}$ chart by substituting $\bar{\bar{x}} = 33.55, \bar{R} = 6.2, A_2 = 0.58$ into Eq. (14.10):

$$\text{LCL}_{\bar{x}} = 33.55 - 0.58 \times 6.2 = 29.95$$

$$\text{UCL}_{\bar{x}} = 33.55 + 0.58 \times 6.2 = 37.15$$

These results differ from those obtained by the computer only in the second decimal place; the difference is due to roundoff errors.

The lower and upper control limits for the R chart are obtained by substituting $\bar{R} = 6.2, D_3 = 0, D_4 = 2.11$ into Eqs. (14.11) and (14.12). This yields the result (rounded to one decimal place)

$$\text{LCL}_R = D_3 \bar{R} = 0$$

$$\text{UCL}_R = D_4 \bar{R} = 2.11 \times 6.2 = 13.1$$

Looking at the $\bar{x}$ chart in Fig. 14.2, we see that three samples (10, 12, and 18) are outside the control limits. There are also two samples (9 and 13) outside the control limits of the R chart. ∎

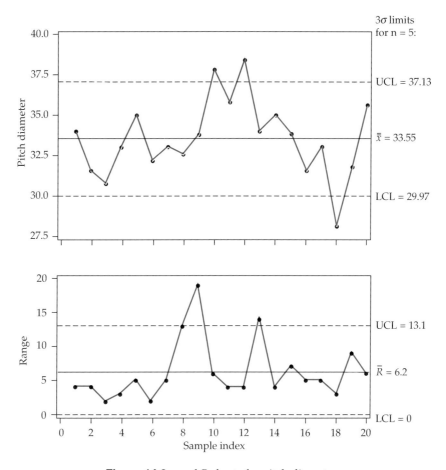

Figure 14.2 x and R charts for pitch diameters

14.2.1   Detecting a Shift in the Process Mean

A control chart is not infallible. It can fail to detect quickly an out-of-control process, or it can incorrectly send an out-of-control signal. The similarity with hypothesis testing should be noted. The null hypothesis is that the process is in a state of statistical control. Failure to detect an out-of-control process is a type I error, and an incorrect out-of-control signal is a type II error.

We begin our analysis of the performance of the $\bar{x}$ chart by computing the expected number of samples taken before an incorrect out-of-control signal is indicated. Under the null hypothesis (that the process is in statistical control), the sample means $\bar{X}_1, \ldots, \bar{X}_k$ are a random sample taken from a normal distribution with mean μ_0 and variance $\sigma/\sqrt{n}$. Let T denote the first time a sample mean falls outside the control limits. This is another example of a waiting time problem of the sort discussed in Chap. 3. Consequently, the random variable T has the geometric distribution

$$P(T = i) = q^{i-1}p, \quad i = 1, 2, \ldots$$

where p (the probability that a single sample falls outside the control limits) is given by

$$p = P\left(|\overline{X} - \mu_0| > \frac{3\sigma}{\sqrt{n}}\right) = 0.0026$$

and $q = 1 - p$. Since T has a geometric distribution, it follows that

$$P(T \leq i) = 1 - (1 - p)^i \tag{14.13}$$

$$E(T) = \frac{1}{p} \tag{14.14}$$

Equation (14.13) gives the probability that an incorrect out-of-control signal occurs at the ith sample or earlier; Eq. (14.14) gives the expected number of samples required. Since $p = 0.0026$, we see that $E(T) = 384.62 \approx 385$. In other words, the average number of samples taken before an incorrect out-of-control signal occurs is approximately 385.

---

example **14.3**

Suppose a process that is assumed to be in control with process mean μ_0 and variance σ is thrown out of control because the process mean has shifted to $\mu_0 + \delta\sigma$, where δ is a constant (positive or negative). What is the probability that the three-sigma control chart will detect this shift no later than the ith sample? What is the expected number of samples required in order for the shift to be detected?

Solution. Under the hypothesis that the process mean has shifted to $\mu_0 + \delta\sigma$, the sample means $\overline{X}_1, \ldots, \overline{X}_k$ are a random sample taken from a normal distribution with mean $\mu_0 + \delta\sigma$ and variance $\sigma/\sqrt{n}$. Let T denote the first time a sample mean falls outside the control limits. Consequently, the distribution of T is again geometric, where p (the probability that a single sample falls outside the control limits) is given by

$$p = P(|\overline{X} - \mu_0| > \frac{3\sigma}{\sqrt{n}} \Big| \mu = \mu_0 + \delta\sigma) \tag{14.15}$$

$$= \Phi(-3 - \delta\sqrt{n}) + [1 - \Phi(3 - \delta\sqrt{n})]$$

The derivation of Eq. (14.15) is left to the reader as Prob. 14.3.

To illustrate the formula, suppose the process mean of Example 14.1 is shifted to 207; the process standard deviation, however, remains fixed at $\sigma = 10$. What is the probability that the shift is detected no later than the third sample? What is the expected number of steps until the shift is detected?

The shift will be detected when a sample mean lies outside the the control limits. Denoting the number of steps required to detect this shift by T, we see that it too has a geometric distribution with probability p given by Eq. (14.15). A shift in the process mean from 200 to 207 means that $200 + \delta\sigma = 207$. Consequently, $\delta = 0.7$ since $\sigma = 10$. Inserting the values $\delta = 0.7, n = 5$ into Eq. (14.15), we obtain the value $p = 0.0764$. Therefore,

$$P(T \leq 3) = 1 - (1 - 0.0764)^3 = 0.2121$$

$$E(T) = \frac{1}{0.0764} = 13.09 \approx 13$$

In other words, it will take 13 samples, on average, for the control chart to detect this shift in the process mean. Even after three samples have been taken, the probability is still approximately 0.79 that the shift in the process mean remains undetected. ∎

14.3 *p* CHARTS AND *c* CHARTS

P *Chart for Proportion of Defective Items*

Another important measure of quality is the proportion of defective (nonconforming) items produced. At regular time intervals $i = 1, \ldots, k$ the quality control inspector takes a sample of size n and records the number of defective items for the ith sample, denoted X_i, and also the proportion of defective items $\hat{p}_i = X_i/n$. The sample proportion is the total number of defective items divided by the total number of items sampled; it is denoted $\bar{p}$, where

$$\bar{p} = \frac{\sum_{1 \leq i \leq k} X_i}{kn}$$

When the process is in statistical control, the distribution of X_i is binomial with parameters n, p, and, when n is large ($n > 30$, say), the ith sample proportion has an approximate normal distribution with mean p and variance $p(1 - p)/n$. Consequently,

$$P\left(p - 3\sqrt{\frac{p(1-p)}{n}} \leq \hat{p}_i \leq p + 3\sqrt{\frac{p(1-p)}{n}}\right) \approx 0.9974$$

When the true proportion of defective items is unknown, we replace it with its estimate $\bar{p}$; it follows that

$$P\left(\bar{p} - 3\sqrt{\frac{\bar{p}(1-\bar{p})}{n}} \leq \hat{p}_i \leq \bar{p} + 3\sqrt{\frac{\bar{p}(1-\bar{p})}{n}}\right) \approx 0.9974$$

Reasoning as in the case of the $\bar{x}$ chart, we see that it is very unlikely for a sample proportion to fall outside the bounds determined by the lower control limit and upper control limit, defined by

$$\text{LCL}_{\bar{p}} = \bar{p} - 3\sqrt{\frac{\bar{p}(1-\bar{p})}{n}} \quad \text{and} \quad \text{UCL}_{\bar{p}} = \bar{p} + 3\sqrt{\frac{\bar{p}(1-\bar{p})}{n}} \quad (14.16)$$

We call $\text{LCL}_{\bar{p}}$ and $\text{UCL}_{\bar{p}}$ the three-sigma control limits. The p chart monitors the proportion of defective items, with the center line drawn at $\bar{p}$ and the lower and upper control limits $\text{LCL}_{\bar{p}}$ and $\text{UCL}_{\bar{p}}$ defined by Eq. (14.16). (*Note:* It sometimes happens that the lower control limit is negative; when this occurs, we set it equal to zero.) The points $(i, \hat{p}_i), i = 1, \ldots, k$ are then plotted. A lack of control is indicated when one or more sample proportions lie outside the control limits.

---

example 14.4   This example is a simulation of Deming's red bead experiment (Example 1.3) described in Chap. 1. We repeat the details. Suppose 4000 beads are placed in a bowl. Of these, 3200 are white and 800 are red. The red beads represent the defective items. Each worker stirs the beads and then, while blindfolded, inserts a special tool into the mixture and draws out exactly 50 beads. The aim of this process is to produce white beads; red beads, as previously noted, are defective and will not be accepted by the customers. When the worker is done, the sampled beads are replaced, and the beads are thoroughly stirred for the next worker, who repeats the process. The results for six workers over a five-day period are shown in Table 14.3. Compute $\bar{p}$, $\text{LCL}_{\bar{p}}$, and $\text{UCL}_{\bar{p}}$; graph the p chart; and determine if the process is in control.

TABLE 14.3 NUMBER OF RED BEADS DRAWN
BY SIX WORKERS OVER FIVE DAYS

	Worker					
Day	1	2	3	4	5	6
1	3	9	10	11	6	10
2	15	9	11	9	10	10
3	15	13	6	10	17	10
4	6	12	9	15	8	9
5	6	14	12	9	12	11

Solution. The total number of items inspected is $kn = 30 \times 50 = 1500$, and the total number of defective items is 307. So $\bar{p} = 307/1500 = 0.204$ (rounded to three decimal places). Therefore, the lower and upper control limits are

$$\text{LCL}_{\bar{p}} = 0.204 - 0.171 = 0.033$$
$$\text{UCL}_{\bar{p}} = 0.204 + 0.171 = 0.375$$

The p chart is shown in Fig. 14.3. No points lie outside the control limits, so we conclude that the process is in control. Notice that worker 1 "produced"

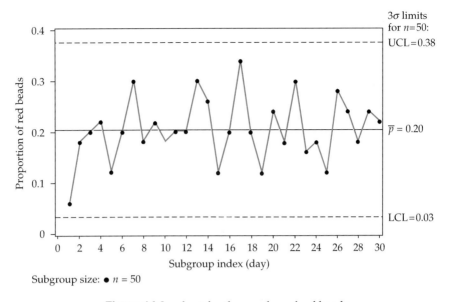

3σ limits
for *n*=50:
UCL=0.38

$\bar{p} = 0.20$

LCL=0.03

Subgroup size: ● *n* = 50

Figure 14.3 *p* chart for the number of red beads

the smallest number of defectives on day 1 but produced the largest number of defectives on day 2. What the control chart tells us is that the excessive proportion of defectives is due to the system and not the workers.

c Chart for the Number of Defects per Unit

The *c* chart is used when it is the number of defects per item that is of interest rather than the proportion of defective items. Consider, for example, an aircraft fuselage assembled with hundreds of rivets. One defective rivet is highly unlikely to compromise aircraft safety, but 10 or more might. The variable of interest here is *C*, the number of defective rivets per fuselage. A *c* chart is a control chart for the number of defects per item. The control limits are based on the assumption that the number of defects per item is a Poisson random variable with parameter λ. This assumption is a reasonable one for an airplane fuselage consisting of hundreds of rivets, each of which has a small probability of being defective.

Control Limits for the *c* **Control Charts.** Let $C_i, i = 1, \ldots, k$, denote the number of defects found on the *i*th item. We assume that C_i has a Poisson distribution with parameter λ. Therefore, it has mean value λ and standard deviation $\sqrt{\lambda}$. The theoretical three-sigma control limits are

$$\text{LCL} = \lambda - 3\sqrt{\lambda} \quad \text{and} \quad \text{UCL} = \lambda + 3\sqrt{\lambda}$$

When λ is unknown, we estimate it by dividing the total number of defects by the total number of items sampled; we denote this quantity by $\bar{c}$. Thus,

$$\bar{c} = \frac{\sum_{1 \le i \le k} C_i}{k}$$

The estimated control limits are given by

$$\text{LCL}_{\bar{c}} = \bar{c} - 3\sqrt{\bar{c}} \quad \text{and} \quad \text{UCL}_{\bar{c}} = \bar{c} + 3\sqrt{\bar{c}} \tag{14.17}$$

example 14.5   The following data are a record of the number of defective rivets (missing, misaligned, etc.) on 14 aircraft fuselages.

$$2, 9, 10, 11, 6, 10, 16, 7, 5, 5, 6, 8, 6, 5$$

Compute $\bar{c}$ and the control limits, and plot the c chart.

Solution. The total number of defects is 106, so $\bar{c} = 106/14 = 7.57$. Therefore, the lower and upper control limits are

$$\text{LCL}_{\bar{c}} = 7.57 - 8.25 = -0.68 \quad \text{and} \quad \text{UCL}_{\bar{c}} = 7.57 + 8.25 = 15.82$$

Since $\bar{c} - 3\sqrt{\bar{c}} < 0$, we set $\text{LCL}_{\bar{c}} = 0$. The c chart is displayed in Fig. 14.4. There is one point (airplane 7, with $C_7 = 16$) outside the control limits, so the system is not in control.

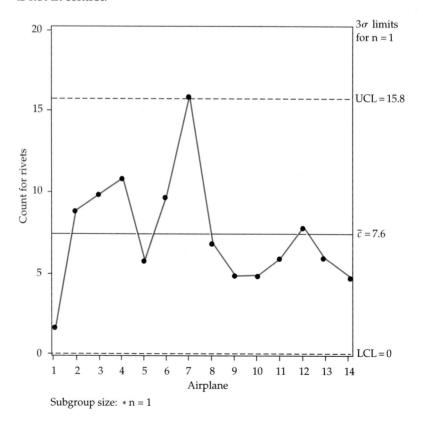

Figure 14.4 c chart for number of defective rivets

PROBLEMS

14.1 The thickness of an epitaxial layer on a silicon wafer is specified to be 14.5 ± 0.5 μm. Assume that the manufacturing process is in statistical control, where the distribution of the thickness is normal with mean $\mu = 14.5$ and standard deviation $\sigma = 0.4$. To monitor the production process, the mean thickness of four wafers is determined at regular time intervals.

(a) Compute the lower and upper control limits $LCL_{\bar{x}}$ and $UCL_{\bar{x}}$ for a three-sigma control chart for the process mean.

(b) Compute the lower and upper control limits LCL_R and UCL_R for a three-sigma control chart for the process variation.

(c) Determine the probability that a sample mean (based on a sample of size $n = 4$) falls outside the tolerance limits 14.5 ± 0.5.

14.2 Suppose the process mean in Prob. 14.1 shifts to $\mu = 15.0$.

(a) What is the probability that this shift will be detected on the first sample after the shift has occurred?

(b) What is the probability that this shift will be detected by the third sample?

(c) What is the expected number of samples required before this shift is detected?

14.3 Derive the formula for p (the probability that a single sample falls outside the control limits) stated in Eq. (14.15).

14.4 The diameters (measured in millimeters) of ball bearings are monitored by $\bar{x}$ and R charts. After 30 samples of size $n = 6$ were taken, the following values were recorded: $\sum_i \bar{x}_i = 150$ mm and $\sum_i \bar{R}_i = 12.0$.

(a) Assuming the process is in statistical control, calculate the $\bar{x}$ and R control limits.

(b) Assuming that the ball bearing diameters are normally distributed, estimate σ.

14.5 The dimension of a rheostat knob was specified to be 0.140 ± 0.003 in.; otherwise the knob would not fit into another part. Five knobs from each each hour's production were inspected, and the dimensions were recorded (in units of 0.001 in.) in the following table. The sample means and sample ranges are recorded in the last two columns.

(a) Draw the histogram of the raw data using the class marks

$$132, 134, 136, 138, 140, 142, 144, 146, 148$$

How many rheostats were nonconforming?

(b) Is the process in statistical control? Draw the $\bar{x}$ and R charts to find out.

Sample	Dimensions (0.0001 in.)					$\bar{x}$	R
1	140	143	137	134	135	137.8	9
2	138	143	143	145	146	143.0	8
3	139	133	147	148	139	141.2	15
4	143	141	137	138	140	139.8	6
5	142	142	145	135	136	140.0	10
6	136	144	143	136	137	139.2	8
7	142	147	137	142	138	141.2	10
8	143	137	145	137	138	140.0	8
9	141	142	147	140	140	142.0	7
10	142	137	145	140	132	139.2	13
11	137	147	142	137	135	139.6	12
12	137	146	142	142	140	141.4	9
13	142	142	139	141	142	141.2	3
14	137	145	144	137	140	140.6	8
15	144	142	143	135	144	141.6	9
16	140	132	144	145	141	140.4	13
17	137	137	142	143	141	140.0	6
18	137	142	142	145	143	141.8	8
19	142	142	143	140	135	140.4	8
20	136	142	140	139	137	138.8	6
21	142	144	140	138	143	141.4	6
22	139	146	143	140	139	141.4	7
23	140	145	142	139	137	140.6	8
24	134	147	143	141	142	141.4	13
25	138	145	141	137	141	140.4	8
26	140	145	143	144	138	142.0	7
27	145	145	137	138	140	141.0	8

Source: E. L. Grant and R. S. Leavenworth, *Statistical Quality Control,* 7th ed., New York, McGraw-Hill, 1996. Used with permission.

14.6 Items are shipped in cartons of several hundred. From each carton 100 items are selected at random and inspected for defects. For a shipment of 12 cartons the following data on the number of defective items found in the sample for each carton were recorded:

$$0, 1, 1, 2, 0, 2, 4, 1, 2, 1, 7, 5$$

Calculate the three-sigma control limits for the p chart. Is the process in statistical control? If not, list the values that are out of control.

14.7 The proportion of defective items produced by a process in control is known to be $p = 0.02$. The process is monitored by taking samples of size 50 and determining the proportion of defective items in each sample.

(a) Compute the lower and upper control limits for the proportion defective.

(b) Suppose the proportion of defective items produced increases to 0.04. Let T denote the first sample for which the proportion defective lies above the upper control limit obtained in part (a). Describe the distribution of T. Name it and identify its parameters. Compute $E(T)$.

14.8 A c chart is used to monitor the number of typos per page in a printed document. Based on past experience, the number of typos is assumed to have a Poisson distribution with $\bar{c} = 1.5$.

(a) Assuming the process is in statistical control, find the three-sigma control limits.

(b) Use Table A.2 (Poisson distribution) to determine the probability that a point on the control chart will fall outside the control limits.

(c) Assuming the process is in statistical control, determine the expected number of pages one has to sample before an incorrect out-of-control signal is given by the c chart.

14.9 The number of defects in a bolt of cloth is to be monitored by means of a c chart. Defects include flaws in the weave and irregularities in the dye. Before the actual implementation of the control chart a trial run using simulated data was carried out. The first six observations in the following data set come from a Poisson distribution with parameter $\lambda_1 = 6$, the next eight observations come from a Poisson distribution with parameter $\lambda_2 = 9$, and the final six observations come from a Poisson distribution with parameter $\lambda_3 = 6.5$. Consequently, the simulated data do not come from a process in statistical control. Calculate the three-sigma control limits using Eq. (14.2) and plot the c control chart. Does the control chart detect a lack of control?

$$1, 5, 6, 7, 3, 6, 11, 9, 8, 10$$
$$12, 10, 9, 8, 4, 6, 5, 6, 4, 7$$

14.10 An alternative method for constructing lower and upper control limits $\text{LCL}_{\bar{c}}$ and $\text{UCL}_{\bar{c}}$ for a c chart is to choose them so that the inequalities

$$P(C < \text{LCL}_{\bar{c}}) < \frac{\alpha}{2} \quad \text{and} \quad P(C > \text{UCL}_{\bar{c}}) < \frac{\alpha}{2}$$

are satisfied. It is assumed that C has a Poisson distribution. When $\text{LCL}_{\bar{c}} = 0$, then an out-of-control signal occurs only when $C > \text{UCL}_{\bar{c}}$.

(a) Find the control limits when $\lambda = 4$ and $\alpha = 0.01$.

(b) Find the control limits when $\lambda = 6$ and $\alpha = 0.01$.

(c) Find the control limits when $\lambda = 6$ and $\alpha = 0.05$.

14.4 CHAPTER SUMMARY

This chapter gave a brief introduction to statistical quality control. No improvements in product quality are possible until the manufacturing process is in a state of statistical control, and this means that one must first identify and eliminate variation due to special causes. Control charts are used to detect a lack of control. There is a similarity between hypothesis testing and statistical control. The null hypothesis is that the process is in a state of statistical control. Failure to detect an out-of-control process is a type I error, and an incorrect out-of-control signal is a type II error.

To Probe Further. The common practice in manufacturing, according to Grant and Leavenworth (see the reference at the end of this paragraph), is to set the control chart limits far enough apart so that type I errors are rare on the grounds that "it seldom pays to hunt for trouble without a strong basis for confidence that trouble is really there." On the other hand, the Shewhart control chart, because it looks at only one sample at a time, may not detect quickly enough a lack of control caused by a shift in the process mean. There are several methods for reducing the type II error probability (failure to detect an out-of-control process). The *cumulative sum control chart* is a useful method for coping with this type of problem. These and other refinements and extensions are discussed in E. L. Grant and R. S. Leavenworth, *Statistical Quality Control*, 7th ed., (New York, McGraw-Hill, 1996), which is the classic text on this subject. Part 4 contains an extensive account of the economic aspects, historical origins, current trends, and future directions of the quality assurance movement in the United States and Japan.

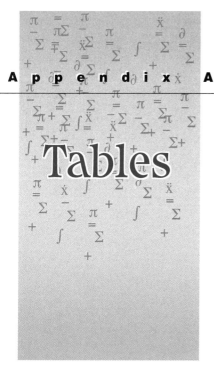

Tables

TABLE A.1   BINOMIAL DISTRIBUTION

$$B(x;n,p) = \sum_{0 \le y \le x} b(y;n,p)$$

The values of $B(x;n,p)$ for $0.5 < p < 1.0$ are obtained by using the formula

$$B(x;n,1-p) = 1 - B(n-1-x;n,p)$$

						p					
n	x	0.05	0.10	0.15	0.20	0.25	0.30	0.35	0.40	0.45	0.50
5	0	0.774	0.590	0.444	0.328	0.237	0.168	0.116	0.078	0.050	0.031
	1	0.977	0.919	0.835	0.737	0.633	0.528	0.428	0.337	0.256	0.188
	2	0.999	0.991	0.973	0.942	0.896	0.837	0.765	0.683	0.593	0.500
	3	1.000	1.000	0.998	0.993	0.984	0.969	0.946	0.913	0.869	0.813
	4	1.000	1.000	1.000	1.000	0.999	0.998	0.995	0.990	0.982	0.969
	5	1.000	1.000	1.000	1.000	1.000	1.000	1.000	1.000	1.000	1.000
10	0	0.599	0.349	0.197	0.107	0.056	0.028	0.013	0.006	0.003	0.001
	1	0.914	0.736	0.544	0.376	0.244	0.149	0.086	0.046	0.023	0.011
	2	0.988	0.930	0.820	0.678	0.526	0.383	0.262	0.167	0.100	0.055
	3	0.999	0.987	0.950	0.879	0.776	0.650	0.514	0.382	0.266	0.172
	4	1.000	0.998	0.990	0.967	0.922	0.850	0.751	0.633	0.504	0.377
	5	1.000	1.000	0.999	0.994	0.980	0.953	0.905	0.834	0.738	0.623
	6	1.000	1.000	1.000	0.999	0.996	0.989	0.974	0.945	0.898	0.828
	7	1.000	1.000	1.000	1.000	1.000	0.998	0.995	0.988	0.973	0.945
	8	1.000	1.000	1.000	1.000	1.000	1.000	0.999	0.998	0.995	0.989
	9	1.000	1.000	1.000	1.000	1.000	1.000	1.000	1.000	1.000	0.999
	10	1.000	1.000	1.000	1.000	1.000	1.000	1.000	1.000	1.000	1.000
15	0	0.463	0.206	0.087	0.035	0.013	0.005	0.002	0.000	0.000	0.000
	1	0.829	0.549	0.319	0.167	0.080	0.035	0.014	0.005	0.002	0.000
	2	0.964	0.816	0.604	0.398	0.236	0.127	0.062	0.027	0.011	0.004
	3	0.995	0.944	0.823	0.648	0.461	0.297	0.173	0.091	0.042	0.018
	4	0.999	0.987	0.938	0.836	0.686	0.515	0.352	0.217	0.120	0.059
	5	1.000	0.998	0.983	0.939	0.852	0.722	0.564	0.403	0.261	0.151
	6	1.000	1.000	0.996	0.982	0.943	0.869	0.755	0.610	0.452	0.304
	7	1.000	1.000	0.999	0.996	0.983	0.950	0.887	0.787	0.654	0.500
	8	1.000	1.000	1.000	0.999	0.996	0.985	0.958	0.905	0.818	0.696
	9	1.000	1.000	1.000	1.000	0.999	0.996	0.988	0.966	0.923	0.849
	10	1.000	1.000	1.000	1.000	1.000	0.999	0.997	0.991	0.975	0.941
	11	1.000	1.000	1.000	1.000	1.000	1.000	1.000	0.998	0.994	0.982
	12	1.000	1.000	1.000	1.000	1.000	1.000	1.000	1.000	0.999	0.996
	13	1.000	1.000	1.000	1.000	1.000	1.000	1.000	1.000	1.000	1.000
	14	1.000	1.000	1.000	1.000	1.000	1.000	1.000	1.000	1.000	1.000
	15	1.000	1.000	1.000	1.000	1.000	1.000	1.000	1.000	1.000	1.000

						p					
n	x	0.05	0.10	0.15	0.20	0.25	0.30	0.35	0.40	0.45	0.50
20	0	0.358	0.122	0.039	0.012	0.003	0.001	0.000	0.000	0.000	0.000
	1	0.736	0.392	0.176	0.069	0.024	0.008	0.002	0.001	0.000	0.000
	2	0.925	0.677	0.405	0.206	0.091	0.035	0.012	0.004	0.001	0.000
	3	0.984	0.867	0.648	0.411	0.225	0.107	0.044	0.016	0.005	0.001
	4	0.997	0.957	0.830	0.630	0.415	0.238	0.118	0.051	0.019	0.006
	5	1.000	0.989	0.933	0.804	0.617	0.416	0.245	0.126	0.055	0.021
	6	1.000	0.998	0.978	0.913	0.786	0.608	0.417	0.250	0.130	0.058
	7	1.000	1.000	0.994	0.968	0.898	0.772	0.601	0.416	0.252	0.132
	8	1.000	1.000	0.999	0.990	0.959	0.887	0.762	0.596	0.414	0.252
	9	1.000	1.000	1.000	0.997	0.986	0.952	0.878	0.755	0.591	0.412
	10	1.000	1.000	1.000	0.999	0.996	0.983	0.947	0.872	0.751	0.588
	11	1.000	1.000	1.000	1.000	0.999	0.995	0.980	0.943	0.869	0.748
	12	1.000	1.000	1.000	1.000	1.000	0.999	0.994	0.979	0.942	0.868
	13	1.000	1.000	1.000	1.000	1.000	1.000	0.998	0.994	0.979	0.942
	14	1.000	1.000	1.000	1.000	1.000	1.000	1.000	0.998	0.994	0.979
	15	1.000	1.000	1.000	1.000	1.000	1.000	1.000	1.000	0.998	0.994
	16	1.000	1.000	1.000	1.000	1.000	1.000	1.000	1.000	1.000	0.999
	17	1.000	1.000	1.000	1.000	1.000	1.000	1.000	1.000	1.000	1.000
	18	1.000	1.000	1.000	1.000	1.000	1.000	1.000	1.000	1.000	1.000
	19	1.000	1.000	1.000	1.000	1.000	1.000	1.000	1.000	1.000	1.000
	20	1.000	1.000	1.000	1.000	1.000	1.000	1.000	1.000	1.000	1.000
25	0	0.277	0.072	0.017	0.004	0.001	0.000	0.000	0.000	0.000	0.000
	1	0.642	0.271	0.093	0.027	0.007	0.002	0.000	0.000	0.000	0.000
	2	0.873	0.537	0.254	0.098	0.032	0.009	0.002	0.000	0.000	0.000
	3	0.966	0.764	0.471	0.234	0.096	0.033	0.010	0.002	0.000	0.000
	4	0.993	0.902	0.682	0.421	0.214	0.090	0.032	0.009	0.002	0.000
	5	0.999	0.967	0.838	0.617	0.378	0.193	0.083	0.029	0.009	0.002
	6	1.000	0.991	0.930	0.780	0.561	0.341	0.173	0.074	0.026	0.007
	7	1.000	0.998	0.975	0.891	0.727	0.512	0.306	0.154	0.064	0.022
	8	1.000	1.000	0.992	0.953	0.851	0.677	0.467	0.274	0.134	0.054
	9	1.000	1.000	0.998	0.983	0.929	0.811	0.630	0.425	0.242	0.115
	10	1.000	1.000	1.000	0.994	0.970	0.902	0.771	0.586	0.384	0.212
	11	1.000	1.000	1.000	0.998	0.989	0.956	0.875	0.732	0.543	0.345
	12	1.000	1.000	1.000	1.000	0.997	0.983	0.940	0.846	0.694	0.500
	13	1.000	1.000	1.000	1.000	0.999	0.994	0.975	0.922	0.817	0.655
	14	1.000	1.000	1.000	1.000	1.000	0.998	0.991	0.966	0.904	0.788
	15	1.000	1.000	1.000	1.000	1.000	1.000	0.997	0.987	0.956	0.885
	16	1.000	1.000	1.000	1.000	1.000	1.000	0.999	0.996	0.983	0.946
	17	1.000	1.000	1.000	1.000	1.000	1.000	1.000	0.999	0.994	0.978
	18	1.000	1.000	1.000	1.000	1.000	1.000	1.000	1.000	0.998	0.993
	19	1.000	1.000	1.000	1.000	1.000	1.000	1.000	1.000	1.000	0.998
	20	1.000	1.000	1.000	1.000	1.000	1.000	1.000	1.000	1.000	1.000
	21	1.000	1.000	1.000	1.000	1.000	1.000	1.000	1.000	1.000	1.000
	22	1.000	1.000	1.000	1.000	1.000	1.000	1.000	1.000	1.000	1.000
	23	1.000	1.000	1.000	1.000	1.000	1.000	1.000	1.000	1.000	1.000
	24	1.000	1.000	1.000	1.000	1.000	1.000	1.000	1.000	1.000	1.000
	25	1.000	1.000	1.000	1.000	1.000	1.000	1.000	1.000	1.000	1.000

Source: Produced using SAS.

TABLE A.2 CUMULATIVE POISSON DISTRIBUTION

$$P(x;\lambda) = \sum_{0 \le y \le x} p(y;\lambda)$$

	λ								
x	0.1	0.2	0.3	0.4	0.5	0.6	0.7	0.8	0.9
0	0.905	0.819	0.741	0.670	0.607	0.549	0.497	0.449	0.407
1	0.995	0.982	0.963	0.938	0.910	0.878	0.844	0.809	0.772
2	1.000	0.999	0.996	0.992	0.986	0.977	0.966	0.953	0.937
3	1.000	1.000	1.000	0.999	0.998	0.997	0.994	0.991	0.987
4	1.000	1.000	1.000	1.000	1.000	1.000	0.999	0.999	0.998
5	1.000	1.000	1.000	1.000	1.000	1.000	1.000	1.000	1.000
6	1.000	1.000	1.000	1.000	1.000	1.000	1.000	1.000	1.000

	λ								
x	1	1.5	2	2.5	3	3.5	4	4.5	5
0	0.368	0.223	0.135	0.082	0.050	0.030	0.018	0.011	0.007
1	0.736	0.558	0.406	0.287	0.199	0.136	0.092	0.061	0.040
2	0.920	0.809	0.677	0.544	0.423	0.321	0.238	0.174	0.125
3	0.981	0.934	0.857	0.758	0.647	0.537	0.433	0.342	0.265
4	0.996	0.981	0.947	0.891	0.815	0.725	0.629	0.532	0.440
5	0.999	0.996	0.983	0.958	0.916	0.858	0.785	0.703	0.616
6	1.000	0.999	0.995	0.986	0.966	0.935	0.889	0.831	0.762
7	1.000	1.000	0.999	0.996	0.988	0.973	0.949	0.913	0.867
8	1.000	1.000	1.000	0.999	0.996	0.990	0.979	0.960	0.932
9	1.000	1.000	1.000	1.000	0.999	0.997	0.992	0.983	0.968
10	1.000	1.000	1.000	1.000	1.000	0.999	0.997	0.993	0.986
11	1.000	1.000	1.000	1.000	1.000	1.000	0.999	0.998	0.995
12	1.000	1.000	1.000	1.000	1.000	1.000	1.000	0.999	0.998
13	1.000	1.000	1.000	1.000	1.000	1.000	1.000	1.000	0.999
14	1.000	1.000	1.000	1.000	1.000	1.000	1.000	1.000	1.000

TABLE A.2   (*continued*)

					λ				
x	6	7	8	9	10	11	12	13	14
0	0.002	0.001	0.000	0.000	0.000	0.000	0.000	0.000	0.000
1	0.017	0.007	0.003	0.001	0.000	0.000	0.000	0.000	0.000
2	0.062	0.030	0.014	0.006	0.003	0.001	0.001	0.000	0.000
3	0.151	0.082	0.042	0.021	0.010	0.005	0.002	0.001	0.000
4	0.285	0.173	0.100	0.055	0.029	0.015	0.008	0.004	0.002
5	0.446	0.301	0.191	0.116	0.067	0.038	0.020	0.011	0.006
6	0.606	0.450	0.313	0.207	0.130	0.079	0.046	0.026	0.014
7	0.744	0.599	0.453	0.324	0.220	0.143	0.090	0.054	0.032
8	0.847	0.729	0.593	0.456	0.333	0.232	0.155	0.100	0.062
9	0.916	0.830	0.717	0.587	0.458	0.341	0.242	0.166	0.109
10	0.957	0.901	0.816	0.706	0.583	0.460	0.347	0.252	0.176
11	0.980	0.947	0.888	0.803	0.697	0.579	0.462	0.353	0.260
12	0.991	0.973	0.936	0.876	0.792	0.689	0.576	0.463	0.358
13	0.996	0.987	0.966	0.926	0.864	0.781	0.682	0.573	0.464
14	0.999	0.994	0.983	0.959	0.917	0.854	0.772	0.675	0.570
15	0.999	0.998	0.992	0.978	0.951	0.907	0.844	0.764	0.669
16	1.000	0.999	0.996	0.989	0.973	0.944	0.899	0.835	0.756
17	1.000	1.000	0.998	0.995	0.986	0.968	0.937	0.890	0.827
18	1.000	1.000	0.999	0.998	0.993	0.982	0.963	0.930	0.883
19	1.000	1.000	1.000	0.999	0.997	0.991	0.979	0.957	0.923
20	1.000	1.000	1.000	1.000	0.998	0.995	0.988	0.975	0.952
21	1.000	1.000	1.000	1.000	0.999	0.998	0.994	0.986	0.971
22	1.000	1.000	1.000	1.000	1.000	0.999	0.997	0.992	0.983
23	1.000	1.000	1.000	1.000	1.000	1.000	0.999	0.996	0.991
24	1.000	1.000	1.000	1.000	1.000	1.000	0.999	0.998	0.995
25	1.000	1.000	1.000	1.000	1.000	1.000	1.000	0.999	0.997
26	1.000	1.000	1.000	1.000	1.000	1.000	1.000	1.000	0.999
27	1.000	1.000	1.000	1.000	1.000	1.000	1.000	1.000	0.999
28	1.000	1.000	1.000	1.000	1.000	1.000	1.000	1.000	1.000

Source: Produced using SAS.

TABLE A.3 AREA Φ(z) UNDER THE STANDARD NORMAL CURVE

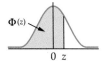

z	0.00	0.01	0.02	0.03	0.04	0.05	0.06	0.07	0.08	0.09
−3.4	0.0003	0.0003	0.0003	0.0003	0.0003	0.0003	0.0003	0.0003	0.0003	0.0002
−3.3	0.0005	0.0005	0.0005	0.0004	0.0004	0.0004	0.0004	0.0004	0.0004	0.0003
−3.2	0.0007	0.0007	0.0006	0.0006	0.0006	0.0006	0.0006	0.0005	0.0005	0.0005
−3.1	0.0010	0.0009	0.0009	0.0009	0.0008	0.0008	0.0008	0.0008	0.0007	0.0007
−3.0	0.0013	0.0013	0.0013	0.0012	0.0012	0.0011	0.0011	0.0011	0.0010	0.0010
−2.9	0.0019	0.0018	0.0018	0.0017	0.0016	0.0016	0.0015	0.0015	0.0014	0.0014
−2.8	0.0026	0.0025	0.0024	0.0023	0.0023	0.0022	0.0021	0.0021	0.0020	0.0019
−2.7	0.0035	0.0034	0.0033	0.0032	0.0031	0.0030	0.0029	0.0028	0.0027	0.0026
−2.6	0.0047	0.0045	0.0044	0.0043	0.0041	0.0040	0.0039	0.0038	0.0037	0.0036
−2.5	0.0062	0.0060	0.0059	0.0057	0.0055	0.0054	0.0052	0.0051	0.0049	0.0048
−2.4	0.0082	0.0080	0.0078	0.0075	0.0073	0.0071	0.0069	0.0068	0.0066	0.0064
−2.3	0.0107	0.0104	0.0102	0.0099	0.0096	0.0094	0.0091	0.0089	0.0087	0.0084
−2.2	0.0139	0.0136	0.0132	0.0129	0.0125	0.0122	0.0119	0.0116	0.0113	0.0110
−2.1	0.0179	0.0174	0.0170	0.0166	0.0162	0.0158	0.0154	0.0150	0.0146	0.0143
−2.0	0.0228	0.0222	0.0217	0.0212	0.0207	0.0202	0.0197	0.0192	0.0188	0.0183
−1.9	0.0287	0.0281	0.0274	0.0268	0.0262	0.0256	0.0250	0.0244	0.0239	0.0233
−1.8	0.0359	0.0351	0.0344	0.0336	0.0329	0.0322	0.0314	0.0307	0.0301	0.0294
−1.7	0.0446	0.0436	0.0427	0.0418	0.0409	0.0401	0.0392	0.0384	0.0375	0.0367
−1.6	0.0548	0.0537	0.0526	0.0516	0.0505	0.0495	0.0485	0.0475	0.0465	0.0455
−1.5	0.0668	0.0655	0.0643	0.0630	0.0618	0.0606	0.0594	0.0582	0.0571	0.0559
−1.4	0.0808	0.0793	0.0778	0.0764	0.0749	0.0735	0.0721	0.0708	0.0694	0.0681
−1.3	0.0968	0.0951	0.0934	0.0918	0.0901	0.0885	0.0869	0.0853	0.0838	0.0823
−1.2	0.1151	0.1131	0.1112	0.1093	0.1075	0.1056	0.1038	0.1020	0.1003	0.0985
−1.1	0.1357	0.1335	0.1314	0.1292	0.1271	0.1251	0.1230	0.1210	0.1190	0.1170
−1.0	0.1587	0.1562	0.1539	0.1515	0.1492	0.1469	0.1446	0.1423	0.1401	0.1379
−0.9	0.1841	0.1814	0.1788	0.1762	0.1736	0.1711	0.1685	0.1660	0.1635	0.1611
−0.8	0.2119	0.2090	0.2061	0.2033	0.2005	0.1977	0.1949	0.1922	0.1894	0.1867
−0.7	0.2420	0.2389	0.2358	0.2327	0.2296	0.2266	0.2236	0.2206	0.2177	0.2148
−0.6	0.2743	0.2709	0.2676	0.2643	0.2611	0.2578	0.2546	0.2514	0.2483	0.2451
−0.5	0.3085	0.3050	0.3015	0.2981	0.2946	0.2912	0.2877	0.2843	0.2810	0.2776
−0.4	0.3446	0.3409	0.3372	0.3336	0.3300	0.3264	0.3228	0.3192	0.3156	0.3121
−0.3	0.3821	0.3783	0.3745	0.3707	0.3669	0.3632	0.3594	0.3557	0.3520	0.3483
−0.2	0.4207	0.4168	0.4129	0.4090	0.4052	0.4013	0.3974	0.3936	0.3897	0.3859
−0.1	0.4602	0.4562	0.4522	0.4483	0.4443	0.4404	0.4364	0.4325	0.4286	0.4247
−0.0	0.5000	0.4960	0.4920	0.4880	0.4840	0.4801	0.4761	0.4721	0.4681	0.4641

TABLE A.3 (*continued*)

z	0.00	0.01	0.02	0.03	0.04	0.05	0.06	0.07	0.08	0.09
0.0	0.5000	0.5040	0.5080	0.5120	0.5160	0.5199	0.5239	0.5279	0.5319	0.5359
0.1	0.5398	0.5438	0.5478	0.5517	0.5557	0.5596	0.5636	0.5675	0.5714	0.5753
0.2	0.5793	0.5832	0.5871	0.5910	0.5948	0.5987	0.6026	0.6064	0.6103	0.6141
0.3	0.6179	0.6217	0.6255	0.6293	0.6331	0.6368	0.6406	0.6443	0.6480	0.6517
0.4	0.6554	0.6591	0.6628	0.6664	0.6700	0.6736	0.6772	0.6808	0.6844	0.6879
0.5	0.6915	0.6950	0.6985	0.7019	0.7054	0.7088	0.7123	0.7157	0.7190	0.7224
0.6	0.7257	0.7291	0.7324	0.7357	0.7389	0.7422	0.7454	0.7486	0.7517	0.7549
0.7	0.7580	0.7611	0.7642	0.7673	0.7704	0.7734	0.7764	0.7794	0.7823	0.7852
0.8	0.7881	0.7910	0.7939	0.7967	0.7995	0.8023	0.8051	0.8078	0.8106	0.8133
0.9	0.8159	0.8186	0.8212	0.8238	0.8264	0.8289	0.8315	0.8340	0.8365	0.8389
1.0	0.8413	0.8438	0.8461	0.8485	0.8508	0.8531	0.8554	0.8577	0.8599	0.8621
1.1	0.8643	0.8665	0.8686	0.8708	0.8729	0.8749	0.8770	0.8790	0.8810	0.8830
1.2	0.8849	0.8869	0.8888	0.8907	0.8925	0.8944	0.8962	0.8980	0.8997	0.9015
1.3	0.9032	0.9049	0.9066	0.9082	0.9099	0.9115	0.9131	0.9147	0.9162	0.9177
1.4	0.9192	0.9207	0.9222	0.9236	0.9251	0.9265	0.9279	0.9292	0.9306	0.9319
1.5	0.9332	0.9345	0.9357	0.9370	0.9382	0.9394	0.9406	0.9418	0.9429	0.9441
1.6	0.9452	0.9463	0.9474	0.9484	0.9495	0.9505	0.9515	0.9525	0.9535	0.9545
1.7	0.9554	0.9564	0.9573	0.9582	0.9591	0.9599	0.9608	0.9616	0.9625	0.9633
1.8	0.9641	0.9649	0.9656	0.9664	0.9671	0.9678	0.9686	0.9693	0.9699	0.9706
1.9	0.9713	0.9719	0.9726	0.9732	0.9738	0.9744	0.9750	0.9756	0.9761	0.9767
2.0	0.9772	0.9778	0.9783	0.9788	0.9793	0.9798	0.9803	0.9808	0.9812	0.9817
2.1	0.9821	0.9826	0.9830	0.9834	0.9838	0.9842	0.9846	0.9850	0.9854	0.9857
2.2	0.9861	0.9864	0.9868	0.9871	0.9875	0.9878	0.9881	0.9884	0.9887	0.9890
2.3	0.9893	0.9896	0.9898	0.9901	0.9904	0.9906	0.9909	0.9911	0.9913	0.9916
2.4	0.9918	0.9920	0.9922	0.9925	0.9927	0.9929	0.9931	0.9932	0.9934	0.9936
2.5	0.9938	0.9940	0.9941	0.9943	0.9945	0.9946	0.9948	0.9949	0.9951	0.9952
2.6	0.9953	0.9955	0.9956	0.9957	0.9959	0.9960	0.9961	0.9962	0.9963	0.9964
2.7	0.9965	0.9966	0.9967	0.9968	0.9969	0.9970	0.9971	0.9972	0.9973	0.9974
2.8	0.9974	0.9975	0.9976	0.9977	0.9977	0.9978	0.9979	0.9979	0.9980	0.9981
2.9	0.9981	0.9982	0.9982	0.9983	0.9984	0.9984	0.9985	0.9985	0.9986	0.9986
3.0	0.9987	0.9987	0.9987	0.9988	0.9988	0.9989	0.9989	0.9989	0.9990	0.9990
3.1	0.9990	0.9991	0.9991	0.9991	0.9992	0.9992	0.9992	0.9992	0.9993	0.9993
3.2	0.9993	0.9993	0.9994	0.9994	0.9994	0.9994	0.9994	0.9995	0.9995	0.9995
3.3	0.9995	0.9995	0.9995	0.9996	0.9996	0.9996	0.9996	0.9996	0.9996	0.9997
3.4	0.9997	0.9997	0.9997	0.9997	0.9997	0.9997	0.9997	0.9997	0.9997	0.9998

Source: Produced using SAS.

TABLE A.4 CRITICAL VALUES $t_\nu(\alpha)$ OF THE t DISTRIBUTION

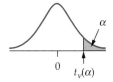

	α				
ν	0.10	0.05	0.025	0.01	0.005
1	3.078	6.314	12.706	31.821	63.657
2	1.886	2.920	4.303	6.965	9.925
3	1.638	2.353	3.182	4.541	5.841
4	1.533	2.132	2.776	3.747	4.604
5	1.476	2.015	2.571	3.365	4.032
6	1.440	1.943	2.447	3.143	3.707
7	1.415	1.895	2.365	2.998	3.499
8	1.397	1.860	2.306	2.896	3.355
9	1.383	1.833	2.262	2.821	3.250
10	1.372	1.812	2.228	2.764	3.169
11	1.363	1.796	2.201	2.718	3.106
12	1.356	1.782	2.179	2.681	3.055
13	1.350	1.771	2.160	2.650	3.012
14	1.345	1.761	2.145	2.624	2.977
15	1.341	1.753	2.131	2.602	2.947
16	1.337	1.746	2.120	2.583	2.921
17	1.333	1.740	2.110	2.567	2.898
18	1.330	1.734	2.101	2.552	2.878
19	1.328	1.729	2.093	2.539	2.861
20	1.325	1.725	2.086	2.528	2.845
21	1.323	1.721	2.080	2.518	2.831
22	1.321	1.717	2.074	2.508	2.819
23	1.319	1.714	2.069	2.500	2.807
24	1.318	1.711	2.064	2.492	2.797
25	1.316	1.708	2.060	2.485	2.787
26	1.315	1.706	2.056	2.479	2.779
27	1.314	1.703	2.052	2.473	2.771
28	1.313	1.701	2.048	2.467	2.763
29	1.311	1.699	2.045	2.462	2.756
30	1.310	1.697	2.042	2.457	2.750
40	1.303	1.684	2.021	2.423	2.704
60	1.296	1.671	2.000	2.390	2.660
∞	1.282	1.645	1.960	2.326	2.576

Source: Produced using SAS.

TABLE A.5 QUANTILES $Q_\nu(p) = \chi_\nu^2(1-p)$ OF THE χ^2 DISTRIBUTION

$$P(\chi_\nu^2 \le Q_\nu(p)) = p$$

					p					
ν	0.005	0.01	0.025	0.05	0.1	0.9	0.95	0.975	0.99	0.995
1	0.000	0.000	0.001	0.004	0.016	2.706	3.841	5.024	6.635	7.879
2	0.010	0.020	0.051	0.103	0.211	4.605	5.991	7.378	9.210	10.597
3	0.072	0.115	0.216	0.352	0.584	6.251	7.815	9.348	11.345	12.838
4	0.207	0.297	0.484	0.711	1.064	7.779	9.488	11.143	13.277	14.860
5	0.412	0.554	0.831	1.145	1.610	9.236	11.070	12.833	15.086	16.750
6	0.676	0.872	1.237	1.635	2.204	10.645	12.592	14.449	16.812	18.548
7	0.989	1.239	1.690	2.167	2.833	12.017	14.067	16.013	18.475	20.278
8	1.344	1.646	2.180	2.733	3.490	13.362	15.507	17.535	20.090	21.955
9	1.735	2.088	2.700	3.325	4.168	14.684	16.919	19.023	21.666	23.589
10	2.156	2.558	3.247	3.940	4.865	15.987	18.307	20.483	23.209	25.188
11	2.603	3.053	3.816	4.575	5.578	17.275	19.675	21.920	24.725	26.757
12	3.074	3.571	4.404	5.226	6.304	18.549	21.026	23.337	26.217	28.300
13	3.565	4.107	5.009	5.892	7.042	19.812	22.362	24.736	27.688	29.819
14	4.075	4.660	5.629	6.571	7.790	21.064	23.685	26.119	29.141	31.319
15	4.601	5.229	6.262	7.261	8.547	22.307	24.996	27.488	30.578	32.801
16	5.142	5.812	6.908	7.962	9.312	23.542	26.296	28.845	32.000	34.267
17	5.697	6.408	7.564	8.672	10.085	24.769	27.587	30.191	33.409	35.718
18	6.265	7.015	8.231	9.390	10.865	25.989	28.869	31.526	34.805	37.156
19	6.844	7.633	8.907	10.117	11.651	27.204	30.144	32.852	36.191	38.582
20	7.434	8.260	9.591	10.851	12.443	28.412	31.410	34.170	37.566	39.997
21	8.034	8.897	10.283	11.591	13.240	29.615	32.671	35.479	38.932	41.401
22	8.643	9.542	10.982	12.338	14.041	30.813	33.924	36.781	40.289	42.796
23	9.260	10.196	11.689	13.091	14.848	32.007	35.172	38.076	41.638	44.181
24	9.886	10.856	12.401	13.848	15.659	33.196	36.415	39.364	42.980	45.559
25	10.520	11.524	13.120	14.611	16.473	34.382	37.652	40.646	44.314	46.928
26	11.160	12.198	13.844	15.379	17.292	35.563	38.885	41.923	45.642	48.290
27	11.808	12.879	14.573	16.151	18.114	36.741	40.113	43.195	46.963	49.645
28	12.461	13.565	15.308	16.928	18.939	37.916	41.337	44.461	48.278	50.993
29	13.121	14.256	16.047	17.708	19.768	39.087	42.557	45.722	49.588	52.336
30	13.787	14.953	16.791	18.493	20.599	40.256	43.773	46.979	50.892	53.672
40	20.707	22.164	24.433	26.509	29.051	51.805	55.758	59.342	63.691	66.766
50	27.991	29.707	32.357	34.764	37.689	63.167	67.505	71.420	76.154	79.490
60	35.534	37.485	40.482	43.188	46.459	74.397	79.082	83.298	88.379	91.952

Source: Produced using SAS.

$$\alpha = 0.05$$

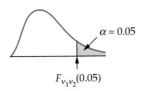

$\alpha = 0.05$

$F_{\nu_1 \nu_2}(0.05)$

					ν_1				
ν_2	1	2	3	4	5	6	7	8	9
1	161.45	199.50	215.71	224.58	230.16	233.99	236.77	238.88	240.54
2	18.51	19.00	19.16	19.25	19.30	19.33	19.35	19.37	19.38
3	10.13	9.55	9.28	9.12	9.01	8.94	8.89	8.85	8.81
4	7.71	6.94	6.59	6.39	6.26	6.16	6.09	6.04	6.00
5	6.61	5.79	5.41	5.19	5.05	4.95	4.88	4.82	4.77
6	5.99	5.14	4.76	4.53	4.39	4.28	4.21	4.15	4.10
7	5.59	4.74	4.35	4.12	3.97	3.87	3.79	3.73	3.68
8	5.32	4.46	4.07	3.84	3.69	3.58	3.50	3.44	3.39
9	5.12	4.26	3.86	3.63	3.48	3.37	3.29	3.23	3.18
10	4.96	4.10	3.71	3.48	3.33	3.22	3.14	3.07	3.02
11	4.84	3.98	3.59	3.36	3.20	3.09	3.01	2.95	2.90
12	4.75	3.89	3.49	3.26	3.11	3.00	2.91	2.85	2.80
13	4.67	3.81	3.41	3.18	3.03	2.92	2.83	2.77	2.71
14	4.60	3.74	3.34	3.11	2.96	2.85	2.76	2.70	2.65
15	4.54	3.68	3.29	3.06	2.90	2.79	2.71	2.64	2.59
16	4.49	3.63	3.24	3.01	2.85	2.74	2.66	2.59	2.54
17	4.45	3.59	3.20	2.96	2.81	2.70	2.61	2.55	2.49
18	4.41	3.55	3.16	2.93	2.77	2.66	2.58	2.51	2.46
19	4.38	3.52	3.13	2.90	2.74	2.63	2.54	2.48	2.42
20	4.35	3.49	3.10	2.87	2.71	2.60	2.51	2.45	2.39
21	4.32	3.47	3.07	2.84	2.68	2.57	2.49	2.42	2.37
22	4.30	3.44	3.05	2.82	2.66	2.55	2.46	2.40	2.34
23	4.28	3.42	3.03	2.80	2.64	2.53	2.44	2.37	2.32
24	4.26	3.40	3.01	2.78	2.62	2.51	2.42	2.36	2.30
25	4.24	3.39	2.99	2.76	2.60	2.49	2.40	2.34	2.28
26	4.23	3.37	2.98	2.74	2.59	2.47	2.39	2.32	2.27
27	4.21	3.35	2.96	2.73	2.57	2.46	2.37	2.31	2.25
28	4.20	3.34	2.95	2.71	2.56	2.45	2.36	2.29	2.24
29	4.18	3.33	2.93	2.70	2.55	2.43	2.35	2.28	2.22
30	4.17	3.32	2.92	2.69	2.53	2.42	2.33	2.27	2.21
40	4.08	3.23	2.84	2.61	2.45	2.34	2.25	2.18	2.12
60	4.00	3.15	2.76	2.53	2.37	2.25	2.17	2.10	2.04
120	3.92	3.07	2.68	2.45	2.29	2.18	2.09	2.02	1.96

TABLE A.6 (*continued*)

$$\alpha = 0.05$$

ν_2	ν_1								
	10	**12**	**15**	**20**	**24**	**30**	**40**	**60**	**120**
1	241.88	243.91	245.95	248.01	249.05	250.10	251.14	252.20	253.25
2	19.40	19.41	19.43	19.45	19.45	19.46	19.47	19.48	19.49
3	8.79	8.74	8.70	8.66	8.64	8.62	8.59	8.57	8.55
4	5.96	5.91	5.86	5.80	5.77	5.75	5.72	5.69	5.66
5	4.74	4.68	4.62	4.56	4.53	4.50	4.46	4.43	4.40
6	4.06	4.00	3.94	3.87	3.84	3.81	3.77	3.74	3.70
7	3.64	3.57	3.51	3.44	3.41	3.38	3.34	3.30	3.27
8	3.35	3.28	3.22	3.15	3.12	3.08	3.04	3.01	2.97
9	3.14	3.07	3.01	2.94	2.90	2.86	2.83	2.79	2.75
10	2.98	2.91	2.85	2.77	2.74	2.70	2.66	2.62	2.58
11	2.85	2.79	2.72	2.65	2.61	2.57	2.53	2.49	2.45
12	2.75	2.69	2.62	2.54	2.51	2.47	2.43	2.38	2.34
13	2.67	2.60	2.53	2.46	2.42	2.38	2.34	2.30	2.25
14	2.60	2.53	2.46	2.39	2.35	2.31	2.27	2.22	2.18
15	2.54	2.48	2.40	2.33	2.29	2.25	2.20	2.16	2.11
16	2.49	2.42	2.35	2.28	2.24	2.19	2.15	2.11	2.06
17	2.45	2.38	2.31	2.23	2.19	2.15	2.10	2.06	2.01
18	2.41	2.34	2.27	2.19	2.15	2.11	2.06	2.02	1.97
19	2.38	2.31	2.23	2.16	2.11	2.07	2.03	1.98	1.93
20	2.35	2.28	2.20	2.12	2.08	2.04	1.99	1.95	1.90
21	2.32	2.25	2.18	2.10	2.05	2.01	1.96	1.92	1.87
22	2.30	2.23	2.15	2.07	2.03	1.98	1.94	1.89	1.84
23	2.27	2.20	2.13	2.05	2.01	1.96	1.91	1.86	1.81
24	2.25	2.18	2.11	2.03	1.98	1.94	1.89	1.84	1.79
25	2.24	2.16	2.09	2.01	1.96	1.92	1.87	1.82	1.77
26	2.22	2.15	2.07	1.99	1.95	1.90	1.85	1.80	1.75
27	2.20	2.13	2.06	1.97	1.93	1.88	1.84	1.79	1.73
28	2.19	2.12	2.04	1.96	1.91	1.87	1.82	1.77	1.71
29	2.18	2.10	2.03	1.94	1.90	1.85	1.81	1.75	1.70
30	2.16	2.09	2.01	1.93	1.89	1.84	1.79	1.74	1.68
40	2.08	2.00	1.92	1.84	1.79	1.74	1.69	1.64	1.58
60	1.99	1.92	1.84	1.75	1.70	1.65	1.59	1.53	1.47
120	1.91	1.83	1.75	1.66	1.61	1.55	1.50	1.43	1.35

$$\alpha = 0.01$$

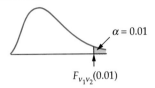

$\alpha = 0.01$

$F_{\nu_1\nu_2}(0.01)$

					ν_1				
ν_2	1	2	3	4	5	6	7	8	9
1	4052	5000	5403	5625	5764	5859	5928	5981	6022
2	98.50	99.00	99.17	99.25	99.30	99.33	99.36	99.37	99.39
3	34.12	30.82	29.46	28.71	28.24	27.91	27.67	27.49	27.35
4	21.20	18.00	16.69	15.98	15.52	15.21	14.98	14.80	14.66
5	16.26	13.27	12.06	11.39	10.97	10.67	10.46	10.29	10.16
6	13.75	10.92	9.78	9.15	8.75	8.47	8.26	8.10	7.98
7	12.25	9.55	8.45	7.85	7.46	7.19	6.99	6.84	6.72
8	11.26	8.65	7.59	7.01	6.63	6.37	6.18	6.03	5.91
9	10.56	8.02	6.99	6.42	6.06	5.80	5.61	5.47	5.35
10	10.04	7.56	6.55	5.99	5.64	5.39	5.20	5.06	4.94
11	9.65	7.21	6.22	5.67	5.32	5.07	4.89	4.74	4.63
12	9.33	6.93	5.95	5.41	5.06	4.82	4.64	4.50	4.39
13	9.07	6.70	5.74	5.21	4.86	4.62	4.44	4.30	4.19
14	8.86	6.51	5.56	5.04	4.69	4.46	4.28	4.14	4.03
15	8.68	6.36	5.42	4.89	4.56	4.32	4.14	4.00	3.89
16	8.53	6.23	5.29	4.77	4.44	4.20	4.03	3.89	3.78
17	8.40	6.11	5.18	4.67	4.34	4.10	3.93	3.79	3.68
18	8.29	6.01	5.09	4.58	4.25	4.01	3.84	3.71	3.60
19	8.18	5.93	5.01	4.50	4.17	3.94	3.77	3.63	3.52
20	8.10	5.85	4.94	4.43	4.10	3.87	3.70	3.56	3.46
21	8.02	5.78	4.87	4.37	4.04	3.81	3.64	3.51	3.40
22	7.95	5.72	4.82	4.31	3.99	3.76	3.59	3.45	3.35
23	7.88	5.66	4.76	4.26	3.94	3.71	3.54	3.41	3.30
24	7.82	5.61	4.72	4.22	3.90	3.67	3.50	3.36	3.26
25	7.77	5.57	4.68	4.18	3.85	3.63	3.46	3.32	3.22
26	7.72	5.53	4.64	4.14	3.82	3.59	3.42	3.29	3.18
27	7.68	5.49	4.60	4.11	3.78	3.56	3.39	3.26	3.15
28	7.64	5.45	4.57	4.07	3.75	3.53	3.36	3.23	3.12
29	7.60	5.42	4.54	4.04	3.73	3.50	3.33	3.20	3.09
30	7.56	5.39	4.51	4.02	3.70	3.47	3.30	3.17	3.07
40	7.31	5.18	4.31	3.83	3.51	3.29	3.12	2.99	2.89
60	7.08	4.98	4.13	3.65	3.34	3.12	2.95	2.82	2.72
120	6.85	4.79	3.95	3.48	3.17	2.96	2.79	2.66	2.56

TABLE A.6   (*continued*)

$$\alpha = 0.01$$

					ν_1				
ν_2	10	12	15	20	24	30	40	60	120
1	6056	6106	6157	6209	6235	6261	6287	6313	6339
2	99.40	99.42	99.43	99.45	99.46	99.47	99.47	99.48	99.49
3	27.23	27.05	26.87	26.69	26.60	26.50	26.41	26.32	26.22
4	14.55	14.37	14.20	14.02	13.93	13.84	13.75	13.65	13.56
5	10.05	9.89	9.72	9.55	9.47	9.38	9.29	9.20	9.11
6	7.87	7.72	7.56	7.40	7.31	7.23	7.14	7.06	6.97
7	6.62	6.47	6.31	6.16	6.07	5.99	5.91	5.82	5.74
8	5.81	5.67	5.52	5.36	5.28	5.20	5.12	5.03	4.95
9	5.26	5.11	4.96	4.81	4.73	4.65	4.57	4.48	4.40
10	4.85	4.71	4.56	4.41	4.33	4.25	4.17	4.08	4.00
11	4.54	4.40	4.25	4.10	4.02	3.94	3.86	3.78	3.69
12	4.30	4.16	4.01	3.86	3.78	3.70	3.62	3.54	3.45
13	4.10	3.96	3.82	3.66	3.59	3.51	3.43	3.34	3.25
14	3.94	3.80	3.66	3.51	3.43	3.35	3.27	3.18	3.09
15	3.80	3.67	3.52	3.37	3.29	3.21	3.13	3.05	2.96
16	3.69	3.55	3.41	3.26	3.18	3.10	3.02	2.93	2.84
17	3.59	3.46	3.31	3.16	3.08	3.00	2.92	2.83	2.75
18	3.51	3.37	3.23	3.08	3.00	2.92	2.84	2.75	2.66
19	3.43	3.30	3.15	3.00	2.92	2.84	2.76	2.67	2.58
20	3.37	3.23	3.09	2.94	2.86	2.78	2.69	2.61	2.52
21	3.31	3.17	3.03	2.88	2.80	2.72	2.64	2.55	2.46
22	3.26	3.12	2.98	2.83	2.75	2.67	2.58	2.50	2.40
23	3.21	3.07	2.93	2.78	2.70	2.62	2.54	2.45	2.35
24	3.17	3.03	2.89	2.74	2.66	2.58	2.49	2.40	2.31
25	3.13	2.99	2.85	2.70	2.62	2.54	2.45	2.36	2.27
26	3.09	2.96	2.81	2.66	2.58	2.50	2.42	2.33	2.23
27	3.06	2.93	2.78	2.63	2.55	2.47	2.38	2.29	2.20
28	3.03	2.90	2.75	2.60	2.52	2.44	2.35	2.26	2.17
29	3.00	2.87	2.73	2.57	2.49	2.41	2.33	2.23	2.14
30	2.98	2.84	2.70	2.55	2.47	2.39	2.30	2.21	2.11
40	2.80	2.66	2.52	2.37	2.29	2.20	2.11	2.02	1.92
60	2.63	2.50	2.35	2.20	2.12	2.03	1.94	1.84	1.73
120	2.47	2.34	2.19	2.03	1.95	1.86	1.76	1.66	1.53

Source: Produced using SAS.

TABLE A.7 CRITICAL VALUES FOR THE STUDENTIZED RANGE

$$q(\alpha; n, \nu)$$

$\alpha = 0.05$

					n				
ν	2	3	4	5	6	7	8	9	10
1	17.97	26.98	32.82	37.08	40.41	43.12	45.40	47.36	49.07
2	6.08	8.33	9.80	10.88	11.74	12.44	13.03	13.54	13.99
3	4.50	5.91	6.82	7.50	8.04	8.48	8.85	9.18	9.46
4	3.93	5.04	5.76	6.29	6.71	7.05	7.35	7.60	7.83
5	3.64	4.60	5.22	5.67	6.03	6.33	6.58	6.80	6.99
6	3.46	4.34	4.90	5.30	5.63	5.90	6.12	6.32	6.49
7	3.34	4.16	4.68	5.06	5.36	5.61	5.82	6.00	6.16
8	3.26	4.04	4.53	4.89	5.17	5.40	5.60	5.77	5.92
9	3.20	3.95	4.41	4.76	5.02	5.24	5.43	5.59	5.74
10	3.15	3.88	4.33	4.65	4.91	5.12	5.30	5.46	5.60
11	3.11	3.82	4.26	4.57	4.82	5.03	5.20	5.35	5.49
12	3.08	3.77	4.20	4.51	4.75	4.95	5.12	5.27	5.39
13	3.06	3.73	4.15	4.45	4.69	4.88	5.05	5.19	5.32
14	3.03	3.70	4.11	4.41	4.64	4.83	4.99	5.13	5.25
15	3.01	3.67	4.08	4.37	4.59	4.78	4.94	5.08	5.20
16	3.00	3.65	4.05	4.33	4.56	4.74	4.90	5.03	5.15
17	2.98	3.63	4.02	4.30	4.52	4.70	4.86	4.99	5.11
18	2.97	3.61	4.00	4.28	4.49	4.67	4.82	4.96	5.07
19	2.96	3.59	3.98	4.25	4.47	4.65	4.79	4.92	5.04
20	2.95	3.58	3.96	4.23	4.45	4.62	4.77	4.90	5.01
24	2.92	3.53	3.90	4.17	4.37	4.54	4.68	4.81	4.92
30	2.89	3.49	3.85	4.10	4.30	4.46	4.60	4.72	4.82
40	2.86	3.44	3.79	4.04	4.23	4.39	4.52	4.63	4.73
60	2.83	3.40	3.74	3.98	4.16	4.31	4.44	4.55	4.65
120	2.80	3.36	3.68	3.92	4.10	4.24	4.36	4.47	4.56
∞	2.77	3.31	3.63	3.86	4.03	4.17	4.29	4.39	4.47

n = number of treatments; ν = degrees of freedom.

TABLE A.7   (*continued*)

$$\alpha = 0.05$$

ν	n									
	11	12	13	14	15	16	17	18	19	20
1	50.59	51.96	53.20	54.33	55.36	56.32	57.22	58.04	58.83	59.56
2	14.39	14.75	15.08	15.38	15.65	15.91	16.14	16.37	16.57	16.77
3	9.72	9.95	10.15	10.35	10.52	10.69	10.84	10.98	11.11	11.24
4	8.03	8.21	8.37	8.52	8.66	8.79	8.91	9.03	9.13	9.23
5	7.17	7.32	7.47	7.60	7.72	7.83	7.93	8.03	8.12	8.21
6	6.65	6.79	6.92	7.03	7.14	7.24	7.34	7.43	7.51	7.59
7	6.30	6.43	6.55	6.66	6.76	6.85	6.94	7.02	7.10	7.17
8	6.05	6.18	6.29	6.39	6.48	6.57	6.65	6.73	6.80	6.87
9	5.87	5.98	6.09	6.19	6.28	6.36	6.44	6.51	6.58	6.64
10	5.72	5.83	5.93	6.03	6.11	6.19	6.27	6.34	6.40	6.47
11	5.61	5.71	5.81	5.90	5.98	6.06	6.13	6.20	6.27	6.33
12	5.51	5.61	5.71	5.80	5.88	5.95	6.02	6.09	6.15	6.21
13	5.43	5.53	5.63	5.71	5.79	5.86	5.93	5.99	6.05	6.11
14	5.36	5.46	5.55	5.64	5.71	5.79	5.85	5.91	5.97	6.03
15	5.31	5.40	5.49	5.57	5.65	5.72	5.78	5.85	5.90	5.96
16	5.26	5.35	5.44	5.52	5.59	5.66	5.73	5.79	5.84	5.90
17	5.21	5.31	5.39	5.47	5.54	5.61	5.67	5.73	5.79	5.84
18	5.17	5.27	5.35	5.43	5.50	5.57	5.63	5.69	5.74	5.79
19	5.14	5.23	5.31	5.39	5.46	5.53	5.59	5.65	5.70	5.75
20	5.11	5.20	5.28	5.36	5.43	5.49	5.55	5.61	5.66	5.71
24	5.01	5.10	5.18	5.25	5.32	5.38	5.44	5.49	5.55	5.59
30	4.92	5.00	5.08	5.15	5.21	5.27	5.33	5.38	5.43	5.47
40	4.82	4.90	4.98	5.04	5.11	5.16	5.22	5.27	5.31	5.36
60	4.73	4.81	4.88	4.94	5.00	5.06	5.11	5.15	5.20	5.24
120	4.64	4.71	4.78	4.84	4.90	4.95	5.00	5.04	5.09	5.13
∞	4.55	4.62	4.68	4.74	4.80	4.85	4.89	4.93	4.97	5.01

TABLE A.7   (continued)

$$\alpha = 0.01$$

ν	2	3	4	5	6	7	8	9	10
					n				
1	90.03	135.0	164.3	185.6	202.2	215.8	227.2	237.0	245.6
2	14.04	19.02	22.29	24.72	26.63	28.20	29.53	30.68	31.69
3	8.26	10.62	12.17	13.33	14.24	15.00	15.64	16.20	16.69
4	6.51	8.12	9.17	9.96	10.58	11.10	11.55	11.93	12.27
5	5.70	6.98	7.80	8.42	8.91	9.32	9.67	9.97	10.24
6	5.24	6.33	7.03	7.56	7.97	8.32	8.61	8.87	9.10
7	4.95	5.92	6.54	7.01	7.37	7.68	7.94	8.17	8.37
8	4.75	5.64	6.20	6.62	6.96	7.24	7.47	7.68	7.86
9	4.60	5.43	5.96	6.35	6.66	6.91	7.13	7.33	7.49
10	4.48	5.27	5.77	6.14	6.43	6.67	6.87	7.05	7.21
11	4.39	5.15	5.62	5.97	6.25	6.48	6.67	6.84	6.99
12	4.32	5.05	5.50	5.84	6.10	6.32	6.51	6.67	6.81
13	4.26	4.96	5.40	5.73	5.98	6.19	6.37	6.53	6.67
14	4.21	4.89	5.32	5.63	5.88	6.08	6.26	6.41	6.54
15	4.17	4.84	5.25	5.56	5.80	5.99	6.16	6.31	6.44
16	4.13	4.79	5.19	5.49	5.72	5.92	6.08	6.22	6.35
17	4.10	4.74	5.14	5.43	5.66	5.85	6.01	6.15	6.27
18	4.07	4.70	5.09	5.38	5.60	5.79	5.94	6.08	6.20
19	4.05	4.67	5.05	5.33	5.55	5.73	5.89	6.02	6.14
20	4.02	4.64	5.02	5.29	5.51	5.69	5.84	5.97	6.09
24	3.96	4.55	4.91	5.17	5.37	5.54	5.69	5.81	5.92
30	3.89	4.45	4.80	5.05	5.24	5.40	5.54	5.65	5.76
40	3.82	4.37	4.70	4.93	5.11	5.26	5.39	5.50	5.60
60	3.76	4.28	4.59	4.82	4.99	5.13	5.25	5.36	5.45
120	3.70	4.20	4.50	4.71	4.87	5.01	5.12	5.21	5.30
∞	3.64	4.12	4.40	4.60	4.76	4.88	4.99	5.08	5.16

TABLE A.7 *(continued)*

$$\alpha = 0.01$$

ν	11	12	13	14	15	16	17	18	19	20
1	253.2	260.0	266.2	271.8	277.0	281.8	286.3	290.4	294.3	298.0
2	32.59	33.40	34.13	34.81	35.43	36.00	36.53	37.03	37.50	37.95
3	17.13	17.53	17.89	18.22	18.52	18.81	19.07	19.32	19.55	19.77
4	12.57	12.84	13.09	13.32	13.53	13.73	13.91	14.08	14.24	14.40
5	10.48	10.70	10.89	11.08	11.24	11.40	11.55	11.68	11.81	11.93
6	9.30	9.48	9.65	9.81	9.95	10.08	10.21	10.32	10.43	10.54
7	8.55	8.71	8.86	9.00	9.12	9.24	9.35	9.46	9.55	9.65
8	8.03	8.18	8.31	8.44	8.55	8.66	8.76	8.85	8.94	9.03
9	7.65	7.78	7.91	8.03	8.13	8.23	8.33	8.41	8.49	8.57
10	7.36	7.49	7.60	7.71	7.81	7.91	7.99	8.08	8.15	8.23
11	7.13	7.25	7.36	7.46	7.56	7.65	7.73	7.81	7.88	7.95
12	6.94	7.06	7.17	7.26	7.36	7.44	7.52	7.59	7.66	7.73
13	6.79	6.90	7.01	7.10	7.19	7.27	7.35	7.42	7.48	7.55
14	6.66	6.77	6.87	6.96	7.05	7.13	7.20	7.27	7.33	7.39
15	6.55	6.66	6.76	6.84	6.93	7.00	7.07	7.14	7.20	7.26
16	6.46	6.56	6.66	6.74	6.82	6.90	6.97	7.03	7.09	7.15
17	6.38	6.48	6.57	6.66	6.73	6.81	6.87	6.94	7.00	7.05
18	6.31	6.41	6.50	6.58	6.65	6.73	6.79	6.85	6.91	6.97
19	6.25	6.34	6.43	6.51	6.58	6.65	6.72	6.78	6.84	6.89
20	6.19	6.28	6.37	6.45	6.52	6.59	6.65	6.71	6.77	6.82
24	6.02	6.11	6.19	6.26	6.33	6.39	6.45	6.51	6.56	6.61
30	5.85	5.93	6.01	6.08	6.14	6.20	6.26	6.31	6.36	6.41
40	5.69	5.76	5.83	5.90	5.96	6.02	6.07	6.12	6.16	6.21
60	5.53	5.60	5.67	5.73	5.78	5.84	5.89	5.93	5.97	6.01
120	5.37	5.44	5.50	5.56	5.61	5.66	5.71	5.75	5.79	5.83
∞	5.23	5.29	5.35	5.40	5.45	5.49	5.54	5.57	5.61	5.65

Source: Reproduced with permission of the Biometrika Trustees from Table 29 of E. S. Pearson and H. O. Hartley, *Biometrika Tables for Statisticians, Vol. 1,* 3d ed., Cambridge: Cambridge University Press, 1966.

Number of observations in subgroup, n	Factor d_2, $d_2 = \dfrac{\bar{R}}{\sigma}$	Factor d_3, $d_3 = \dfrac{\sigma_R}{\sigma}$	Factor c_2, $c_2 = \dfrac{\bar{\sigma}_{\text{RMS}}}{\sigma}$	Factor c_4, $c_4 = \dfrac{\bar{s}}{\sigma}$
2	1.128	0.8525	0.5642	0.7979
3	1.693	0.8884	0.7236	0.8862
4	2.059	0.8798	0.7979	0.9213
5	2.326	0.8641	0.8407	0.9400
6	2.534	0.8480	0.8686	0.9515
7	2.704	0.8332	0.8882	0.9594
8	2.847	0.8198	0.9027	0.9650
9	2.970	0.8078	0.9139	0.9693
10	3.078	0.7971	0.9227	0.9727
11	3.173	0.7873	0.9300	0.9754
12	3.258	0.7785	0.9359	0.9776
13	3.336	0.7704	0.9410	0.9794
14	3.407	0.7630	0.9453	0.9810
15	3.472	0.7562	0.9490	0.9823
16	3.532	0.7499	0.9523	0.9835
17	3.588	0.7441	0.9551	0.9845
18	3.640	0.7386	0.9576	0.9854
19	3.689	0.7335	0.9599	0.9862
20	3.735	0.7287	0.9619	0.9869
21	3.778	0.7242	0.9638	0.9876
22	3.819	0.7199	0.9655	0.9882
23	3.858	0.7159	0.9670	0.9887
24	3.895	0.7121	0.9684	0.9892
25	3.931	0.7084	0.9696	0.9896
30	4.086	0.6926	0.9748	0.9914
35	4.213	0.6799	0.9784	0.9927
40	4.322	0.6692	0.9811	0.9936
45	4.415	0.6601	0.9832	0.9943
50	4.498	0.6521	0.9849	0.9949
55	4.572	0.6452	0.9863	0.9954
60	4.639	0.6389	0.9874	0.9958
65	4.699	0.6337	0.9884	0.9961
70	4.755	0.6283	0.9892	0.9964
75	4.806	0.6236	0.9900	0.9966
80	4.854	0.6194	0.9906	0.9968
85	4.898	0.6154	0.9912	0.9970
90	4.939	0.6118	0.9916	0.9972
95	4.978	0.6084	0.9921	0.9973
100	5.015	0.6052	0.9925	0.9975

Estimate of $\sigma = \bar{R}/d_2$ or $\bar{s}/c_4$ or $\bar{\sigma}_{\text{RMS}}/c_2$; $\sigma_R = \bar{R}/d_3$. These factors assume sampling from a normal distribution.

Source: Reproduced from E. L. Grant and R. S. Leavenworth, *Statistical Quality Control*, 7th ed., New York, McGraw-Hill, 1996. Used with permission.

TABLE A.9 FACTORS FOR DETERMINING FROM $\overline{R}$ THE THREE-SIGMA CONTROL LIMITS FOR $\overline{X}$ AND R CHARTS

Number of observations in subgroup, n	Factor for $\overline{X}$ chart, A_2	Factors for R chart	
		Lower control limit, D_3	Upper control limit, D_4
2	1.88	0	3.27
3	1.02	0	2.57
4	0.73	0	2.28
5	0.58	0	2.11
6	0.48	0	2.00
7	0.42	0.08	1.92
8	0.37	0.14	1.86
9	0.34	0.18	1.82
10	0.31	0.22	1.78
11	0.29	0.26	1.74
12	0.27	0.28	1.72
13	0.25	0.31	1.69
14	0.24	0.33	1.67
15	0.22	0.35	1.65
16	0.21	0.36	1.64
17	0.20	0.38	1.62
18	0.19	0.39	1.61
19	0.19	0.40	1.60
20	0.18	0.41	1.59

Upper control limit for $\overline{X}$ = $UCL_{\overline{x}}$ = $\overline{\overline{X}} + A_2\overline{R}$

Lower control limit for $\overline{X}$ = $LCL_{\overline{x}}$ = $\overline{\overline{X}} - A_2\overline{R}$

(If aimed-at or standard value $\overline{X}_0$ is used rather than $\overline{\overline{X}}$ as the central line on the control chart, $\overline{X}_0$ should be substituted for $\overline{\overline{X}}$ in the preceding formulas.)

Upper control limit for R = UCL_R = $D_4\overline{R}$

Lower control limit for R = LCL_R = $D_3\overline{R}$

All factors in the table are based on the normal distribution.

There is no lower control limit for R chart where n is less than 7.

Source: Reproduced from E. L. Grant and R. S. Leavenworth, *Statistical Quality Control,* 7th ed., New York, McGraw-Hill, 1996. Used with permission.

TABLE A.10 FACTORS FOR DETERMINING FROM σ THE THREE-SIGMA CONTROL LIMITS FOR $\bar{X}$, R, AND s OR σ_{RMS} CHARTS

Number of observations in subgroup, n	Factors for $\bar{X}$ chart, A	Factors for R chart		Factors for σ_{RMS} chart		Factors for s chart	
		Lower control limit, D_1	Upper control limit, D_2	Lower control limit, B_1	Upper control limit, B_2	Lower control limit, B_5	Upper control limit, B_6
2	2.12	0	3.69	0	1.84	0	2.61
3	1.73	0	4.36	0	1.86	0	2.28
4	1.50	0	4.70	0	1.81	0	2.09
5	1.34	0	4.92	0	1.76	0	1.96
6	1.22	0	5.08	0.03	1.71	0.03	1.87
7	1.13	0.20	5.20	0.10	1.67	0.11	1.81
8	1.06	0.39	5.31	0.17	1.64	0.18	1.75
9	1.00	0.55	5.39	0.22	1.61	0.23	1.71
10	0.95	0.69	5.47	0.26	1.58	0.28	1.67
11	0.90	0.81	5.53	0.30	1.56	0.31	1.64
12	0.87	0.92	5.59	0.33	1.54	0.35	1.61
13	0.83	1.03	5.65	0.36	1.52	0.37	1.59
14	0.80	1.12	5.69	0.38	1.51	0.40	1.56
15	0.77	1.21	5.74	0.41	1.49	0.42	1.54
16	0.75	1.28	5.78	0.43	1.48	0.44	1.53
17	0.73	1.36	5.82	0.44	1.47	0.46	1.51
18	0.71	1.43	5.85	0.46	1.45	0.48	1.50
19	0.69	1.49	5.89	0.48	1.44	0.49	1.48
20	0.67	1.55	5.92	0.49	1.43	0.50	1.47
21	0.65			0.50	1.42	0.52	1.46
22	0.64			0.52	1.41	0.53	1.45
23	0.63			0.53	1.41	0.54	1.44
24	0.61			0.54	1.40	0.55	1.43
25	0.60			0.55	1.39	0.56	1.42

TABLE A.10   (*continued*)

Number of observations in subgroup, *n*	Factors for $\overline{X}$ chart, A	Factors for R chart		Factors for σ_{RMS} chart		Factors for s chart	
		Lower control limit, D_1	Upper control limit, D_2	Lower control limit, B_1	Upper control limit, B_2	Lower control limit, B_5	Upper control limit, B_6
30	0.55			0.59	1.36	0.60	1.38
35	0.51			0.62	1.33	0.63	1.36
40	0.47			0.65	1.31	0.66	1.33
45	0.45			0.67	1.30	0.68	1.31
50	0.42			0.68	1.28	0.69	1.30
55	0.40			0.70	1.27	0.71	1.28
60	0.39			0.71	1.26	0.72	1.27
65	0.37			0.72	1.25	0.73	1.26
70	0.36			0.74	1.24	0.74	1.25
75	0.35			0.75	1.23	0.75	1.24
80	0.34			0.75	1.23	0.76	1.24
85	0.33			0.76	1.22	0.77	1.23
90	0.32			0.77	1.22	0.77	1.22
95	0.31			0.77	1.21	0.78	1.22
100	0.30			0.78	1.20	0.78	1.21

$\mathrm{UCL}_{\overline{X}} = \mu + A\sigma$ $\mathrm{LCL}_{\overline{X}} = \mu - A\sigma$

(If actual average is to be used rather than standard or aimed-at average, $\overline{\overline{X}}$ should be substituted for μ in the preceding formulas.)

$$\mathrm{UCL}_R = D_2\sigma \qquad \mathrm{UCL}_s = B_6\sigma \qquad \mathrm{UCL}_{\sigma_{\mathrm{RMS}}} = B_2\sigma$$
$$\text{Central line}_R = d_2\sigma \qquad \text{Central line}_s = c_4\sigma \qquad \text{Central line}_{\sigma_{\mathrm{RMS}}} = c_2\sigma$$
$$\mathrm{LCR}_R = D_1\sigma \qquad \mathrm{LCL}_s = B_5\sigma \qquad \mathrm{LCL}_{\sigma_{\mathrm{RMS}}} = B_1\sigma$$

There is no lower control limit for dispersion chart where *n* is less than 6.

Source: E. L. Grant and R. S. Leavenworth, *Statistical Quality Control,* 7th ed., New York, McGraw-Hill, 1996. Used with permission.

ANSWERS TO ODD-NUMBERED PROBLEMS

Chapter 1

1.1 (a)
```
0 114555799999
1 00000000022222255
2 0058
3 000023
4 05
5
6 0
```

(b) Use six classes with class marks 0.05, 0.15, 0.25, 0.35, 0.45, 0.55

1.3 Use six classes with class marks $-2.6, -2.2, -1.8, -1.4, -1.0, -0.6$

1.5 (a) $146 < 147 < 148 < 149 < 151 < 152 < 153 \leq 153 < 154 < 157$ **(b)** $\hat{F}_{10}(145) = 0$, $\hat{F}_{10}(150) = 0.4, \hat{F}_{10}(150.5) = 0.4, \hat{F}_{10}(152.9) = 0.6, \hat{F}_{10}(153.1) = 0.8, \hat{F}_{10}(158) = 1$

1.7 (a) The order statistics are

7.29040	7.31960	7.44720	7.49650
7.57190	7.63570	7.71835	7.72270
7.74010	7.75750	7.77490	7.77925
7.79810	7.87640	7.87785	7.89525
7.91120	7.94310	7.96195	8.01705
8.06055	8.07070	8.07360	8.19250

(b) 50%; 50% **(c)** 25%; 75% **(d)** 75%; 25% **(e)** $k = 5$ classes; class marks are 7.3, 7.5, 7.7, 7.9, 8.1

1.9 (a)
```
48 079         3
49 09          2
50 135         3
51 3379        4
52 28          2
53 1134466889 10
54 89          2
55 2778        4
56 18          2
57 37          2
58 8           1
59 9           1
```

(b) Use six classes with class marks 49, 51, 53, 55, 57, 59

(c)
```
58 3           1
59 6           1
60
61 08          2
62 003888      6
63 334678      6
64 368         3
65 13444466    8
66 0134489     7
67 0           1
68 1           1
```

(d) Use seven classes with class marks 58.5, 60.0, 61.5, 63.0, 64.5, 66.0, 67.5 **(e)** The daughters' heights are symmetrically distributed about a single well-defined peak. The mothers' heights are skewed to the left.

1.11 (a)
```
13 00          2
14 5569        4
15 0004        4
16 003569      6
17 01355       5
18 00035       5
19 00067       5
20 04          2
21
22 0           1
23 55          2
```

(b) Use six classes with class marks 130, 150, 170, 190, 210, 230 **(c)** The scatter plot shows that the father's weight tends to increase with the father's height.

1.13 (a)
```
180 0          1      196 0000000    7
182                   198 00000000   8
184                   200 00000       5
186                   202 00          2
188 000        3      204 00          2
190 000        3      206 0           1
192 00000000   8      208 0           1
194 0000000    7
```

565

(b) Use six classes with class marks 182.5, 187.5, 192.5, 197.5, 202.5, 207.5   **(c)** The level of cadmium dust is greater than or equal to the federal standard 22.92% = (11/48) × 100% of the time. Clearly, this is not a safe workplace.

1.15 Use six classes with class marks 1800, 3000, 4200, 5400, 6600, 7800.   **(a)** The distribution is skewed right, with most of the observations concentrated around \$1800.   **(b)** $\tilde{x} = 2315$, $Q_1 = 2101, Q_3 = 3191, \text{IQR} = 1090$. There are four outliers: 5746, 5878, 6169, 7787

1.17 (a) $Q(0.20) = 27, Q(0.40) = 35, Q(0.60) = 45, Q(0.80) = 65$ **(b)** $Q_1 = 31, Q_3 = 55$

1.19 (a) Range = 105, $\tilde{x} = 173, Q_1 = 152, Q_3 = 190, \text{IQR} = 38$ **(b)** No outliers

1.21 (a) Range = 31, $Q_1 = 26, Q_3 = 37.5, \text{IQR} = 11.5$

1.25 (a) $Q(0.10) = 4.6$ **(b)** Range = 1799.9, IQR = 435.3   **(c)** There are five outliers: 1108.2, 1148.5, 1569.3, 1750.6, 1802.1

1.27 (a) Range = 260, $\tilde{x} = 22$, $Q_1 = 12$, $Q_3 = 87$, IQR = 75; outliers are 225, 246, 261   **(b)** Range = 215, $\tilde{x} = 63$, $Q_1 = 18$, $Q_3 = 111$, IQR = 93; no outliers   **(c)** Range = 207, $\tilde{x} = 41.5$, $Q_1 = 18.5$, $Q_3 = 92.5$, IQR = 74; one outlier, 200   **(d)** Range = 216, $\tilde{x} = 60$, $Q_1 = 33, Q_3 = 118, \text{IQR} = 85$; no outliers

1.29 $Q_1 = 4, Q_3 = 25, \text{IQR} = 21$

1.31 (a) Net revenue $R = \$2$ **(b)** Net revenue $R = -\$2$

1.33 (a) Commercial Lab (BOD): $\bar{x} = 25.27, s^2 = 387.42, s = 19.68$; State Lab (BOD): $\bar{x} = 34.64$, $s^2 = 109.25, s = 10.45$ **(b)** Commercial Lab (SS): $\bar{x} = 46.45, s^2 = 1014.07, s = 31.84$; State Lab (SS): $\bar{x} = 33.18, s^2 = 363.76, s = 19.07$

1.35 $\bar{x} = 16.46, s^2 = 0.34, s = 0.58$

1.37 $\bar{x} = 70.19, s = 2.88$

1.39 $\bar{x} = 8568, s^2 = 19771046, s = 4446.46$

1.41 $\bar{x} = 151, s = 3.46$

1.43 (a) *Exposed:* $\bar{x} = 31.85$, $s^2 = 207.57$, $s = 14.40$ *Control:* $\bar{x} = 15.88$, $s^2 = 20.60$, $s = 4.54$ **(b)** Because the mean lead level in the exposed group is approximately twice that of the control group, we infer that the exposed group has higher lead levels than the control group.

1.45 $\bar{x} = 196.13, s^2 = 25.69, s = 5.07$

1.47 (a) The summary statistics for each diet are given by Diet A: $\bar{x} = 9.04$, $s = 3.67$; Diet B: $\bar{x} = 5.86$, $s = 3.44$; Diet C: $\bar{x} = 10.79$, $s = 3.27$. There are two outliers in Diet A: 2.1 and 15.3 pounds.   **(c)** Diet C is most effective, followed by diets A and B (least effective). However, Diet C does not appear to be much better than Diet A.

Chapter 2

2.1 (a) A_0 = TTTT; $P(A_0) = 1/16$ **(b)** A_1 = HTTT, THTT, TTHT, TTTH; $P(A_1) = 4/16$ **(c)** B_3 = HHHT, HHTH, HTHH, THHH, HHHH; $P(B_3) = 5/16$ **(d)** B_4 = HHHH; $P(B_4) = 1/16$ **(e)** A_4 = HHHH; $P(A_4) = 1/16$

2.3 (a) B_2 **(b)** A_1 **(c)** A_4 **(d)** A_0

2.5 (a) $c = 0.2013$ **(b)** $P(\omega_3) = 0.2013, P(\omega_3, \omega_4, \omega_5) = 0.4717$

2.7 (a) $A_3 \cup A_5 \cup A_7 \cup A_9 \cup A_{11}$ **(b)** $A_2 \cup A_3 \cup A_4 \cup A_5 \cup A_6$ **(c)** $A_7 \cup A_8 \cup A_9 \cup A_{10} \cup A_{11} \cup A_{12}$

2.9 (a) 0.9   **(b)** $1 - 0.9 = 0.1$ **(c)** $1 - P(A \cap B) = 1 - 0.2 = 0.8$ **(d)** 0.2   **(e)** 0.5   **(f)** 0.5   **(g)** 0.8

2.11 (a) T   **(b)** F   **(c)** F

2.15 (a) $56 + 28 = 84$ **(b)** $10 + 5 = 15$

2.17 $10 \times 10 \times 10 \times 26 \times 26 \times 26 = 17{,}576{,}000$

2.19 34,650

2.21 (a) 0.3   **(b)** 0.5   **(c)** 0.83

2.23 (a) The sample space consists of 10^5 quintuples of integers $(x_1, x_2, x_3, x_4, x_5)$, where each x_i is an integer from 1 to 10.   **(b)** $P_{10,5}/10^5 = 0.3024$

2.25 (i) $P(\text{at least one ace}) = 0.6651$ **(ii)** $P(\text{at least two aces}) = 0.6186$ So A has the greater chance of winning.

2.27 **(a)** (*Notation:* N = nickel, D = dime). There are $\binom{5}{2}$ = 10 subsets of size 2 taken from five objects. The subsets are (NN), (ND), (ND), (ND), (ND), (ND), (ND), (DD), (DD), (DD). That is, there is only one way to get two nickels, 6 ways to get one nickel and one dime, and three ways to get two dimes. **(b)** The monetary value of (NN), (ND), and (DD) are 10, 15, and 20 cents respectively; so P(total value coins = 10) = 1/10; P(total value coins = 15) = 6/10; P(total value coins = 20) = 3/10

2.29 **(a)** 3/105 = 0.029 **(b)** 28/105 = 0.267 **(c)** 21/105 = 0.20 **(d)** 56/105 = 0.533 **(e)** P(at least one is good) = 1 − P(neither is good) = 1 − 0.2 = 0.8 **(f)** P(at most one is good) = P(none are good) + P(exactly one is good) = 0.20 + 0.533 = 0.733. **(g)** 6/105 = 0.057

2.31 Let M, W, and E denote the event that the student selected is a man, a woman, and an engineering student, respectively; P(M | E) = 0.61

2.33 P(Class A risk | accident occurred) = 0.10 P(Class B risk | accident occurred) = 0.67 P(Class C risk | accident occurred) = 0.23

2.35 P(both are defective | at least one is defective) = $\frac{1}{8}$

2.37 **(a)** 0.33 **(b)** 0.5

2.39 **(a)** $0.98 \times 0.95 = 0.931$ **(b)** $1 - 0.931 = 0.069$

Chapter 3

3.1 **(a)** $P(X \le 0) = 10/15$ **(b)** $P(X < 0) = 5/15$ **(c)** $P(X \le 1) = 14/15$ **(d)** $P(X \le 1.5) = 14/15$ **(e)** $P(|X| \le 1) = 13/15$ **(f)** $P(|X| < 1) = 5/15$

3.3 **(a)** $c = 1/30$ **(b)** $F(-2.5) = 0$, $F(-0.5) = 15/30$, $F(0) = 21/30$, $F(1.5) = 26/30$, $F(1.7) = 26/30$, $F(3) = 1$

3.5

x	1	5	10	15
$f(x)$	0.4	0.3	0.2	0.1

3.7 There are $C_{5,2} = 10$ sample points given by (12), (13), (14), (15), (23), (24), (25), (34), (35), (45)

w	3	4	5	6	7	8	9
$f(w)$	0.1	0.1	0.2	0.2	0.2	0.1	0.1

3.9 $f(x) = [\binom{4}{x}\binom{46}{5-x}]/\binom{50}{5}$, $x = 0, 1, 2, 3, 4$

3.13 **(a)** $c = 0.438$ **(b)** $E(X) = 1.1899$, $E(X^2) = 3.1898$, $V(X) = 1.7739$

3.15 **(a)** $E(X) = 1.5$ **(b)** $E(X^2) = 3$ so $V(X) = 3 - (1.5)^2 = 0.75$

3.17 $E(X) = 1.25$, $V(X) = E(X^2) - 1.25^2 = 0.8633$

3.19 **(a)** $E(X) = 3$ **(b)** $E(X^2) = 11$ **(c)** $V(X) = 11 - 3^2 = 2$ **(d)** $E(2^X) = 12.4$ **(e)** $E(\sqrt{X}) = 1.68$ **(f)** $E(2X + 5) = 2E(X) + 5 = 2 \times 3 + 5 = 11$ **(g)** $V(2X + 5) = 2^2 V(X) = 4 \times 2 = 8$

3.21 **(a)** $E(X) = 0$ **(b)** $E(X^2) = 16/15$ **(c)** $V(X) = 16/15$ **(d)** $E(2^X) = 1.28$ **(e)** $E(2^{-X}) = 1.28$ **(f)** $E(\cos(\pi X)) = -1/15$

3.23 **(a)** $E(X) = -0.33$ **(b)** $E(X^2) = 2$ **(c)** $V(X) = 1.89$ **(d)** $E(2^X) = 1.32$ **(e)** $E(2^{-X}) = 1.85$ **(f)** $E(\cos(\pi X)) = 0.2$

3.25 $E(X) = (2n + 1)/3$

3.29 **(a)** See the answers to part (b). **(b)** $h(0) = \binom{14}{4}/\binom{20}{4} = 0.2066$; $R_h(x) = [(4 - x)(6 - x)]/[(x + 1)(10 + x + 1)]$, $x = 0, 1, 2, 3$, so $h(1) = R_h(0) \times h(0) = 0.4508$; $h(2) = R_h(1) \times h(1) = 0.2817$; $h(3) = R_h(2) \times h(2) = 0.0578$; $h(4) = R_h(3) \times h(3) = 0.0031$

3.31 **(a)** $n = 5$, $N = 50$, $D = 1$ $P(X = 0) = 0.90$ **(b)** $n = 10$, $N = 100$, $D = 2$ $P(X = 0) = 0.8091$ **(c)** $n = 20$, $N = 200$, $D = 4$ $P(X = 0) = 0.6539$

3.33 $P(Y = 1) = \frac{2}{10}$, $P(Y = x) = (P_{8,x-1}/P_{10,x-1}) \times 2/[10 - (x - 1)]$, $x = 2, \ldots, 9$

3.35 $P(U = i) = (i - 1)/15$, $i = 2, 3, 4, 5, 6$

3.37 **(a)** $B(4; 10, 0.1) = 0.998$; $B(4; 10, 0.4) = 0.663$; $B(4; 10, 0.6) = 0.166$; $B(4; 10, 0.8) = 0.006$ **(b)** $1 - B(8; 20, 0.2) = 0.01$; $1 - B(8; 20, 0.5) = 0.748$; $1 - B(8; 20, 0.7) = B(11; 20, 0.3) = 0.995$; $1 - B(8; 20, 0.9) = B(11; 20, 0.1) \approx 1.00$

3.41 0.02

3.43 **(a)** Y is binomial with parameters $n = 20, p = 0.10$ **(b)** $E(Y) = 20 \times 0.10 = 2$ and $V(Y) = 20 \times 0.1 \times 0.9 = 1.8$ **(c)** $P(Y > 4) = 1 - P(Y \leq 3) = 1 - 0.867 = 0.133$ **(d)** The probability that 6 or more batteries have lifetimes less than 3.5 hours equals $P(Y \geq 6) = 1 - B(5; 20, 0.10) = 1 - 0.989 = 0.011$. The probability of this event occurring is quite small; it would appear to be inconsistent with the assumption that 1 in 10 batteries has a lifetime less than 3.5 hours.

3.45 The number of patients who recover is given by the binomially distributed random variable X with parameters $n = 20$ and $p = 0.4$ (if the drug is worthless) and $p = 0.8$ if the recovery rate for patients given the drug is indeed 80 percent.   **(a)** $P(X \geq 12) = 0.057$ **(b)** $P(X \geq 12) = 0.99$

3.47 0.9375

3.49 **(a)** 0.21   **(b)** 0.0159

3.51 16,000

3.55 **(a)** $\exp(-1.5) = 0.2231$; $1.5 \times \exp(-1.5) = 0.3347$; $(1.5)^2/2 \times \exp(-1.5) = 0.2510$; $(1.5)^3/6 \times \exp(-1.5) = 0.1255$ **(b)** $p(0; 1.5) = 0.223$; $p(1; 1.5) = P(1; 1.5) - P(0; 1.5) = 0.558 - 0.223 = 0.335$; $p(2; 1.5) = P(2; 1.5) - P(1; 1.5) = 0.809 - 0.558 = 0.251$; $p(3; 1.5) = P(3; 1.5) - P(2; 1.5) = 0.934 - 0.809 = 0.125$

3.57 **(a)** 0.6065   **(b)** 0.3033   **(c)** 0.0902

3.59 **(a)** The expected net revenue is \$15.63.   **(b)** The expected net revenue is \$16.23.   **(c)** A five-seat van has greater expected net revenue, so it is more profitable to operate.

Chapter 4

4.1 **(a)** $P(1 < Y < 2) = 0.2492$ **(b)** $P(Y > 3) = 0.1054$ **(c)** Solve $1 - \exp(-0.75\eta_{(0.95)}) = 0.95$: $\eta_{(0.95)} = 3.9943$

4.3 **(b)** $0.0146, 0.9730$ **(c)** $f(y) = F'(y) = 3y^2/8, 0 < y < 2; f(y) = 0$ elsewhere

4.5 **(a)** $k = 3$ **(c)** $F(x) = 0, x \leq 0; F(x) = 1 - (1-x)^3, 0 < x < 1; F(x) = 1, x \geq 1$ **(e)** $P(X \leq 0.25) = 0.5781; P(X \geq 0.75) = 0.0156$ **(f)** $\eta_{0.5} = 0.2063$

4.7 **(b)** $F(x) = 0, x \leq -1; F(x) = x + (x^2/2) + \frac{1}{2}, -1 < x \leq 0; F(x) = x - (x^2/2) + \frac{1}{2}, 0 \leq x < 1; F(x) = 1, x \geq 1$ **(c)** $P(-0.4 < X < 0.6) = 0.74$ **(d)** median $= 0$ **(e)** $\eta_{(0.10)} = -0.5528; \eta_{(0.90)} = 0.5528$ [*Note:* The percentiles are obtained by solving $F(\eta_p) = p$.]

4.9 **(a)** $k = \frac{3}{4}$ **(b)** $P(0.5 < S < 1.5) = 0.6880$ **(c)** $P(S > 1) = 0.6875$

4.11 $P(T > 1) = \exp(-1/43.3) = 0.9770$

4.13 **(a)** F is a continuous df because $f(x) = F'(x) = 0, x \leq 0; f(x) = F'(x) = [1/(1+x)^2] > 0, x > 0$, so $f(x) \geq 0; \int_{-\infty}^{\infty} f(t)\, dt = \int_0^\infty [1/(1+t)^2]\, dt = \lim_{x \to \infty} \int_0^x [1/(1+t)^2]\, dt = \lim_{x \to \infty} [x/(x+1)] = 1$ **(b)** F is a continuous df because $f(x) = F'(x) = 0, x \leq 2; f(x) = F'(x) = 3/x^2 > 0, x > 2$, so $f(x) \geq 0; \int_{-\infty}^{\infty} f(t)\, dt = \int_2^\infty (3/t^2)\, dt = \lim_{x \to \infty} \int_2^x (3/t^2)\, dt = \lim_{x \to \infty} [1 - (4/x^2)] = 1$ **(c)** F is not a continuous df because $F(x) = \sin x < 0, -\pi/2 < x < 0$

4.15 **(a)** $E(S) = 1.2$ **(b)** $V(S) = 0.6933$

4.17 **(a)** $E(\cos \Theta) = 0, V(\cos \Theta) = 0.5$ **(b)** $E(\sin \Theta) = 0, V(\sin \Theta) = 0.5$ **(c)** $E(|\cos \Theta|) = 2/\pi$ **(d)** No. $E(\cos \Theta) = 0$, but $\cos (E(\Theta)) = \cos 0 = 1$

4.21 The expected profit is given by Eq. (4.22).   **(a)** $K = 700$, so $E(R) = 0.9795$; $K = 900$, so $E(R) = 0.4394$ **(b)** $K = 1098$ hours.

4.23 **(a)** $P(|Z| > 1) \leq 1$; $P(|Z| > 1) = 0.3174$ **(b)** $P(|Z| > 1.5) \leq 1/(1.5)^2 = 0.4444$; $P(|Z| > 1.5) = 0.1336$ **(c)** $P(|Z| > 2) \leq 1/2^2 = 0.25$; $P(|Z| > 2) = 0.0456$

4.25 **(a)** $P(X \geq 24.5) = 0.9525$ **(b)** $P(X \leq 34) = 0.9332$

4.27 Safe life $= 2.988$ hours

4.29 **(a)** $P(X < 0) = 0.0228; P(X > 20) = 0.0918$ **(b)** $P(|X - 12| < c = 9.87) = 0.9; P(|X - 12| < c = 11.76) = 0.95; P(|X - 12| < c = 15.48) = 0.99$

4.33 $c = 100.5$

4.35 The following answers were obtained without the continuity correction. **(a)** $P(X \geq 160) = 0.124$ **(b)** $P(X \leq 140) = 0.124$ **(c)** $P(|X - 150| \geq 20) = 0.0208$ **(d)** $P(135 < X < 165) = 0.8948$

4.37 We use the normal approximation without the continuity correction. **(a)** $P(Y = y) = b(100; 0.7, y), \quad y = 0, 1, \ldots, 100$ **(b)** $P(Y \geq 75) \approx 0.1379$ **(c)** $P(Y \geq 80) \approx 0.0146$ **(d)** $P(Y \leq 65) \approx 0.1379$

4.39 Since V is $N(0, 1)$ distributed, it follows that $E(V^2) = 1$ and $E(V^4) = 3$. Consequently, $E(K) = E(mV^2/2) = m/2E(V^2) = m/2$, $E(K^2) = (m/2)^2E(V^4) = 3m^2/4$, so $V(K) = 3m^2/4 - m^2/4 = m^2/2$

4.41 **(a)** $F(1) = 1 - \exp(-(1/3)^2) = 0.1052$; $F(2) = 1 - \exp(-(2/3)^2) = 0.3588$; $F(3) = 1 - \exp(-1) = 0.6321$ **(b)** Using Eq. (4.22), we have $C_1 = 50$, $C_2 = 100$, $K = 1$; so $E(R) = 50P(T > 1) = 50 \times \exp(-(1/3)^2) = \$44.74$

4.43 **(a)** $B(1, 1) = 1$ **(b)** $B(1, 2) = 1/2$ **(c)** $B(2, 3) = 1/12$ **(d)** $B(3, 2) = 1/12$

4.45 **(a)** $F(x) = 0, x \leq 0; F(x) = 4x^3 - 3x^4, 0 \leq x \leq 1; F(x) = 1, x \geq 1$ **(b)** 0.0508

4.47 The range of $X = U^2$ is the interval $[0, 1]$. **(a)** $F(x) = 0, x < 0; F(x) = P(X \leq x) = P(U^2 \leq x); P(-\sqrt{x} \leq U \leq \sqrt{x}) = \sqrt{x}, 0 \leq x \leq 1; F(x) = 1, x > 1$ **(b)** $f(x) = 1/(2\sqrt{x}), 0 < x < 1$ $f(x) = 0$, elsewhere

4.51 **(a)** $G(y) = 0, y \leq 0; G(y) = \int_0^y [dt/(1+t)^2] = y/(1+y), 0 \leq y < \infty$

4.53 The condition $1 \leq T < \infty$ implies that $0 < X = T^{-1} \leq 1$. Therefore, for $0 < x \leq 1$ we have $F(x) = P(0 < X \leq x) = P(0 < T^{-1} \leq x) = P(T \geq 1/x) = x^2, 0 \leq x \leq 1; f(x) = F'(x) = 2x$, $0 \leq x \leq 1; f(x) = 0$, otherwise

4.55 $F(y) = 0,\ y \leq 0;\ F(y) = P(Y \leq y) = P(\sqrt{X} \leq y) = P(X \leq y^2) = 2\int_0^{y^2}(1-x)dx = 1 - (1-y^2)^2,\ 0 \leq y \leq 1;\ F(y) = 1,\ y > 1;\ f(y) = F'(y) = 2y(1-y^2),\ 0 \leq y \leq 1;\ f(y) = 0$, elsewhere

4.57 All three statements are false; the counterexamples follow. **(a)** If X is $N(0, 1)$, then $E(X)^2 = 0^2 = 0 \neq E(X^2) = 1$ **(b)** If X is uniform on the interval $[1, 2]$, then $E(X) = 1.5$ and $E(1/X) = \int_1^2 x^{-1}dx = \ln 2 = 0.6931 \neq 1/1.5 = 0.6667 = 1/E(X)$ **(c)** If X is $N(0, 1)$, then $E(X) = 0$ but $P(X = 0) = 0$

Chapter 5

5.1 **(a)**

	Y				
X	2	3	4	5	$f_X(x)$
0	1/24	3/24	1/24	1/24	6/24
1	1/12	1/12	3/12	1/12	6/12
2	1/12	1/24	1/12	1/24	6/24
$f_Y(y)$	5/24	6/24	9/24	4/24	

(b) X and Y are not independent. Note that $f(0, 2) = 1/24 \neq f_X(0)f_Y(2) = (6/24) \times (5/24)$.

5.3 **(a)** For $x = 0, 1, 2$ $P(Y \leq 1 \mid X = x) = 12/15, 1, 1$ **(b)** For $y = 0, 1, 2$ $P(Y = y \mid X = 0) = 1/5, 3/5, 1/5$

5.5 **(a)**

	X_2	
X_1	5	10
5	1/10	3/10
10	3/10	3/10

(b)

t	5	15	20
$f_T(t)$	1/10	6/10	3/10

5.7 The faces of the three six–sided dice are: Die 1 $= \{5, 7, 8, 9, 10, 18\}$, Die 2 $= \{2, 3, 4, 15, 16, 17\}$, Die 3 $= \{1, 6, 11, 12, 13, 14\}$ The sample space corresponding to throwing die 1 and die 2 consists of 36 sample points (each assigned probability 1/36):

Sample Space of (D_1, D_2)					
(5, 2)	(5, 3)	(5, 4)	(5, 15)	(5, 16)	(5, 17)
(7, 2)	(7, 3)	(7, 4)	(7, 15)	(7, 16)	(7, 17)
(8, 2)	(8, 3)	(8, 4)	(8, 15)	(8, 16)	(8, 17)
(9, 2)	(9, 3)	(9, 4)	(9, 15)	(9, 16)	(9, 17)
(10, 2)	(10, 3)	(10, 4)	(10, 15)	(10, 16)	(10, 17)
(18, 2)	(18, 3)	(18, 4)	(18, 15)	(18, 16)	(18, 17)

The set $\{D_1 > D_2\}$ consists of 21 sample points given by $\{(5, 2)$ $(5, 3)$ $(5, 4)$ $(7, 2)$ $(7, 3)$ $(7, 4)$ $(8, 2)$ $(8, 3)$ $(8, 4)$ $(9, 2)$ $(9, 3)$ $(9, 4)$ $(10, 2)$ $(10, 3)$ $(10, 4)$ $(18, 2)$ $(18, 3)$ $(18, 4)$ $(18, 15)$ $(18, 16)$ $(18, 17)\}$. So $P\{D_1 > D_2\} = 21/36$. The remaining assertions are derived by the same method.

5.11 $Cov(X, Y) = -6400$

5.13 (a)

X_1	X_2					$f_{X_1}(x_1)$
	0	1	2	3	4	
0	0.12	0.09	0.03	0.03	0.03	0.3
1	0.08	0.06	0.02	0.02	0.02	0.2
2	0.08	0.06	0.02	0.02	0.02	0.2
3	0.08	0.06	0.02	0.02	0.02	0.2
4	0.04	0.03	0.01	0.01	0.01	0.1
$f_{X_2}(x_2)$	0.40	0.30	0.1	0.1	0.1	1.00

(b) P (M_1 has more breakdowns than M_2) $= P(X_1 > X_2) = 0.47$ **(c)** P (same number of breakdowns) $= P(X_1 = X_2) = 0.11$

(d)

T	0	1	2	3	4
$f_T(t)$	0.12	0.17	0.17	0.19	0.17

T	5	6	7	8
$f_T(t)$	0.09	0.05	0.03	0.01

5.15 (a) $E(3X - 2Y) = 12$ **(b)** $V(3X) = 81$ **(c)** $V(-2Y) = 64$ **(d)** $V(3X - 2Y) = 81 + 64 = 145$

5.19 (a) $P(M_n \leq x) = P(X_1 \leq x \cap \cdots \cap X_n \leq X) = (x/N)^n$ **(b)** $P(L_n > x) = P(X_1 > x \cap \cdots \cap X_n > x) = ((N - x)/N))^n$, so $P(L_n \leq x) = 1 - ((N - x)/N))^n$

5.21 (a) $E(Y_1) = 15, E(Y_2) = 4, E(Y_3) = 1$ **(b)** $P(Y_1 = 15, Y_2 = 4, Y_3 = 1) = 0.0829$

5.23 0.0612

5.25 (a) $P(X(2) \leq 3) = 0.857$ **(b)** $P(X(2) = 2, X(4) = 6) = 0.0376$

5.27 (a) The system function is given by $X = 1 - (1 - X_1X_2)(1 - X_3X_4)$ **(b)** The reliability is given by $E(X) = 1 - E(1 - X_1X_2)(1 - X_3X_4)) = 1 - (1 - p^2)(1 - p^2) = 1 - (1 - p^2)^2$

5.29 (a) $r = 0.8123$ **(b)** $r \approx 1.00$

5.31 (a) $c = 2/3$ **(b)** $f_X(x) = \frac{2}{3}(x + 1)$, $0 \leq x \leq 1$; $f_X(x) = 0$ elsewhere; $f_Y(y) = \frac{1}{3}(4y + 1)$, $0 \leq y \leq 1$; $f_Y(y) = 0$ elsewhere **(c)** $\mu_X = 0.56$, $\mu_Y = 0.61$ **(d)** $\sigma_X = 0.2833$, $\sigma_Y = 0.2644$

5.33 (a) $P(X^2 + Y^2 \leq 1/4) = \pi(1/2)^2/\pi = 0.25$ **(b)** $\{(x, y) : x > y, x^2 + y^2 \leq 1\} = $ semicircle, so $P(X > Y) = 0.5$ **(c)** $P(X = Y) = 0$ **(d)** $\{(x, y) : y < 2x, x^2 + y^2 \leq 1\} = $ semicircle, so $P(Y < 2X) = 0.5$ **(e)** $F_R(r) = 0$, $r \leq 0$; $F_R(r) = P(R \leq r) = P(\sqrt{R} \leq \sqrt{r}) = (\pi r)/\pi = r, 0 \leq r \leq 1$; $F_R(r) = 1, r \geq 1$

5.35 (a) $P(X + Y \leq 1) = 0.2143$ **(b)** $P(Y > X) = 0.2857$ **(c)** $f_X(x) = \frac{2}{7}([1 + x]^3 - x^3), 0 \leq x \leq 1$; $f_X(x) = 0$ elsewhere; $f_Y(y) = \frac{2}{7}([1 + y]^3 - y^3), 0 \leq y \leq 1$; $f_Y(y) = 0$ elsewhere

5.37 (a) $P(Y < 1/2 \mid X < 1/2) = 0.196$ **(b)** $f(y \mid x) = [6(xy + y^2)]/(3x + 2), 0 \leq x \leq 2, 0 \leq y \leq 1 = 0$ elsewhere **(c)** $E(Y \mid X = x) = \int_0^1 y[6(xy + y^2)]/(3x + 2) \, dy = (2x + 1.5)/(3x + 1)$

5.39 (a) $P(Y < 1) = 0.50$ **(b)** $P(Y < 1 \mid X = 0) = 0.7764$ **(c)** $E(Y \mid X = 0) = -1$ **(d)** $V(X + Y) = 43$

5.41 (a) $P(X < 4) = 0.7486$ **(b)** $P(X < 4 \mid Y = 1) = 0.8438$ **(c)** $E(X \mid Y = 1) = 2$ **(d)** $V(X + Y) = 9$

Chapter 6

6.1 (a) $E(X) = 50$ **(b)** $V(X) = 180$ **(c)** $P(|X - 50| \le 25) = 0.9372$ **(d)** $\eta_{(0.9)} = 67.173$

6.3 Suppose the time (measured in minutes) to repair a component is normally distributed with $\mu = 65, \sigma = 10$. **(a)** $P(X < 60) = 0.3085$ **(b)** $T = N(520, 800); 0.0793$

6.5 (a) $\overline{X} - \overline{Y} \sim N(1, 25/36)$ **(b)** $P(\overline{X} > \overline{Y}) = 0.8849$

6.7 (a) $\overline{X} - \overline{Y} \sim N(0, (1347.22)^2)$ **(b)** $P(|\overline{X} - \overline{Y}| > 2500) = 0.0628$ **(c)** $P(\overline{X} - \overline{Y} < -2500) = 0.0314$

6.9 (a) $E(W_n) = n \times 175, V(W_n) = n \times 400$ **(b)** $W_n \sim N(n \times 175, n \times 400)$ **(c)** $P(W_{18} > 3000) = 0.9616$ **(d)** $N = 16$

6.11 (a) $E(T) = 40 \times 0.4 = 16$, $V(T) = 25.6$ **(b)** Largest value is 40; smallest value is -40 **(c)** The central limit theorem asserts that the distribution of T is approximately $N(16, 25.6)$ $P(T > 25) \approx 0.0325; P(T < 0) \approx 0.0008$

6.13 (b) $\Gamma(3/2) = \sqrt{\pi}/2; \Gamma(5/2) = 3\sqrt{\pi}/4$

6.15 (b) $\alpha = 4$ and $\beta = 0.5$

6.17 (a) $a = 3.94, b = 18.307$ **(b)** $a = 7.261, b = 24.996$ **(c)** $a = 6.262, b = 27.488$ **(d)** $a = 9.591, b = 34.170$

6.19 (a) $a = b^3/2$ **(b)** The df of the kinetic energy $Y = (m/2) \times V^2$ is given by $G(y) = F(\sqrt{2y/m})$, where $F(v) = P(V \le v) = \int_0^v ay^2 e^{-by^2} dy$ **(c)** $g(y) = [(b^3\sqrt{y})/m^{3/2}] \exp(-b^2 y/m)$, $0 \le y < \infty$

6.21 (a) $t_9(0.05) = 1.833$ **(b)** $t_9(0.01) = 2.821$ **(c)** $t_{18}(0.025) = 2.101$ **(d)** $t_{18}(0.01) = 2.552$

6.23 (a) $F_{3,20}(0.05) = 3.10$ **(b)** $F_{3,20}(0.01) = 4.94$ **(c)** $F_{4,30}(0.05) = 2.69$ **(d)** $F_{4,30}(0.01) = 4.02$

Chapter 7

7.1 (a) $75 \pm 1.47 = [73.53, 76.47]$ **(b)** $[73.86, \infty)$ **(c)** $75 \pm 1.75 = [73.25, 75.75]$ **(d)** $[73.53, \infty)$

7.3 (a) 15.88 ± 1.34 (we use the value $t_{32}(0.05) \approx 1.69$) **(b)** 15.88 ± 2.17 (we use the value $t_{32}(0.005) \approx 2.74$)

7.7 (a) $d = 1.96 \times (4/5) = 1.568$; the confidence interval is $148.3 \pm 1.568 = [146.73, 149.87]$ **(b)** $d = 2.064 \times (4/5) = 1.6512$ [$t_{24}(0.025) = 2.064$]; the confidence interval is $148.3 \pm 1.6512 = [146.65, 149.95]$

7.9 $(n - 1)s^2 = 5 \times 51.2 = 256, \chi_5^2(0.05) = 11.07, \chi_5^2(0.95) = 1.145$; the confidence interval for σ^2 is $[23.13, 223.58]$

7.11 (a) Because the sample size is large, we can assume that the distribution of the sample mean is approximately normal. **(b)** $38,000 - 2.33 \times (3200/\sqrt{100}) = [37,254.40, \infty)$

7.13 (a) $\overline{x} = 0.152, s = 0.0043; t_9(0.025) = 2.262$. Consequently, the endpoints of the confidence interval are $L = 0.1493, U = 0.1547$ (0.152 ± 0.0027). **(b)** $\overline{x} = 0.152, s = 0.0043; t_9(0.005) = 3.25$. Consequently, the endpoints of the confidence interval are $L = 0.148, U = 0.156$ (0.152 ± 0.0044).

7.15 (a) $196.13 \pm 2.690 \times (5.07/\sqrt{48}) = [194.16, 198.10]$ **(b)** $196.13 + 1.679 \times (5.07/\sqrt{48}) = (-\infty, 197.36]$

7.19 (a) $[141.4, 395.6]$ **(b)** $[14.03, 39.26]$

7.21 The confidence interval for σ^2 is $[0.0485, 0.1790]$.

7.23 The confidence interval is $[-0.001419, 0.005619]$.

7.25 The confidence interval is $[-5299.26, 299.26]$.

7.27 (a) $[-1.9136, -1.1736]$ **(b)** $[-2.0658, -1.0214]$

7.29 $[-0.013416, -0.002244]$

7.31 The upper 99 percent confidence interval is $[0, 0.0217]$.

7.33 (a) $[0.0286, 0.0714]$ **(b)** $(0, 0.068]$

7.35 (a) $n = 2401$ **(b)** $n = 865$

7.37 **(a)** $= [0, 0.054]$ **(b)** $[0, 0.083]$ **(c)** $[-0.053, -0.0004]$ Because the confidence interval does not contain 0, it is reasonable to conclude that machine 2 produces a higher proportion of defective chips than machine 1.

7.39 The 95 percent confidence interval is $[-0.096, 0.029]$. Because the confidence interval includes 0, it follows that the true difference between the proportions of defective items could be positive or negative. So the evidence suggesting that machine 2 produces more defective items than machine 1 is rather weak.

7.41 **(a)** $\hat{\mu} = 3.72, \hat{\sigma}^2 = 0.0316, \hat{\sigma} = 0.1778$. **(b)** $\hat{\mu}$ is unbiased; the others are not.

7.45 $\hat{p} = \frac{185}{351} = 0.527$ and $\hat{p}(1 - \hat{p}) = 0.249$; $\hat{p}$ is an unbiased estimator, but $\hat{p}(1 - \hat{p})$ is biased; see Prob. 7.30.

Chapter 8

8.1 **(b)** $\alpha = 0.5; 0.052; \beta(205) = 0.206$ **(c)** $\alpha = 0.0516; \beta(205) = 0.7939$

8.3 **(a)** $\alpha = 0.0668; \beta(25) = 0.0122$ **(b)** $\alpha = 0.03; \beta(25) = 0.03$

8.5 **(a)** $c = 12.44; \beta(12.5) = 0.5120$ **(b)** $c = 11.71; \beta(12.5) = 0.653$

8.7 **(a)** $H_0: \mu = 0.1$ against $H_1: \mu < 0.1$; Rejection region: $\mathscr{C} = (\bar{x} : \bar{x} < 0.08\%)$ **(b)** $\alpha = 0.1151$
(c) $\beta(0.09) = 0.2743$ **(d)** 0.2743

8.9 **(a)** $[\bar{x} - 4.128, \bar{x} + 4.128]$ **(b)** Reject if $|\bar{x} - 50| > 4.128$ **(c)** $P(t_{24} > 1.5); P(t_{24} > 2)$

8.13 **(a)** Reject if $\bar{X} < c = 0.2 - t_6(\alpha)(s/\sqrt{7})$ **(b)** The cutoff value $c = 0.2 - 1.943(0.03/\sqrt{7}) = 0.1780$. Since $\bar{x} = 0.17 < 0.178$, we reject H_0. **(c)** P-value $= P(t_6 < -\sqrt{7}) = P(t_6 < -2.65)$
(d) Upper confidence interval $= (-\infty, 0.192]$, which does not contain 0.20. Reject H_0. The concentration of arsenic could be as high as 0.192 percent.

8.15 **(a)** Because of the large sample size ($n = 100$), the distribution of $\bar{X}$ is approximately normal. **(b)** Reject H_0 if $\sqrt{n}(\bar{X} - 40{,}000)/3200 < -z(\alpha)$. **(c)** P-value $= P(\bar{X} < 39{,}360 \mid \mu = 40{,}000) = 0.0228$, which is significant at the 5 percent level.

8.21 **(a)** $P(t_{11} < -1.155)$ **(b)** $P(t_{11} < -1.732)$ **(c)** $P(t_{11} < -2.079)$ **(d)** (a) and (b) are not significant at the 5 percent level, whereas (c) is significant.

8.23 $\pi(\mu) = \Phi(21 - 0.75\mu)$

8.25 $n = 21$

8.27 **(a)** The 95 percent confidence interval $[-0.0013, 0.0057]$ contains 0, so the null hypothesis of no difference between the means is not rejected. **(b)** P-value $= P(t_{10} > 1.4113) > P(t_{10} > 1.812) = 0.10$. Consequently, the P-value is greater than 0.10.

8.29 $[(\bar{X}_1 - \bar{X}_2)/(s_p \sqrt{(1/12) + (1/12)})] = -0.5399; P$-value $= P(|t_{22}| > 0.5399) > P(|t_{22}| > 1.321) = 0.20$. We do not reject H_0.

8.31 **(a)** $[(\bar{X}_1 - \bar{X}_2)/(s_p \sqrt{(1/7) + (1/6)})] = -2.8419;$ P-value $= P(|t_{11}| > 2.8419) < P(|t_{11}| > 2.718) = 0.02$. Test is significant at the 5 percent level.

8.33 $[\bar{D}/(s_D/\sqrt{n})] = 2.4196; P(|t_{14}| > 2.4196) \leq P(|t_{14}| > 2.145) = 0.05$; The P-value is less than 0.05.

8.35 **(a)** $[(\bar{X}_1 - \bar{X}_2)/s(\bar{X}_1 - \bar{X}_2)] = -3.915; P(|t_{23}| > 3.915) \leq P(|t_{23}| > 2.807) = 0.01$; The P-value is less than 0.01.

8.37 **(a)** $[(\bar{X}_1 - \bar{X}_2)/s(\bar{X}_1 - \bar{X}_2)] = -1.8074; 0.025 = P(t_{16} < -2.120) \leq P(t_{16} < -1.8074) \leq P(t_{16} < -1.746) = 0.05$ Consequently, $0.025 < P$-value < 0.05. **(b)** The upper 99 percent one-sided confidence interval $(-\infty, 0.6008]$ contains 0, so we do not reject H_0.

8.39 $s_1^2/s_2^2 \sim F_{5,9}$. Accept H_0 because $0.0984 = [1/F_{9,5}(0.01)] \leq (s_1^2/s_2^2) = 0.3235 \leq F_{5,9}(0.01) = 6.06$. Since we do not reject the null hypothesis that $\sigma_1^2 = \sigma_2^2$, one is justified in using the two-sample t test.

8.41 We list the order statistics $x_{(j)}$ (column EXPOSED) and normal scores z_j (column Z). The task of plotting $(z_j, x_{(j)})$ is left to the reader.

Exposed	Z	Exposed	Z	Exposed	Z
10	−2.16611	23	−0.38941	38	0.47279
13	−1.69062	24	−0.30867	39	0.55959
14	−1.43420	25	−0.22988	39	0.65084
15	−1.24775	27	−0.15251	41	0.74786
16	−1.09680	31	−0.07603	43	0.85250
17	−0.96742	34	0.00000	44	0.96742
18	−0.85250	34	0.07603	45	1.09680
20	−0.74786	35	0.15251	48	1.24775
21	−0.65084	35	0.22988	49	1.43420
22	−0.55959	36	0.30867	62	1.69062
23	−0.47279	37	0.38941	73	2.16611

8.43 We list the order statistics $x_{(j)}$ (column D) and normal scores z_j (column Z). The task of plotting $(z_j, x_{(j)})$ is left to the reader.

D	Z	D	Z	D	Z
−9	−2.16611	7	−0.38941	23	0.47279
−6	−1.69062	9	−0.30867	23	0.55959
−3	−1.43420	13	−0.22988	25	0.65084
−3	−1.24775	14	−0.15251	25	0.74786
0	−1.09680	14	−0.07603	25	0.85250
1	−0.96742	15	0.00000	30	0.96742
1	−0.85250	16	0.07603	32	1.09680
2	−0.74786	16	0.15251	36	1.24775
4	−0.65084	17	0.22988	42	1.43420
5	−0.55959	18	0.30867	47	1.69062
6	−0.47279	22	0.38941	60	2.16611

8.45 We list the order statistics $x_{(j)}$ (column X) and normal scores z_j (column Z). The task of plotting $(z_j, x_{(j)})$ is left to the reader.

X	Z	X	Z	X	Z	X	Z
181	−2.31099	193	−0.64206	196	0.02611	199	0.70764
189	−1.86273	193	−0.57913	197	0.07841	200	0.77642
189	−1.62498	193	−0.51842	197	0.13093	200	0.84909
189	−1.45441	194	−0.45956	197	0.18380	200	0.92654
190	−1.31801	195	−0.40225	197	0.23720	201	1.00999
190	−1.20251	195	−0.34623	198	0.29129	201	1.10115
190	−1.10115	195	−0.29129	198	0.34623	202	1.20251
192	−1.00999	195	−0.23720	198	0.40225	202	1.31801
192	−0.92654	195	−0.18380	198	0.45956	204	1.45441
192	−0.84909	195	−0.13093	198	0.51842	204	1.62498
192	−0.77642	196	−0.07841	199	0.57913	206	1.86273
193	−0.70764	196	−0.02611	199	0.64206	209	2.31099

Chapter 9

9.1 (a) $\alpha = 0.078$, $\beta(0.30) = 0.850$, $\beta(0.5) = 0.377$ **(b)** $\alpha = 0.004$, $\beta(0.3) = 0.953$, $\beta(0.5) = 0.623$

9.3 $c = 11, B(0.6) = 0.2173$

9.5 (a) $\pi(p) = 1 - B(9; 15, p)$ (b) $\alpha = 0.095$ (c) $B(0.6) = 0.39$

9.7 H_0: $p = 0.50$ against H_1: $p < 0.50$. Reject H_0 if $\hat{p} < 0.4772$. However, $\hat{p} = 625/1300 = 0.4808 > 0.4772$; we do not reject. P-value $= 0.0823$

9.9 $n = 87$

9.11 The two-sided 95 percent confidence interval for $p_A - p_B$ is $[-0.0562, 0.0162]$; it contains 0, so the difference is not statistically significant.

9.13 $\chi^2 = 0.8 < \chi^2_{5,0.05} = 11.07$; consequently the null hypothesis is not rejected.

9.17 $\chi^2 = 34.076 > 12.592 = \chi^2_6(0.05)$. We conclude that the three classes of voters differ with respect to the importance of these four public policy issues.

9.19 $\chi^2 = 18.812 > \chi^2_1(0.01) = 6.635$; therefore, we reject the hypothesis that lateral deflection and range are independent.

9.21 $\chi^2 = 1.207 < 9.210 = \chi^2_2(0.01)$. We conclude that there is no statistically significant difference between the two processes.

Chapter 10

10.1 (c) $\hat{\beta} = 0.667$ and $\hat{\beta}_1 = 1.000$

10.3 (b) $\hat{\beta}_0 = -0.008, \hat{\beta}_1 = 0.244$ (c) $r = 0.9856, R^2 = 0.9715, \text{SSE} = 5.87$

(d)

Obs	Dep Var Y	Predict Value	Residual
1	2.0000	3.4043	-1.4043
2	5.0000	3.8918	1.1082
3	7.0000	6.5731	0.4269
4	9.0000	10.2295	-1.2295
5	10.0000	9.4982	0.5018
6	13.0000	12.1795	0.8205
7	20.0000	20.2235	-0.2235

10.5 (b) $\hat{\beta}_0 = 31.446, \hat{\beta}_1 = -1.186$ (c) $r = -0.6998, R^2 = 0.4897, \text{SSE} = 35.64$ (d) $\hat{y}(1.6) = 29.5487, y - \hat{y} = -0.5487; \hat{y}(4.6) = 25.9909, y - \hat{y} = -0.9909$ (e) $R^2(\text{weight}) = 0.6919 > 0.4897 = R^2(\text{displacement})$

10.7 (b) $\hat{\beta}_0 = 3.133, \hat{\beta}_1 = 0.157$ (c) $r = 0.7160, R^2 = 0.5127, \text{SSE} = 0.411$ (d) $\hat{y}(1.6) = 3.3848, y - \hat{y} = 0.0522; \hat{y}(4.6) = 3.8566, y - \hat{y} = 0.1434$ (e) $R^2(\text{weight}) = 0.7052 > R^2(\text{displacement}) = 0.5127$

10.13 (a)

Analysis of Variance

Source	DF	Sum of Squares	Mean Square	F Value	Prob > F
Model	1	250708.69157	250708.69157	25.121	0.0010
Error	8	79841.30843	9980.16355		
C Total	9	330550.00000			

	Root MSE	99.90077	R-square	0.7585

(b) $R^2 = 0.7585$ (c) 95 percent confidence interval for $\hat{\beta}_0$ is $[-1283.79, 144.85]$; 95 percent confidence interval for $\hat{\beta}_1$ is $[3.72, 10.07]$. (d) $\hat{y}(220) = 948.1$; 95 percent confidence interval is $[874.2, 1022.0]$.

10.15 (a) 95 percent confidence interval for $\hat{\beta}_0$ is $[33.54, 44.03]$; 95 percent confidence interval for $\hat{\beta}_1$ is $[-5.05, -1.82]$. (b) $[28.73, 31.46]$ (c) $[27.01, 33.19]$

10.17 (b)

Analysis of Variance

Source	DF	Sum of Squares	Mean Square	F Value	Prob > F
Model	1	816.44062	816.44062	47.052	0.0001
Error	11	190.87015	17.35183		
C Total	12	1007.31077			

	Root MSE	4.16555	R-square	0.8105

(c) $R^2 = 0.8105$ (d) 95 percent confidence interval for $\hat{\beta}_0$ is [22.485, 31.311]; 95 percent confidence interval for $\hat{\beta}_1$ is [0.0114, 0.0226].

10.19 (b)

Analysis of Variance

Source	DF	Sum of Squares	Mean Square	F Value	Prob > F
Model	1	22426.66667	22426.66667	235.512	0.0001
Error	10	952.25000	95.22500		
C Total	11	23378.91667			

Root MSE	9.75833	R-square	0.9593

(c) $\hat{\beta}_0 = 153.917$, $\hat{\beta}_1 = 2.417$ (d) $R^2 = 0.9593$ (e) 95 percent confidence interval for $\hat{\beta}_0$ is [135.944, 171.890]; 95 percent confidence interval for $\hat{\beta}_1$ is [2.066, 2.768].   (f) 269.9

10.21 The least squares estimates for Model($Y = \ln P \mid x = \ln D$) are $\hat{\beta}_0 = 0.000016833$, $\hat{\beta}_1 = 1.499701$. Consequently, $a = \exp(0.000016833) = 1.000$ and $b = 1.4997015 \approx 1.5$. So these results are consistent with Kepler's predicted values.

10.23 (c) The least squares estimates for Model($Y = \ln \mid x = \ln v$) are $\hat{\beta}_0 = 3.175244$, $\hat{\beta}_1 = -1.002195$.   **(d)** The equation $pv = k$ implies that $p = kv^{-1}$. Thus Boyle's law implies that $b = -1$, which is consistent with the least squares estimate $\hat{\beta}_1 = -1.002195$.

10.25 Refer back to Prob. 10.19.   **(a)** The predicted values $\hat{y}_i$ and studentized residuals e_i^* are listed in the following table.

Observation	Predicted value	Studentized residual
1	231.250	−0.13894
2	269.917	−0.84735
3	327.917	−0.57543
4	308.583	−1.17635
5	269.917	−1.59658
6	192.583	0.81560
7	250.583	−0.27905
8	269.917	0.97222
9	269.917	−0.31218
10	211.917	0.24383
11	347.250	1.49350
12	289.250	1.70131

(b) The order statistics of the studentized residuals and the corresponding z scores are listed in the next table.

Observation	Order stats	z
1	−1.59658	−1.73166
2	−1.17635	−1.15035
3	−0.84735	−0.81222
4	−0.57543	−0.54852
5	−0.31218	−0.31864
6	−0.27905	−0.10463
7	−0.13894	0.10463
8	0.24383	0.31864
9	0.81560	0.54852
10	0.97222	0.81222
11	1.49350	1.15035
12	1.70131	1.73166

10.29 $r = 0.9165$

10.31 (b) $r(X_2, X_3) = 0.695$ **(d)** $r(X_3, X_4) = -0.602$

10.33 The scatter plot suggests a weak positive correlation between the two variables. The sample correlation coefficient $r = 0.45908$.

Chapter 11

11.3 **(a)**
$$\begin{pmatrix} Y_1 \\ Y_2 \\ Y_3 \\ Y_4 \\ Y_5 \\ Y_6 \\ Y_7 \end{pmatrix} = \begin{pmatrix} 1 & 14 \\ 1 & 16 \\ 1 & 27 \\ 1 & 42 \\ 1 & 39 \\ 1 & 50 \\ 1 & 83 \end{pmatrix} + \begin{pmatrix} \epsilon_1 \\ \epsilon_2 \\ \epsilon_3 \\ \epsilon_4 \\ \epsilon_5 \\ \epsilon_6 \\ \epsilon_7 \end{pmatrix}$$

(b) $\begin{pmatrix} 7 & 271 \\ 271 & 13{,}855 \end{pmatrix}\begin{pmatrix} \hat{\beta}_0 \\ \hat{\beta}_1 \end{pmatrix} = \begin{pmatrix} 66 \\ 3375 \end{pmatrix}$

(d) $s^2(\mathbf{X'X})^{-1} = \begin{pmatrix} 0.6907 & -0.0135 \\ -0.0135 & 0.0003 \end{pmatrix}$

11.9 **(a)** $y = -28.3654 + 0.6013x_1 + 1.0279x_2$ **(b)** 65.2695   **(c)** $\hat{y}(841, 48) = 526.7$, residual $= 8.3028$ **(d)**

Analysis of Variance

Source	DF	Sum of Squares	Mean Square	F Value	Prob > F
Model	2	171328.66023	85664.33012	1304.326	0.0001
Error	7	459.73977	65.67711		
C Total	9	171788.40000			

	Root MSE	8.10414	R-square	0.9973

$R^2 = 0.9973$, and adjusted $R^2 = 0.9966$.   **(e)** $\text{SSE}(r) = 1166.75898$, $\text{DF}(r) = 8$; $\text{SSE}(f) = 459.73977$, $\text{DF}(f) = 7$. The F ratio $F = 10.7651 > F_{1,7}(0.05) = 5.59$. On the other hand, $F = 10.7651 < F_{1,7}(0.01) = 12.25$. The P-value is 0.0135.

11.11 **(b)** $y = -26{,}219 + 189.204551x - 0.331194x^2$ **(d)** $\text{SSE}(r) = 22508.10547$, $DF(r) = 6$; $\text{SSE}(f) = 10213.02766$, $\text{DF}(f) = 5$. The F ratio is $F = 6.0172 > F_{1,5}(0.05) = 6.61$. The P-value is 0.0577.   **(e)** The following table lists the predicted values and the corresponding studentized residuals. Using this information, you can easily plot the residuals against the predicted values.

OBS	PREDICT	YSRESID
1	792.542	-0.87628
2	802.187	-0.05864
3	789.683	1.27463
4	774.065	0.92301
5	752.485	-0.45056
6	678.948	-0.97745
7	637.497	-1.20873
8	517.593	2.00390

The following table lists the order statistics of the studentized residuals and the z (normal) scores. Using this information you can graph the normal probability plot.

OBS	SRESID	z
1	-1.20873	-1.53412
2	-0.97745	-0.88715
3	-0.87628	-0.48878
4	-0.45056	-0.15731
5	-0.05864	0.15731
6	0.92301	0.48878
7	1.27463	0.88715
8	2.00390	1.53412

11.13 Looking at the ANOVA table, we see that the F ratio is 1.774 with a P-value of 0.1925. The coefficient of multiple determination is 0.2496, which is quite low. The model is not a good one.

```
                        Analysis of Variance
                             Sum of          Mean
        Source        DF     Squares        Square    F Value    Prob > F
        Model          3     0.02685        0.00895    1.774      0.1925
        Error         16     0.08072        0.00505
        C Total       19     0.10757
            Root MSE        0.07103     R-square        0.2496
```

Chapter 12

12.5 (a) $\bar{y}_{1.} = 221.67$, $\bar{y}_{2.} = 202.67$, $\bar{y}_{3.} = 177.00$, $\bar{y}_{4.} = 249.00$, $\bar{y}_{..} = 212.58$ **(b)** SSTr $= 8319.58$, SSE $= 3523.33$, SST $= 11842.92$

(c) Dependent Variable: Y   Burst Strength (psi)

```
                              Sum of           Mean
        Source          DF    Squares         Square   F Value  Pr > F
        Model            3   8319.5833333  2773.1944444   6.30   0.0168
        Error            8   3523.3333333   440.4166667
        Corrected Total 11  11842.9166667
```

$F = 6.30 > 4.07 = F_{3,8}(0.05)$. Reject H_0.

12.7

```
                              Sum of           Mean
        Source          DF    Squares         Square    F Value  Pr > F
        Model            2   76.85405000   38.42702500    74.21   0.0001
        Error            9    4.66045000    0.51782778
        Corrected Total 11   81.51450000
```

$F = 74.21 > 4.26 = F_{2,9}(0.05)$. Reject H_0.

12.9 (c) Dependent Variable: POUNDS

```
                              Sum of           Mean
        Source          DF    Squares         Square    F Value  Pr > F
        Model            2   155.1450286   77.5725143     6.01    0.0083
        Error           22   284.1725714   12.9169351
        Corrected Total 24   439.3176000
```

$F = 6.01 > 3.44 = F_{2,22}(0.05)$. Reject H_0.

12.13 (a) LSD $= t_{12}(0.025)\sqrt{\text{MSE}/J} = 122.21$. The four 95 percent confidence intervals are $[2755.79, 3000.21]$, $[3031.79, 3276.21]$, $[2815.29, 3059.71]$, $[2551.79, 2796.21]$.

(b)

Comparison	Confidence interval
$\mu_2 - \mu_3$	$[44.42, 390.08]$
$\mu_2 - \mu_1$	$[103.67, 449.33]$
$\mu_2 - \mu_4$	$[307.67, 653.33]$
$\mu_3 - \mu_1$	$[-113.58, 231.08]$
$\mu_3 - \mu_4$	$[90.42, 436.08]$

(c)

$\bar{y}_{4.}$	$\bar{y}_{1.}$	$\bar{y}_{3.}$	$\bar{y}_{2.}$
2674.00	2878.00	2937.50	3154.00

12.15 (a) LSD $= t_{21}(0.025)\sqrt{\text{MSE}/J} = 3.4055$. The three 95 percent confidence intervals are $[12.34, 19.16]$, $[11.22, 18.03]$, $[14.22, 221.03]$ **(b)** HSD $= 4.1275$

$\bar{y}_{2.}$	$\bar{y}_{1.}$	$\bar{y}_{3.}$
14.625	15.750	16.625

There are no significant differences among the means.

12.19 (a)

Parameter	95 percent confidence interval
μ_1	$[6.73, 11.45]$
μ_2	$[1.97, 7.30]$
μ_3	$[8.49, 13.76]$

(b) We use Eq. (12.23). The differences among the means are significant when the confidence interval does not contain 0.

Comparison	95 percent confidence interval
$\mu_3 - \mu_1$	[2.035, 6.318], significant
$\mu_3 - \mu_2$	[1.667, 11.012], significant
$\mu_1 - \mu_2$	[−0.145, 8.754], not significant

12.21 (b) Dependent Variable: Y   Toxic level (ppm)

Source	DF	Sum of Squares	Mean Square	F Value
Model	3	95.73458333	31.91152778	4.85
Error	20	131.66166667	6.58308333	
Corrected Total	23	227.39625000		

$P(F_{3,20} > 4.85) = 0.0108$. We reject H_0.   **(c)** $\hat{\sigma}^2 = 6.5831, \hat{\sigma}_A^2 = 4.2214$

Chapter 13

13.1

Source	DF	Sum of Squares	Mean Square	F Value	Pr > F
Model	2	219.5000000	109.7500000	0.49	0.6299
Error	9	2029.5000000	225.5000000		
Corrected Total	11	2249.0000000			

13.3 (a) The grand mean is $\hat{\mu} = 102.56$. The treatment means are $\hat{\mu}_1 = 97.67$, $\hat{\mu}_2 = 103.33$, $\hat{\mu}_3 = 106.67$. Block means are 100.67, 100.33, 101.00, 104.00, 103.00, 106.33.

(b)

Source	DF	Sum of Squares	Mean Square	F Value	Pr > F
Model	7	330.8888889	47.2698413	18.50	0.0001
Error	10	25.5555556	2.5555556		
Corrected Total	17	356.4444444			

13.5 There are three treatments corresponding to the three configurations. There are five blocks corresponding to the workloads.

(b)

Source	DF	Sum of Squares	Mean Square	F Value	Pr > F
Model	6	13165.60000	2194.26667	74.13	0.0001
Error	8	236.80000	29.60000		
Corrected Total	14	13402.40000			

(c) The three estimated treatment means are 51, 52, 113.6. HSD = 9.8322 and MSE = 29.6. Consequently, the mean execution time for the no-cache memory is significantly different from the first two mean execution times. The difference between the two-cache memory and the one-cache memory is not significant.

13.7 (a) Dependent Variable: YIELD

Source	DF	Sum of Squares	Mean Square	F Value	Pr > F
Model	8	6427.398095	803.424762	10.18	0.0003
Error	12	947.433333	78.952778		
Corrected Total	20	7374.831429			

13.9 (a) Dependent Variable: Y1

Source	DF	Sum of Squares	Mean Square	F Value	Pr > F
Model	3	1.94196306	0.64732102	7.78	0.0003
Error	44	3.66207788	0.08322904		
Corrected Total	47	5.60404094			

13.15 (a)

Estimated cell means

	Factor B			
Factor *A*	1	2	3	Row mean
1	5340.00	4990.00	4815.00	5048.33
2	5045.00	5345.00	4515.00	4968.33
3	5675.00	5415.00	4917.50	5335.83
Column mean	5353.33	5250.00	4749.17	$\bar{y}_{..} = 5117.50$

(c) Dependent Variable: STRENGTH

Source	DF	Sum of Squares	Mean Square	F Value	Pr > F
Model	8	4066400.000	508300.000	1.85	0.1108
Error	27	7415475.000	274647.222		
Corrected Total	35	11481875.000			

Source	DF	Type I SS	Mean Square	F Value	Pr > F
GAUGER	2	896450.000	448225.000	1.63	0.2142
BREAKER	2	2506116.667	1253058.333	4.56	0.0196
GAUGER*BREAKER	4	663833.333	165958.333	0.60	0.6629

13.17 (a)

Estimated cell means

	Factor B				
Factor *A*	40	60	80	100	Row mean
50	18.5	18.5	23.0	27.5	21.875
75	10.5	15.5	14.5	29.0	17.375
100	14.0	19.5	24.0	26.5	21.0
125	19.0	22.0	22.5	30.0	23.375
Column mean	15.5	18.875	21.0	28.25	$\bar{y}_{..} = 20.9063$

(c) Dependent Variable: WARPING

Source	DF	Sum of Squares	Mean Square	F Value	Pr > F
Model	15	968.2187500	64.5479167	9.52	0.0001
Error	16	108.5000000	6.7812500		
Corrected Total	31	1076.7187500			

Source	DF	Type I SS	Mean Square	F Value	Pr > F
TEMP	3	156.0937500	52.0312500	7.67	0.0021
COPPER	3	698.3437500	232.7812500	34.33	0.0001
TEMP*COPPER	9	113.7812500	12.6423611	1.86	0.1327

13.19 (a) $L_{X1}/8 = 0.352, L_{X8}/8 = 0.5658, L_{X1X8}/8 = 0.125$ **(b)** SSX1 = $2.817^2/16 = 0.496$, SSX1 = $(-4.526)^2/16 = 1.280$, SSX1X8 = $1^2/16 = 0.0625$ **(c)** The sample mean of Y2 at X8 = 0 is -0.3656; the sample mean of Y2 at X8 = 1 is -0.9313. Setting the nozzle position at X8 = 1 produces a smaller variance, since $\exp(-0.9313) < \exp(-0.3656)$.

13.21 (a)

X4 = 0, X8 = 0	14.821, 14.757, 14.415, 14.932
X4 = 1, X8 = 0	13.880, 13.860, 13.972, 13.907
X4 = 0, X8 = 1	14.888, 14.921, 14.843, 14.878
X4 = 1, X8 = 1	14.037, 14.165, 14.032, 13.914

(b) Dependent Variable: Y1   Epitaxial thickness

Source	DF	Sum of Squares	Mean Square	F Value	Pr > F
Model	3	2.87631725	0.95877242	60.30	0.0001
Error	12	0.19080650	0.01590054		
Corrected Total	15	3.06712375			

Source	DF	Anova SS	Mean Square	F Value	Pr > F
X8	1	0.08037225	0.08037225	5.05	0.0441
X4	1	2.79558400	2.79558400	175.82	0.0001
X8*X4	1	0.00036100	0.00036100	0.02	0.8827

(c) $X4 = 0,\ X8 = 0$ $-0.4425,\ -0.3267,\ -0.3131,\ -0.2292$
$X4 = 1,\ X8 = 0$ $-0.6505,\ -0.4969,\ -0.3467,\ -0.1190$
$X4 = 0,\ X8 = 1$ $-1.1989,\ -0.6270,\ -0.4369,\ -0.6154$
$X4 = 1,\ X8 = 1$ $-1.4307,\ -1.4230,\ -0.8663,\ -0.8625$

(d) Dependent Variable: Y2   Log of s-square

Source	DF	Sum of Squares	Mean Square	F Value	Pr > F
Model	3	1.65055071	0.55018357	7.97	0.0034
Error	12	0.82812271	0.06901023		
Corrected Total	15	2.47867342			

Source	DF	Anova SS	Mean Square	F Value	Pr > F
X8	1	1.28034883	1.28034883	18.55	0.0010
X4	1	0.24897605	0.24897605	3.61	0.0818
X8*X4	1	0.12122583	0.12122583	1.76	0.2097

13.23

Contrasts	SS_{effect}	SS_{effect}/SST	SS_{effect}/SST (complete factorial)
$l_A = 339$	9576.75	0.4588	0.4463
$l_B = 313$	8164.083	0.3911	0.4463
$l_C = 183$	2790.750	0.1337	0.0735

Source	DF	Sum of Squares	Mean Square	F Value	Pr > F
Model	3	20531.58333	6843.86111	160.40	0.0001
Error	8	341.33333	42.66667		
Corrected Total	11	20872.91667			

Source	DF	Type I SS	Mean Square	F Value	Pr > F
A	1	9576.750000	9576.750000	224.46	0.0001
B	1	8164.083333	8164.083333	191.35	0.0001
C	1	2790.750000	2790.750000	65.41	0.0001

13.25

Contrasts	SS_{effect}	SS_{effect}/SST	SS_{effect}/SST (complete factorial)
$l_A = 37.5$	175.78	0.8953	0.5302
$l_B = 10.3$	13.26	0.0675	0.0054
$l_C = 3.1$	1.20	0.0061	0.0023

Source	DF	Sum of Squares	Mean Square	F Value	Pr > F
Model	3	190.2437500	63.4145833	41.69	0.0018
Error	4	6.0850000	1.5212500		
Corrected Total	7	196.3287500			

Source	DF	Type I SS	Mean Square	F Value	Pr > F
A	1	175.7812500	175.7812500	115.55	0.0004
B	1	13.2612500	13.2612500	8.72	0.0419
C	1	1.2012500	1.2012500	0.79	0.4244

Chapter 14

14.1 **(a)** $\text{LCL}_{\overline{X}} = 13.9$, $\text{UCL}_{\overline{X}} = 15.1$ **(b)** $\text{LCL}_R = 0$, $\text{UCL}_R = 1.88$ **(c)** $P(|\overline{X} - 14.5| > 0.5) = 0.0155$

14.5 **(a)** Forty-two rheostats are nonconforming; that is, approximately 31 percent failed to meet the specifications.   **(b)** $\text{LCL}_{\overline{X}} = 135.67$, $\text{UCL}_{\overline{X}} = 145.62$, $\overline{\overline{x}} = 140.64$. $\text{LCL}_R = 0$, $\text{UCL}_R = 18.2$, $\overline{R} = 8.6$. The $\overline{x}$ and R control charts do not show a lack of control.

14.7 **(a)** $\text{LCL}_{\overline{p}} = 0$, $\text{UCL}_{\overline{p}} = 0.0794$ **(b)** $P(\hat{p} > 0.0794 \mid p = 0.04) = 0.0778$. T has a geometric distribution with parameter $p = 0.0778$. $E(T) = 12.85$.

14.9 $\text{LCL}_{\overline{c}} = 0$, $\text{UCL}_{\overline{c}} = 14.702$. The maximum number of observed defects is 12, so the control chart does not detect a lack of control.

INDEX